Toshiyuki Fukada • Taiho Kambe
Editors

Zinc Signaling

Second Edition

Springer

Editors
Toshiyuki Fukada
Molecular and Cellular Physiology
Faculty of Pharmaceutical Sciences
Tokushima Bunri University
Tokushima, Japan

Taiho Kambe
Division of Integrated Life Science
Graduate School of Biostudies
Kyoto University
Kyoto, Japan

ISBN 978-981-15-0556-0 ISBN 978-981-15-0557-7 (eBook)
https://doi.org/10.1007/978-981-15-0557-7

This Springer imprint is published by the registered company Springer Nature Singapore Pte Ltd.
The registered company address is: 152 Beach Road, #21-01/04 Gateway East, Singapore 189721, Singapore

Foreword

Seventy years ago (in the 1950s), it was first recognized that zinc is an essential metal in humans and the zinc-binding protein metallothionein was first described. Two decades later, acrodermatitis enteropathica was shown to be genetic disease of zinc deficiency in humans, and soon thereafter (in the 1980s), the first mammalian metallothionein genes were cloned, their dramatic transcriptional induction by zinc was discovered, and the first mammalian zinc transporter was cloned. During the next 30 years, we have come to understand that zinc can dynamically modulate many cellular signaling cascades and multiple aspects of human cellular physiology and that disruption of zinc metabolism causes diseases.

Revealing the mechanisms that control zinc metabolism is a major focus of the field that is advancing rapidly. Since the discovery that MTF-1 functions as a zinc sensor that regulates metallothionein gene transcription, multiple other zinc-sensing processes have been discovered. Two multimember families of zinc transporters (24 genes) are being studied in great detail, and members have been shown to play cell-type-specific and organelle-specific roles in zinc homeostasis. Mouse models with mutations in most of these genes are available and have begun to help elucidate physiological functions of many zinc transporters and have revealed previously unknown roles of specific zinc transporters in human disease. Zinc fluxes have recently been shown to modify multiple kinase signal transduction cascades, as well as affect transcription and protein translation, localization, processing, and turnover. It is now clear that zinc signals play important roles in fertilization, early development, and cell cycle, as well as in cancer progression and other diseases.

The second addition of this book contains a wide range of current studies which continue to address mechanisms of zinc metabolism. The creation of zinc-sensing fluorescent probes, zinc transporter expression vectors, specific antibodies, and genetic mouse models has supported studies of the mechanistic aspects of zinc metabolism in great depth. Studies of structure-function relationships in zinc transport proteins continue to progress, and studies of zinc-protein and protein-protein interaction and modifications are beginning to reveal multiple mechanisms of cellular and organismal zinc signaling and its role in health and disease.

University of Kansas Medical Center
Kansas City, Kansas, USA
Glen K. Andrews

Contents

1 **Opening the Second Era of Zinc Signaling Study** 1
Toshiyuki Fukada

2 **Regulation of Cellular Zinc Ions and Their Signaling Functions**. . . . 5
Wolfgang Maret

3 **Zinc Transporter Proteins: A Review and a New View from Biochemistry**. 23
Taiho Kambe, Eisuke Suzuki, and Taiki Komori

4 **The Metallothionein-Zinc Landscape: How It Shapes Antimicrobial Immunity**. 57
Debabrata Chowdhury, George S. Deepe Jr, and Kavitha Subramanian Vignesh

5 **Role of Zinc Signaling in Mast Cell, Basophil, T Cell, and B Cell** . . . 79
Keigo Nishida, Michiko Kato, and Ryota Uchida

6 **Review: The Role of Zinc Signaling in Reproduction** 99
Atsuko Kageyama, Takafumi Namiki, Junya Ito, and Naomi Kashiwazaki

7 **Zinc Signaling in Skeletal Muscle** 123
Kristyn Gumpper and Jianjie Ma

8 **Zinc Signaling in Aging Heart Function**. 139
Belma Turan, Deniz Billur, and Yusuf Olgar

9 **Zinc Signaling in the Life and Death of Neurons**. 165
Elias Aizenman

10 **Possible Therapeutic Roles of Metallothionein-3 and Zinc in Endosome-Autophagosome-Lysosome Pathway (EALP) Dysfunction in Astrocytes**. 187
Ha Na Kim, Bo-Ra Seo, Sook-Jeong Lee, and Jae-Young Koh

11 **Zinc in Neurodegeneration** 201
Stuart D. Portbury, Ashley I. Bush, and Paul A. Adlard

12 **Role of Zinc Transporters in Type 2 Diabetes and Obesity** 229
Ayako Fukunaka and Yoshio Fujitani

13 **Zinc Signals in Immunology** 243
Martina Maywald and Lothar Rink

14 **Zinc Signals in Inflammation** 279
Sannette C. Hall and Daren L. Knoell

15 **Zinc Transporters and Zinc Signaling in Skin Formation and Diseases** 305
Bum-Ho Bin, Mi-Gi Lee, Takafumi Hara, Teruhisa Takagishi, and Toshiyuki Fukada

16 **Post-translational Mechanisms of Zinc Signalling in Cancer** 319
Thirayost Nimmanon and Kathryn M. Taylor

17 **Zinc Signaling (Zinc'ing) in Intestinal Function** 347
Michal Hershfinkel, Johnny Iv, and Shannon L. Kelleher

18 **In Situ Imaging of Zinc with Synthetic Fluorescent Probes** 365
Jiyao Yu and Christoph J. Fahrni

19 **Zinc Signals in Biology** 389
Hak Chung and Amanda J. Bird

Index 411

Chapter 1
Opening the Second Era of Zinc Signaling Study

Toshiyuki Fukada

Abstract Almost half a decade has passed since the publication of the first edition of *Zinc Signaling*, entitled "*Zinc Signals in Cellular Functions and Disorders*", which was the first exclusive book dedicated to "zinc signaling" at that time. Since then, a number of novel findings have been reported, which have updated our knowledge of zinc-related biology and provided new insights into many cellular functions, physiology, and pathophysiology associated with zinc. This book, *Zinc Signaling, 2nd Edition* has been planned to be published with a primary goal of updating the available information about the role of zinc signaling in biological processes at both molecular and physiological levels, as well as highlighting the new achievements in the field in the last 5 years after the publication of the first edition. This book will certainly assist in addressing the questions underlying this unique phenomenon and discerning its future direction, and would help in knowing what is new and what remains to be solved in the next era.

Keywords Zinc · Zinc signaling · Zinc transporter · Metallothionein · Physiology · Disease

1.1 Introduction

Zinc, an essential trace element, plays indispensable roles in multiple cellular processes. It regulates many protein functions including those that involve transcription factors, enzymes, adapters, and growth factors, in the capacity of a structural and/or

T. Fukada (✉)
Molecular and Cellular Physiology, Faculty of Pharmaceutical Sciences,
Tokushima Bunri University, Tokushima, Japan
e-mail: fukada@ph.bunri-u.ac.jp

T. Fukada, T. Kambe (eds.), *Zinc Signaling*,
https://doi.org/10.1007/978-981-15-0557-7_1

catalytic factor. Recent studies have highlighted yet another function of zinc as an intra- and intercellular signaling mediator, which is now recognized as the "zinc signal." Indeed, zinc regulates the cellular signaling pathways, which enables the conversion of extracellular stimuli into the intracellular signals as well as controls the various intracellular and extracellular events (Hara 2017; Takagishi 2017). The zinc signal is essential for physiology and its dysregulation causes a variety of diseases such as diabetes, cancer, osteoarthritis, dermatitis, and dementia. This indicates that "zinc signal" is an emerging topic, which will assist our understanding of the nature of physiology and pathophysiology.

Five years ago, Dr. Taiho Kambe and I had edited and published the first book titled, *Zinc Signals in Cellular Functions and Disorders* (Fukada and Kambe 2014) about zinc signal. Since then, with the advent of new techniques and knowledge about this field, many novel findings have accumulated rapidly (Kambe 2017; Hara 2017; Takagishi 2017). This made us realize that the study of zinc signaling has become an emerging life science field, and that there is a necessity to update our knowledge about zinc signaling; hence, the decision was made to publish the second edition.

1.2 Progresses, Questions, and Directions

To reiterate, this second edition is to update the current information available about the crucial role of zinc signaling at both molecular and physiological levels. This will help in addressing the specific questions and discerning the future direction through 18 such chapters by the invited authors that have been included in this book. Each chapter highlights the progress of a particular research area as well as the relevant new questions and future direction.

From a molecular and biochemical point of view, the first article in Chap. 2 by Maret provides an overview of the regulatory functions of zinc signaling through zinc transporters and metallothionein, etc., which enable the transmission of information within and between the cells. The article by Kambe et al. summarizes various zinc transporters, that is, the family of zinc transporters (ZNTs) and Zrt- and Irt-like proteins (ZIPs), in addition to discussing the roles of these transporters from a biochemical viewpoint in Chap. 3.

The following chapters address the current understanding of zinc signaling in physiology. In Chap. 4, Chowdhury et al. review the updates about the role of metallothionein in antimicrobial immunity. They elaborate on the recent progress in immunological roles of metallothionein with a focus on zinc regulation in response to pathogen invasion. Nishida et al., in Chap. 5, describe the role of zinc signaling in immune cells, particularly focusing on mast cells, basophil, T cells, and B cells. Kageyama et al., in Chap. 6, summarize the current knowledge about the role of zinc signaling in reproduction and the role of "zinc sparks" in fertilization. Gumpper and Ma address how zinc regulates the function of skeletal muscles in Chap. 7, with a focus on the TRIM family proteins that modulate the integrity of muscle fibers as well as regeneration. Turan et al. focus on the role of zinc transporters in mammalian heart function in Chap. 8. Neurology is one of the intensive fields in zinc signaling-related studies, and Aizenman comprehensively discusses the role of zinc signaling in neuronal cell death in Chap. 9.

This book also contains the updated information about the involvement of zinc signaling in several diseases. In Chap. 10, Kim et al. describe new insights into zinc and metallothionein in the context of diverse neurodegenerative diseases. Portbury et al., in Chap. 11, summarize the advances in our understanding of the role of zinc signaling in Alzheimer's disease, Parkinson's disease, and head trauma/traumatic brain injury. Fukunaka and Fujitani focus on the role of zinc transporters in lifestyle-related diseases, particularly in diabetes and obesity, in Chap. 12. Maywald and Rink discusses the updated information on the role of zinc signaling in both innate and acquired immunity in Chap. 13. Hall and Knoell discuss the roles of zinc signaling in infection and inflammation in Chap. 14.

In Chap. 15, Bin et al. elaborate on the role of zinc signaling in skin formation and skin diseases caused by systemic or topical imbalance in zinc homeostasis. Nimmanon and Taylor focus on the role of zinc signaling in cancers and its effect on post-translational modifications such as phosphorylation, in Chap. 16. Hershfinkel et al., in Chap. 17, provide new insights into zinc signaling in intestinal functions by focusing on zinc transporters and channels as well as on the role of the zinc receptor ZnR/GPR39, a G-protein coupled receptor that senses changes in the concentration of extracellular zinc. Yu and Fahrni review the updated information on zinc imaging using chemical fluorescent zinc probes in Chap. 18. Finally, in Chap. 19, Chung and Bird comprehensively discuss the underlying mechanisms that control zinc signaling and its biological significance in bacteria, fungi, nematodes, and fish.

Among these, skin, skeletal muscle, heart, intestine, and chemical fluorescent zinc probe are new topics, which were not included in the first edition. Hence, we must emphasize that this second edition will provide updated and novel insights into the role of zinc signaling, mediated by zinc transporters, channels, and zinc-binding proteins, in health and diseases, from molecular as well as physiological perspectives. We hope that this will present our readers with novel opportunities to develop new ideas and connections to resolve persisting questions in the future.

Finally, we would like to express our heartfelt gratitude to all the authors for their tremendous efforts in supporting this publication. Without their valuable assistance, this timely and successful publication with its useful updates on zinc signaling biology would not have been possible.

Conflicts of Interest The authors declare no conflict of interest.

References

Fukada T, Kambe T (eds) (2014) Zinc signals in cellular functions and disorders. Springer Nature, Tokyo

Hara T, Takeda TA, Takagishi T, Fukue K, Kambe T, Fukada T (2017) Physiological roles of zinc transporters: molecular and genetic importance in zinc homeostasis. J Physiol Sci 67(2):283–301

Kambe T, Matsunaga M, Takeda TA (2017) Understanding the contribution of zinc transporters in the function of the early secretory pathway. Int J Mol Sci 18(10):E2179

Takagishi T, Hara T, Fukada T (2017) Recent advances in the role of SLC39A/ZIP zinc transporters in vivo. Int J Mol Sci 18(12):E2708

Chapter 2
Regulation of Cellular Zinc Ions and Their Signaling Functions

Wolfgang Maret

Abstract Zinc signaling adds a similarly complex regulatory system to calcium signaling and extends signaling with metal ions to lower concentrations. Zinc ions have intracellular and extracellular regulatory functions in addition to their catalytic and structural functions in several thousand human proteins. Cellular zinc metabolism and Zn^{2+} signaling are under exquisite homeostatic control. In order to ascertain specificity in functions without interference by other metal ions, zinc ions are controlled in a defined range of concentrations and re-distributed subcellularly. Several dozen proteins, zinc transporters of two major families, metallothioneins, and proteins that sense zinc ion concentrations participate in this control. Total cellular zinc concentrations are relatively high. They are buffered in such a way that "free" zinc ion concentrations are very low but not negligible. Key features of cellular zinc biology are subcellular storage, release, and, in some cells, exocytosis of zinc ions for a purpose. The chemical nature of zinc ion signals is not known because the biological ligands of zinc when not bound to proteins have not been identified. How zinc ion signals are generated, shaped, and transduced and how effector proteins decode the signal involves remarkable mechanisms of coordination dynamics in proteins. Affecting the proteins that control cellular zinc homeostasis by nutritional deficiencies, xenobiotics, and mutations in their genes interferes with healthy growth and development, is a cause of disease, and influences the course of virtually any disease.

Keywords Zinc · Zn^{2+} signaling · Zinc transporters · Metallothioneins · Cellular zinc homeostasis

At the time a most comprehensive review was published on "the biochemical basis of zinc physiology," the basis was the then known structures and functions of zinc proteins (Vallee and Falchuk 1993). A quarter of a century later, we can record a

W. Maret (✉)
Metal Metabolism Group, Department of Nutritional Sciences, School of Life Course Sciences, Faculty of Life Sciences and Medicine, King's College London, London, UK
e-mail: wolfgang.maret@kcl.ac.uk

T. Fukada, T. Kambe (eds.), *Zinc Signaling*,
https://doi.org/10.1007/978-981-15-0557-7_2

heightened impact of zinc on biology and significant advances that include a further impressive increase in the number of zinc proteins, roles of zinc ions in signal transduction, and major insights into how cellular zinc is controlled homeostatically (Rink 2011). How zinc is regulated cannot be discussed without considering how the other essential metal ions are regulated. Each metal ion is controlled in a certain range of affinities so that it can function specifically without interference with the others. Zinc is the second-most competitive divalent metal ion next to copper, which is tightly controlled by cellular metallochaperones. Critical for the biological actions of zinc is its high affinity to binding sites in proteins, leaving very little zinc unbound. The affinities are in the picomolar or even femtomolar range (Maret 2004a; Kluska et al. 2018). Based on simple mass equilibria, the resulting steady-state "free" zinc ion concentrations are expected to be picomolar as indeed confirmed experimentally (Krężel and Maret 2006). These low concentrations are not negligible. They are functionally important and tightly controlled. Total cellular zinc concentrations are remarkably high. They are hundreds of micromolar, putting them in the range of those metabolites such as ATP. Therefore, zinc is not a trace element in the cell. In this chapter, I shall describe how biological control of zinc is achieved, why it is so complex, and how zinc regulates a multitude of cellular processes at several hierarchical levels.

Zinc biology is based on the zinc ion (Zn^{2+}), which is redox-inert in biology, and its interaction with proteins. Hence, we need to be concerned only with one valence state, which is often referred to as "zinc" although the word stands for the element (Zn^{o}). Despite not participating directly in the redox biology of living organisms, zinc affects redox metabolism and redox signaling indirectly (Maret 2019). Zinc coordination occurs with the nitrogen, oxygen, and sulfur donors of the side chains of mainly four amino acids: His, Glu, Asp, and Cys, and only occasionally others. Coordination is flexible as zinc does not underlie the geometric constraints governing the coordination of the transition metal ions. Therefore, a variation of the number and the types of coordinating donors can modulate the chemical characteristics so that zinc can serve many different functions. A remarkable property is that the sulfur donor of cysteine confers redox activity on zinc coordination environments. It makes this redox-inert metal ion part of redox metabolism (Maret and Vallee 1998). Oxidoreduction of the sulfur allows biological control of protein functions through reversible zinc binding and mobilization of zinc from its tight binding sites (Maret 2006). It is not the only way of conferring mobility on zinc, though. While zinc is bound permanently during the lifetime of many proteins, it binds transiently to some proteins that have sites with coordination dynamics to enhance dissociation rates and move zinc ions (Maret and Li 2009; Maret 2011a). Thus, coordination environments also have a role in biological time and determine whether zinc resides in the protein for a very long time, for example, carbonic anhydrase with a half-life of about half a year, or is moved relatively fast during transfer to other proteins or transport through membrane proteins (Maret 2012). For example, transport rates of zinc transporters are on a 10–100 ms time scale, many orders of magnitude faster than the zinc dissociation rates of zinc from zinc metalloproteins (Chao and Fu 2004). This mobility is achieved by protein dynamics and reducing or not having scaffolding of sites, a term that refers to fastening donor ligands in the primary coordination sphere through interactions with a secondary coordination sphere. The coordination environments of the zinc ion when not bound to proteins are not known. Because of the abundance of low molecular weight molecules that could potentially serve as ligands, the coordination chemistry in the cell is much more complex than simple aqueous solution chemistry (Krężel and Maret 2016). The roles of non-protein ligands in the control of cellular zinc are not known either, but glutathione is a major candidate (Marszałek et al. 2018).

Zinc has functions in cellular signaling pathways that include proliferation and differentiation of cells (Beyersmann and Haase 2001). At the molecular level, its functions are based on catalytic, structural, and regulatory roles in proteins. It includes binding sites between proteins or their subunits (Maret 2004b). The first two functions are well characterized in numerous zinc metalloproteins. Owing to its ability to organize protein domains, zinc contributes significantly to the diversity of protein architecture and biomolecular interactions. Although zinc regulation has been mentioned in the literature for a long time, the details and the extent of regulatory functions of zinc were unknown, mainly because there was no knowledge about how such regulation with a metal ion that is tightly bound to proteins is possible. The sheer number of zinc proteins is impressive and makes zinc the most important metal ion for protein structure and function. Bioinformatics approaches defined the zinc proteome (Andreini et al. 2006). Humans have an estimated 3000 zinc proteins. This count does not include sites with regulatory functions, the number of which is unknown as will be discussed later (Maret 2008).

All of these cellular functions of zinc require an efficient regulatory system. A quantitative concept of how multicellular eukarya regulate cellular zinc and how zinc can regulate protein functions is the subject of two recent articles. In the previous edition of this book, the molecular aspects of how cellular zinc and zinc ion transients are handled were discussed (Maret 2014). The cellular aspects involve a high degree of compartmentalization, intracellular trafficking, and different types of zinc ion signals (Maret 2017). The present article builds on the content of these articles and provides an update on the cellular and molecular mechanisms of how zinc is regulated and how it can serve as a signaling ion.

2.1 Proteins Regulating Zinc: Buffering and Muffling in Cellular Zinc Homeostasis

Key to understanding the control of zinc is the affinity of proteins for zinc and the resulting available "free" zinc ion concentrations. The control of "free" metal ion concentrations is commonly described by metal ion buffering in analogy to proton buffering. The pK_a of acid-base pairs and the ratio of their concentrations determine the pH value. Likewise, the pK_d of ligands for zinc and the ratio of bound and unbound ligands determine the pZn value. With the relatively high affinities of cellular zinc-binding sites in zinc proteins, the resulting "free" zinc ion concentrations are very low, yielding pZn values of ten or higher (Krężel and Maret 2006). Another parameter that is important for control is the zinc buffering capacity. It describes how resistant the pZn is to change. It is in the micromolar range in cells (Krężel and Maret 2006).

In biology, in contrast to simple solution chemistry, removal or addition of zinc by transport processes also contributes to metal buffering of a cell. This property is referred to as muffling (Colvin et al. 2010). Minimally, two dozen zinc transporters, a dozen metallothioneins, and at least one zinc sensor, metal-regulatory element (MRE) binding transcription factor-1 (MTF-1), participate in cellular zinc homeostasis. This remarkably large number of proteins is needed to regulate zinc ion transients (fluctuations) in addition to ascertaining proper zinc concentrations and to distribute zinc subcellularly for its functions in organelles. An important aspect of the subcellular biology of zinc is its storage in vesicles (zincosomes), and the role of zinc-containing vesicles in cellular exocytosis of zinc ions for extracellular functions.

2.2 Metallothioneins

Metallothionein (MT) has been known for 70 years (Margoshes and Vallee 1957). The protein, thionein (T), obtained its name from being sulfur-rich. It binds metal ions, hence the name metallothionein for the metal-bound form. Its function was thought to be elusive for a long time. In fact, it could not have been defined because knowledge about how cellular zinc is regulated and what exactly the affinities of MT for zinc are did not exist. Zinc coordination in MT is entirely based on interactions with the sulfur donors of cysteines in 2 "clusters," 1 binding 3 zinc ions to 9 cysteines and the other 4 zinc ions to 11 cysteines (Robbins et al. 1991). Inorganic biochemists pondered long over what purpose such a binding in a unique structure might have in biology. The answer came from determining the affinities of MT for its seven zinc ions and relating them to the affinities of other proteins for zinc as well as to the "free" zinc ion concentrations. The answer is that MT is a regulated biological zinc (and copper) buffer (Krężel and Maret 2017).

MT has not just one high affinity for zinc as previously thought and which would make it a storage molecule. Instead, its affinities for zinc vary over a wider range, exactly the range needed for buffering cellular zinc (Krężel and Maret 2007). The dissociation constants (10^{-11} and 10^{-9} M) are commensurate with the measured "free" zinc ion concentrations of a few hundred picomolar. The affinities of some sites of MT for zinc are weaker than those of zinc for zinc-requiring metalloproteins, thus leaving the large number of zinc proteins requiring zinc for function in the zinc-bound state in the presence of MT. MT can be compared to a "clean sweeper," keeping zinc away from spurious, weaker binding sites, but at the same time ascertaining that genuine zinc metalloproteins obtain and retain their zinc. On the other hand, the induction of T, changing the MT/T (bound/unbound ligand) ratio, removes zinc from proteins that have weaker zinc-binding sites than genuine zinc metalloproteins and can bind zinc when zinc ion concentrations increase (Krężel and Maret 2008). The fact that the binding sites of MT are not saturated with zinc under normal physiological conditions, as expressed by an MT/T ratio that varies as a function of zinc availability, supports a function as a zinc buffer (Krężel and Maret 2008).

Zinc buffering of MT is regulated. On the protein level it is redox regulation (Maret 2011b). The thiols in thionein are coupled to redox systems (Sagher et al. 2006; Maret 2006). In this way, redox reactions can control the zinc chelating capacity of MT, making more zinc available under oxidizing conditions when "T_{ox}" is formed and making less zinc available under reducing conditions when "T_{red}" is formed (Fig. 2.1) (Maret 2000). Regulation also occurs on the gene level. In humans, about a dozen genes code for MTs and they are differentially regulated (Li and Maret 2008). A host of signaling pathways, in particular stress signals, control the total concentration of MT and indicate the large number of processes associated with managing cellular zinc. A significant number of transcription factors controlling MT gene expression are zinc proteins (Krężel and Maret 2017).

One could posit that other proteins and ligands with zinc-binding capacity buffer zinc as well. They certainly do but tighter or weaker binding is not relevant for the physiological range of buffering zinc for signaling, just as, for example, only one of the three pK_a values of phosphoric acid is relevant for pH buffering in a particular range. MT does not work in isolation. It is a "rheostat" for zinc, linking cellular signal transduction pathways to MTF-1 sensing and gene expression (Hardyman et al. 2016). Only if zinc buffering in MT is exhausted, MTF-1 is activated for making more thionein, exporting zinc ions, and for other purposes.

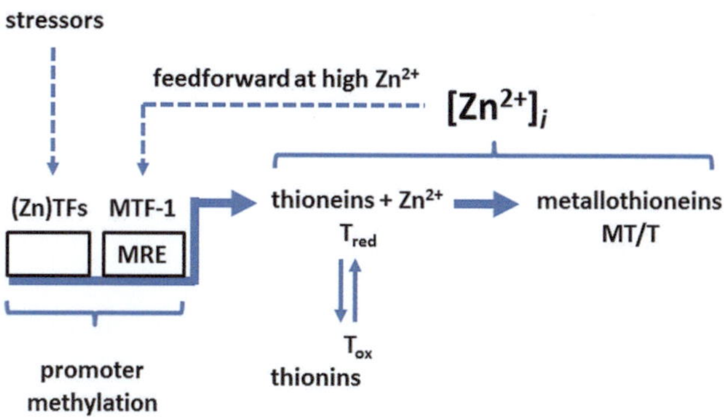

Fig. 2.1 Zinc buffering. At least a dozen human genes code for thioneins that bind zinc ions – and other ions such as cuprous ions – in their reduced form (T_{red}) and form metallothioneins (MT). MT is usually not fully saturated with zinc ions as required for a buffer, a property described with an MT/T ratio. The protein is coupled to redox reactions. Oxidation forming partially oxidized thionin (T_{ox}) reduces the zinc-binding capacity of the protein and makes zinc available, while reduction of thionin forming T_{red} reduces zinc availability. Many different pathways regulate the expression of the thionein genes, in particular many chemical and physical stressors, and include a large number of zinc-containing transcription factors (ZnTFs). One feedforward regulation is the induction of thionein at increased zinc ion concentrations via the MTF-1 transcription factor. It generates zinc chelating and reducing capacity. Further regulation is afforded via methylation of the promoters of thionins

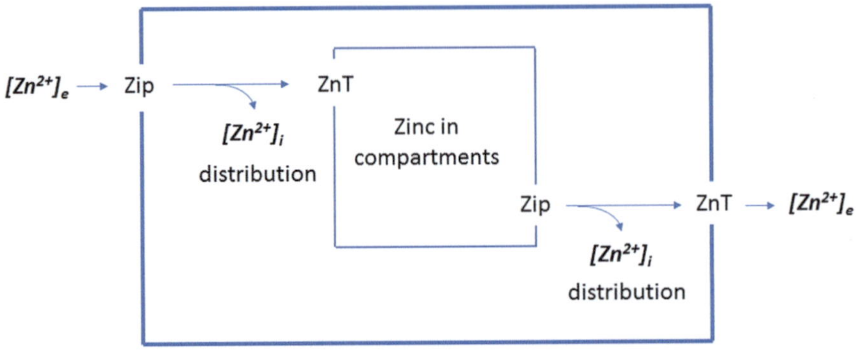

Fig. 2.2 Co-operation of zinc transporters. Plasma membrane transporters of the Zip family import extracellular zinc, $[Zn^{2+}]_e$, while transporters on intracellular membranes of the ZnT family distribute intracellular zinc, $[Zn^{2+}]_i$, subcellularly. In this directional presentation of zinc flux, the use of the two types of transporters in the export pathway is opposite to the one in the import pathway. Both import from outside the cell and import from cellular compartments increase "free" zinc ion concentrations via Zip transporters while ZnT transporters decrease them

MT is also involved in the cellular translocation of zinc. During the cell cycle, it moves from the cytosol to the nucleus (Nagel and Vallee 1995). It also translocates from the cytosol to the mitochondrial inter membrane space (Ye et al. 2001).

In addition to this dynamic zinc buffering by MT (Fig. 2.1), muffling involves transport of zinc out of the cell or into a cellular compartment (Fig. 2.2). This process is also rather complex with multiple importers and exporters and their regulation. It is required for compartmentalization and control of zinc ion signals in addition to overall homeostatic control of cellular zinc.

2.3 Zinc Transporters

In contrast to MT, the zinc coordination chemistry in the transport sites of two families of zinc transporters, Zips and ZnTs, does not involve sulfur donors. 3D structures of the mammalian proteins are unknown and all inferences about their functions are made from the 3D structures of bacterial homologues. While all of them transport zinc, their specificity is poorly characterized as some family members can transport other metal ions, in particular iron and manganese.

The impressive number of 24 zinc transporters illustrates how important it is to regulate cellular zinc (Fukada and Kambe 2011; Kambe et al. 2015). Zinc transporters need to regulate zinc ion transients in addition to maintaining the correct levels of zinc: some Zips control the formation of zinc ion signals; ZnTs restore the

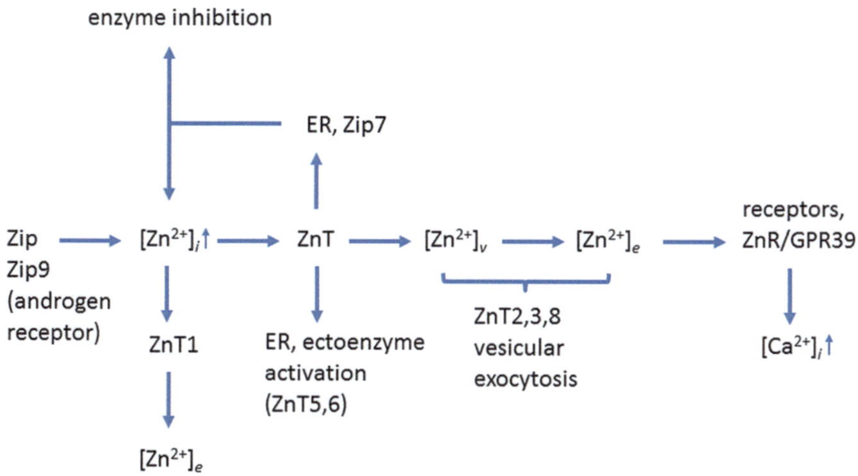

Fig. 2.3 Zinc muffling. In addition to the regulation of zinc ions via the metallothionein/thionein buffering system (Fig. 2.1), two families of zinc transporters participate in regulating cellular zinc by import, distribution, and export, and in providing zinc-dependent proteins with zinc. The transporters have different activities, such as being an androgen receptor (Zip9), loading enzymes of the secretory pathway with zinc (ZnT5,6), or providing exocytotic vesicles with zinc, $[Zn^{2+}]_v$, (ZnT2,3,8). In addition, zinc release from the endoplasmic reticulum (ER) via Zip7 can induce intracellular signaling leading to enzyme inhibition. Extracellular zinc signaling to another cell is achieved through exocytosis and binding to a zinc receptor (ZnR, GPR39), which stimulates intracellular calcium $[Ca^{2+}]_i$ release

steady-state (Fig. 2.3). Most of the Zip transporters are located on the plasma membrane though some can localize intracellularly. With the exception of ZnT1 and the evolutionary related ZnT10, ZnTs localize to intracellular membranes. Different cellular compartments require specific acquisition and delivery of zinc and transport against various concentration gradients. To accomplish this, zinc transporters may use different ions such as bicarbonate, calcium, or protons in symport or antiport (Gaither and Eide 2000; Girijashanker et al. 2008; Levy et al. 2019). The transporters are secondary active transporters; none of them are ATP-dependent. At least for one ZnT, it has been shown that it can couple to a proton gradient generating pump. The high concentration of zinc in exocytotic vesicles requires an active process as indeed shown for ZnT2 being coupled to the vacuolar ATPase (Lee et al. 2017). Tissue-specific expression of zinc transporters and removal from the plasma membrane and degradation affords further regulation.

2.3.1 *ZnTs (Solute Carrier Family SLC30A1-10)*

ZnTs belong to the cation diffusion facilitator (CDF) family of proteins. 3D structures have been reported for the *E. coli* protein YiiP and a related protein from *Shewanella oneidensis* (Lu and Fu 2007; Lopez-Redondo et al. 2018). These bacterial proteins have additional metal-binding sites believed to sense zinc ions for export. For mammalian proteins, zinc sensing mechanisms are unknown, however.

The bacterial proteins are dimers with six α-helices in the transmembrane domain (TMD) in the monomer and with a more variable, relatively large C-terminal cytoplasmic domain (CTD) with an αββαβ structure, except for ZnT9. The CTD has the capacity to bind several metal ions and to serve as a hub for interactions with other proteins. Whether zinc binding to the CTD is permanent for maintaining its structure or transient for either delivering zinc ions to the TMD or sensing them is not known. Bacterial CDFs work by conformational changes with an alternative access mechanism and proton antiport and serve as a model for mammalian ZnTs, but significant differences exist (Lu et al. 2009). While the bacterial proteins export an excess of zinc ions from cells, eukaryotic cells require zinc to be re-distributed intracellularly when there is no known excess of zinc ions. How the intracellular transporters acquire zinc from the low steady-state picomolar "free" zinc ion concentrations is unknown. The majority of ZnTs are within organellar membranes and serve diverse functions, such as filling vesicular stores and supplying organelles and secreted proteins with zinc and loading exocytotic vesicles with zinc for fundamental biological processes, among which neurotransmission (ZnT3), insulin storage (ZnT8), and lactation (ZnT2) are the best characterized.

Overall, the core structure of the TMD and CTD is relatively well conserved but significant differences exist in the different members regarding extensions at the N- and C-termini and additional loops between TM helices. ZnT5 has twice the number of TM helices. The phylogenetic tree shows clades consisting of the vesicular transporters (ZnT2-4, ZnT8), ZnT1 and ZnT10, ZnT5 and ZnT7, and ZnT6 and ZnT9 (Hogstrand and Fu 2014). ZnT6 does not transport zinc but heterodimerizes with ZnT5 when supplying zinc for zinc proteins of the secretory pathway (Tsuji et al. 2017).

Characterization of the individual CTD of human ZnT8 shows that it folds independently and that its metal binding is different from that of the bacterial transporters, binding only two instead of the four or six metal ions at the dimer interface in the bacterial transporters (Parsons et al. 2018).

2.3.2 *ZIPs (Zrt/Irt-Like Proteins) (Solute Carrier Family SLC39A1-14)*

A 3D structure has been determined for a ZIP transporter from *Bordetella bronchiseptica*. The TM helices have a novel 3+2+3 arrangement (Zhang et al. 2017). The zinc transport site is a binuclear metal site. Human Zip2, however, is predicted to have a mononuclear metal transport site and shows pH and voltage modulation of metal transport (Gyimesi et al. 2019). The 3D structure of the extracellular domain of human Zip4 was determined in an approach to understanding the entire mammalian protein by piecing together the structures of individual domains (Zhang et al. 2016). It forms two subdomains with a linker, which is characteristic of all nine LIV1 members with the so-called PAL motif. The extracellular domain forms a dimer. Some Zips form heteromers (Taylor et al. 2016). Two other parts of the Zip4 molecule were investigated regarding zinc binding, and they are believed to participate in zinc sensing, an extracellular loop with histidines (Chun et al. 2019) and an intracellular loop with histidines (Bafaro et al. 2019). The intracellular loop is thought to sense zinc ion concentrations that are sufficiently high and signal no further need for zinc import. This sensing is linked to removal of Zip4 from the plasma membrane and degradation (Shakenabat et al. 2015).

The phylogenetic tree encompasses one major clade, the LIV1 subfamily, containing 12/4; 8/14; 5 & 10/6; 13/7 and another clade containing Zip1-3. Zip11 and Zip9 are more distantly related. Zip9 is an androgen receptor and zinc transporter (Thomas et al. 2018).

2.4 MTF-1

MTF-1, metal regulatory element (MRE)-binding transcription factor-1, controls zinc-dependent gene expression. The protein has six zinc fingers which are implicated in sensing increased zinc ion concentrations (Laity and Andrews 2007). It responds to stressors such as heavy metals, hypoxia, and oxidative stress and induces the expression of thionein and ZnT1 to maintain cellular zinc homeostasis. It is essential for embryonic liver development (Günther et al. 2012). Another emerging aspect of MTF-1 is that it regulates miRNAs via MREs and thus another level of regulation of zinc metabolism (Francis and Grider 2019).

2.5 Signaling with Inorganic Ions: Ca^{2+} and Zn^{2+}

Phosphate metabolism is critical for energy metabolism. Because the strong interaction of phosphate with calcium can cause precipitation of calcium phosphate, free calcium concentrations must be kept very low in the cell compared to the environment. Zinc ions, because of their generally higher affinity for ligands, are

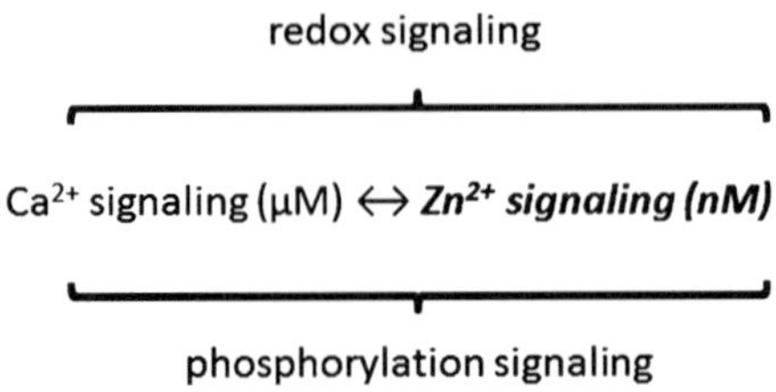

Fig. 2.4 Signaling with redox-inert metal ions. Zinc signaling complements calcium signaling but occurs in a separate, much lower range of concentrations. Both signaling systems interact with redox and phosphorylation signaling

Table 2.1 Ca^{2+} vs Zn^{2+} as carriers of signals

Property	Calcium (Ca^{2+})	Zinc (Zn^{2+})
Steady-state concentration	10–100 nM	10–100 pM
Transients	>1 μM	>1 nM
Concentrations across plasma membrane	1 mM outside vs 100 nM inside (free calcium)	15 μM outside vs 250 μM (total zinc)[a]
Signal type	Puffs, sparks, and waves	Sparks and waves
Main function of signal	Activation	Inhibition
Buffering	Calmodulin and others	Metallothionein and others
Coordination chemistry	Oxygen	Oxygen, nitrogen, sulfur
Motifs	EF hands and others	Zinc fingers and others
Transporters	Active transport, channels, exchangers	Secondary active transport, channels
Intracellular stores	ER, mitochondria, acidocalcisomes	ER, mitochondria, zincosomes
Second messenger causing release from store	IP_3	cAMP(?)[b]
Extracellular metal-sensing receptor for first messenger	CaR	ZnR (GPR39)

[a]The extracellular "free" zinc concentration is not known
[b]Observed in fungal biochemistry (Kjellerup et al. 2018)

kept at even lower concentrations, but they must be acquired from an environment low in zinc. It has been stated that signaling substances occur at rather low concentrations for energetic reasons (Carafoli and Krebs 2016). For both zinc and calcium, it is achieved through buffering by proteins and exquisite regulation. Uncontrolled increases are a major pathway for cytotoxicity and cell death. "Free" zinc ion concentrations are about three orders of magnitude lower than the already low steady state free calcium concentrations, although the total concentrations of the two ions differ at the most only by one order of magnitude: calcium a few millimolar, zinc hundreds of micromolar. This difference allows the two signaling systems to co-exist and complement each other (Fig. 2.4). Measuring the lower concentrations of "free" zinc with sufficiently selective probes and sensors has been experimentally challenging. The biological functions of the two metal ions have a lot in common (Table 2.1).

And the two signaling systems co-operate. Zinc ions inhibit the Ca^{2+}-ATPase on the plasma membrane with a K_i value of 80 pM and thus exert control over cellular-free calcium concentrations (Hogstrand et al. 1999). Another target of zinc is the cardiac ryanodine receptor/calcium channel (RyR2), which leaks calcium when cellular "free" zinc ion concentrations increase above 2 nM (Reilly-O'Donnell et al. 2017). Together with Mg^{2+}, the three redox-inert metal ions can regulate processes over many orders of magnitude in concentrations. Its redox inertness and flexible coordination environments make zinc ideally suited for a role as a messenger and for transmitting specific signals to sites on proteins.

2.6 Zinc Regulating Proteins (Zinc/Zn^{2+} Signaling)

2.6.1 Intracellular Regulation

How can the very low concentrations of available "free" zinc ions regulate and serve as signaling ions in a zinc-buffered environment? When fluctuations of cellular zinc ion concentrations were measured, it became apparent that zinc ions indeed regulate biological functions (Fukada et al. 2011). Specific mechanisms exist for signal generation, transduction, and decoding. Two pathways are known to elicit cellular zinc ion transients. One is the release of zinc ions from an intracellular store (Yamasaki et al. 2007; Taylor et al. 2012a) and the other is the release of zinc from zinc/thiolate sites in proteins such as in metallothioneins. Release of zinc from a store in the ER occurs in response to the hormone stimulation of cells, which results in casein kinase-2-dependent phosphorylation and opening of the Zip7 channel (Taylor et al. 2012b). The released zinc is required for skin (dermis) and B-cell development through controlling ER zinc, and hence ER stress, and cytosolic zinc (Bin et al. 2017; Woodruff et al. 2018; Anzilotti et al. 2019) (Fig. 2.3). Casein kinase-2 is a zinc protein with zinc finger-mediated dimerization of the regulatory subunit (Chantalat et al. 1998). Whether other Zip transporters are regulated in a similar way is not known. Stimulation also occurs with zinc ions, indicating that there is a zinc-induced zinc release just as in the case of calcium (Kjellerup et al. 2018). Release of zinc from zinc/thiolate coordination environments occurs under conditions of redox signaling and oxidative or carbonyl stress (Hao and Maret 2006). It can lead to Zn^{2+}-dependent gene expression, which is part of a feedback loop as it includes proteins involved in chelating zinc and in an antioxidant response. Aside from zinc-dependent gene expression, other molecular targets of the zinc ions have not been linked directly to the events that cause a zinc transient. In contrast to *activation* of gene expression, the typical response is an *inhibition* of enzymes (Maret et al. 1999). The overall effect can be confusing because zinc inhibition on the molecular level can result in the activation of a process. For example, a general observation is that an increase of cellular free zinc ions enhances phosphorylation signaling (Haase and Maret 2003). One cause for this activation is an inhibition of protein tyrosine phosphatases (PTPs), which are the main regulators of phosphorylation signaling. In PTP1B, the phosphatase controlling the insulin and leptin receptors, zinc inhibits only the enzyme in the phosphorylated form through the generation of a zinc-binding site in the closed conformation (Bellomo et al. 2014, 2016). The inhibition constant for receptor protein tyrosine phosphatase β (RPTPβ) is 21 pM, suggesting that some PTPs are inhibited tonically and need to be activated by zinc removal (Wilson et al. 2012). Many other enzymes with inhibition constants in the nanomolar range have been reported (Maret 2013a). One common feature is that catalytic residues positioned in proximity can bind a metal ion, that is, cysteine-dependent enzymes with a nearby histidine. Such diads and triads of amino acids in the active site have dual functions: They serve as catalytic groups and they provide the donors for zinc binding and inhibition. Inhibitory zinc-binding sites on several enzymes, including caspases, cathepsins, and kallikreins, have been characterized structurally (Maret 2013a; Eron et al. 2018). The inhibition involves specific mechanisms of binding and regulation such as allosterism and binding one or several zinc ions. The inhibition of many proteinases indicates a role of zinc in proteostasis. Since these zinc-inhibited enzymes do not need zinc for activity they are not considered to be zinc enzymes although their zinc binding is almost as strong as that for genuine zinc enzymes. Because of the strong metal inhibition, the enzymes are often supplied commercially with the chelating agent ethylenediaminetetraacetic acid (EDTA) to keep them active, thus masking the zinc inhibition. While most zinc sites in zinc proteins are readily recognized by signatures/motifs with characteristic sets of donors linked by relatively short non-coordinating amino acid spacers, this is not the case in inhibitory sites. Therefore, these sites are not amenable to the database mining approach that has been so successful in predicting the total number of zinc metalloproteins in the human genome (Andreini et al. 2006). Consequently, we do not have estimates of the number of regulatory sites. An even less understood facet is how the zinc-inhibited proteins are activated. At least two ways can be envisaged. One is lowering the free zinc ion concentrations by muffling, removing zinc ions with transporters, and the other is buffering, binding zinc by a protein such as thionein. In the first case, the re-activation would depend on the rate of zinc dissociation, which could be limiting unless assisted by coordination dynamics of the proteins as in the zinc transport sites of zinc transporters. In the second case, zinc dissociation would be fast if a chelating agent accelerates it but slow if gene expression such as induction of thionein is involved. To distinguish the two possibilities, one would need to know the duration of the cellular zinc ion transients in addition to their amplitudes, which are in the order of a few nanomolar when measured globally. The link between zinc ion signals and their targets is often tenuous and it remains unknown whether local zinc ion transients or a global change of the pZn value is responsible for the effect. Linking zinc ion transients with specific targets is one of the pressing issues in the field of zinc signaling to explain the great number of functional readouts that have been reported.

An exception to the rule that the action of each metal is well separated from that of others is the binding of zinc to magnesium sites. Work dating back to 1961 showed that zinc inhibits Mg-dependent phosphoglucomutase with an inhibition constant (IC_{50}) of about 2 pM (Milstein 1961). The tight inhibition was then used to set an upper limit to the "free" zinc ion concentration possible without inhibiting the enzyme (Peck and Ray 1971). Investigations into whether zinc is indeed a physiological modulator of the activity of this enzyme have not been pursued. Also, Mg^{2+} activates PTP1B while Zn^{2+} inhibits it (Bellomo et al. 2018). Another relationship between the two essential minerals is that some kinases prefer a Zn-ATP complex over the typically employed Mg-ATP complex (McCormick 2002).

2.6.2 Extracellular Regulation

In addition to intracellular signaling, there is extracellular zinc ion signaling. It was first described in neurotransmission and established zinc ions stored in synaptic vesicles (boutons) as a neuromodulator (Frederickson et al. 2005). The best described target is the N-methyl-D-aspartate (NMDA) receptor. Its structure shows zinc in the high affinity (5 nM) binding site in the N2A subunit bound to His-44, His-128, Glu-266, and Asp-282. Only two ligands bind zinc at the low affinity site (7.5 μM): His-127 and Glu-284 (Romero-Hernandez et al. 2016).

In addition to the zinc-glutamatergic neurons, many other cells have the capacity to secrete zinc ions. They do so for different purposes with paracrine, autocrine, or even endocrine effects. One target is ZnR/GPR39, a G-protein coupled receptor on the plasma membrane activating phospholipase β, releasing Ca^{2+} from the ER, and stimulating extracellular signal-regulated kinases (ERK1/2) and protein kinase B (Akt) (Hershfinkel 2018). Fertilized mammalian embryos emit zinc sparks. The exocytotically released zinc ions are coordinated with cellular calcium transients and modify the structure of the zona pellucida to avoid polyspermy (Que et al. 2019). The ZnT2,3 and 8 zinc transporters are involved in loading exocytotic vesicles with zinc. The exocytosed zinc ions in other tissues and their targets have received comparatively less attention. Zinc affinities for extracellular proteins probably do not need to be as low as for those in the cytosol. Zinc binds to and affects the function of a large number of membrane channels, but only few of them have affinities in the nanomolar range (Peralta and Huidobru-Toro 2016). The conditions for buffering zinc ions are different in the extracellular environment, the vesicles, and the cytosol. Therefore, intracellular sites on membrane proteins may need higher affinities for zinc modulation than extracellular sites.

2.7 Zinc Functions in Health and Disease

Zinc is involved in virtually all aspects of cellular functions and is important for the control of proliferation, differentiation, and programmed cell death (Maret 2013b). On an organismal level, it is essential for growth, development, maintenance of function, and protection against various physical and chemical insults. The implications of perturbation of zinc metabolism for disease are huge. The role of zinc in diseases can originate from mutations, the number of which is quite high in zinc transporters and MTs regulating cellular zinc (Hogstrand and Maret 2016; Raudenska et al. 2014). Several diseases are specifically caused by a disruption of zinc transporter function and thus perturbation of zinc homeostasis: acrodermatitis enteropathica (Zip4), neonatal transient zinc deficiency (ZnT2), and SCD-Ehlers-Danlos Syndrome (Zip13). In many more diseases or syndromes, zinc is involved causally. Nutritional zinc deficiency and interference of xenobiotics with zinc metabolism are also causes of diseases. Discovery of the numerous molecular functions of zinc has been a boon for biochemistry but a bane for a wider acceptance of the overall importance of zinc for the cell biology. Reasons for the latter are the lack of a biomarker for the cellular functions of zinc and the overall issue that pleiotropic functions of zinc may give the impression of a lack of specificity thus concealing its true importance.

Acknowledgment I thank my colleague Dr. Ron Jacob at King's College London for stimulating discussions.

References

Andreini C, Banci L, Bertini I, Rosato A (2006) Counting the zinc-proteins encoded in the human genome. J Proteome Res 5:196–201

Anzilotti C, Swan DJ, Boisson B, Deobagkar-Lele M, Oliveira C, Chabosseau P, Engelhardt KR, Xu X, Chen R, Alvarez L, Berlinguer-Palmini R, Bull KR, Cawthorne E, Cribbs AP, Crockford TL, Dang TS, Fearn A, Fenech EJ, de Jong SJ, Lagerholm BC, Ma CS, Sims D, van děn Berg B, Xu Y, Cant AJ, Kleiner G, Leahy TR, de la Morena MT, Puck JM, Shapiro RS, van der Burg M, Chapman JR, Christianson JC, Davies B, McGrath JA, Przyborski S, Santibanez Koref M, Tangye SG, Werner A, Rutter GA, Padilla-Parra S, Casanova JL, Cornall RJ, Conley ME, Hambleton S (2019) An essential role for the Zn_{2+} transporter ZIP7 in B cell development. Nat Immunol 20:350–361

Bafaro EM, Maciejewski MW, Hoch JC, Dempski RE (2019) Concomitant disorder and high affinity zinc binding in the hZIP4 intracellular loop. Protein Sci 28:868–880

Bellomo E, Massarotti A, Hogstrand C, Maret W (2014) Zinc ions modulate protein tyrosine phosphatase 1B activity. Metallomics 6:1229–1239

Bellomo E, Singh BK, Massarotti A, Hogstrand C, Maret W (2016) The metal face of PTP-1B. Coord Chem Rev 327–328:70–83

Bellomo E, Abro A, Hogstrand C, Maret W, Domene C (2018) Role of zinc and magnesium ions in the modulation of phosphoryl transfer in protein tyrosine phosphatase 1B. J Am Chem Soc 140:4446–4454

Beyersmann D, Haase H (2001) Functions of zinc in signaling, proliferation and differentiation of mammalian cells. Biometals 14:331–341

Bin BH, Bhin J, Seo J, Kim SY, Lee E, Park K, Choi DH, Takagishi T, Hara T, Hwang D, Koseki H, Asada Y, Shimoda S, Mishima K, Fukada T (2017) Requirement of zinc transporter SLC39A7/ZIP7 for dermal development to fine-tune endoplasmic reticulum function by regulating protein disulphide isomerase. J Invest Dermatol 137:1682–1691

Carafoli E, Krebs J (2016) Why calcium? How calcium became the best communicator. J Biol Chem 291:20849–20857

Chantalat L, Leroy D, Filhol O, Nueda A, Benitez MJ, Chambaz EM, Cochet C, Dideberg O (1998) Crystal structure of the human protein kinase CK2 regulatory subunit reveals its zinc finger-mediated dimerization. EMBO J 18:2930–2940

Chao Y, Fu D (2004) Kinetic study of the antiport mechanism of an *Escherichia coli* zinc transporter ZitB. J Biol Chem 279:12043–12050

Chun H, Korolnek T, Lee C-J, Coyne HJ III, Winge DR, Kim B-E, Petris MJ (2019) An extracellular histidine-containing motif in the zinc transporter ZIP4 plays a role in zinc sensing and zinc-induced endocytosis in mammalian cells. J Biol Chem 294:2815–2826

Colvin RA, Holmes WR, Fontaine CP, Maret W (2010) Cytosolic zinc buffering and muffling: their role in intracellular zinc homeostasis. Metallomics 2:306–317

Eron SJ, MacPherson DJ, Dagbay KB, Hardy JA (2018) Multiple mechanisms of zinc-mediated inhibition for the apoptotic caspases -3, -6, -7, and -8. ACS Chem Biol 13:1279–1290

Francis M, Grider A (2019) MiRNA-target interactions in osteogenic signalling pathways involving zinc via the metal regulatory element. Biometals 13:111–121

Frederickson CJ, Koh J-Y, Bush AI (2005) The neurobiology of zinc in health and disease. Nat Rev Neurosci 6:449–462

Fukada T, Kambe T (2011) Molecular and genetic features of zinc transporters in physiology and pathogenesis. Metallomics 3:662–674

Fukada T, Yamasaki S, Nishida K, Murakami M, Hirano T (2011) Zinc homeostasis and signaling in health and diseases. J Biol Inorg Chem 16:1123–1134

Gaither LA, Eide DJ (2000) Functional expression of the human hZIP2 zinc transporter. J Biol Chem 275:5560–5564

Girijashanker K, He L, Soleimani M, Reed JM, Li H, Liu Z, Wang B, Dalton TP, Nebert DW (2008) Slc39a14 gene encodes ZIP14, a metal/bicarbonate symporter: similarities to the ZIP8 transporter. Mol Pharmacol 73:1413–1423

Günther V, Lindert U, Schaffner W (2012) The taste of heavy metals: gene regulation by MTF-1. Biochim Biophys Acta 1823:1416–1425

Gyimesi G, Albano G, Fuster DG, Hediger MA, Pujol-Giménez J (2019) Unraveling the structural elements of pH sensitivity and substrate binding in the human zinc transporter SLC39A2 (ZIP2). J Biol Chem 294:8046–8063

Haase H, Maret W (2003) Intracellular zinc fluctuations modulate protein tyrosine phosphatase activity in insulin/insulin-like growth factor-1 signaling. Exp Cell Res 291:289–298

Hao Q, Maret W (2006) Aldehydes release zinc from proteins. A pathway from oxidative stress/lipid peroxidation to cellular functions of zinc. FEBS J 273:4300–4310

Hardyman JE, Tyson J, Jackson KA, Aldridge C, Cockell SJ, Wakeling LA, Valentine RA, Ford D (2016) Zinc sensing by metal-responsive transcription factor 1 (MTF1) controls metallothionein and ZnT1 expression to buffer the sensitivity of the transcriptome response to zinc. Metallomics 8:337–343

Hershfinkel M (2018) The zinc sensing receptor, ZnR/GPR39, in health and disease. Int J Mol Sci 19:E439

Hogstrand C, Fu D (2014) Zinc. In: Maret W, Wedd AG (eds) Binding, transport and storage of metal ions in biological cells. Royal Society of Chemistry, Cambridge, pp 666–694

Hogstrand C, Maret W (2016) Genetics of human zinc deficiencies. In: eLS (encyclopedia of life sciences), Wiley, Chichester

Hogstrand C, Verbost PM, Wendelaar Bonga SE (1999) Inhibition of human Ca^{2+}-ATPase by Zn^{2+}. Toxicology 133:1139–1145

Kambe T, Tsuji T, Hashimoto A, Itsumura N (2015) The physiological, biochemical, and molecular roles of zinc transporters in zinc homeostasis and metabolism. Physiol Rev 95:749–784

Kjellerup L, Winther AL, Wilson D, Fuglsang AT (2018) Cyclic AMP pathway activation and extracellular zinc induce rapid intracellular zinc mobilization in *Candida albicans*. Front Microbiol 9:502. https://doi.org/10.3389/fmicb.2018.00502. eCollection 2018

Kluska K, Adamczyk J, Krężel A (2018) Metal binding properties, stability, and reactivity of zinc fingers. Coord Chem Rev 367:18–64
Krężel A, Maret W (2006) Zinc buffering capacity of a eukaryotic cell at physiological pZn. J Biol Inorg Chem 11:1049–1062
Krężel A, Maret W (2007) The nanomolar and picomolar zinc binding properties of metallothionein. J Am Chem Soc 129:10911–10921
Krężel A, Maret W (2008) Thionein/metallothionein control Zn(II) availability and the activity of enzymes. J Biol Inorg Chem 13:401–409
Krężel A, Maret W (2016) The biological inorganic chemistry of zinc ions. Arch Biochem Biophys 611:3–19
Krężel A, Maret W (2017) The functions of metamorphic metallothioneins in zinc and copper metabolism. Int J Mol Sci 18:1237
Laity JH, Andrews JK (2007) Understaning the mechanisms of zinc-sensing by metal-responsive element binding transcription factor-1 (MTF-1). Arch Biochem Biophys 463:201–210
Lee S, Rivera OC, Kelleher SL (2017) Zinc transporter 2 interacts with vacuolar ATPase and is required for polarization, vesicle acidification, and secretion in mammary epithelial cells. J Biol Chem 292:21598–21613
Levy M, Elkoshi N, Barber-Zucker S, Hoch E, Zarivatch R, Hershfinkel M, Sekler I (2019) Zinc transporter 10 (ZnT10)-dependent extrusion of cellular Mn^{2+} is driven by an active Ca^{2+}-coupled exchange. J Biol Chem 294:5879–5889
Li Y, Maret W (2008) Human metallothionein metallomics. J Anal At Spectrom 23:1055–1062
Lopez-Redondo ML, Coudray N, Zhang Z, Alexopoulos J, Stokes DL (2018) Structural basis for the alternating access mechanism of the cation diffusion facilitator YiiP. Proc Natl Acad Sci USA 115:3042–3047
Lu M, Fu D (2007) Structure of the zinc transporter YiiP. Science 317:1746–1748
Lu M, Chai J, Fu D (2009) Structural basis for autoregulation of the zinc transporter YiiP. Nat Struct Mol Biol 16:1063–1067
Maret W (2000) The function of zinc metallothionein: a link between cellular zinc and redox state. J Nutr 130:1455S–1458S
Maret W (2004a) Zinc and sulfur: a critical biological partnership. Biochemistry 43:3301–3309
Maret W (2004b) Protein interface zinc sites: a role of zinc in the supramolecular assembly of proteins and in transient protein-protein interactions. In: Messerschmidt A, Bode W, Cygler M (eds) Handbook of metalloproteins, vol 3. Wiley, Chichester, pp 432–441
Maret W (2006) Zinc coordination environments in proteins as redox sensors and signal transducers. Antioxid Redox Signal 8:1419–1441
Maret W (2008) Zinc proteomics and the annotation of the human zinc proteome. Pure Appl Chem 80:2679–2687
Maret W (2011a) Metals on the move: zinc ions in cellular regulation and in the coordination dynamics of zinc proteins. Biometals 24:411–418
Maret W (2011b) Redox biochemistry of mammalian metallothioneins. J Biol Inorg Chem 16:1079–1086
Maret W (2012) New perspectives of zinc coordination environments in proteins. J Inorg Biochem 111:110–116
Maret W (2013a) Inhibitory zinc sites in enzymes. Biometals 26:197–204
Maret W (2013b) Zinc biochemistry: from a single zinc enzyme to a key element of life. Adv Nutr 4:82–91
Maret W (2014) Molecular aspects of zinc signals. In: Fukuda T, Kambe T (eds) Zinc signals in cellular functions and disorders. Springer Japan KK, Tokyo, pp 7–26
Maret W (2017) Zinc in cellular regulation: the nature and significance of "zinc signals". Int J Mol Sci 18:2285
Maret W (2019) The redox biology of redox-inert zinc ions. Free Radic Biol Med 134:311–326
Maret W, Li Y (2009) Coordination dynamics of zinc in proteins. Chem Rev 109:4682–4707

Maret W, Vallee BL (1998) Thiolate ligands in metallothionein confer redox activity on zinc clusters. Proc Natl Acad Sci USA 95:3478–3482
Maret W, Jacob C, Vallee BL, Fischer EH (1999) Inhibitory sites in enzymes: zinc removal and reactivation by thionein. Proc Natl Acad Sci USA 96:1936–1940
Margoshes M, Vallee BL (1957) A cadmium protein from equine kidney cortex. J Am Chem Soc 79:4813
Marszałek I, Goch W, Bal W (2018) Ternary Zn(II) complexes of FluoZin-3 and the low molecular weight component of the exchangeable cellular zinc pool. Inorg Chem 57:9826–9838
McCormick D (2002) Micronutrient cofactor research with extensions to applications. Nutr Res Rev 15:245–262
Milstein C (1961) Inhibition of phosphoglucomutase by trace metals. Biochem J 79:591–596
Nagel WW, Vallee BL (1995) Cell cycle regulation of metallothionein in human colonic cancer cells. Proc Natl Acad Sci USA 92:579–583
Parsons D, Hogstrand C, Maret W (2018) The C-terminal cytosolic domain of the human zinc transporter ZnT8 and its diabetes risk variant. FEBS J 285:1237–1250
Peck EJ, Ray WJ (1971) Metal complexes of phosphoglucomutase in vivo. J Biol Chem 246:1160–1167
Peralta FA, Huidobro-Toro JP (2016) Zinc as allosteric ion channel modulator: ionotropic receptors as metalloproteins. Int J Mol Sci 17:1059
Que EL, Duncan FE, Lee HC, Hornick JE, Vogt S, Fissore RA, O'Halloran TV, Woodruff TK (2019) Bovine eggs release zinc in response to parthenogenetic and sperm-induced egg activation. Theriogenology 127:41–48
Raudenska M, Gumulec J, Podlaha O, Sztalmachova M, Babula P, Eckschlager T, Adam V, Kizek R, Masarik M (2014) Metallothionein polymorphisms in pathological processes. Metallomics 6:55–68
Reilly-O'Donnell B, Robertson GB, Karumbi A, McIntyre C, Bal W, Nishi M, Takeshima H, Stewart AJ, Pitt SJ (2017) Dysregulated Zn^{2+} homeostasis impairs cardiac type-2 ryanodine receptor and mitsugumin 23 functions, leading to sarcoplasmic reticulum Ca^{2+} leakage. J Biol Chem 292:13361–13373
Rink L (ed) (2011) Zinc in human health, Biomedical and Health Research, vol 76. IOS Press, Amsterdam
Robbins AH, McRee DE, Williamson M, Collett SA, Xuong NH, Furey WF, Wang BC, Stout CD (1991) Refined crystal structure of Cd, Zn metallothionein at 2.0 Å resolution. J Mol Biol 221:1269–1293
Romero-Hernandez A, Simorowski N, Karakas E, Furukawa H (2016) Molecular basis for subtype specificity and high-affinity zinc inhibition in the GluN1-GluN2A NMDA receptor amino-terminal domain. Neuron 92:1324–1336
Sagher D, Brunell D, Hejmancik JF, Kantorov M, Brot N, Weissbach H (2006) Thionein can serve as a reducing agent for the methionine sulfoxide reductases. Proc Natl Acad Sci USA 103:8656–8661
Shakenabat S, Din J, Maret W (2015) Genetic causes and gene-nutrient interactions in mammalian zinc deficiencies: acrodermatitis enteropathica and transient neonatal zinc deficiency as examples. J Trace Elem Biol Med 29:47–62
Taylor KM, Kille P, Hogstrand C (2012a) Protein kinase CK2 opens the gate for zinc signaling. Cell Cycle 11:1863–1864
Taylor KM, Hiscox S, Nicholson RI, Hogstrand C, Kille P (2012b) Protein kinase CK2 triggers cytosolic zinc signaling pathways by phosphorylation of zinc channel ZIP7. Sci Signal 5(210):ra11
Taylor KM, Muraina IA, Brethour D, Schmitt-Ulms G, Nimmanon T, Ziliotto S, Kille P, Hogstrand C (2016) Zinc transporter ZIP10 forms a heteromer with ZIP6 which regulates embryonic development and cell migration. Biochem J 473:2531–2544
Thomas P, Converse A, Berg HA (2018) ZIP9, a novel membrane androgen receptor and zinc transporter protein. Gen Comp Endocrinol 257:130–136

Tsuji T, Kurokawa Y, Chiche J, Pouysségur J, Sato H, Fukuzawa H, Nagao M, Kambe T (2017) Dissecting the process of activation of cancer-promoting zinc-requiring ectoenzymes by zinc metalation mediated by ZNT transporters. J Biol Chem 292:2159–2173

Vallee BL, Falchuk KH (1993) The biochemical basis of zinc physiology. Physiol Rev 73:79–118

Wilson M, Hogstrand C, Maret W (2012) Picomolar concentrations of free zinc(II) ions regulate receptor protein tyrosine phosphatase beta activity. J Biol Chem 287:9322–9326

Woodruff G, Bouwkamp CG, de Vrij FM, Lovenberg T, Bonaventure P, Kushner SA, Harrington AW (2018) The zinc transporter SLC39A7 (ZIP7) is essential for regulation of cytosolic zinc levels. Mol Pharmacol 94:1092–1100

Yamasaki S, Sakata-Sogawa K, Hasegawa A, Suzuki T, Kabu K, Sato E, Kurosaki T, Yamashita S, Tokunaga M, Nishida K, Hirano T (2007) Zinc is a novel intracellular second messenger. J Cell Biol 177:637–645

Ye B, Maret W, Vallee BL (2001) Zinc metallothionein imported into liver mitochondria modulates respiration. Proc Natl Acad Sci USA 98:2317–2322

Zhang T, Sui D, Hu J (2016) Structural insights of ZIP4 extracellular domain critical for optimal zinc transport. Nature Commun 7:11979

Zhang T, Liu J, Fellner M, Zhang C, Sui D, Hu J (2017) Crystal structure of a ZIP zinc transporter reveal a binuclear center in the transport pathway. Sci Adv 3:e1700344

Chapter 3
Zinc Transporter Proteins: A Review and a New View from Biochemistry

Taiho Kambe, Eisuke Suzuki, and Taiki Komori

Abstract Zinc is an essential biological metal found in approximately 10% of the human proteome. Zinc regulates a large number of proteins and their functions, including transcription factors, enzymes, adapters, receptors, and growth factors, acting as a structural or catalytic cofactor or as a signaling mediator. Increasing evidence indicates that the transport of zinc across biological membranes plays a pivotal role in its biological functions. Zinc transport is mostly mediated by two zinc transporter proteins, ZNT and ZIP. Members of both transporter families are involved in a variety of biological events, which in humans are often associated with health and disease. In this chapter, we review the current understanding of the biochemical functions of both transporter protein families with a particular focus on their biological subgroupings.

Keywords Zrt, Irt-like protein (ZIP) · Zn transporter (ZNT) · Solute carrier family (SLC) · Membrane transport · Cation diffusion facilitator (CDF)

3.1 Introduction

Under physiological conditions, zinc is present as a divalent cation that does not have a redox potential. In contrast to copper and iron, it therefore does not require a redox reaction during membrane transport (Kambe 2013). Thus, precise spatiotemporal control of the expression of zinc transport proteins is critically important for net zinc transport across biological membranes. Increasing evidence suggests that the cellular concentration and distribution of zinc, which are mediated through zinc mobilization across cellular membranes, are highly involved in a large number of cellular responses, such as activation/repression of transcription and translation, post-translational controls including membrane trafficking and protein stability, and

T. Kambe (✉) · E. Suzuki · T. Komori
Division of Integrated Life Science, Graduate School of Biostudies,
Kyoto University, Kyoto, Japan
e-mail: kambe1@kais.kyoto-u.ac.jp

T. Fukada, T. Kambe (eds.), *Zinc Signaling*,
https://doi.org/10.1007/978-981-15-0557-7_3

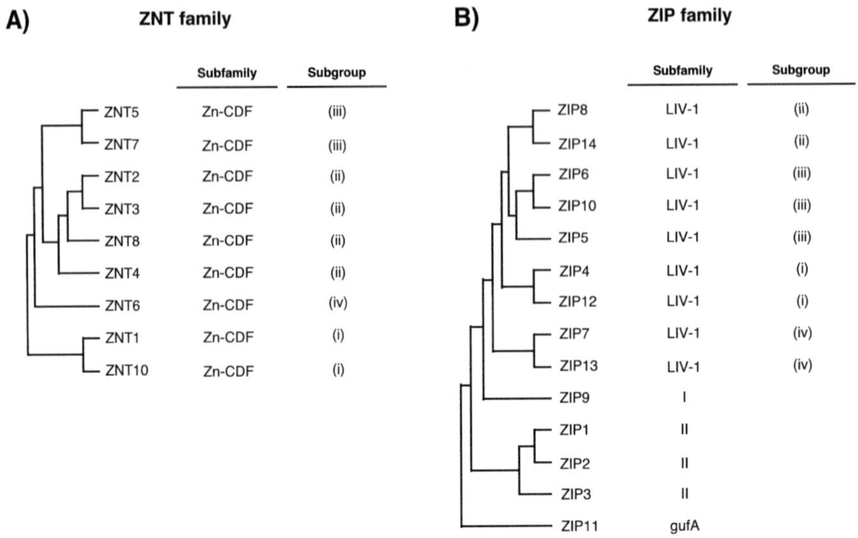

Fig. 3.1 Phylogeny of ZNT and ZIP family proteins. A neighbor-joining phylogenetic tree was generated with ClustalW, based on protein sequence alignments. (**a**) ZNT family members. (**b**) ZIP family members. In (**A**) and (**B**), the subfamily and subgroup classification assigned in this chapter are shown on the right

activation/inactivation of many enzymes related to cellular signaling (Fukada and Kambe 2011; Maret and Li 2009). Therefore, understanding the physiological/pathological regulations and biochemical properties of zinc transport is essential for understanding the biological roles of zinc. In metazoans, cellular zinc transport is primarily mediated by two zinc transporter proteins, the Zn transporter (ZnT)/solute carrier family 30 (ZNT/SLC30A) and Zrt, Irt-like protein (ZIP)/SLC39A (Fig. 3.1), that function as exporters and importers, respectively. In this chapter, the biochemical properties of these zinc transporter proteins are discussed. Their physiological and pathological roles are referred to in other chapters of this book.

3.2 History of ZNT and ZIP Family Proteins

In 1995, the first mammalian zinc transporter was identified by expression cloning using zinc-sensitive mutant Baby hamster kidney cells (Palmiter and Findley 1995). Since then, more than 20 zinc transporters, including ZNTs and ZIPs, have been identified (Gaither and Eide 2001; Kambe et al. 2004), all of which can function as secondary active transporters. The importance of zinc transporters is unquestionable, with a growing number of studies showing their involvement in both physiological and pathological processes. Before outlining their biochemical properties, we briefly discuss their historical backgrounds, which may give us important clues that would enable us to discover new functions of zinc transporter protein.

The first zinc transporter protein of the ZNT family, named Zrc1p, was identified in *Saccharomyces cerevisiae* and confers resistance to high zinc concentrations (Kamizono et al. 1989). Subsequently, Cot1p and CzcD were identified in *S. cerevisiae* and the *Cupriavidus metallidurans* (formerly *Alcaligenes eutrophus*) strain CH34, both of which were also shown to confer resistance to high levels of zinc and other metals such as cobalt (Conklin et al. 1992; Nies 1992). Due to their sequence homology, these three proteins were grouped together and named cation diffusion facilitator (CDF) proteins (Gaither and Eide 2001). However, it was later discovered that, despite their name, CDF proteins do not serve as diffusion facilitators but function as secondary active transporters. Proteins within this family are conserved at all phylogenetic levels (Gaither and Eide 2001). The first mammalian ZNT family member, rat Znt1, was identified as a member of the CDF family (Palmiter and Findley 1995). The second ZNT transporter, Znt2, was identified by expression cloning (Palmiter et al. 1996a) with other ZNT family proteins identified by other methods, including EST and genome database searches (Chimienti et al. 2004; Huang et al. 2002; Kambe et al. 2002; Kirschke and Huang 2003; Palmiter et al. 1996b). To date, there are ten members of the ZNT family, ZNT1-ZNT10; however, ZNT9 has been shown to act as a nuclear receptor coactivator (GRIP1-associated coactivator 63 (GAC 63)) (Chen et al. 2007), although there are studies reporting its involvement in zinc metabolism (Perez et al. 2017). Since the zinc transport ability of ZNT9 is yet to be fully elucidated, we have excluded it from our ZNT family discussion (Figs. 3.2 and 3.3).

Homologs of the human ZNT family proteins have been found in the genome sequences of rats, mice, chickens, zebrafish, fruit flies, nematodes, yeast, and plants (*Arabidopsis thaliana*) and are shown in Table 3.1. This table highlights that ZNT family proteins are highly conserved in vertebrates, and in nematodes, yeast, and plants. However, species such as nematodes, yeast, and plants have family members that produce ZNT family proteins, which are not homologous to vertebrate members. CDF family members are generally classified into three subfamilies, Zn-CDF, Zn/Fe-CDF, or Mn-CDF, based on their phylogenetic relationships and metal substrate specificities (Kambe 2012). All vertebrate ZNT family members belong to the Zn-CDF subgroup (Fig. 3.2). CDF members with low homology to vertebrates (Table 3.1) belong to Zn/Fe-CDF or Mn-CDF subfamilies and have been shown to be involved in other divalent cation transport in addition to zinc, which suggests that these proteins play diverse roles in metal transport.

The identification of ZIP family members is somewhat confusing. Among the ZIP family, ZIP6 was the first to be identified. However, it was originally identified as a highly expressed gene in breast cancer and named LIV-1 (Manning et al. 1994), with its zinc transport activity not evident at that time (Taylor 2000). Numerous studies have since shown that members of the ZIP family, including ZIP6, are associated with cancer development and metastasis (Bafaro et al. 2017) (see Chap. 16), which is consistent with the initial identification of ZIP6 in breast cancer. Zrt1p and Zrt2p zinc transporters were subsequently identified in *S. cerevisiae* (Zhao and Eide 1996a, b) simultaneous to the identification of IRT1 iron transporter in *A. thaliana* (Eide et al. 1996), all of which were identified from excellent genetic studies in

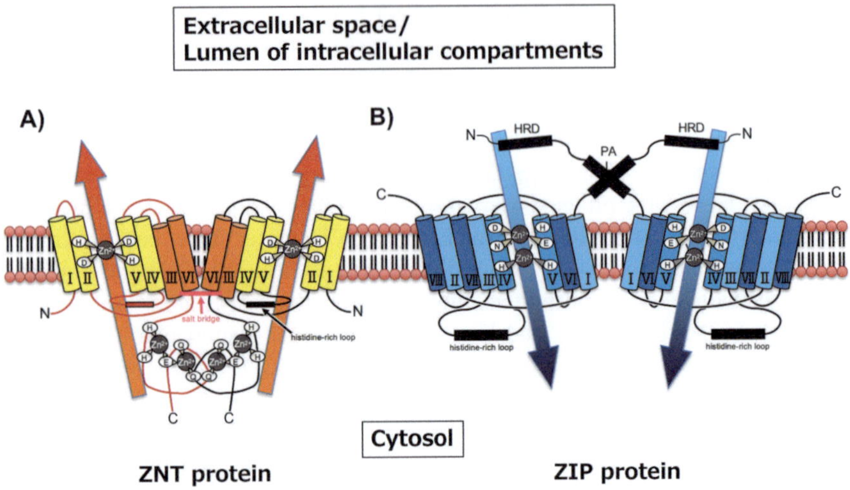

Fig. 3.2 Cartoon of predicted structures of ZNT and ZIP family proteins. (**A**). Predicted topology of the ZNT protein is shown. ZNT protomers most likely have six transmembrane domains (TMDs), in which TMDs, I, II, IV, and V form a compact four-helical bundle and the remaining TMDs, III and VI, form a two-helical pair outside the bundle. TMD II and V of the compact four-helical bundle create a transmembranous zinc binding site, which is formed by conserved aspartic acid (D) and histidine (H) residues. Predictions based on the structural properties of the bacterial homolog, YiiP (Coudray et al. 2013; Gupta et al. 2014; Lopez-Redondo et al. 2018; Lu et al. 2009; Lu and Fu 2007). ZNT proteins transport zinc from the cytosol into the extracellular space or intracellular compartments. (**B**) Predicted topology of the ZIP protein is shown. ZIP protomers are likely to have eight transmembrane domains (TMDs) consisting of a novel 3+2+3 TMD architecture, in which the first three TMDs (I to III) are symmetrically related to the last three TMDs (VI to VIII) by a pseudo-twofold axis. The conserved amphipathic amino acid residues in TMD IV (histidine, asparagine (N), and aspartic acid) and in TMD V (two histidines and one glutamic acid (E) in the potential metalloprotease motif (**HE**XP**H**EXGD) in LIV-1 subfamily, which are indicated in bold) form a binuclear metal center within the TMDs. However, the amphipathic amino acid residues in TMD V are only partially conserved in other subfamilies. Prediction based on the structure of the bacterial ZIP protein homolog, BbZIP (Zhang et al. 2017). ZIP4's long extracellular N-terminal region is divided into two structural domains, the helix-rich domain (HRD), and the PAL motif-domain (see text) (Zhang et al. 2016). ZIP family members transport zinc in the opposite direction to the ZNT family

yeast. These findings lead to the name "ZIP" (ZRT, IRT-like protein) (Eng et al. 1998), with the name suggesting that ZIP members can transport both zinc and iron, and possibly other divalent cations through a broader specificity. In fact, many recent studies show that mammalian ZIP family members can also transport manganese and cadmium (see Sect. 3.4.4.2) (Aydemir and Cousins 2018; Jenkitkasemwong et al. 2012). After the identification of ZIP1, ZIP2, and ZIP3 from EST database searches, an important discovery for the ZIP family in mammals was the identification of the *SLC39A4/ZIP4* gene, whose mutated form is responsible for *acrodermatitis enteropathica*, an inherited zinc deficiency disorder in humans. This finding

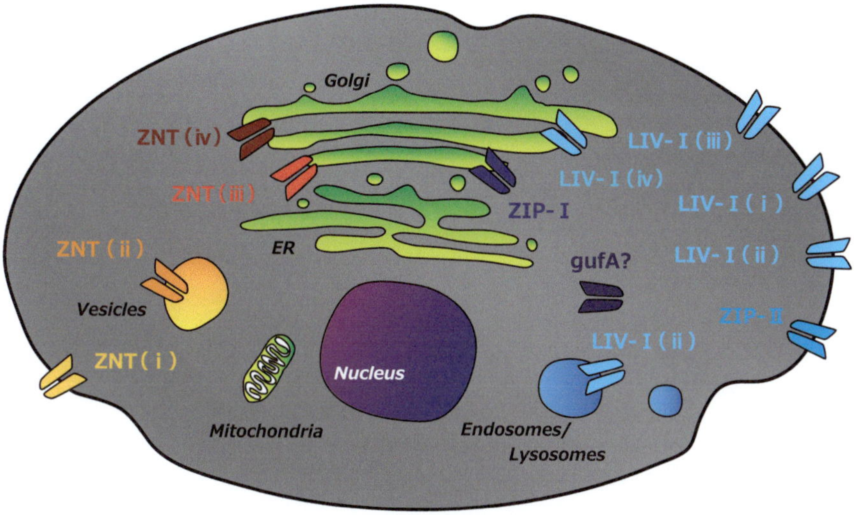

Fig. 3.3 Subcellular localization of ZNT and ZIP proteins from the view of subfamily and subgroup. The subcellular localizations of ZNT and ZIP protein members are shown according to their subfamily and subgroup assignments described in the main text and Fig. 3.1. ZNT members are shown according to their subgroups: subgroup (i) ZNT1 and ZNT10; subgroup (ii) ZNT2, ZNT3, ZNT4, and ZNT8; subgroup (iii) ZNT5 and ZNT7; and subgroup (iv) ZNT6. ZIP family members, except for LIV-1, are shown according to their subfamily: subfamily (ZIP-I) ZIP9; subfamily (ZIP-II) ZIP1, ZIP2, and ZIP3; and subfamily (gufA) ZIP11. LIV-1 subfamily members are assigned into subgroups: subgroup (i) ZIP4 and ZIP12; subgroup (ii) ZIP8 and ZIP14; subgroup (iii) ZIP5, ZIP6, and ZIP10; and subgroup (iv) ZIP7 and ZIP13

highlights the importance of the ZIP family for human and vertebrate physiopathology. Other ZIP family proteins have been discovered in genome databases, with 14 members now identified (Fig. 3.1) (Eide 2004; Jeong and Eide 2013).

Homologs of the human ZIP family have also been found in the genome sequences of rats, mice, chickens, zebrafish, fruit flies, nematodes, yeast, and plants (*A. thaliana*) and are shown in Table 3.2. This table shows that all ZIP family members are highly conserved in mammals, as in the case of the ZNT family. ZIP family members are generally classified into four subfamilies, ZIP-I, ZIP-II, LIV-1, and gufA (Fig. 3.1) (see Sect. 3.4.2) (Dempski 2012; Gaither and Eide 2001; Kambe et al. 2004; Taylor and Nicholson 2003), based on their phylogenetic relationships. ZIP4 is essential for zinc absorption in mammals, but a homologous gene is not present in chickens.

A number of ZNT and ZIP transporters have been shown to be involved in human genetic disorders. Moreover, numerous mutant and knockout animals have been generated for most of their orthologues. The data in Tables 3.1 and 3.2 may prove to be useful for reorganizing the orthologue functions of both the protein families.

Table 3.1 Amino acid sequence similarity among ZNT family members found in representative model organisms

H. sapiens	*R. norvegicus*	*M. musculus*	*G. gallus*	*D. rerio*	*D. melanogaster*	*C. elegans*	*S. cerevisiae*	*A. thaliana*
ZNT1 (SLC30A1) 507 A.A.	455/511 (89%)	455/508 (89%)	379/532 (71%)	351/495 (70%)	252/474 (53%)	61/100 (61%)	–	–
NP_067017.2	NP_074044.1	NP_033605.1	BBM28213.1	NP_957173.1	NP_647801.1	NP_509095.1 (Cdf1)	–	–
ZNT2 (SLC30A2) 372 A.A.	334/376 (88%)	333/373 (89%)	300/374 (80%)	260/341 (76%)	120/167 (71%)	229/331 (69%)	176/315 (55%)	180/315 (57%)
NP_001004434.1	NP_001076591	NP_001034766.1	XP_003642614.2	NP_001038485.1	NP_609741.3	NP_510091 (Cdf2)	NP_014961 (Cot1p)	NP_191753.1 (AtMTP2)
ZNT3 (SLC30A3) 388 A.A.	355/388 (91%)	359/388 (92%)	236/321 (73%)	–	201/404 (49%)	–	-	181/359 (50%)
NP_003450.2	NP_001013261.1	NP_035903.2	XP_025004924.1	–	NP_723901.1	–	–	NP_850459.1 (AtMTP1)
ZNT4 (SLC30A4) 429 A.A.	407/430 (94%)	408/430 (94%)	362/437 (82%)	222/310 (71%)	210/371 (57%)	173/274 (63%)	159/353 (45%)	183/312 (58%)
NP_037441.2	NP_742063.1	NP_035904.2	NP_001280212.1	NP_956937.1	NP_610185.1	NP_499691.1 (ttm-1)[a]	NP_013970.1 (Zrc1p)	NP_191440.2 (AtMTP3)
ZNT5 (SLC30A5) 765 A.A.	715/744 (96%)	736/765 (96%)	714/775 (92%)	675/781 (86%)	–	424/730 (58%)	159/291 (54%)	215/399 (53%)
NP_075053.2	NP_001099874.1	NP_075023.2	NP_001026590.2	NP_001002322.1	–	NP_740931 (CDF5)	NP_010491 (Msc2p)	NP_178539.2 (AtMTP12)
ZNT6 (SLC30A6) 501 A.A	437/501 (87%)	439/501 (87%)	433/501 (86%)	375/511 (73%)	–	157/330 (47%)	27/80 (33%)	112/247 (45%)
NP_001180442.1	NP_001264208.1	NP_659047.2	NP_001006402.1	NP_991214.1	–	NP_001033385.1 (Toc1)[b]	NP_014437 (Zrg17p)[b]	NP_187817.2 (AtMTP5)[b]
ZNT7 (SLC30A7) 376 A.A	364/378 (96%)	365/378 (96%)	339/378 (89%)	316/387 (81%)	270/374 (72%)	–	–	169/269 (63%)
NP_598003.2	NP_001178644.1	NP_075703.1	NP_001008788.1	NP_001093556.1	NP_650049.1	–	–	NP_178539.2

ZNT8 (SLC30A8) 369 A.A	288/322 (89%)	337/369 (91%)	285/378 (75%)	243/351 (69%)	205/351 (58%)	–	–	173/339 (51%)
NP_7762 50.2	NP_001124010.1	NP_766404.1	XP_025003910.1	XP_021323959.1	NP_723732.2[c]	–	–	NP_180502.2 (AtMTP4)
ZNT10 (SLC30A10) 485 A.A	407/486 (83%)	410/485 (84%)	329/490 (67%)	259/422 (61%)	194/455 (42%)	–	–	–
NP_061183.2	NP_001099455.1	NP_001028458.1	XP_015139382.1	NP_001121706.2	NP_649233.2[d]	–	–	–
Not classified[b]						NP_509164[e]	NP_013902 (Mft1p)[f]	OAP07976 (ZAT1)
						NP_498611[e]	NP_015100 (Mft2p)[f]	NP_182304.2 (AtMTP6)
						NP_492028[e]		NP_564594.1 (AtMTP7)
						NP_001024066		NP_191365.2 (AtMTP8)[g]
						NP_509279		NP_178070.2 (AtMTP9)
								NP_173081.2 (AtMTP10)[g]
								NP_181477.1 (AtMTP11)[g]

ZNT protein members and their homologs in humans (*Homo sapiens*), rats (*Rattus norvegicus*), mice (*Mus musculus*), chickens (*Gallus gallus*), zebrafish (*Danio rerio*), fruit flies (*Drosophila melanogaste*r), nematodes (*Caenorhabditis elegans*), yeast (*Saccharomyces cerevisiae*), and plants (*Arabidopsis thaliana*) are arranged according to the sequence homology using NCBI BLAST tool (http://www.ncbi.nlm.nih.gov/BLAST/). The numbers indicate the similarity (positives, %) between the human sequence and the corresponding sequence, whose accessions are described below

[a]Located to the plasma membrane (Roh et al. 2013)

[b]Not found by sequence alignment but are assigned according to their functional homology

[c]Has a long extra N-terminal portion

[d]Has HNED motif instead of NDHD motif in the intramembranous metal binding site

[e]Shows weak similarity to yeast Mft1p and Mft2p

[f]Known to transport iron

[g]Known to transport manganese

Table 3.2 Amino acid sequence similarity among ZIP family members found in representative model organisms

H. sapiens	*R. norvegicus*	*M. musculus*	*G. gallus*	*D. rerio*	*D. melanogaster*	*C. elegans*	*S. cerevisiae*	*A. thaliana*
ZIP1 (SLC39A1) 324A.A. NP_001258886.1	313/324 (96%) NP_001128049.1	311/324 (95%) NP_038929.2	141/246 (57%) XP_015154092.2	172/271 (63%) NP_997748.2	149/295 (50%) NP_610231.2	165/338 (48%) NP_493626.1 (ZIPT-1)	133/393 (33%) NP_011259.1 (Zrt1p)[a]	98/239 (41%) NP_567590.3 (IRT1)
								103/249 (41%) NP_180786.1
								111/267 (41%) NP_055252.2 (ZIP5)
								101/241 (41%) NP_178488.1
								104/245 (42%) OAO90686.1 (ZIP8)
								103/285 (36%) OAO97412.1 (ZIP9)
								82/176 (46%) NP_564766.1 (IRT3)
ZIP2 (SLC39A2) 309A.A. NP_055394.2	262/309 (84%) NP_001100730.1	265/309 (85%) NP_001034765.2	–	–	148/334 (44%) NP_525107.1	147/385 (38%) NP_001026796.1 (ZIPT-2.1)	137/429 (31%) NP_013231.1 (Zrt2p)[a]	134/327 (40%) NP_001031670.1 (IRT2)
						150/370 (40%) NP_500517.1 (ZIPT-2.2)		135/359 (37%) NP_187881.1
						148/339 (43%) NP_496876.1 (ZIPT-2.3)		
						164/498 (32%) NP_001249972.1 (ZIPT-2.4)		

ZIP3 (SLC39A3) 314A.A. NP_653165	292/317 (92%) NP_001008357.1	292/317 (92%) NP_598896.2	277/315 (87%) XP_015155452.1	–	147/297 (49%) NP_525107.1	146/377 (38%) NP_495126.1 (ZIPT-3)	–	122/310 (39%) OAP13793.1 (ZIP11)
ZIP4 (SLC39A4) 647A.A. NP_570901.2	527/657 (80%) NP_001071137.1	524/661 (79%) NP_082340.1	–	367/659 (55%) XP_017207743.1	–	–	–	–
ZIP5 (SLC39A5) 540A.A. NP_775867.2	468/539 (86%) NP_001102198.1	471/539 (87%) NP_082368.1	300/525 (57%) XP_025001341.1	286/520 (55%) XP_690258.3	–	–	–	–
ZIP6 (SLC39A6) 755A.A. NP_036451.4	694/760 (91%) NP_001019916.1	709/766 (92%) NP_631882.2	580/762 (76%) XP_419071.2	443/751 (58%) NP_001001591.1	305/757 (40%) NP_001303364	–	–	–
ZIP7 (SLC39A7) 469A.A. NP_008910.2	416/475 (87%) NP_001008885.1	427/478 (89%) NP_001071177.1	–	271/405 (66%) NP_571006.3	225/354 (63%) NP_524931.1 (Catsup)	191/348 (54%) NP_503070.2 (ZipT-7.1) 289/512 (56%) NP_510563.2 (ZIPT-7.2)	193/485 (39%) NP_012241.1 (yKE4p)	179/373 (47%) OAP17350.1 (IAR1)
ZIP8 (SLC39A8) 460A.A. NP_001128618.1	426/462 (92%) NP_001011952.1	431/462 (93%) NP_080504.3	355/436 (81%) XP_427970.3	289/452 (63%) XP_009305480.1	–	–	–	–
ZIP9 (SLC39A9) 307A.A. NP_060845.2	260/266 (97%) NP_001030101.1	297/308 (96%) NP_080520.2	292/307 (95%) NP_001007934.1	280/309 (90%) NP_001013558.1	201/310 (64%) NP_651919.3	210/389 (53%) NP_506393.1 (ZIPT-9)	142/361 (39%) NP_014722.1 (Atx2p)	71/179 (39%) NP_187881.1
ZIP10 (SLC39A10) 831A.A. NP_065075.1	764/834 (91%) NP_001102266.2	769/833 (92%) NP_766241.2	669/863 (77%) XP_426562.2	501/896 (55%) NP_956965.1	272/526 (51%) NP_523974.3 (FOI)	–	–	176/842 (20%) NP_174411.2 (ZIP10)

(continued)

Table 3.2 (continued)

H. sapiens	*R. norvegicus*	*M. musculus*	*G. gallus*	*D. rerio*	*D. melanogaster*	*C. elegans*	*S. cerevisiae*	*A. thaliana*
ZIP11 (SLC39A11) 342A.A.	306/342 (89%)	315/343 (91%)	276/342 (80%)	86/102 (84%)	230/346 (66%)	237/345 (68%)	151/526 (28%)	129/354 (36%)
NP_001153242.1	NP_001013060.1	NP_001159975.1	XP_024997602.1	XP_021335877.1	NP_610712.1	NP_491614.1 (ZIPT11)	NP_012746.1 (Zrt3p)	NP_001326769.1 (ZTP29)
ZIP12 (SLC39A12) 691A.A.	481/558 (86%)	583/693 (84%)	541/704 (76%)	–	–	200/743 (26%)	–	–
NP_001138667.1	NP_001099594.1	NP_001012305.1	XP_015137490.1			NP_503096.2 (ZIPT-17)		
ZIP13 (SLC39A13) 371A.A.	321/348 (92%)	346/371 (93%)	270/346 (78%)	233/327 (71%)	198/339 (58%)	180/412 (43%)	–	–
NP_509719.2 (ZIPT-13)	NP_001034285.1	NP_080997.1	NP_001008471.1	NP_001005306.3	NP_001189315	NP_001121697.2		
ZIP14 (SLC39A14) 492A.A.	441/487 (90%)	455/487 (93%)	417/457 (91%)	360/469 (76%)	189/406 (46%)	–	–	–
NP_001121903.1	NP_001100745.1	NP_659057.2	XP_024998709.1	NP_001313628.1	NP_001097608.1			

ZIP protein members and their homologs in humans (*Homo sapiens*), rats (*Rattus norvegicus*), mice (*Mus musculus*), chickens (*Gallus gallus*), zebrafish (*Danio rerio*), fruit flies (*Drosophila melanogaste*r), nematodes (*Caenorhabditis elegans*), yeast (*Saccharomyces cerevisiae*), and plants (*Arabidopsis thaliana*) are arranged as in Table 3.1. The numbers indicate the similarity (positives, %) between the human sequence and the corresponding sequence, whose accessions are described below

[a]Zrt1p is a high-affinity while Zrp2p is a low-affinity zinc transporter

3.3 ZNT Transporters

3.3.1 Biochemical and Structural Properties of Bacterial ZNT Homologs

ZNT transporters function as zinc efflux proteins, transporting zinc from the cytosol into intracellular compartments or into the extracellular space. Based on high resolution structures of the bacterial homolog YiiP (*E. coli* and *S. oneidensis*), which belongs to Zn/Fe-CDF subfamily (Coudray et al. 2013; Gupta et al. 2014; Lopez-Redondo et al. 2018; Lu et al. 2009; Lu and Fu 2007), ZNT transporters are thought to form homodimers, enabling them to transport zinc across biological membranes. The structure of YiiP reveals a topology in which each protomer most likely has six transmembrane domains (TMDs) with cytosolic N- and C-termini, as predicted by hydrophobicity plots (Gaither and Eide 2001; Paulsen and Saier 1997). The TMDs are grouped into a compact four-helical bundle consisting of TMDs, I, II, IV, and V, and a two-helical pair outside the bundle, consisting of TMDs III and VI. The compact four-helical bundle creates a channel in which the intramembranous tetrahedral zinc-binding site of TMDs II and V is located. This zinc-binding site consists of one histidine (H) and three aspartic acid (D) residues (DDHD core motif). Homodimerization of YiiP is stabilized by intermolecular salt-bridges which ensure the correct orientation of TMDs III and VI by interlocking the TMDs at the dimer interface (Lu et al. 2009). Another zinc-binding site is formed in the cytosolic C-terminal region, which exhibits a binuclear zinc-coordination site. YiiP has a highly conserved metallochaperone-like structure, with a characteristic αββα structure despite a high degree of sequence variability (Cherezov et al. 2008; Higuchi et al. 2009; Uebe et al. 2018). Recently, a ZNT homolog lacking the C-terminal region was reported in a marine bacterium (Kolaj-Robin et al. 2015), suggesting that this region may not be required for zinc transport activity.

YiiP is functional as a proton-zinc exchanger, in which an alternative access mechanism is in operation. TMDs of YiiP can adopt cytosolic-facing (inward-facing) and periplasmic-facing (outward-facing) conformations, both of which can bind zinc or protons (Gupta et al. 2014; Lopez-Redondo et al. 2018). Zinc binding in the cytosolic C-terminal region may induce conformational changes in the TMDs, facilitating zinc transport by the alternative access mechanism (Coudray et al. 2013; Gupta et al. 2014; Lopez-Redondo et al. 2018) in which the extracellular proton provides a driving force for exporting zinc from the cytosol. This information provides the framework for exploring the biochemical and structural properties of ZNT transporters.

3.3.2 Properties of ZNT Transporter Proteins

ZNT family proteins are predicted to have a similar topology to YiiP and, therefore, are thought to form homodimers (Fukunaka et al. 2009; Golan et al. 2016; Itsumura et al., 2013; Lasry et al. 2014; Murgia et al. 2008). The conserved motif of (F/Y) G(W/Y/F)XRXE, which is positioned on the first cytosolic loop between TMDs II and III, is thought to be involved in homodimer formation (Fig. 3.4) (Lasry et al. 2012). The conserved arginine (R) in the motif is thought to be involved in the formation of intermolecular salt-bridges on the cytoplasmic membrane surface which are important for zinc transport (Figs. 3.2 and 3.4) (Fukue et al. 2018). In addition to homodimer formation, ZNT5 and ZNT6 form heterodimers (Fukunaka et al. 2009; Golan et al. 2015; Lasry et al. 2014; Suzuki et al. 2005b). Other ZNT members could also form heterodimers (Golan et al. 2015; Zhao et al. 2016), the significance of which is unclear. Similar to YiiP, ZNT proteins are functional as zinc-proton exchangers (Ohana et al. 2009; Shusterman et al. 2014). This mechanism of zinc transport is reasonable, particularly in one subgroup (ZNT2, ZNT3, ZNT4, and ZNT8) (see Sect. 3.3.3.2) which are localized to acidic compartments, such as endosomes, lysosomes, or intracellular vesicles. The activity of ZNT proteins may be tunable and could be controlled by the lipid composition of the vesicles where they are localized (Merriman et al. 2016). Recent computational simulations, based on energy calculations, of the ZNT zinc permeation pathway shows a favorable zinc translocation via the alternative access mechanism, consistent with the model of YiiP (Golan et al. 2018)

Interestingly, the DDHD core motif of YiiP changes to HDHD within TMDs II and V in most of the ZNT members (Fig. 3.3). Numerous biochemical studies reveal that this core motif is essential for zinc transport, as substitution of histidine or aspartic acid residues to alanine (A) abolishes zinc transport (Fujimoto et al. 2013; Ohana et al. 2009; Tsuji et al. 2017). In addition, ZNT10 has an altered motif, NDHD, within TMDs II and V (Fig. 3.3), that enables ZNT10 to transport manganese (Leyva-Illades et al. 2014; Nishito et al. 2016) (see Sect. 3.3.3.1). Moreover, replacing the histidine residue in TMD II with aspartic acid (DDHD, *i.e.*, the YiiP motif) in ZNT5 and ZNT8 allows it to transport cadmium as well as zinc (Hoch et al. 2012). Thus, this position in TMD II, which constitutes the intramembranous zinc-binding site, is critical for regulating metal substrate specificity. ZNT6 has no zinc transport activity, with two amino acids of the HDHD motif replaced by hydrophobic residues (Fig. 3.4). Instead, ZNT6 is functional as an auxiliary protomer of

Fig. 3.4 (continued) external loops (EL), and C-terminal cytosolic α-helices and β-sheets based on the crystal structure of YiiP are indicated in yellow, pink, turquoise, and lavender, respectively. Residues highlighted in red indicate the histidine (H) and aspartic acid (D) residues constituting the intramembranous zinc-binding site. Residues highlighted in black and gray are highly conserved and semi-conserved, respectively. The asparagine residue (N) in TMD II of ZNT10, which has been speculated to be involved in the recognition of manganese, is indicated in green. Residues highlighted in light blue indicate the amino acid residues forming the cytosolic binuclear zinc-coordination site. Residues highlighted in blue indicate the highly conserved arginine (R) and well-conserved glutamic acid (E) or glutamine (Q) likely involved in the formation of salt-bridges on the cytosolic side. The amino acid sequences at the C-terminal end of ZNT6 (30 aa) and ZNT10 (4 aa) are not displayed in the alignment. The TMDs indicated in ZNT5 correspond to TMDs between X and XV. "-" denotes a gap in the alignment. Blue circles indicate the amino acid residues involved in zinc binding. The (F/Y)G(W/Y/F)XRXE sequence, which is proposed to be involved in dimerization, is indicated in lavender. Blue, green, red, or honey circles below the sequences indicate the amino acid residues involved in zinc binding (different color means the coordination of different zinc ions, and honey color means the coordination of zinc ion in the neighboring subunit). This figure is used and modified from Kambe et al. 2014 with permission

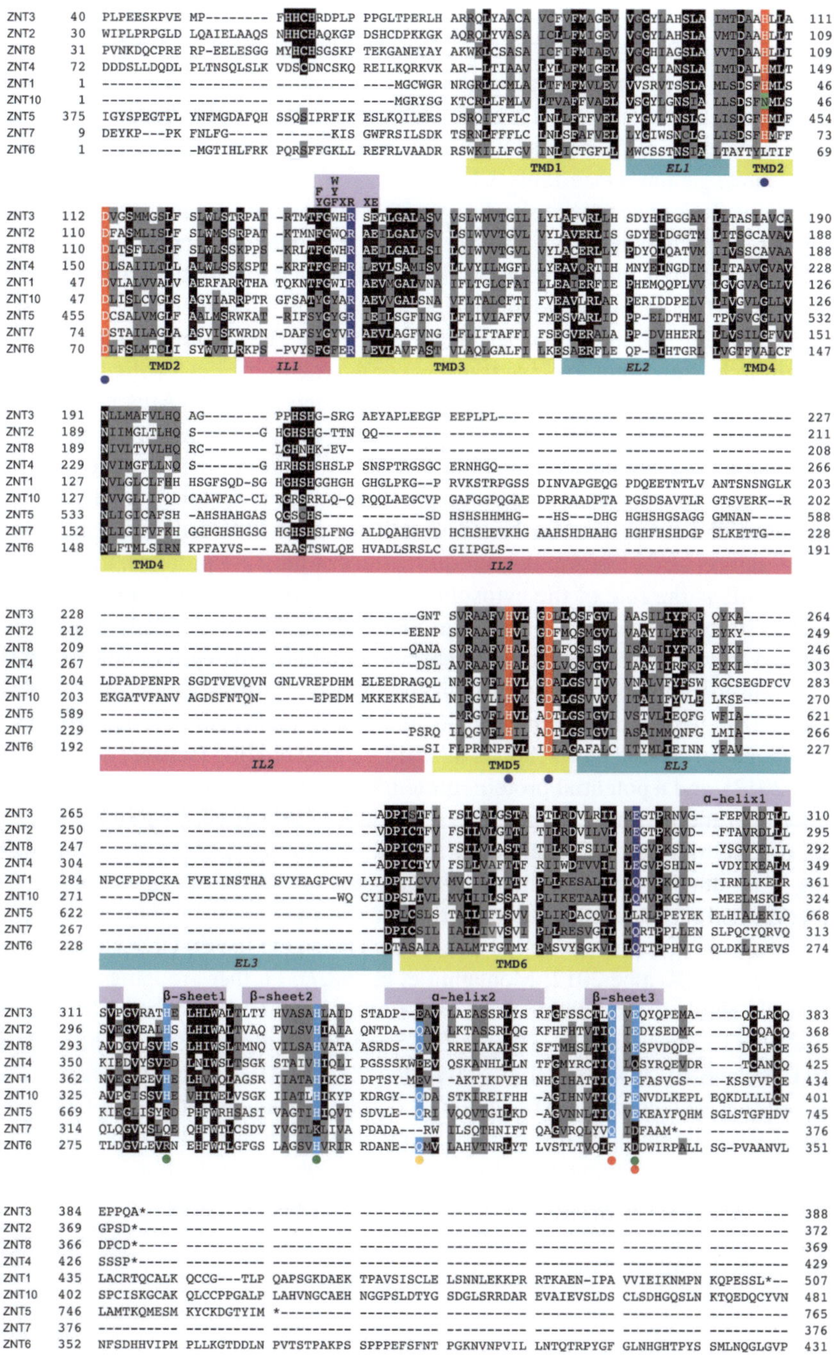

Fig. 3.4 Sequence alignment of ZNT family proteins. The alignment is ordered according to similarities among subfamilies. The putative transmembrane domains (TMD), intracel lular loops (IL),

the ZNT5 and ZNT6 heterodimer (Fukunaka et al. 2009) (see Sect. 3.3.3.4). Interestingly, several plant ZNT homologs (*e.g.*, MTP8, MTP9, and MTP11), which are known to be manganese transporters and are grouped within the Mn-CDF subfamily of CDF proteins (Gustin et al. 2011; Pedas et al. 2014; Tsunemitsu et al. 2018; Ueno et al. 2015), have a DDDD core motif. No orthologues have been found in vertebrate genomes (see Table 3.1).

Due to a lack of sequence similarity with YiiP, several unique features of ZNT transporter proteins have been determined in a number of biochemical studies. One such feature is the unique cytosolic histidine-rich loop with variable lengths between TMDs IV and V, which is also found in the plant ZNT homologs (Blindauer and Schmid 2010; Kambe et al. 2014). The histidine-rich loop was thought to be important for zinc transport or as a sensor for cytosolic zinc levels by coordinating cytosolic zinc through its histidine residues (Arus et al. 2013; Kawachi et al. 2008; Suzuki et al. 2005b; Tanaka et al. 2015; Tanaka et al. 2013). However, a ZNT mutant, in which all histidine residues in the loop are mutated to alanine, still possesses zinc transport activity, although this activity is decreased, indicating that the histidine residues are not essential for zinc transport (Fukue et al. 2018). Another unique feature of the ZNT proteins, in comparison to YiiP, is the role of the cytosolic N-terminal region, which, in YiiP, is too short to study. A recent study indicates that ZNT members are likely to be functional even when the N-terminal is absent, although this region could regulate zinc transport via an interaction with the cytosolic histidine-rich loop (Fukue et al. 2018). Thus far, the functions of the N-terminal region are reported as a mitochondrial sorting motif (Seo et al. 2011), a zinc binding (sensor) motif (Arus et al. 2013), and a potential protein–protein interaction motif resembling the leucine zipper motif (Murgia et al. 1999), in addition to participating in the regulation of zinc transport (Kawachi et al. 2012). Interestingly, ZNT5 has a uniquely long N-terminal sequence containing nine potential TMDs (Kambe et al. 2002), however its functional importance is not yet known.

The nine ZNT members described above all belong to the Zn-CDF subfamily of CDF proteins (Kambe 2012; Montanini et al. 2007). Further groupings, based on sequence similarity, subdivide them into four groups: (i) ZNT1 and ZNT10, (ii) ZNT2, ZNT3, ZNT4, and ZNT8, (iii) ZNT5 and ZNT7, and (iv) ZNT6 (Gustin et al. 2011; Kambe 2012; Kambe et al. 2006) (Figs. 3.1 and 3.3). Subgroup (i) contains the cell surface localized ZNTs, although, despite it belonging to the Zn-CDF subfamily, ZNT10 transports manganese. Subgroup (ii) contains transporters involved in intracellular compartments and vesicles, while subgroup (iii) contains transporters involved in zinc transport in the early secretory pathway. Subgroup (iv) contains only ZNT6, which is functional as an auxiliary protomer without zinc transport activity, as described above.

3.3.3 *Biochemical Characterization of the ZNT Subgroups*

Here, we provide a brief summary of each of the ZNT subgroups. For more detailed information about their physiopathological functions, we refer the reader to the following reviews, Bowers and Srai (2018), Hara et al. (2017), and Kambe et al. (2015). Additionally, the details of mice phenotypes have been described in other chapters of this book.

3.3.3.1 ZNT1 and ZNT10 Subgroup

Both ZNT1 and ZNT10 are known to be functional at the plasma membrane as efflux transporters of cytosolic zinc and manganese, respectively (see above) (Palmiter and Findley 1995; Leyva-Illades et al. 2014; Nishito and Kambe 2019), although their intracellular localization is also reported. As mentioned above, their difference in metal substrate specificity is due to differences in their metal binding motifs (HDHD in ZNT1, NDHD in ZNT10). ZNT1 expression increases in response to excess zinc through binding of metal-response element-binding transcription factor-1 (MTF-1) to the metal response element (MRE) in its promoter, in a fashion similar to metallothionein (Langmade et al. 2000). This is consistent with its cellular function, which is to reduce the toxicity of excess zinc. In contrast, there are no reports describing manganese-induced expression of ZNT10, although such regulation would be important as mutations of *SL C30A10/ZNT10* gene result in parkinsonism with hypermanganesemia (Quadri et al. 2012; Tuschl et al. 2012). ZNT1 is thought to be mostly localized to the basolateral membrane in polarized cells (McMahon and Cousins 1998), while ZNT10 is reported to be localized to the apical membrane (Taylor et al. 2019). Overall, they have a sequence similarity of 37% (Nishito et al. 2016), therefore, the differing amino acids may be responsible for their unique subcellular localizations, although the regulatory mechanisms for their trafficking to the cell surface have not yet been elucidated. Both have a very short cytosolic N-terminal region (Fukue et al. 2018; Kambe et al. 2014) (Fig. 3.3). Recently, ZNT10 was reported to be functional as a manganese-calcium exchanger (Levy et al. 2019), the mode of which needs to be further investigated.

3.3.3.2 ZNT2, ZNT3, ZNT4, and ZNT8 Subgroup

These ZNT subgroup members are localized to the membranes of cytosolic secretory vesicles and play essential roles in transporting zinc into their lumens (Hennigar and Kelleher 2012; Kambe 2011). This subgroup functions as zinc-proton exchangers and is able to do so due to the acidic environment of the vesicular lumen they transport to. Apart from ZNT4, these transporters are expressed in a tissue-specific manner. Zinc transport into cytosolic secretory vesicles is crucial as a high concentration of zinc, which is required for many physiological responses. Examples include the insulin granules in pancreatic β cells, which accumulate high levels of zinc mediated by ZNT8, and the presynaptic vesicles in a subset of glutamatergic neurons, which also accumulate high levels of zinc mediated by ZNT3. This zinc accumulation is essential for good health as single nucleotide polymorphisms in *SLC30A8/ZNT8* are shown to be associated with an increased susceptibility to type 2 diabetes (see Chap. 12), while ZNT3 is crucial for neuronal activity (see Chaps. 9 and 11). High levels of zinc in breast milk are secreted through the vesicles of the secretory mammary epithelial cells (Lee and Kelleher 2016), which in humans is mediated by ZNT2. Thus, mutations of *SLC30A2/ZNT2* gene cause *transient neonatal zinc deficiency*, another inherited zinc deficiency in humans (Golan et al. 2017). A similar phenotype (*i.e.*, low levels of zinc secretion) was observed in mice with a loss of function Znt4 mutant (McCormick et al. 2016). ZNT2 plays a role in zinc transport into the cell granules of Paneth cells in the crypts of Lieberkühn of the small intestine (Podany et al. 2016). There are other vesicles in the body which accumulate high levels of zinc, such as cytosolic vesicles in epithelial cells of the lateral prostate, pigment epithelial cells in the retina, and mast cells (Kambe et al. 2014), whose zinc transport may also be controlled by ZNT family members within this subgroup. ZNT4 has been shown to localize to endosomes, lysosomes, and the *trans*-Golgi network, in addition to cytosolic vesicles. Consistent with this, ZNT4 is likely to play other roles, including in the regulation of several enzymes (McCormick and Kelleher 2012; Tsuji et al. 2017).

3.3.3.3 ZNT5 and ZNT7 Subgroup

ZNT5 and ZNT7 are localized to the early secretory pathway, including the endoplasmic reticulum (ER) and the Golgi apparatus, both of which require efficient zinc transport. Impairment of zinc transport by both ZNT5 and ZNT7 most likely results in the misfolding or incomplete assembly of some zinc-containing proteins. This leads to a decrease in the activity of zinc-dependent ectoenzymes and zinc-dependent chaperone proteins that monitor and assist protein folding, which can trigger the unfolded protein response (UPR) (Ellis et al. 2004; Fukunaka et al. 2011; Ishihara et al. 2006; Takeda et al. 2018). Thus, both ZNTs are important for homeostasis in the early secretory pathway by mediating the zinc supply to the nascent proteins (Kambe et al. 2016, 2017). Consistent with this, the expression of *ZNT5* mRNA is transcriptionally upregulated by the UPR (Ishihara et al. 2006). Both ZNT5 and ZNT7 have a unique di-proline motif (PP-motif) in luminal loop 2, which is essential for the activation of ectoenzymes, such as alkaline phosphatases (Fujimoto et al. 2016). This unique motif is highly conserved in most orthologues suggesting an important role in enzyme activation.

3.3.3.4 ZNT6 Subgroup

Co-immunoprecipitation, bimolecular fluorescence complementation, and genetic experiments show the heterodimerization of ZNT6 and ZNT5 (Fukunaka et al. 2009; Lasry et al. 2014; Suzuki et al. 2005a). The heterodimer formation of ZNT5 and ZNT6 is a unique feature among the ZNT family and is highly conserved among ZNT5 and ZNT6 functional orthologues, such as those in nematode (Cdf5 and Toc1) (Fujimoto et al. 2016), yeast (Msc2p and Zrg17p in *S. cerevisiae*, and Cis4 and Zrg17 in *Schizosaccharomyces pombe*) (Choi et al. 2018; Ellis et al. 2005; Ellis et al. 2004), and plant (MTP12 and MTP5 in *A. thaliana*) (Fujiwara et al., 2015). Each of these functional orthologues is also localized to the early secretory pathway. Considering the conservation of heterodimer formation, it is interesting to note that the fruit fly does not possess orthologues of ZNT5 and ZNT6 (see Table 3.1), which may suggest a unique regulatory mechanism of zinc transport and physiology in the fruit fly.

3.4 ZIP Transporters

3.4.1 Biochemical and Structural Properties of Bacterial ZIP Homologs

As with ZNT proteins, recent structural studies reveal that ZIP homologs (*Bordetella bronchiseptica* ZIP (BbZIP)) form homodimers. These studies reveal that each protomer has eight TMDs with extracellular N- and C-termini (Zhang et al. 2017), as predicted in ZIP family proteins (Gaither and Eide 2001; Kambe et al. 2004). BbZIP has a novel 3+2+3 transmembrane architecture, in which the first three TMDs (TMDs I to III) are symmetrically related to the last three TMDs (TMDs VI to VIII) by a pseudo-twofold axis. The TMDs IV and V, which are symmetrically related by the same axis, are sandwiched between the three TMD repeats (Zhang et al. 2017). BbZIP has a binuclear metal center, which coordinates zinc and is formed by several conserved amphipathic amino acid residues, including histidine, asparagine (N), and aspartic acid, within TMD IV, and two histidine residues and one glutamic acid (E) in TMD V (Figs. 3.2 and 3.5). The latter amino acids correspond to those in the potential metalloprotease motif (HEXPHEXGD) of LIV-1 subfamily (see 3.4.4). An in vitro study using reconstituted proteoliposomes proposes that the zinc transport of BbZIP (reported as ZIPB (Lin et al. 2010)) may be by a selective electrodiffusional channel. However, a definitive zinc transport mechanism for ZIP family proteins has not yet been clarified.

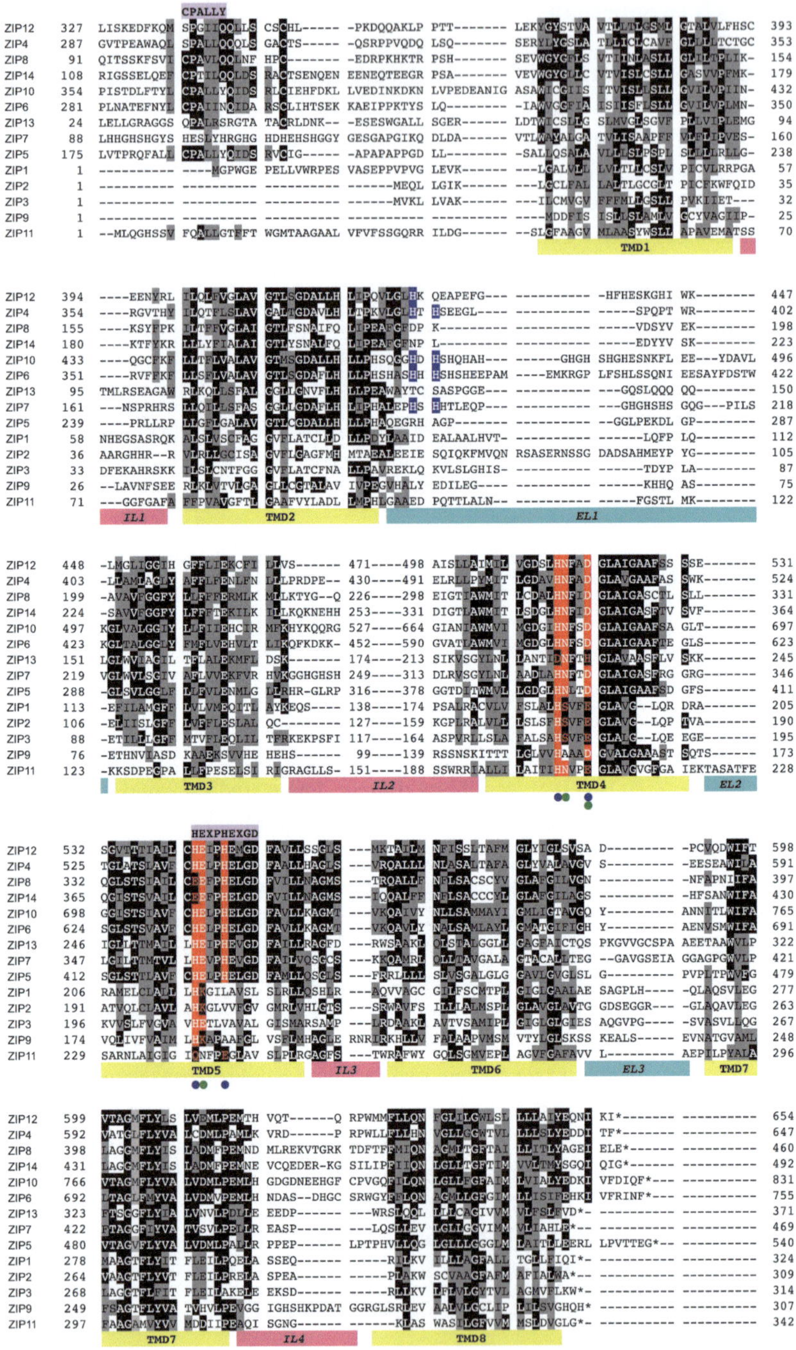

Fig. 3.5 Sequence alignment of ZIP family proteins. The alignment is ordered according to similarities among subfamilies. The putative transmembrane domains (TMD), intracellular loops (IL),

3.4.2 Properties of ZIP Family Proteins

Proteins of the ZIP family form homodimers which enable them to transport zinc across membranes (Bin et al. 2011), as shown in BbZIP (Lin et al. 2010; Zhang et al. 2017). Moreover, recent studies show that they can also form heterodimers (Taylor et al. 2016) (see Sect. 3.4.4.3). ZIP family proteins are thought to have eight TMDs, in which the N- and C-terminal regions are located outside the plasma membrane or in the lumen of intracellular compartments, as predicted by hydrophobicity plots (Taylor and Nicholson 2003) and computational modeling (Antala et al. 2015). This predicted topology is consistent with that of BbZIP (Zhang et al. 2017). The region, which is conserved the most among the ZIP family, is found in TMDs IV and V where numerous amphipathic amino acids, including a conserved histidine, asparagine, aspartic acid, and glutamic acid residues, are found (Fig. 3.2). These amino acid residues constitute an intramembranous binuclear zinc-binding site, and thus form a pore through which zinc passes, as shown in the BbZIP structure. A symport mechanism, whereby zinc is transported alongside bicarbonate ions, is suggested (Gaither and Eide 2000; Girijashanker et al. 2008) but has not been confirmed experimentally (Franz et al. 2018). Recently, a proton-mediated regulatory mechanism was proposed for regulating the velocity and directionality of zinc transport by the ZIP family (Franz et al. 2018; Franz et al. 2014). Most ZIP family proteins have a variable cytosolic loop that is rich in histidine residues between TMDs III and IV (Blindauer and Schmid 2010; Kambe et al. 2015). Its physiological function has not yet been elucidated but it may play a regulatory role in trafficking (Bowers and Srai 2018; Huang and Kirschke 2007) or contribute to binding and sensing of cytosolic zinc (Bafaro et al. 2015), as suggested for the ZNT family. Moreover, histidine residues within the loop may have a unique function, as complete substitution of all histidine residues to alanine in ZIP4 causes a loss of zinc-induced ubiquitination and degradation, although this has no effect on zinc-stimulated endocytosis (Mao et al. 2007). The histidine residues (HXH motif) in the extracellular loop between TMD II and III, which are conserved in a number of ZIP family members, are required for zinc sensing and zinc-induced endocytosis of ZIP4 (Chun et al. 2018) (Fig. 3.5). These results suggest that endocytosis of ZIP proteins, in

Fig. 3.5 (continued) and external loops (EL) are shown below the alignment in yellow, pink, and turquoise, respectively. Residues highlighted in black and gray are highly conserved and semi-conserved, respectively. Residues highlighted in red indicate the positions of residues important for zinc binding. Residues highlighted in blue indicate the positions of residues likely required for zinc sensing and zinc-induced endocytosis of a number of LIV-1 members. Conserved sequences in ZIP family proteins (PAL and HEXPHEXGD motifs) are indicated in lavender. The amino acid sequences of the N-terminal region of the LIV-1 subfamily members (the first nine proteins) are not displayed in the alignment, and the IL2 loop between TMD3 and TMD4 is omitted from the figure. "-" denotes a gap in the alignment. Blue or green circles below the sequences indicate the amino acid residues involved in zinc binding (different color means the coordination of different zinc ions). This figure is used and modified from Kambe et al. 2014 with permission response to excess zinc, is mediated via zinc binding to conserved sensory sequences. The long extracellular region of some ZIP family members is known to be cleaved during zinc deficiency and in response to other stimuli (Ehsani et al. 2012; Hogstrand et al. 2013; Kambe and Andrews 2009), which may be important for the regulation of their zinc transport activity or their cellular trafficking.

As described above, ZIP family members are classified into four subfamilies based on their phylogenetic relationships (Gaither and Eide 2001; Kambe et al. 2004). The 14 mammalian members are classified as ZIP-I (ZIP9), ZIP-II (ZIP1-ZIP3), gufA (ZIP11), and LIV-1 (ZIP4–8, ZIP10, ZIP12-ZIP14) (Figs. 3.1 and 3.3) (Dempski 2012; Jeong and Eide 2013; Taylor and Nicholson 2003). Among them, the LIV-1 subfamily, whose name arises from its original identification in breast cancer, is the largest. LIV-1 subfamily members have a potential metalloprotease motif (HEXPHEXGD) in TMD V and an extracellular CPALLY motif (hereafter referred to as a PAL motif, see below) immediately preceding TMD I (Taylor et al., 2007) (Figs. 3.2 and 3.5). LIV-1 subfamily proteins are further divided into four subgroups: (i) ZIP4 and ZIP12; (ii) ZIP8 and ZIP14; (iii) ZIP5, ZIP6, and ZIP10; and (iv) ZIP7 and ZIP13 (Kambe et al. 2006; Zhang et al. 2016) (Figs. 3.1 and 3.3), each of which has unique sequence similarities. ZIP4 and ZIP12 of subgroup (i) possess a helix rich domain (HRD) in addition to the PAL motif-domain in the extracellular N-terminal region (Zhang et al. 2016). Subgroup (ii) members have a unique ability to transport manganese and iron, in addition to zinc. Subgroup (iii) contains ZIPs whose extracellular region, proximal to the membrane, has a unique domain called a prion fold. Subgroup (iv) contains members which are localized to the early secretory pathway and transport zinc from the lumen to the cytosol.

3.4.3 *Biochemical Characterization of the ZIP Subfamilies*

Here, we provide a brief summary of each of the ZIP subfamilies with the exception of LIV-1 which is further discussed in Sect. 3.4.4. For a more detailed discussion about the physiopathological functions of ZIP families, we refer the reader to the following reviews, Bowers and Srai (2018), Hara et al. (2017), and Kambe et al. (2015). Additionally, the details of mice phenotypes have been described in other chapters of this book.

3.4.3.1 ZIP-I Subfamily

ZIP9 is the only member belonging to this subfamily in vertebrates (Matsuura et al. 2009). ZIP9 is described as a dual-functioning protein because in addition to transporting zinc across cellular membranes, it can also function as a high affinity membrane androgen receptor through which testosterone activates G proteins thereby inducing cell signaling (Thomas et al. 2014). Thus, steroid and zinc signaling pathways cooperate to regulate physiological functions in mammalian cells through ZIP9 (Thomas et al. 2018). ZIP9 is localized to the plasma membrane and intracellular compartments, such as the Golgi (Berg et al. 2014; Matsuura et al. 2009). It may regulate its function by altering its subcellular localization.

3.4.3.2 ZIP–II Subfamily

ZIP-II subfamily includes ZIP1, ZIP2, and ZIP3. These proteins are homologous to each other, both in amino acid sequence and gene structure. In particular, regions of TMD IV are highly conserved (Dufner-Beattie et al. 2003; Kambe et al. 2014), with the aspartic acid residue observed in LIV-1 subfamily, which is expected to form a binuclear metal center, replaced by glutamic acid (Fig. 3.5). Moreover, the highly conserved histidine in TMD V, which is also expected to form a binuclear metal center, is replaced by a hydrophobic amino acid (valine (V) or leucine (L)) (Fig. 3.5), suggesting that the mechanism of zinc coordination in this subfamily may be different from other ZIP homologs including LIV-1 subfamily. Consistent with their homology, the ZIP-II subfamily do have similar zinc transport mechanisms (Dufner-Beattie et al. 2003). However, their expression is differentially regulated in a tissue-specific manner. ZIP1 has been extensively investigated and its endocytosis and degradation are shown to be mediated through a di-leucine motif (LL motif, not shown in Fig. 3.5) within the variable cytosolic loop between TMD III and IV (Huang and Kirschke 2007). This motif is not conserved in ZIP2 and ZIP3. Mice containing knockouts of each individual gene show zinc-sensitive phenotypes during pregnancy, similar to the phenotypes observed in triple knockout mice (Dufner-Beattie et al. 2006; Kambe et al. 2008; Peters et al. 2007). This suggests that each protein has a unique cell-specific function that is required for adaptation to a zinc deficiency during development.

3.4.3.3 gufA Subfamily

ZIP11 is the only vertebrate protein of the gufA subfamily. This subfamily includes the bacterial and archaeal ZupT (Yu et al. 2013) and *S. cerevisiae* Zrt3p, suggesting that the gufA subfamily proteins arose from an ancient zinc transporter. The biological functions or expression profiles of gufA subfamily members remains unclear, although its subcellular localization to the nucleus or Golgi apparatus is reported (Kelleher et al. 2012; Martin et al. 2013). Interestingly, ZIP11 almost lacks histidine and cysteine residues, which suggests that ZIP11 binds zinc in a different manner than other ZIP family members (Kambe et al. 2015; Yu et al. 2013).

3.4.4 Biochemical Characterization of LIV-1 Subfamily

As mentioned above, LIV-1 subfamily members are further divided into four subgroups, (i) ZIP4 and ZIP12, (ii) ZIP8 and ZIP14, (iii) ZIP5, ZIP6, and ZIP10, and (iv) ZIP7 and ZIP13 (Kambe et al. 2006; Zhang et al. 2016). Recent reports indicate unique physiopathological roles for each of the members, this is discussed in detail in other reviews (Bowers and Srai 2018; Hara et al. 2017; Kambe et al. 2015) and in other chapters of this book. Here, we provide a brief summary of each of the LIV-1 subgroups.

3.4.4.1 ZIP4 and ZIP12 Subgroup

While both ZIP4 and ZIP12 contain HRD and PAL motif-domains in the extracellular N-terminal region (Zhang et al. 2016), their physiological functions are very different. The biochemistry, physiopathology, and structure of ZIP4 have been extensively studied, while much less is known about ZIP12. ZIP12 has been identified as an important molecule for neurulation and neurite extension (Chowanadisai et al. 2013), and as a major regulator of hypoxia-induced pulmonary vascular remodeling (Zhao et al. 2015). The long extracellular N-terminal region of ZIP4 forms a homodimer without the need for TMDs (Zhang et al. 2017), in which the PAL motif is crucial (Zhang et al. 2016). This region is cleaved during zinc deficiency, raising the question as to how the cleavage is regulated. ZIP4 protein expression is regulated in response to zinc levels, while the regulation of ZIP12 expression has not yet been elucidated. Overall sequence similarity between ZIP4 and ZIP12 is 46%.

3.4.4.2 ZIP8 and ZIP14 Subgroup

A prominent feature of this subgroup is that both proteins can transport iron and manganese as well as zinc (Aydemir and Cousins 2018; Jenkitkasemwong et al. 2012). Their involvement in iron and manganese transport is crucial as mutations within both genes result in disease, such as severe dysglycosylation due to a type II congenital disorder of glycosylation (CDG) (Boycott et al. 2015; Park et al. 2015) or childhood-onset parkinsonism-dystonia (Tuschl et al. 2016). Their unique metal-binding specificities are most likely related to their sequences; ZIP8 and ZIP14 have glutamic acid in place of histidine in the HEXPHEXGD motif of TMD V (i.e., EEXPHELGD in both) (Fig. 3.5). However, this has not yet been tested at the molecular level. Both proteins are primarily localized to the plasma membrane, with some reports suggesting that their subcellular localization includes endosomes and lysosomes (Aydemir et al. 2009; Guthrie et al. 2015). Their physiopathological importance for zinc physiology is reported in many articles (Hojyo et al. 2011; Kim et al. 2014; Liu et al. 2013; Liuzzi et al. 2005).

3.4.4.3 ZIP5, ZIP6, and ZIP10 Subgroup

These three ZIPs have a unique extracellular domain in the region proximal to the membrane, called a prion fold, which may indicate an evolutionary link between them and prion proteins (Ehsani et al. 2011; Ehsani et al. 2012; Pocanschi et al. 2013).

This may also indicate their involvement in the etiology of prion diseases. Among these three, ZIP6 and ZIP10 share the highest homology and have been shown to form functional heterodimers in an interactome analysis (Brethour et al. 2017; Taylor et al. 2016), which could explain the overlap in ZIP6- and ZIP10-dependent phenotypes, such as cancer development and metastasis (Brethour et al. 2017) or oocyte-to-embryo transitions (Kong et al. 2014). An important aspect is that *Zip10*-knockout mice show significant phenotypes (Bin et al. 2017b; Gao et al. 2017; Hojyo et al. 2014; Miyai et al. 2014), which seem to be independent of ZIP6. In contrast to their extended expression, ZIP5 is expressed in a tissue-specific manner and is uniquely localized to the basolateral membrane in polarized cells, such as intestinal epithelial cells (Dufner-Beattie et al. 2004; Wang et al. 2004).

3.4.4.4 ZIP7 and ZIP13 Subgroup

The N-terminal region of this subgroup is shorter than those of other LIV-1 subfamily members, therefore sequence alignment of their PAL motifs is difficult (Fig. 3.5). Compared with other LIV-1 subfamily members, both proteins are uniquely localized to the early secretory pathway, including the ER and Golgi apparatus. Zinc release from the early secretory pathway, mediated by these proteins, plays an important role in regulating cell signaling (Fukada et al. 2008; Fukunaka et al. 2017; Nimmanon et al. 2017; Taylor et al. 2012; Tuncay et al. 2017). Moreover, their zinc transport activity contributes to the homeostatic maintenance of the secretory pathway (Bin et al. 2017a; Jeong et al. 2012; Ohashi et al. 2016). Notably, ZIP13 and its orthologues have a unique zinc coordination motif in TMD IV, in which the conserved histidine is replaced by aspartic acid (Dxx**H**NFx**D** sequence changes to Nxx**D**NFx**H**) (Xiao et al. 2014) (Fig. 3.5). Considering the evidence for ZNT1 and ZNT10 subgroup (see Sect. 3.3.3.1) and ZIP8 and ZIP14 subgroup in the LIV-1 subfamily (see Sect. 3.4.4.2), this change may alter the metal substrate specificity. Interestingly, the ZIP13 orthologue in the fruit fly is reported as an iron transporter, which mediates iron transport from the cytosol to the lumen of the ER or that of the Golgi (Xiao et al. 2019; Xiao et al. 2014), which is the opposite direction assumed for mammalian ZIP13. These findings raise the necessity for further investigation in ZIP13.

Compared with ZIP13, ZIP7 is conserved within other species (Table 3.2). Their physiological functions are extremely diverse but are related to the early secretory pathway, in particular the ER, and have been shown in yeast (yKE4p), drosophila (Catsup), nematode (ZipT-7.1), and plant (IAR1) (Groth et al. 2013; Kumanovics et al. 2006; Lasswell et al. 2000; Zhao et al. 2018), indicating its biological significance. The histidine-rich sequence in the N-terminal region is highly conserved among ZIP7 orthologues (Adulcikas et al 2018), suggesting that a unique zinc-sensing mechanism may be used. In mammals, ZIP7 expression is upregulated by the UPR and its loss results in disruption of ER functions (Ohashi et al. 2016; Tuncay et al. 2017), similar to ZNT5 and ZNT7 (see Sect. 3.3.3.3) (Ishihara et al.

2006; Tuncay et al. 2017). Thus, luminal zinc homeostasis in the early secretory pathway is regulated by both ZIPs and ZNTs.

3.5 Concluding Remarks and Perspectives

A number of studies reveal that many ZIP and ZNT family proteins are involved in human genetic diseases. Moreover, various phenotypes of numerous knockout mice unveil the fundamental importance of ZNTs and ZIPs (see other chapters). This chapter summarized the biochemical features of both protein families, with a particular focus on their biological subgroupings. To our knowledge, this is the first time such a review has been attempted and is therefore useful in providing a comprehensive overview of both ZNT and ZIP families. Future studies should aim to elucidate the molecular mechanisms that enable both families to control their spatiotemporal zinc transport. These studies would be facilitated by a comprehensive resource in which their classification and subgroups are described. This is required in order to further understand zinc signaling in physiopathological processes.

Acknowledgments This work was supported by an endowment from Renatech Co. Ltd.

References

Adulcikas J, Norouzi S, Bretag L, Sohal SS, Myers S (2018) The zinc transporter SLC39A7 (ZIP7) harbours a highly-conserved histidine-rich N-terminal region that potentially contributes to zinc homeostasis in the endoplasmic reticulum. Comput Biol Med 100:196–202. https://doi.org/10.1016/j.compbiomed.2018.07.007

Antala S, Ovchinnikov S, Kamisetty H, Baker D, Dempski RE (2015) Computation and functional studies provide a model for the structure of the zinc transporter hZIP4. J Biol Chem 290(29):17796–17805. https://doi.org/10.1074/jbc.M114.617613

Arus D, Dancs A, Nagy NV, Gajda T (2013) A comparative study on the possible zinc binding sites of the human ZnT3 zinc transporter protein. Dalton Trans 42(33):12031–12040. https://doi.org/10.1039/c3dt50754h

Aydemir TB, Cousins RJ (2018) The multiple faces of the metal transporter ZIP14 (SLC39A14). J Nutr 148(2):174–184. https://doi.org/10.1093/jn/nxx041

Aydemir TB, Liuzzi JP, McClellan S, Cousins RJ (2009) Zinc transporter ZIP8 (SLC39A8) and zinc influence IFN-gamma expression in activated human T cells. J Leukoc Biol 86(2):337–348. https://doi.org/10.1189/jlb.1208759

Bafaro EM, Antala S, Nguyen TV, Dzul SP, Doyon B, Stemmler TL, Dempski RE (2015) The large intracellular loop of hZIP4 is an intrinsically disordered zinc binding domain. Metallomics 7(9):1319–1330. https://doi.org/10.1039/c5mt00066a

Bafaro E, Liu Y, Xu Y, Dempski RE (2017) The emerging role of zinc transporters in cellular homeostasis and cancer. Signal Transduct Target Ther 2. https://doi.org/10.1038/sigtrans.2017.29

Berg AH, Rice CD, Rahman MS, Dong J, Thomas P (2014) Identification and characterization of membrane androgen receptors in the ZIP9 zinc transporter subfamily: I. Discovery in female Atlantic croaker and evidence ZIP9 mediates testosterone-induced apoptosis of ovarian follicle cells. Endocrinology:en20141198. https://doi.org/10.1210/en.2014-1198

Bin BH, Fukada T, Hosaka T, Yamasaki S, Ohashi W, Hojyo S, Miyai T, Nishida K, Yokoyama S, Hirano T (2011) Biochemical characterization of human ZIP13 protein: a homo-dimerized zinc transporter involved in the spondylocheiro dysplastic Ehlers-Danlos syndrome. J Biol Chem 286(46):40255–40265. https://doi.org/10.1074/jbc.M111.256784

Bin BH, Bhin J, Seo J, Kim SY, Lee E, Park K, Choi DH, Takagishi T, Hara T, Hwang D, Koseki H, Asada Y, Shimoda S, Mishima K, Fukada T (2017a) Requirement of zinc transporter SLC39A7/ZIP7 for dermal development to fine-tune endoplasmic reticulum function by regulating protein disulfide isomerase. J Invest Dermatol 137(8):1682–1691. https://doi.org/10.1016/j.jid.2017.03.031

Bin BH, Bhin J, Takaishi M, Toyoshima KE, Kawamata S, Ito K, Hara T, Watanabe T, Irie T, Takagishi T, Lee SH, Jung HS, Rho S, Seo J, Choi DH, Hwang D, Koseki H, Ohara O, Sano S, Tsuji T, Mishima K, Fukada T (2017b) Requirement of zinc transporter ZIP10 for epidermal development: implication of the ZIP10-p63 axis in epithelial homeostasis. Proc Natl Acad Sci U S A 114(46):12243–12248. https://doi.org/10.1073/pnas.1710726114

Blindauer CA, Schmid R (2010) Cytosolic metal handling in plants: determinants for zinc specificity in metal transporters and metallothioneins. Metallomics 2(8):510–529. https://doi.org/10.1039/c004880a

Bowers K, Srai SKS (2018) The trafficking of metal ion transporters of the Zrt- and Irt-like protein family. Traffic 19(11):813–822. https://doi.org/10.1111/tra.12602

Boycott KM, Beaulieu CL, Kernohan KD, Gebril OH, Mhanni A, Chudley AE, Redl D, Qin W, Hampson S, Kury S, Tetreault M, Puffenberger EG, Scott JN, Bezieau S, Reis A, Uebe S, Schumacher J, Hegele RA, DR ML, Galvez-Peralta M, Majewski J, Ramaekers VT, Care4Rare Canada C, Nebert DW, Innes AM, Parboosingh JS, Abou Jamra R (2015) Autosomal-recessive intellectual disability with cerebellar atrophy syndrome caused by mutation of the manganese and zinc transporter gene SLC39A8. Am J Hum Genet 97(6):886–893. https://doi.org/10.1016/j.ajhg.2015.11.002

Brethour D, Mehrabian M, Williams D, Wang X, Ghodrati F, Ehsani S, Rubie EA, Woodgett JR, Sevalle J, Xi Z, Rogaeva E, Schmitt-Ulms G (2017) A ZIP6-ZIP10 heteromer controls NCAM1 phosphorylation and integration into focal adhesion complexes during epithelial-to-mesenchymal transition. Sci Rep 7:40313. https://doi.org/10.1038/srep40313

Chen YH, Yang CK, Xia M, Ou CY, Stallcup MR (2007) Role of GAC63 in transcriptional activation mediated by beta-catenin. Nucleic Acids Res 35(6):2084–2092. https://doi.org/10.1093/nar/gkm095

Cherezov V, Hofer N, Szebenyi DM, Kolaj O, Wall JG, Gillilan R, Srinivasan V, Jaroniec CP, Caffrey M (2008) Insights into the mode of action of a putative zinc transporter CzrB in Thermus thermophilus. Structure 16(9):1378–1388

Chimienti F, Devergnas S, Favier A, Seve M (2004) Identification and cloning of a {beta}-cell-specific zinc transporter, ZnT-8, localized into insulin secretory granules. Diabetes 53(9):2330–2337

Choi S, Hu YM, Corkins ME, Palmer AE, Bird AJ (2018) Zinc transporters belonging to the cation diffusion facilitator (CDF) family have complementary roles in transporting zinc out of the cytosol. PLoS Genet 14(3):e1007262. https://doi.org/10.1371/journal.pgen.1007262

Chowanadisai W, Graham DM, Keen CL, Rucker RB, Messerli MA (2013) Neurulation and neurite extension require the zinc transporter ZIP12 (slc39a12). Proc Natl Acad Sci U S A 110(24):9903–9908. https://doi.org/10.1073/pnas.1222142110

Chun H, Korolnek T, Lee CJ, Coyne HJ 3rd, Winge DR, Kim BE, Petris MJ (2018) An extracellular histidine-containing motif in the zinc transporter ZIP4 plays a role in zinc sensing and zinc-induced endocytosis in mammalian cells. J Biol Chem. https://doi.org/10.1074/jbc.RA118.005203

Conklin DS, McMaster JA, Culbertson MR, Kung C (1992) COT1, a gene involved in cobalt accumulation in Saccharomyces cerevisiae. Mol Cell Biol 12(9):3678–3688

Coudray N, Valvo S, Hu M, Lasala R, Kim C, Vink M, Zhou M, Provasi D, Filizola M, Tao J, Fang J, Penczek PA, Ubarretxena-Belandia I, Stokes DL (2013) Inward-facing conformation of the zinc transporter YiiP revealed by cryoelectron microscopy. Proc Natl Acad Sci U S A 110(6):2140–2145. https://doi.org/10.1073/pnas.1215455110

Dempski RE (2012) The cation selectivity of the ZIP transporters. Curr Top Membr 69:221–245. https://doi.org/10.1016/B978-0-12-394390-3.00009-4

Dufner-Beattie J, Langmade SJ, Wang F, Eide D, Andrews GK (2003) Structure, function, and regulation of a subfamily of mouse zinc transporter genes. J Biol Chem 278(50):50142–50150

Dufner-Beattie J, Kuo YM, Gitschier J, Andrews GK (2004) The adaptive response to dietary zinc in mice involves the differential cellular localization and zinc regulation of the zinc transporters ZIP4 and ZIP5. J Biol Chem 279(47):49082–49090

Dufner-Beattie J, Huang ZL, Geiser J, Xu W, Andrews GK (2006) Mouse ZIP1 and ZIP3 genes together are essential for adaptation to dietary zinc deficiency during pregnancy. Genesis 44(5):239–251. https://doi.org/10.1002/dvg.20211

Ehsani S, Huo H, Salehzadeh A, Pocanschi CL, Watts JC, Wille H, Westaway D, Rogaeva E, St George-Hyslop PH, Schmitt-Ulms G (2011) Family reunion – the ZIP/prion gene family. Prog Neurobiol 93(3):405–420. https://doi.org/10.1016/j.pneurobio.2010.12.001

Ehsani S, Salehzadeh A, Huo H, Reginold W, Pocanschi CL, Ren H, Wang H, So K, Sato C, Mehrabian M, Strome R, Trimble WS, Hazrati LN, Rogaeva E, Westaway D, Carlson GA, Schmitt-Ulms G (2012) LIV-1 ZIP Ectodomain shedding in prion-infected mice resembles cellular response to transition metal starvation. J Mol Biol 422(4):556–574. https://doi.org/10.1016/j.jmb.2012.06.003

Eide DJ (2004) The SLC39 family of metal ion transporters. Pflugers Arch 447(5):796–800

Eide D, Broderius M, Fett J, Guerinot ML (1996) A novel iron-regulated metal transporter from plants identified by functional expression in yeast. Proc Natl Acad Sci U S A 93(11):5624–5628

Ellis CD, Wang F, MacDiarmid CW, Clark S, Lyons T, Eide DJ (2004) Zinc and the Msc2 zinc transporter protein are required for endoplasmic reticulum function. J Cell Biol 166(3):325–335

Ellis CD, Macdiarmid CW, Eide DJ (2005) Heteromeric protein complexes mediate zinc transport into the secretory pathway of eukaryotic cells. J Biol Chem 280(31):28811–28818

Eng BH, Guerinot ML, Eide D, Saier MH Jr (1998) Sequence analyses and phylogenetic characterization of the ZIP family of metal ion transport proteins. J Membr Biol 166(1):1–7

Franz MC, Simonin A, Graeter S, Hediger MA, Kovacs G (2014) Development of the first fluorescence screening assay for the SLC39A2 zinc transporter. J Biomol Screen 19(6):909–916. https://doi.org/10.1177/1087057114526781

Franz MC, Pujol-Gimenez J, Montalbetti N, Fernandez-Tenorio M, DeGrado TR, Niggli E, Romero MF, Hediger MA (2018) Reassessment of the transport mechanism of the human zinc transporter SLC39A2. Biochemistry 57(26):3976–3986. https://doi.org/10.1021/acs.biochem.8b00511

Fujimoto S, Itsumura N, Tsuji T, Anan Y, Tsuji N, Ogra Y, Kimura T, Miyamae Y, Masuda S, Nagao M, Kambe T (2013) Cooperative functions of ZnT1, Metallothionein and ZnT4 in the cytoplasm are required for full activation of TNAP in the early secretory pathway. PLoS One 8(10):e77445. https://doi.org/10.1371/journal.pone.0077445

Fujimoto S, Tsuji T, Fujiwara T, Takeda TA, Merriman C, Fukunaka A, Nishito Y, Fu D, Hoch E, Sekler I, Fukue K, Miyamae Y, Masuda S, Nagao M, Kambe T (2016) The PP-motif in luminal loop 2 of ZnT transporters plays a pivotal role in TNAP activation. Biochem J 473(17):2611–2621. https://doi.org/10.1042/BCJ20160324

Fujiwara T, Kawachi M, Sato Y, Mori H, Kutsuna N, Hasezawa S, Maeshima M (2015) A high molecular mass zinc transporter MTP12 forms a functional heteromeric complex with MTP5 in the Golgi in Arabidopsis thaliana. FEBS J 282(10):1965–1979. https://doi.org/10.1111/febs.13252

Fukada T, Kambe T (2011) Molecular and genetic features of zinc transporters in physiology and pathogenesis. Metallomics 3(7):662–674. https://doi.org/10.1039/c1mt00011j

Fukada T, Civic N, Furuichi T, Shimoda S, Mishima K, Higashiyama H, Idaira Y, Asada Y, Kitamura H, Yamasaki S, Hojyo S, Nakayama M, Ohara O, Koseki H, Dos Santos HG, Bonafe L, Ha-Vinh R, Zankl A, Unger S, Kraenzlin ME, Beckmann JS, Saito I, Rivolta C, Ikegawa S,

Superti-Furga A, Hirano T (2008) The zinc transporter SLC39A13/ZIP13 is required for connective tissue development; its involvement in BMP/TGF-beta signaling pathways. PLoS One 3(11):e3642

Fukue K, Itsumura N, Tsuji N, Nishino K, Nagao M, Narita H, Kambe T (2018) Evaluation of the roles of the cytosolic N-terminus and his-rich loop of ZNT proteins using ZNT2 and ZNT3 chimeric mutants. Sci Rep 8(1):14084. https://doi.org/10.1038/s41598-018-32372-8

Fukunaka A, Suzuki T, Kurokawa Y, Yamazaki T, Fujiwara N, Ishihara K, Migaki H, Okumura K, Masuda S, Yamaguchi-Iwai Y, Nagao M, Kambe T (2009) Demonstration and characterization of the heterodimerization of ZnT5 and ZnT6 in the early secretory pathway. J Biol Chem 284(45):30798–30806

Fukunaka A, Kurokawa Y, Teranishi F, Sekler I, Oda K, Ackland ML, Faundez V, Hiromura M, Masuda S, Nagao M, Enomoto S, Kambe T (2011) Tissue nonspecific alkaline phosphatase is activated via a two-step mechanism by zinc transport complexes in the early secretory pathway. J Biol Chem 286(18):16363–16373. https://doi.org/10.1074/jbc.M111.227173

Fukunaka A, Fukada T, Bhin J, Suzuki L, Tsuzuki T, Takamine Y, Bin BH, Yoshihara T, Ichinoseki-Sekine N, Naito H, Miyatsuka T, Takamiya S, Sasaki T, Inagaki T, Kitamura T, Kajimura S, Watada H, Fujitani Y (2017) Zinc transporter ZIP13 suppresses beige adipocyte biogenesis and energy expenditure by regulating C/EBP-beta expression. PLoS Genet 13(8):e1006950. https://doi.org/10.1371/journal.pgen.1006950

Gaither LA, Eide DJ (2000) Functional expression of the human hZIP2 zinc transporter. J Biol Chem 275(8):5560–5564

Gaither LA, Eide DJ (2001) Eukaryotic zinc transporters and their regulation. Biometals 14(3–4):251–270

Gao H, Zhao L, Wang H, Xie E, Wang X, Wu Q, Yu Y, He X, Ji H, Rink L, Min J, Wang F (2017) Metal transporter Slc39a10 regulates susceptibility to inflammatory stimuli by controlling macrophage survival. Proc Natl Acad Sci U S A 114(49):12940–12945. https://doi.org/10.1073/pnas.1708018114

Girijashanker K, He L, Soleimani M, Reed JM, Li H, Liu Z, Wang B, Dalton TP, Nebert DW (2008) Slc39a14 gene encodes ZIP14, a metal/bicarbonate symporter: similarities to the ZIP8 transporter. Mol Pharmacol 73(5):1413–1423. https://doi.org/10.1124/mol.107.043588

Golan Y, Berman B, Assaraf YG (2015) Heterodimerization, altered subcellular localization, and function of multiple zinc transporters in viable cells using bimolecular fluorescence complementation. J Biol Chem 290(14):9050–9063. https://doi.org/10.1074/jbc.M114.617332

Golan Y, Itsumura N, Glaser F, Berman B, Kambe T, Assaraf YG (2016) Molecular basis of transient neonatal zinc deficiency: NOVEL ZnT2 MUTATIONS DISRUPTING ZINC BINDING AND PERMEATION. J Biol Chem 291(26):13546–13559. https://doi.org/10.1074/jbc.M116.732693

Golan Y, Kambe T, Assaraf YG (2017) The role of the zinc transporter SLC30A2/ZnT2 in transient neonatal zinc deficiency. Metallomics 9(10):1352–1366. https://doi.org/10.1039/c7mt00162b

Golan Y, Alhadeff R, Glaser F, Ganoth A, Warshel A, Assaraf YG (2018) Demonstrating aspects of multiscale modeling by studying the permeation pathway of the human ZnT2 zinc transporter. PLoS Comput Biol 14(11):e1006503. https://doi.org/10.1371/journal.pcbi.1006503

Groth C, Sasamura T, Khanna MR, Whitley M, Fortini ME (2013) Protein trafficking abnormalities in Drosophila tissues with impaired activity of the ZIP7 zinc transporter catsup. Development 140(14):3018–3027. https://doi.org/10.1242/dev.088336

Gupta S, Chai J, Cheng J, D'Mello R, Chance MR, Fu D (2014) Visualizing the kinetic power stroke that drives proton-coupled zinc(II) transport. Nature 512(7512):101–104. https://doi.org/10.1038/nature13382

Gustin JL, Zanis MJ, Salt DE (2011) Structure and evolution of the plant cation diffusion facilitator family of ion transporters. BMC Evol Biol 11:76. https://doi.org/10.1186/1471-2148-11-76

Guthrie GJ, Aydemir TB, Troche C, Martin AB, Chang SM, Cousins RJ (2015) Influence of ZIP14 (slc39A14) on intestinal zinc processing and barrier function. Am J Physiol Gastrointest Liver Physiol 308(3):G171–G178. https://doi.org/10.1152/ajpgi.00021.2014

Hara T, Takeda TA, Takagishi T, Fukue K, Kambe T, Fukada T (2017) Physiological roles of zinc transporters: molecular and genetic importance in zinc homeostasis. J Physiol Sci 67(2):283–301. https://doi.org/10.1007/s12576-017-0521-4

Hennigar SR, Kelleher SL (2012) Zinc networks: the cell-specific compartmentalization of zinc for specialized functions. Biol Chem 393(7):565–578. https://doi.org/10.1515/hsz-2012-0128. /j/bchm.2012.393.issue-7/hsz-2012-0128/hsz-2012-0128.xml [pii]

Higuchi T, Hattori M, Tanaka Y, Ishitani R, Nureki O (2009) Crystal structure of the cytosolic domain of the cation diffusion facilitator family protein. Proteins 76(3):768–771. https://doi.org/10.1002/prot.22444

Hoch E, Lin W, Chai J, Hershfinkel M, Fu D, Sekler I (2012) Histidine pairing at the metal transport site of mammalian ZnT transporters controls Zn^{2+} over Cd^{2+} selectivity. Proc Natl Acad Sci U S A 109(19):7202–7207. https://doi.org/10.1073/pnas.1200362109

Hogstrand C, Kille P, Ackland ML, Hiscox S, Taylor KM (2013) A mechanism for epithelial-mesenchymal transition and anoikis resistance in breast cancer triggered by zinc channel ZIP6 and STAT3 (signal transducer and activator of transcription 3). Biochem J 455(2):229–237. https://doi.org/10.1042/BJ20130483

Hojyo S, Fukada T, Shimoda S, Ohashi W, Bin BH, Koseki H, Hirano T (2011) The zinc transporter SLC39A14/ZIP14 controls G-protein coupled receptor-mediated signaling required for systemic growth. PLoS One 6(3):e18059. https://doi.org/10.1371/journal.pone.0018059

Hojyo S, Miyai T, Fujishiro H, Kawamura M, Yasuda T, Hijikata A, Bin BH, Irie T, Tanaka J, Atsumi T, Murakami M, Nakayama M, Ohara O, Himeno S, Yoshida H, Koseki H, Ikawa T, Mishima K, Fukada T (2014) Zinc transporter SLC39A10/ZIP10 controls humoral immunity by modulating B-cell receptor signal strength. Proc Natl Acad Sci U S A 111(32):11786–11791. https://doi.org/10.1073/pnas.1323557111

Huang L, Kirschke CP (2007) A di-leucine sorting signal in ZIP1 (SLC39A1) mediates endocytosis of the protein. FEBS J 274(15):3986–3997. https://doi.org/10.1111/j.1742-4658.2007.05933.x

Huang L, Kirschke CP, Gitschier J (2002) Functional characterization of a novel mammalian zinc transporter, ZnT6. J Biol Chem 277(29):26389–26395

Ishihara K, Yamazaki T, Ishida Y, Suzuki T, Oda K, Nagao M, Yamaguchi-Iwai Y, Kambe T (2006) Zinc transport complexes contribute to the homeostatic maintenance of secretory pathway function in vertebrate cells. J Biol Chem 281(26):17743–17750

Itsumura N, Inamo Y, Okazaki F, Teranishi F, Narita H, Kambe T, Kodama H (2013) Compound heterozygous mutations in SLC30A2/ZnT2 results in low Milk zinc concentrations: a novel mechanism for zinc deficiency in a breast-fed infant. PLoS One 8(5):e64045. https://doi.org/10.1371/journal.pone.0064045

Jenkitkasemwong S, Wang CY, Mackenzie B, Knutson MD (2012) Physiologic implications of metal-ion transport by ZIP14 and ZIP8. Biometals 25(4):643–655. https://doi.org/10.1007/s10534-012-9526-x

Jeong J, Eide DJ (2013) The SLC39 family of zinc transporters. Mol Asp Med 34(2–3):612–619. https://doi.org/10.1016/j.mam.2012.05.011

Jeong J, Walker JM, Wang F, Park JG, Palmer AE, Giunta C, Rohrbach M, Steinmann B, Eide DJ (2012) Promotion of vesicular zinc efflux by ZIP13 and its implications for spondylocheiro dysplastic Ehlers-Danlos syndrome. Proc Natl Acad Sci U S A 109(51):E3530–E3538. https://doi.org/10.1073/pnas.1211775110

Kambe T (2011) An overview of a wide range of functions of ZnT and Zip zinc transporters in the secretory pathway. Biosci Biotechnol Biochem 75(6):1036–1043. doi:JST.JSTAGE/bbb/110056 [pii]

Kambe T (2012) Molecular architecture and function of ZnT transporters. Curr Top Membr 69:199–220. https://doi.org/10.1016/B978-0-12-394390-3.00008-2

Kambe T (2013) Regulation of zinc transport. In: Culotta V, Scott RA (eds) Encyclopedia of inorganic and bioinorganic chemistry. Wiley, Hoboken, pp 301–309

Kambe T, Andrews GK (2009) Novel proteolytic processing of the ectodomain of the zinc transporter ZIP4 (SLC39A4) during zinc deficiency is inhibited by acrodermatitis enteropathica mutations. Mol Cell Biol 29(1):129–139

Kambe T, Narita H, Yamaguchi-Iwai Y, Hirose J, Amano T, Sugiura N, Sasaki R, Mori K, Iwanaga T, Nagao M (2002) Cloning and characterization of a novel mammalian zinc transporter, zinc transporter 5, abundantly expressed in pancreatic beta cells. J Biol Chem 277(21):19049–19055

Kambe T, Yamaguchi-Iwai Y, Sasaki R, Nagao M (2004) Overview of mammalian zinc transporters. Cell Mol Life Sci 61(1):49–68

Kambe T, Suzuki T, Nagao M, Yamaguchi-Iwai Y (2006) Sequence similarity and functional relationship among eukaryotic ZIP and CDF transporters. Genomics Proteomics Bioinformatics 4(1):1–9

Kambe T, Geiser J, Lahner B, Salt DE, Andrews GK (2008) Slc39a1 to 3 (subfamily II) zip genes in mice have unique cell-specific functions during adaptation to zinc deficiency. Am J Physiol Regul Integr Comp Physiol 294(5):R1474–R1481

Kambe T, Tsuji T, Fukue K (2014) In: Fukada T, Kambe T (eds) Zinc transport proteins and zinc signaling. Zinc signals in cellular functions and disorders. Springer, Tokyo, pp 27–53

Kambe T, Tsuji T, Hashimoto A, Itsumura N (2015) The physiological, biochemical, and molecular roles of zinc transporters in zinc homeostasis and metabolism. Physiol Rev 95(3):749–784. https://doi.org/10.1152/physrev.00035.2014

Kambe T, Takeda TA, Nishito Y (2016) Activation of zinc-requiring ectoenzymes by ZnT transporters during the secretory process: biochemical and molecular aspects. Arch Biochem Biophys 611:37–42. https://doi.org/10.1016/j.abb.2016.03.035

Kambe T, Matsunaga M, Takeda TA (2017) Understanding the contribution of zinc transporters in the function of the early secretory pathway. Int J Mol Sci 18(10):E2179. https://doi.org/10.3390/ijms18102179

Kamizono A, Nishizawa M, Teranishi Y, Murata K, Kimura A (1989) Identification of a gene conferring resistance to zinc and cadmium ions in the yeast Saccharomyces cerevisiae. Mol Gen Genet 219(1–2):161–167

Kawachi M, Kobae Y, Mimura T, Maeshima M (2008) Deletion of a histidine-rich loop of AtMTP1, a vacuolar Zn(2+)/H(+) antiporter of Arabidopsis thaliana, stimulates the transport activity. J Biol Chem 283(13):8374–8383

Kawachi M, Kobae Y, Kogawa S, Mimura T, Kramer U, Maeshima M (2012) Amino acid screening based on structural modeling identifies critical residues for the function, ion selectivity and structure of Arabidopsis MTP1. FEBS J 279(13):2339–2356. https://doi.org/10.1111/j.1742-4658.2012.08613.x

Kelleher SL, Velasquez V, Croxford TP, McCormick NH, Lopez V, Macdavid J (2012) Mapping the zinc-transporting system in mammary cells: molecular analysis reveals a phenotype-dependent zinc-transporting network during lactation. J Cell Physiol 227(4):1761–1770. https://doi.org/10.1002/jcp.22900

Kim JH, Jeon J, Shin M, Won Y, Lee M, Kwak JS, Lee G, Rhee J, Ryu JH, Chun CH, Chun JS (2014) Regulation of the catabolic cascade in osteoarthritis by the zinc-ZIP8-MTF1 axis. Cell 156(4):730–743. https://doi.org/10.1016/j.cell.2014.01.007

Kirschke CP, Huang L (2003) ZnT7, a novel mammalian zinc transporter, accumulates zinc in the Golgi apparatus. J Biol Chem 278(6):4096–4102

Kolaj-Robin O, Russell D, Hayes KA, Pembroke JT, Soulimane T (2015) Cation diffusion facilitator family: structure and function. FEBS Lett 589(12):1283–1295. https://doi.org/10.1016/j.febslet.2015.04.007

Kong BY, Duncan FE, Que EL, Kim AM, O'Halloran TV, Woodruff TK (2014) Maternally-derived zinc transporters ZIP6 and ZIP10 drive the mammalian oocyte-to-egg transition. Mol Hum Reprod 20(11):1077–1089. https://doi.org/10.1093/molehr/gau066

Kumanovics A, Poruk KE, Osborn KA, Ward DM, Kaplan J (2006) YKE4 (YIL023C) encodes a bidirectional zinc transporter in the endoplasmic reticulum of Saccharomyces cerevisiae. J Biol Chem 281(32):22566–22574

Langmade SJ, Ravindra R, Daniels PJ, Andrews GK (2000) The transcription factor MTF-1 mediates metal regulation of the mouse ZnT1 gene. J Biol Chem 275(44):34803–34809

Lasry I, Seo YA, Ityel H, Shalva N, Pode-Shakked B, Glaser F, Berman B, Berezovsky I, Goncearenco A, Klar A, Levy J, Anikster Y, Kelleher SL, Assaraf YG (2012) A dominant negative heterozygous G87R mutation in the zinc transporter, ZnT-2 (SLC30A2), results in transient neonatal zinc deficiency. J Biol Chem 287(35):29348–29361. https://doi.org/10.1074/jbc.M112.368159

Lasry I, Golan Y, Berman B, Amram N, Glaser F, Assaraf YG (2014) In situ dimerization of multiple wild type and mutant zinc transporters in live cells using bimolecular fluorescence complementation. J Biol Chem 289(11):7275–7292. https://doi.org/10.1074/jbc.M113.533786

Lasswell J, Rogg LE, Nelson DC, Rongey C, Bartel B (2000) Cloning and characterization of IAR1, a gene required for auxin conjugate sensitivity in Arabidopsis. Plant Cell 12(12):2395–2408

Lee S, Kelleher SL (2016) Molecular regulation of lactation: the complex and requisite roles for zinc. Arch Biochem Biophys 611:86–92. https://doi.org/10.1016/j.abb.2016.04.002

Levy M, Elkoshi N, Barber-Zucker S, Hoch E, Zarivatch R, Hershfinkel M, Sekler I (2019) Zinc transporter 10 (ZnT10)-dependent extrusion of cellular Mn2+ is driven by an active Ca2+–coupled exchange. J Biol Chem. https://doi.org/10.1074/jbc.RA118.006816

Leyva-Illades D, Chen P, Zogzas CE, Hutchens S, Mercado JM, Swaim CD, Morrisett RA, Bowman AB, Aschner M, Mukhopadhyay S (2014) SLC30A10 is a cell surface-localized manganese efflux transporter, and parkinsonism-causing mutations block its intracellular trafficking and efflux activity. J Neurosci 34(42):14079–14095. https://doi.org/10.1523/JNEUROSCI.2329-14.2014

Lin W, Chai J, Love J, Fu D (2010) Selective electrodiffusion of zinc ions in a Zrt-, Irt-like protein, ZIPB. J Biol Chem 285(50):39013–39020. https://doi.org/10.1074/jbc.M110.180620

Liu MJ, Bao S, Galvez-Peralta M, Pyle CJ, Rudawsky AC, Pavlovicz RE, Killilea DW, Li C, Nebert DW, Wewers MD, Knoell DL (2013) ZIP8 regulates host defense through zinc-mediated inhibition of NF-kappaB. Cell Rep 3(2):386–400. https://doi.org/10.1016/j.celrep.2013.01.009

Liuzzi JP, Lichten LA, Rivera S, Blanchard RK, Aydemir TB, Knutson MD, Ganz T, Cousins RJ (2005) Interleukin-6 regulates the zinc transporter Zip14 in liver and contributes to the hypozincemia of the acute-phase response. Proc Natl Acad Sci U S A 102(19):6843–6848

Lopez-Redondo ML, Coudray N, Zhang Z, Alexopoulos J, Stokes DL (2018) Structural basis for the alternating access mechanism of the cation diffusion facilitator YiiP. Proc Natl Acad Sci U S A 115(12):3042–3047. https://doi.org/10.1073/pnas.1715051115

Lu M, Fu D (2007) Structure of the zinc transporter YiiP. Science 317(5845):1746–1748

Lu M, Chai J, Fu D (2009) Structural basis for autoregulation of the zinc transporter YiiP. Nat Struct Mol Biol 16(10):1063–1067

Manning DL, Robertson JF, Ellis IO, Elston CW, McClelland RA, Gee JM, Jones RJ, Green CD, Cannon P, Blamey RW et al (1994) Oestrogen-regulated genes in breast cancer: association of pLIV1 with lymph node involvement. Eur J Cancer 30A(5):675–678

Mao X, Kim BE, Wang F, Eide DJ, Petris MJ (2007) A histidine-rich cluster mediates the ubiquitination and degradation of the human zinc transporter, hZIP4, and protects against zinc cytotoxicity. J Biol Chem 282(10):6992–7000

Maret W, Li Y (2009) Coordination dynamics of zinc in proteins. Chem Rev 109(10):4682–4707

Martin AB, Aydemir TB, Guthrie GJ, Samuelson DA, Chang SM, Cousins RJ (2013) Gastric and colonic zinc transporter ZIP11 (Slc39a11) in mice responds to dietary zinc and exhibits nuclear localization. J Nutr 143(12):1882–1888. https://doi.org/10.3945/jn.113.184457

Matsuura W, Yamazaki T, Yamaguchi-Iwai Y, Masuda S, Nagao M, Andrews GK, Kambe T (2009) SLC39A9 (ZIP9) regulates zinc homeostasis in the secretory pathway: characterization of the ZIP subfamily I protein in vertebrate cells. Biosci Biotechnol Biochem 73(5):1142–1148

McCormick NH, Kelleher SL (2012) ZnT4 provides zinc to zinc-dependent proteins in the trans-Golgi network critical for cell function and Zn export in mammary epithelial cells. Am J Physiol Cell Physiol 303(3):C291–C297. https://doi.org/10.1152/ajpcell.00443.2011

McCormick NH, Lee S, Hennigar SR, Kelleher SL (2016) ZnT4 (SLC30A4)-null ("lethal milk") mice have defects in mammary gland secretion and hallmarks of precocious involution during lactation. Am J Physiol Regul Integr Comp Physiol 310(1):R33–R40. https://doi.org/10.1152/ajpregu.00315.2014

McMahon RJ, Cousins RJ (1998) Regulation of the zinc transporter ZnT-1 by dietary zinc. Proc Natl Acad Sci U S A 95(9):4841–4846

Merriman C, Huang Q, Rutter GA, Fu D (2016) Lipid-tuned zinc transport activity of human ZnT8 protein correlates with risk for Type-2 diabetes. J Biol Chem 291(53):26950–26957. https://doi.org/10.1074/jbc.M116.764605

Miyai T, Hojyo S, Ikawa T, Kawamura M, Irie T, Ogura H, Hijikata A, Bin BH, Yasuda T, Kitamura H, Nakayama M, Ohara O, Yoshida H, Koseki H, Mishima K, Fukada T (2014) Zinc transporter SLC39A10/ZIP10 facilitates antiapoptotic signaling during early B-cell development. Proc Natl Acad Sci U S A 111(32):11780–11785. https://doi.org/10.1073/pnas.1323549111

Montanini B, Blaudez D, Jeandroz S, Sanders D, Chalot M (2007) Phylogenetic and functional analysis of the cation diffusion facilitator (CDF) family: improved signature and prediction of substrate specificity. BMC Genomics 8:107

Murgia C, Vespignani I, Cerase J, Nobili F, Perozzi G (1999) Cloning, expression, and vesicular localization of zinc transporter Dri 27/ZnT4 in intestinal tissue and cells. Am J Phys 277(6 Pt 1):G1231–G1239

Murgia C, Devirgiliis C, Mancini E, Donadel G, Zalewski P, Perozzi G (2008) Diabetes-linked zinc transporter ZnT8 is a homodimeric protein expressed by distinct rodent endocrine cell types in the pancreas and other glands. Nutr Metab Cardiovasc Dis 19:431

Nies H (1992) CzcR and CzcD, gene products affecting regulation of resistance to cobalt, zinc, and cadmium (czc system) in Alcaligenes eutrophus. J Bacteriol 174(24):8102–8110

Nimmanon T, Ziliotto S, Morris S, Flanagan L, Taylor KM (2017) Phosphorylation of zinc channel ZIP7 drives MAPK, PI3K and mTOR growth and proliferation signalling. Metallomics 9(5):471–481. https://doi.org/10.1039/c6mt00286b

Nishito Y, Kambe T (2019) Zinc transporter 1 (ZNT1) expression on the cell surface is elaborately controlled by cellular zinc levels. J Biol Chem 294(43):15686–15697. https://doi.org/10.1074/jbc.RA119.010227

Nishito Y, Tsuji N, Fujishiro H, Takeda T, Yamazaki T, Teranishi F, Okazaki F, Matsunaga A, Tuschl K, Rao R, Kono S, Miyajima H, Narita H, Himeno S, Kambe T (2016) Direct comparison of manganese detoxification/efflux proteins and molecular characterization of ZnT10 as a manganese transporter. J Biol Chem 291(28):14773–14787. https://doi.org/10.1074/jbc.M116.728014

Ohana E, Hoch E, Keasar C, Kambe T, Yifrach O, Hershfinkel M, Sekler I (2009) Identification of the Zn^{2+} binding site and mode of operation of a mammalian Zn^{2+} transporter. J Biol Chem 284(26):17677–17686

Ohashi W, Kimura S, Iwanaga T, Furusawa Y, Irie T, Izumi H, Watanabe T, Hijikata A, Hara T, Ohara O, Koseki H, Sato T, Robine S, Mori H, Hattori Y, Watarai H, Mishima K, Ohno H, Hase K, Fukada T (2016) Zinc transporter SLC39A7/ZIP7 promotes intestinal epithelial self-renewal by resolving ER stress. PLoS Genet 12(10):e1006349. https://doi.org/10.1371/journal.pgen.1006349

Palmiter RD, Findley SD (1995) Cloning and functional characterization of a mammalian zinc transporter that confers resistance to zinc. EMBO J 14(4):639–649

Palmiter RD, Cole TB, Findley SD (1996a) ZnT-2, a mammalian protein that confers resistance to zinc by facilitating vesicular sequestration. EMBO J 15(8):1784–1791

Palmiter RD, Cole TB, Quaife CJ, Findley SD (1996b) ZnT-3, a putative transporter of zinc into synaptic vesicles. Proc Natl Acad Sci U S A 93(25):14934–14939

Park JH, Hogrebe M, Gruneberg M, DuChesne I, von der Heiden AL, Reunert J, Schlingmann KP, Boycott KM, Beaulieu CL, Mhanni AA, Innes AM, Hortnagel K, Biskup S, Gleixner EM, Kurlemann G, Fiedler B, Omran H, Rutsch F, Wada Y, Tsiakas K, Santer R, Nebert DW, Rust

S, Marquardt T (2015) SLC39A8 deficiency: a disorder of manganese transport and glycosylation. Am J Hum Genet 97(6):894–903. https://doi.org/10.1016/j.ajhg.2015.11.003

Paulsen IT, Saier MH Jr (1997) A novel family of ubiquitous heavy metal ion transport proteins. J Membr Biol 156(2):99–103

Pedas P, Schiller Stokholm M, Hegelund JN, Ladegard AH, Schjoerring JK, Husted S (2014) Golgi localized barley MTP8 proteins facilitate Mn transport. PLoS One 9(12):e113759. https://doi.org/10.1371/journal.pone.0113759

Perez Y, Shorer Z, Liani-Leibson K, Chabosseau P, Kadir R, Volodarsky M, Halperin D, Barber-Zucker S, Shalev H, Schreiber R, Gradstein L, Gurevich E, Zarivach R, Rutter GA, Landau D, Birk OS (2017) SLC30A9 mutation affecting intracellular zinc homeostasis causes a novel cerebro-renal syndrome. Brain 140(4):928–939. https://doi.org/10.1093/brain/awx013

Peters JL, Dufner-Beattie J, Xu W, Geiser J, Lahner B, Salt DE, Andrews GK (2007) Targeting of the mouse Slc39a2 (Zip2) gene reveals highly cell-specific patterns of expression, and unique functions in zinc, iron, and calcium homeostasis. Genesis 45(6):339–352. https://doi.org/10.1002/dvg.20297

Pocanschi CL, Ehsani S, Mehrabian M, Wille H, Reginold W, Trimble WS, Wang H, Yee A, Arrowsmith CH, Bozoky Z, Kay LE, Forman-Kay JD, Rini JM, Schmitt-Ulms G (2013) The ZIP5 Ectodomain Co-localizes with PrP and may acquire a PrP-like fold that assembles into a dimer. PLoS One 8(9):e72446. https://doi.org/10.1371/journal.pone.0072446

Podany AB, Wright J, Lamendella R, Soybel DI, Kelleher SL (2016) ZnT2-mediated zinc import into Paneth cell granules is necessary for coordinated secretion and Paneth cell function in mice. Cell Mol Gastroenterol Hepatol 2(3):369–383. https://doi.org/10.1016/j.jcmgh.2015.12.006

Quadri M, Federico A, Zhao T, Breedveld GJ, Battisti C, Delnooz C, Severijnen LA, Di Toro ML, Mignarri A, Monti L, Sanna A, Lu P, Punzo F, Cossu G, Willemsen R, Rasi F, Oostra BA, van de Warrenburg BP, Bonifati V (2012) Mutations in SLC30A10 cause parkinsonism and dystonia with hypermanganesemia, polycythemia, and chronic liver disease. Am J Hum Genet 90(3):467–477. https://doi.org/10.1016/j.ajhg.2012.01.017

Roh HC, Collier S, Deshmukh K, Guthrie J, Robertson JD, Kornfeld K (2013) Ttm-1 encodes CDF transporters that excrete zinc from intestinal cells of C. elegans and act in a parallel negative feedback circuit that promotes homeostasis. PLoS Genet 9(5):e1003522. https://doi.org/10.1371/journal.pgen.1003522

Seo YA, Lopez V, Kelleher SL (2011) A histidine-rich motif mediates mitochondrial localization of ZnT2 to modulate mitochondrial function. Am J Physiol Cell Physiol 300(6):C1479–C1489. https://doi.org/10.1152/ajpcell.00420.2010

Shusterman E, Beharier O, Shiri L, Zarivach R, Etzion Y, Campbell CR, Lee IH, Okabayashi K, Dinudom A, Cook DI, Katz A, Moran A (2014) ZnT-1 extrudes zinc from mammalian cells functioning as a Zn(2+)/H(+) exchanger. Metallomics 6(9):1656–1663. https://doi.org/10.1039/c4mt00108g

Suzuki T, Ishihara K, Migaki H, Matsuura W, Kohda A, Okumura K, Nagao M, Yamaguchi-Iwai Y, Kambe T (2005a) Zinc transporters, ZnT5 and ZnT7, are required for the activation of alkaline phosphatases, zinc-requiring enzymes that are glycosylphosphatidylinositol-anchored to the cytoplasmic membrane. J Biol Chem 280(1):637–643

Suzuki T, Ishihara K, Migaki H, Nagao M, Yamaguchi-Iwai Y, Kambe T (2005b) Two different zinc transport complexes of cation diffusion facilitator proteins localized in the secretory pathway operate to activate alkaline phosphatases in vertebrate cells. J Biol Chem 280(35):30956–30962

Takeda TA, Miyazaki S, Kobayashi M, Nishino K, Goto T, Matsunaga M, Ooi M, Shirakawa H, Tani F, Kawamura T, Komai M, Kambe T (2018) Zinc deficiency causes delayed ATP clearance and adenosine generation in rats and cell culture models. Commun Biol 1:113. https://doi.org/10.1038/s42003-018-0118-3

Tanaka N, Kawachi M, Fujiwara T, Maeshima M (2013) Zinc-binding and structural properties of the histidine-rich loop of Arabidopsis thaliana vacuolar membrane zinc transporter MTP1. FEBS Open Bio 3:218–224. https://doi.org/10.1016/j.fob.2013.04.004

Tanaka N, Fujiwara T, Tomioka R, Kramer U, Kawachi M, Maeshima M (2015) Characterization of the histidine-rich loop of Arabidopsis vacuolar membrane zinc transporter AtMTP1 as a sensor of zinc level in the cytosol. Plant Cell Physiol 56(3):510–519. https://doi.org/10.1093/pcp/pcu194

Taylor KM (2000) LIV-1 breast cancer protein belongs to new family of histidine-rich membrane proteins with potential to control intracellular Zn2+ homeostasis. IUBMB Life 49(4):249–253

Taylor KM, Nicholson RI (2003) The LZT proteins; the LIV-1 subfamily of zinc transporters. Biochim Biophys Acta 1611(1–2):16–30

Taylor KM, Morgan HE, Smart K, Zahari NM, Pumford S, Ellis IO, Robertson JF, Nicholson RI (2007) The emerging role of the LIV-1 subfamily of zinc transporters in breast cancer. Mol Med 13(7–8):396–406

Taylor KM, Hiscox S, Nicholson RI, Hogstrand C, Kille P (2012) Protein kinase CK2 triggers cytosolic zinc signaling pathways by phosphorylation of Zinc Channel ZIP7. Sci Signal 5(210):ra11. https://doi.org/10.1126/scisignal.2002585

Taylor KM, Muraina IA, Brethour D, Schmitt-Ulms G, Nimmanon T, Ziliotto S, Kille P, Hogstrand C (2016) Zinc transporter ZIP10 forms a heteromer with ZIP6 which regulates embryonic development and cell migration. Biochem J 473(16):2531–2544. https://doi.org/10.1042/BCJ20160388

Taylor CA, Hutchens S, Liu C, Jursa T, Shawlot W, Aschner M, Smith DR, Mukhopadhyay S (2019) SLC30A10 transporter in the digestive system regulates brain manganese under basal conditions while brain SLC30A10 protects against neurotoxicity. J Biol Chem 294(6):1860–1876. https://doi.org/10.1074/jbc.RA118.005628

Thomas P, Pang Y, Dong J, Berg AH (2014) Identification and characterization of membrane androgen receptors in the ZIP9 zinc transporter subfamily: II. Role of human ZIP9 in testosterone-induced prostate and breast cancer cell apoptosis. Endocrinology 155(11):4250–4265. https://doi.org/10.1210/en.2014-1201

Thomas P, Converse A, Berg HA (2018) ZIP9, a novel membrane androgen receptor and zinc transporter protein. Gen Comp Endocrinol 257:130–136. https://doi.org/10.1016/j.ygcen.2017.04.016

Tsuji T, Kurokawa Y, Chiche J, Pouyssegur J, Sato H, Fukuzawa H, Nagao M, Kambe T (2017) Dissecting the process of activation of Cancer-promoting zinc-requiring Ectoenzymes by zinc metalation mediated by ZNT transporters. J Biol Chem 292(6):2159–2173. https://doi.org/10.1074/jbc.M116.763946

Tsunemitsu Y, Genga M, Okada T, Yamaji N, Ma JF, Miyazaki A, Kato SI, Iwasaki K, Ueno D (2018) A member of cation diffusion facilitator family, MTP11, is required for manganese tolerance and high fertility in rice. Planta 248(1):231–241. https://doi.org/10.1007/s00425-018-2890-1

Tuncay E, Bitirim VC, Durak A, Carrat GRJ, Taylor KM, Rutter GA, Turan B (2017) Hyperglycemia-induced changes in ZIP7 and ZnT7 expression cause Zn^{2+} release from the Sarco(endo)plasmic reticulum and mediate ER stress in the heart. Diabetes 66(5):1346–1358. https://doi.org/10.2337/db16-1099

Tuschl K, Clayton PT, Gospe SM Jr, Gulab S, Ibrahim S, Singhi P, Aulakh R, Ribeiro RT, Barsottini OG, Zaki MS, Del Rosario ML, Dyack S, Price V, Rideout A, Gordon K, Wevers RA, Chong WK, Mills PB (2012) Syndrome of hepatic cirrhosis, dystonia, polycythemia, and hypermanganesemia caused by mutations in SLC30A10, a manganese transporter in man. Am J Hum Genet 90(3):457–466. https://doi.org/10.1016/j.ajhg.2012.01.018

Tuschl K, Meyer E, Valdivia LE, Zhao N, Dadswell C, Abdul-Sada A, Hung CY, Simpson MA, Chong WK, Jacques TS, Woltjer RL, Eaton S, Gregory A, Sanford L, Kara E, Houlden H, Cuno SM, Prokisch H, Valletta L, Tiranti V, Younis R, Maher ER, Spencer J, Straatman-Iwanowska A, Gissen P, Selim LA, Pintos-Morell G, Coroleu-Lletget W, Mohammad SS, Yoganathan S, Dale RC, Thomas M, Rihel J, Bodamer OA, Enns CA, Hayflick SJ, Clayton PT, Mills PB, Kurian MA, Wilson SW (2016) Mutations in SLC39A14 disrupt manganese homeostasis and cause childhood-onset parkinsonism-dystonia. Nat Commun 7:11601. https://doi.org/10.1038/ncomms11601

Uebe R, Keren-Khadmy N, Zeytuni N, Katzmann E, Navon Y, Davidov G, Bitton R, Plitzko JM, Schuler D, Zarivach R (2018) The dual role of MamB in magnetosome membrane assembly and magnetite biomineralization. Mol Microbiol 107(4):542–557. https://doi.org/10.1111/mmi.13899

Ueno D, Sasaki A, Yamaji N, Miyaji T, Fujii Y, Takemoto Y, Moriyama S, Che J, Moriyama Y, Iwasaki K, Ma JF (2015) A polarly localized transporter for efficient manganese uptake in rice. Nat Plants 1:15170. https://doi.org/10.1038/nplants.2015.170

Wang F, Kim BE, Petris MJ, Eide DJ (2004) The mammalian Zip5 protein is a zinc transporter that localizes to the basolateral surface of polarized cells. J Biol Chem 279(49):51433–51441

Xiao G, Wan Z, Fan Q, Tang X, Zhou B (2014) The metal transporter ZIP13 supplies iron into the secretory pathway in Drosophila melanogaster. elife 3:e03191. https://doi.org/10.7554/eLife.03191

Xiao G, Liu ZH, Zhao M, Wang HL, Zhou B (2019) Transferrin 1 functions in Iron trafficking and genetically interacts with ferritin in Drosophila melanogaster. Cell Rep 26(3):748–758. e745. https://doi.org/10.1016/j.celrep.2018.12.053

Yu Y, Wu A, Zhang Z, Yan G, Zhang F, Zhang L, Shen X, Hu R, Zhang Y, Zhang K, Wang F (2013) Characterization of the GufA subfamily member SLC39A11/Zip11 as a zinc transporter. J Nutr Biochem 24(10):1697–1708. https://doi.org/10.1016/j.jnutbio.2013.02.010

Zhang T, Sui D, Hu J (2016) Structural insights of ZIP4 extracellular domain critical for optimal zinc transport. Nat Commun 7:11979. https://doi.org/10.1038/ncomms11979

Zhang T, Liu J, Fellner M, Zhang C, Sui D, Hu J (2017) Crystal structures of a ZIP zinc transporter reveal a binuclear metal center in the transport pathway. Sci Adv 3(8):e1700344. https://doi.org/10.1126/sciadv.1700344

Zhao H, Eide D (1996a) The yeast ZRT1 gene encodes the zinc transporter protein of a high- affinity uptake system induced by zinc limitation. Proc Natl Acad Sci U S A 93(6):2454–2458

Zhao H, Eide D (1996b) The ZRT2 gene encodes the low affinity zinc transporter in Saccharomyces cerevisiae. J Biol Chem 271(38):23203–23210

Zhao L, Oliver E, Maratou K, Atanur SS, Dubois OD, Cotroneo E, Chen CN, Wang L, Arce C, Chabosseau PL, Ponsa-Cobas J, Frid MG, Moyon B, Webster Z, Aldashev A, Ferrer J, Rutter GA, Stenmark KR, Aitman TJ, Wilkins MR (2015) The zinc transporter ZIP12 regulates the pulmonary vascular response to chronic hypoxia. Nature. https://doi.org/10.1038/nature14620

Zhao Y, Feresin RG, Falcon-Perez JM, Salazar G (2016) Differential targeting of SLC30A10/ZnT10 heterodimers to Endolysosomal compartments modulates EGF-induced MEK/ERK1/2 activity. Traffic 17(3):267–288. https://doi.org/10.1111/tra.12371

Zhao Y, Tan CH, Krauchunas A, Scharf A, Dietrich N, Warnhoff K, Yuan Z, Druzhinina M, Gu SG, Miao L, Singson A, Ellis RE, Kornfeld K (2018) The zinc transporter ZIPT-7.1 regulates sperm activation in nematodes. PLoS Biol 16(6):e2005069. https://doi.org/10.1371/journal.pbio.2005069

Chapter 4
The Metallothionein-Zinc Landscape: How It Shapes Antimicrobial Immunity

Debabrata Chowdhury, George S. Deepe Jr, and Kavitha Subramanian Vignesh

Abstract Nutritional immunity refers to the ability of the host to sequester nutrients from pathogens during infection. Metal ions are important for microbial survival and pathogenesis as well as for host defenses. For example, while zinc (Zn^{2+}) is crucial for microbial fitness within the host, immune cells deprive microbes of these metal ions and retain it for their own defense. However, excess Zn^{2+} may be toxic to both the host and the pathogen. Therefore, Zn^{2+} regulation is a central component of host-pathogen interactions and antimicrobial immunity. Metallothioneins (MTs) are a family of highly conserved cysteine-rich proteins that are ubiquitously expressed in most organisms. Immune cells express MTs in response to a variety of stimuli including cytokines, chemokines, and infectious agents. They regulate intracellular Zn^{2+} homeostasis, protect from oxidative stress, and modulate host immunity during infection. Although Zn^{2+} signals are well known to alter immunological processes, our knowledge of how the MTs-Zn^{2+} axis affects immune response to infections is relatively scarce. Emerging evidence points to a significant role for MTs in regulating host immunity. Thus, this chapter discusses immunomodulatory roles of MTs with a focus on Zn^{2+} regulation in response to pathogen attack.

Keywords Metallothionein · Zinc · Macrophages · Innate Immunity · Infection · Inflammation

---

D. Chowdhury · G. S. Deepe Jr · K. Subramanian Vignesh (✉)
Division of Infectious Diseases, College of Medicine, University of Cincinnati, Cincinnati, OH, USA
e-mail: Kavitha.Subramanian@uc.edu

T. Fukada, T. Kambe (eds.), *Zinc Signaling*,
https://doi.org/10.1007/978-981-15-0557-7_4

4.1 Introduction

Nutritional immunity is a mechanism of defense by which the host restricts nutrient access to pathogens to inhibit their growth. Metal ions are micronutrients that are essential for life. They regulate important functions of immune cells as well as that of invading microbes. One form of nutritional immunity is limitation of metal ions by modulating their availability, concentration, and distribution inside the cell. Zn^{2+} is an important metal ion whose concentration in circulation (plasma and serum) rapidly declines during infection (Besecker et al. 2011; Uttra et al. 2011). This phenomenon is postulated to deprive invading pathogens of Zn^{2+}, control their survival, and prevent dissemination (Hennigar and McClung 2016). Zn^{2+} deficiency or excess alters the number and activities of immune cells thereby modulating host susceptibility to infection. A decrease in dietary Zn^{2+} uptake increases the risk of infectious diseases such as tuberculosis, shigellosis, pneumonia, measles, human immunodeficiency virus (HIV), acute cutaneous leishmaniosis, and malaria (Maywald et al. 2017). On the other hand, dietary Zn^{2+} supplementation aids in improving immune defenses against some of the microbes that cause the aforementioned infections as well as diarrhea, leprosy, chronic hepatitis C, and acute lower respiratory infection (Overbeck et al. 2008). However, there may be a very narrow window within which Zn^{2+} exerts beneficial effects. Recent data show that excess dietary Zn^{2+} uptake increases susceptibility to *Clostridioides difficile* and intensifies disease severity suggesting that exogenous Zn^{2+} administration may adversely impact the clearance of some pathogens (Zackular et al. 2016).

Effects of Zn^{2+} on immunity are complex and have been under investigation for many years. Zn^{2+} deprivation or supplementation in vitro and in vivo affects the expression of several genes in immune cells associated with Zn^{2+} homeostasis, cytokine response, stress responses, reactive oxygen species and reactive nitrogen species (ROS and RNS) signaling, metabolism, and survival (Beck et al. 2006; Cousins 1998; Cousins et al. 2003; Haase et al. 2007). Zn^{2+} deficiency inhibits differentiation, proliferation, and survival of monocytes, polymorphonuclear leukocytes (PMN), natural killer cells, and T and B cells (Bonaventura et al. 2015). However, an excess of Zn^{2+} may lead to toxicity. Therefore, Zn^{2+} homeostasis is tightly regulated in immune cells by Zn^{2+} binding proteins such as metallothioneins (MTs), glutathione, the Zn^{2+}-responsive transcription factor metal-response element-binding transcription factor-1 (MTF-1), Zn^{2+}-transporters, Zn^{2+}-permeable ion channels such as transient receptor potential mucolipin 1 (TRPML1), and Zn^{2+} storage organelles such as zincosomes (Andrews 2001; Crawford et al. 2018; Eide 2004; Inoue et al. 2015; Liu et al. 2012; Palmiter 2004; Palmiter and Huang 2004; Vallee 1995). Taken together, regulation of Zn^{2+} is necessary for adequate immune function and aberrant homeostasis of these metal ions may have adverse effects on the host's ability to defend microbial invaders.

Microorganisms require Zn^{2+} for survival. Thus, restriction of Zn^{2+} during infection may be an effective strategy that immune cells utilize to inhibit microbial growth. Several recent studies have brought to light the importance of Zn^{2+} limitation by MTs in immune cells. Zn^{2+} is bound to MTs through seven binding sites with picomolar binding affinity. This attribute facilitates controlled Zn^{2+} exchange between proteins and promotes nutritional immunity in host cells, where accessibility of Zn^{2+} to pathogens must be restricted. Our work revealed that MT1 and MT2 inhibit fungal growth via sequestration of Zn^{2+} in infected macrophages (Subramanian Vignesh et al. 2013). The MT3 isoform has a very distinct role: it facilitates Zn^{2+} uptake by intracellular fungi (Subramanian Vignesh et al. 2016). Nonetheless, the finding that MTs are an important component of the antimicrobial defense arsenal raises interesting questions about their mechanisms of action. In sum, MTs are expressed in immune cells, regulate inflammatory responses, and control host-pathogen interactions indicating that they may be at the forefront of immunological fitness in the host. Given the role of Zn^{2+} in pathogen virulence and the emerging importance of MTs in immune responses, this chapter focuses on the roles of Zn^{2+} and the MT-Zn^{2+} landscape in antimicrobial immunity.

4.2 Low Zinc Spells a High Infection Risk

Zn^{2+} has vital roles in biochemical processes and is crucial for maintaining the structure, stability, and adequate activity of macromolecules, such as proteins and nucleic acids. About 10% of the human proteome consists of Zn^{2+} binding proteins which require this ion for proper physiological function (Andreini et al. 2006). Zn^{2+} is the second most abundant (total concentration in the human body 2–3 g) transition metal after Fe^{2+} in humans (Kehl-Fie and Skaar 2010).

Prasad et al. first discovered Zn^{2+} deficiency in human male dwarfs in the Middle East in 1963 (Prasad et al. 1963). Zn^{2+} deficiency is associated with susceptibility to infections, memory impairment, and growth retardation. Zn^{2+} deficient animals have decreased immunity to viral infections such as *Herpes simplex* (Feiler et al. 1982) and *Semliki forest* (Singh et al. 1992); bacterial infections such as *Listeria monocytogenes* (Carlomagno et al. 1986; Coghlan et al. 1988), *Francisella tularensis* (Pekarek et al. 1977), *Mycobacterium tuberculosis* (McMurray et al. 1990), and *Salmonella enteritidis* (Kidd et al. 1994); parasitic infections such as *Trypanosoma cruzi* (Fraker et al. 1982), *T. musculi* (Lee et al. 1983), *Toxoplasma gondii* (Tasci et al. 1995), and *Plasmodium yoelii* (Shankar et al. 1995); fungal infections such as *Candida albicans* (Salvin et al. 1987); and helminthic infections such as *Fasciola hepatica* (Flagstad et al. 1972), *Heligmosomoides polygyrus* (Minkus et al. 1992; Shi et al. 1994), *Strongyloides ratti* (Fenwick et al. 1990b), *Schistosoma mansoni* (Nawar et al. 1992), and *Trichinella spiralis* (Fenwick et al. 1990a). Therefore, deficiency of Zn^{2+} cripples the host's ability to clear infections and has a considerable impact on human health.

4.3 The Zinc Pill: To Take or Not to Take?

Zn^{2+} supplementation in general has a beneficial role in antimicrobial immunity. Acrodermatitis enteropathica (caused by mutations in the Zn^{2+} importer, ZIP4) is a rare and severe genetic autosomal recessive disorder characterized by acral and periorificial dermatitis, alopecia, and diarrhea. Weakened resistance to fungi, bacteria, and viruses is observed in bovine acrodermatitis enteropathica during Zn^{2+} deficiency. In humans, oral Zn^{2+} supplementation ameliorates symptoms associated with this disorder (Ciampo et al. 2018; Hambidge et al. 1977). Moreover, Zn^{2+} supplementation reduces the incidence of acute and chronic persistent diarrhea, dysentery (DD), acute lower respiratory infections (Ruel et al. 1997; Sazawal et al. 1995, 1996, 1998), malaria (Bates et al. 1993; Shankar et al. 2000), recurrent furunculosis (Brody 1977), and infection caused by the parasite *S. mansoni* (Friis et al. 1997).

Zn^{2+} lozenges reduce the duration of common cold (Mossad et al. 1996) by blocking the binding of HRV14 on the viral surface to the adhesion molecule ICAM-1 on the nasal mucosal surface, ultimately leading to reduced viral uptake (Novick et al. 1996). Low Zn^{2+} levels commonly detected in the plasma and serum of human HIV patients is associated with disease progression (Bogden et al. 1990; Khalili et al. 2008). Zn^{2+} supplementation partially reverses these effects (Falutz et al. 1988; Shankar and Prasad 1998), not by curtailing viral load but by dampening the frequency of diarrheal episodes and delaying immunological failure. This evidence suggests that adequate Zn^{2+} supplementation may be used as an adjunct therapy in HIV-infected adults (Baum et al. 2010). However, poor survival has also been reported in HIV-infected patients with high Zn^{2+} intake (Tang et al. 1996). In addition, as noted above, excess dietary Zn^{2+} intake may adversely affect *C. difficile* clearance. Clearly, determining accurate Zn^{2+} dosage is an indispensable step in the success of Zn^{2+} therapy for infection control. While optimal Zn^{2+} availability promotes resistance to infection, Zn^{2+} excess may be an Achilles' heel in maximizing the therapeutic potential of this metal ion.

4.4 Zinc: A Prominent Driver on the Road to Innate Defense

Innate immunity delivers a rapid first line of defense against invading pathogens. Activation of myeloid cells with pathogen-associated molecular patterns (PAMPs) or danger-associated molecular patterns (DAMPs) triggers phagocytosis, cytokine release, antigen presentation to T cells, and bolsters antimicrobial defenses. In the following sections, we discuss the molecular mechanisms of Zn^{2+} regulation in innate immune cells and counter-defense mechanisms utilized by pathogens.

Polymorphonuclear leucocytes (PMNs) or granulocytes such as neutrophils, eosinophils, and basophils exert robust antimicrobial functions. Neutrophils are the most abundant PMNs. Shortly after phagocytosis of microbes, PMNs migrate into the infected tissue via adhesion and chemotaxis. They generate reactive oxygen species (ROS) through nicotinamide adenine dinucleotide phosphate-oxidase (NADPH oxidase) and kill the invading pathogens. Oxidative burst is an important defense mechanism of activated PMNs that is impaired by Zn^{2+} chelation. Conversely, Zn^{2+} is redox-inert, it acts as an antioxidant via catalytic action of the Cu^{2+} (copper)/Zn^{2+}-superoxide dismutase. Moreover, Zn^{2+} dampens inflammatory responses that would otherwise augment oxidative stress (Lee 2018).

Zn^{2+} has an important role in the regulation of number and activities of myeloid cells. Rats fed a Zn^{2+} deficient diet have increased total white blood cells and granulocytes (neutrophils, eosinophils, and basophils) in blood (Someya et al. 2009). Decreased dietary Zn^{2+} intake or absorption diminishes eosinophil numbers in blood, liver, and lungs of mice and impairs their ability to clear *H. polygyrus* or *Ascaris suum*, a parasitic nematode (large roundworm) that causes ascariasis in pigs (Laubach 1990; Scott and Koski 2000). In contrast, Zn^{2+} deficiency augments eosinophilic allergic inflammation, and dietary Zn^{2+} supplementation reduces its intensity (Richter et al. 2003). Eosinophil cationic protein (ECP) is a potent secretory cytotoxic granule that has bactericidal and antiviral activities. Zn^{2+} inhibits the release of ECP from eosinophils in culture (Winqvist et al. 1985). Likewise, Zn^{2+} inhibits the release of granular protein histamine from basophils and mast cells in the human lung (Marone et al. 1986).

Zn^{2+} regulates important processes in immune cells that are crucial to infection control. Physiological Zn^{2+} concentration (10^{-3}–10^{-2} mmol/L) in culture medium facilitates serum opsonic activity and ROS-generating capability in human neutrophils whereas excess Zn^{2+} (10 mmol/L) subdues it (Hasegawa et al. 2000). Reduced Zn^{2+} levels in the serum and in neutrophils is associated with impaired phagocytosis and diminished T cell-mediated immunity (Karzakova 2005). Voltage-gated proton (Hv1) channels transport H^+ across the phagosomal membrane and regulate NADPH oxidase function. Hv1-mediated proton efflux balances the negative charge translocated by NADPH oxidase and provides substrate H^+ for the formation of hydrogen peroxide, hypochlorous acid, and ROS crucial to killing pathogens (Decoursey 2012). Zn^{2+} inhibits Hv1 via two plausible mechanisms: it binds at low concentrations to one site on the channel which prevents the opening of the Hv1 pore, thereby inhibiting proton conduction. At high concentrations, Zn^{2+} binds to a second site and thwarts the outward movement of Hv1 voltage sensor (Qiu et al. 2016). In sum, Zn^{2+} exerts antioxidant functions by mitigating superoxide burst (Maret 2006; Prasad 2014), which may explain increased superoxide stress in Zn^{2+}-deficient cells.

Antimicrobial peptides produced by some myeloid cells can trigger nutritional immunity through Zn^{2+} restriction. Neutrophils release calprotectin, a heterodimer of S100A8 and S100A9, and S100A12 (calgranulin C) peptides that sequester Zn^{2+} from pathogens to impair their growth. Calprotectin and calgranulin C restrict Zn^{2+} access to *Staphylococcus aureus*, *C. albicans*, and *Helicobacter pylori* thus stalling their growth (Besold et al. 2018; Corbin et al. 2008; Kehl-Fie and Skaar 2010). Likewise, S100A7 (psoriasin) released by keratinocytes kills *Escherichia coli* by withholding Zn^{2+} (Gläser et al. 2005).

NETosis, defined as the development and secretion of neutrophil extracellular traps (NETs), is a cell death mechanism of neutrophils in response to infections. NETs are composed of DNA, chromatin, and granular proteins which entrap and subsequently kill extracellular bacteria (Brinkmann et al. 2004). Zn^{2+} acts as a signaling molecule to facilitate NETosis (Hasan et al. 2013). Activated neutrophils elevate intracellular free Zn^{2+} via a protein-kinase C-ROS-dependent mechanism. While the precise source of this Zn^{2+} pool in neutrophils is not known, ROS may trigger the release of Zn^{2+} bound to sulfur on proteins such as MTs or glutathione (Maret 1994, 2000). Perhaps, Zn^{2+} is released from multiple reservoirs that may include Zn^{2+}-bound proteins, Zn^{2+} storage organelles, or may be imported from the extracellular milieu. Deciphering the origin of the Zn^{2+} pool may provide clues to the mechanisms by which this ion prepares the neutrophil defense armor that is rapidly deployed during infection. Zn^{2+} also regulates neutrophil chemotaxis. These cells migrate into the host infected tissues in response to chemoattractants such as the bacterial product, N-formyl-l-methionyl-l-leucyl-l-phenylalanine (fMLF). In vitro, Zn^{2+} promotes chemotaxis of neutrophils to fMLF (Hasan et al. 2016; Hujanen et al. 1995; Vruwink et al. 1991). In sum, Zn^{2+} alters several functions of neutrophils such as chemotaxis, phagocytosis, NETosis, and pathogen killing.

Monocytes and macrophages phagocytose microbes, present antigen, and secrete cytokines to shape the immune response. Rats fed a Zn^{2+}-deficient diet have increased number of total monocytes in blood (Someya et al. 2009). A lack of Zn^{2+} promotes differentiation and maturation of monocytes into macrophages by augmenting cAMP production by adenylate cyclase in vitro (Dubben et al. 2010). Zn^{2+} depletion in human monocytes improves the clearance of *E. coli*, *S. aureus*, and *Streptococcus pneumoniae* via phagocytosis and oxidative burst (Mayer et al. 2014).

Studies investigating the role of Zn^{2+} on cytokine expression and secretion have produced varying results depending on Zn^{2+} concentration, duration of Zn^{2+} depletion, experimental conditions, and model system. Supplementation of Zn^{2+} in serum-free media enhances the expression of interleukin 1 beta (IL-1β) and TNF-α in human peripheral blood mononuclear cells (PBMC) (Wellinghausen et al. 1996). Increasing amounts of Zn^{2+} dose-dependently inhibit monocyte activation caused by the superantigens, staphylococcal enterotoxins A and E (SEA, SEE), the *Mycoplasma arthritidis*–derived superantigen (MAS), but not toxic shock syndrome toxin-1 (TSST-1). Zn^{2+} interferes with the interactions between SEA, SEE, and MAS and their major histocompatibility complex class II (MHC-II)-binding sites. These data demonstrate that Zn^{2+} levels control the secretion of cytokines and response to superantigen challenge (Driessen et al. 1995).

Macrophages deploy two divergent Zn^{2+}-associated defense mechanisms against intracellular pathogens. On the one hand, these cells may intoxicate *M. tuberculosis* or *E. coli* with excess Zn^{2+} to kill it (Botella et al. 2011). On the other hand, macrophages sequester Zn^{2+} from *Histoplasma capsulatum* residing within phagosomes (Subramanian Vignesh et al. 2013). Thus, one may speculate that Zn^{2+} acts as a double-edged sword: inadequate amounts arrest microbial growth, while an excess intoxicates them. What mechanisms influence the immune system's decision to "withhold" versus "intoxicate" to overcome microbial pathogenesis remains a conundrum.

T cell–derived IFNγ is important for macrophage activation (Prasad 2014). Insufficient dietary Zn^{2+} intake compromises IFNγ production by T helper type 1 (Th1) cells resulting in impaired activation of monocytes/macrophages (Agnello et al. 2003; Prasad 2000). Moreover, Zn^{2+} deficiency reduces the production of IL-12 by monocytes/macrophages, yielding poor Th1 differentiation (Bao et al. 2011; Langrish et al. 2004). Thus, changes in the Zn^{2+} status not only affect differentiation and activation of monocytes/macrophages but may also compromise adaptive immunity.

4.5 The MT-Zinc Immune-Landscape: An Old Axis with a New Tale

In 1957, Margoshe and Vallee discovered MTs as Cd^{2+} (cadmium)-binding proteins from the horse renal cortex (Margoshes and Vallee 1957). MTs sense intracellular as well as environmental cues and regulate cellular Zn^{2+} homeostasis through sequestration, mobilization, and release (Subramanian Vignesh and Deepe Jr 2017). MTs have a higher Zn^{2+}-binding constant ($K_{Zn} = 3.2 \times 10^{13}$ M^{-1} at pH 7.4) than most Zn^{2+}-binding proteins. Despite this property, they facilitate controlled Zn^{2+} release to proteins with a lower stability constant for Zn^{2+} (Jacob et al. 1998). This metal is readily released from only one of the sites on MT1 and MT2 through interactions with adenosine triphosphate (ATP), guanosine triphosphate (GTP), or glutathione (Maret 2000).

There are 4 MT isoforms in mice and over 16 in humans (Quaife et al. 1994; Uchida et al. 1991). MTs are present in immune cells including those in the bone marrow (Liu et al. 2004), axillary lymph nodes (Haerslev et al. 1994), spleen (Huang et al. 2019; Mita et al. 2002), and thymus (Savino et al. 1984). In immune cells, MTs are induced by Zn^{2+}, Cu^{2+}, and Cd^{2+} (Aydemir et al. 2006; Huber and Cousins 1993; Makhijani 1998; Thorvaldsen et al. 1995); cytokines such as GM-CSF, TNFα, IFNγ, IL-1, IL-4, IL-6, and IL-27 (Cousins and Leinart 1988; Schroeder and Cousins 1990; Sciavolino and Vilček 1995; Ullio et al. 2015; Subramanian Vignesh et al. 2013, 2016; Chuan Wu et al. 2013a, b); and microbial ligands such as lipopolysaccharide (LPS) (Arizono et al. 1995; Leibbrandt and Koropatnick 1994), ROS (Dalton et al. 1994; Nourani et al. 2011; Tate et al. 1995), and nitric oxide (NO) (Arizono et al. 1995). In turn, MTs may regulate the activity of some of these immune modulators. For example, MTs scavenge ROS and regulate the function of GM-CSF and IL-4-polarized macrophages (Li et al. 2004; Subramanian Vignesh et al. 2013, 2016). The interrelationship between immune modulators and MTs is schematically represented in Fig. 4.1.

MTs influence a variety of immune responses in vivo and in vitro. For example, IL-27 induces MT1 and MT2 that prevent type 1 regulatory T (Tr1) cell development. This effect of MTs is due to negative feedback inhibition of signal transducer and activator of transcription (STAT)1 and STAT3 phosphorylation, resulting in

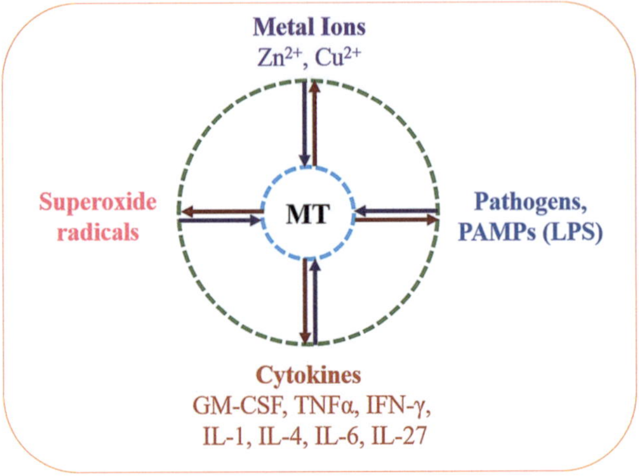

Fig. 4.1 The association between MTs and immune mediators. MTs are induced by metal ions (Zn^{2+}, Cu^{2+}), infection, pathogen associate molecular patterns (PAMPs), cytokines, and superoxide radicals. The protein family in turn controls responses to each of these stimuli, thereby establishing a feedback loop between immune mediators and MTs

diminished Tr1 differentiation and IL-10 production. Thus, the dynamic balance between STATs and MTs calibrates the development and suppressive function of Tr1 cells. The control of Zn^{2+} within the intracellular milieu may arm MTs with the ability to control STAT phosphorylation. This is plausible because Zn^{2+} inhibits the function of phosphatases that downmodulate STAT signaling. By sequestering the ion, MTs may render phosphatases active, leading to increased STAT dephosphorylation (Supasai et al. 2017; Chuan Wu et al. 2013a, b). From an antimicrobial immunity standpoint, this attribute of MTs in Tr1 cells may benefit the host in clearing infection rapidly, before suppressive immunity emerges to subdue inflammation and promote tissue repair.

Mt1$^{-/-}$*Mt2*$^{-/-}$ mice exhibit stronger humoral responses through the elevation of nuclear factor-kappaB (NF-κB) transcription factor activity in splenocytes. These knockout mice display higher circulatory immunoglobulin levels, enhanced B cell differentiation upon OVA challenge, and lymphoproliferative responses to mitogenic stimulation (Crowthers et al. 2000). Exogenous administration of MTs into these mice dampens humoral immunity (Lynes et al. 1993). Thus, MTs temper antibody production by B cells. MTs also influence cytokine production by basophils. Stimulation of the Fc epsilon receptor 1 (FcεRI) induces MT1 and MT2 in mouse basophils to regulate intracellular Zn^{2+}. Lack of MTs increases intracellular free Zn^{2+} which inhibits calcineurin (CaN) phosphatase activity and thus impacts FcεRI-induced nuclear factor of activated T-cell (NFAT)-dependent IL-4 production (Ugajin et al. 2015).

4.6 The MT-Zinc Axis in Infection

Several studies have brought to light the complex functions of MTs and the MT-Zn^{2+} axis in antimicrobial defense. Below, we discuss how the control of signaling pathways, Zn^{2+} homeostasis, and inflammatory responses by different MTs converge to dictate the outcome of host–pathogen interactions.

4.6.1 *Bacterial Infection*

NF-κB is essential for adequate innate immunity to infection. Zn^{2+} is an important negative regulator of NF-κB, while ROS is a positive regulator. MTs subvert the action of Zn^{2+}, possibly by sequestering the intracellular free Zn^{2+} pool indicating that MT is an important intracellular modulator of NF-κB activation (Kim et al. 2003).

In polymicrobial sepsis in mice, deficiency of Zn^{2+} promotes systemic infection and NF-κB activation leading to elevated inflammation, lung injury, and mortality. Zn^{2+} supplementation prior to initiation of sepsis effectively reverses these effects (Bao et al. 2010). NF-κB induces the expression of the Zn^{2+} importer SLC39A8 (ZIP8) that imports Zn^{2+} to inhibit IκB kinase (Iκκ) activity. These findings identify a negative feedback loop that directly regulates a master transcription factor via coordination of Zn^{2+} metabolism (Liu et al. 2013). *Salmonella typhimurium* is a causative agent for inflamed gut. Macrophages infected with this pathogen exhibit elevated levels of free cytoplasmic Zn^{2+} that downmodulates NF-κB activity, as a result affecting the expression of reactive species (ROS and RNS) -forming enzymes phos47 (an NADPH oxidase subunit), inducible NO synthase (iNOS), and proinflammatory cytokines. Macrophages counter this change in Zn^{2+} homeostasis by augmenting MT1 and MT2 that scavenge free Zn^{2+} and restore ROS and RNS production to kill the pathogen. Thus, the limitation of free Zn^{2+} by MTs facilitates the control of intestinal colonization by *S. typhimurium* (Wu et al. 2017). In contrast, *M. tuberculosis*–infected macrophages rapidly increase free Zn^{2+} to poison this intracellular pathogen. This phenomenon is associated with an increase in MTs, MTF-1, and ZnT1, an exporter of Zn^{2+}. MTF-1 translocates to the nucleus upon infection to induce MTs and ZnT1, suggesting that the host mounts a direct and quick response to protect itself from Zn^{2+} intoxication (Botella et al. 2011). Together, these studies suggest that macrophages possess two opposing mechanisms to exert antimicrobial immunity: Zn^{2+} depletion and Zn^{2+} poisoning. The effect of MTs on NF-κB activation is paradoxical depending on whether MTs scavenge Zn^{2+} or ROS. By scavenging ROS, MTs stabilize Iκκ, an inhibitor of NF-κB, ultimately downmodulating activation of this transcription factor. In gastric cells of $Mt1^{-/-}Mt2^{-/-}$ mice with *Helicobacter pylori* infection, NF-κB activation and downstream production of macrophage inflammatory protein (MIP)-1α and monocyte chemoattractant protein (MCP)-1 is increased. The absence of MT1 and MT2 leads to erosive lesions and elevates infiltration of inflammatory leukocytes in the gastric mucosa. Thus, MTs protect against gastric ulceration during *H. pylori* infection by negatively regulating NF-κB activation (Mita et al. 2008).

4.6.2 Fungal Infection

How macrophages utilize the Zn^{2+} pool to resolve mycobacterial versus fungal infection presents an interesting paradox. Granulocyte macrophage-colony stimulating factor (GM-CSF) augments antimicrobial defenses against the intracellular fungus *H. capsulatum*. GM-CSF activated, infected macrophages increase MT1 and MT2 expression via activation of STAT3 and STAT5 transcriptional factors. These MTs bind to the macrophage free Zn^{2+} pool, denying Zn^{2+} access to the pathogen residing within phagosomes. In fact, Zn^{2+} is mobilized into the Golgi apparatus in association with an increase in the Zn^{2+} exporters, ZnT4 and ZnT7 that are expressed on the Golgi membrane. The Zn^{2+} sequestration "feat" by MTs simultaneously boosts ROS to stall fungal growth. Intriguingly, GM-CSF also elevates Zn^{2+} import via the importer Zip2, perhaps, to support an increased demand for Zn^{2+}-dependent host processes during pathogen insult (Subramanian Vignesh et al. 2013).

IL-4 and IL-13 are cytokines that shape macrophage polarization to the M(IL-4) and M(IL-13) phenotypes, respectively. Studies on how Zn^{2+} levels influence macrophage polarization in rodents and human cell lines have produced distinct results. Zn^{2+} deficiency in rodents diminishes IL-4 production by Th2 cells and the proportion of M(IL-4) polarized macrophages in the spleen (Kido et al. 2019). In contrast, in human THP1 monocyte-derived macrophages, Zn^{2+} deficiency inhibits M1 polarization by IFNγ and LPS but does not affect M2 polarization by IL-4. Exogenous addition of Zn^{2+} suppresses the emergence of M(IL-4) macrophages in vitro (Dierichs et al. 2018). The use of different experimental models (rodents versus human cell line and dietary Zn^{2+} deficiency versus Zn^{2+} depletion/supplementation in culture media) may explain some of these findings. Nonetheless, these studies suggest that Zn^{2+} impacts macrophage polarization and can regulate the balance between M1 and M2 polarization states.

M2 macrophages aid in parasite clearance but harbor a permissive milieu for persistence of intracellular pathogens. Recent literature has demonstrated that IL-4 augments intracellular free Zn^{2+} in bone marrow–derived macrophages, microglia, and human monocyte-derived macrophages (Aratake et al. 2018; Subramanian Vignesh et al. 2016). M(IL-4) macrophages from the bone marrow specifically upregulate the MT3 isoform via STAT6 and interferon regulatory factor (IRF)4 signaling. The relationship between MT1/MT2 and MT3 is dichotomous, in that the latter expands the free Zn^{2+} pool while the former shrinks it in macrophages (Subramanian Vignesh et al. 2013, 2016). The action of cathepsin proteases enhances Zn^{2+} release from MT3 (Subramanian Vignesh et al. 2016). This is notable

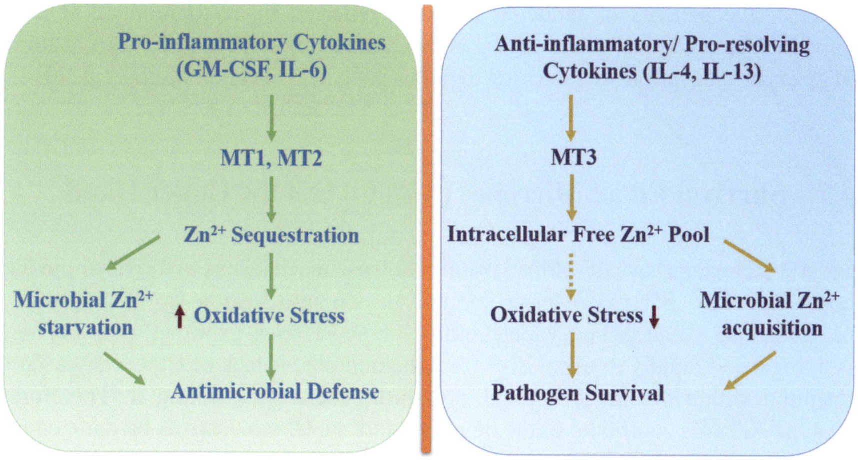

Fig. 4.2 MT isoforms have distinct roles in host–pathogen interactions. GM-CSF and IL-6 elevate MT1 and MT2 that promote intracellular Zn^{2+} sequestration. Limitation of free Zn^{2+} in the host inhibits intracellular microbial growth via the induction of oxidative burst and restriction of Zn^{2+} access to the microbes. In contrast, IL-4 and IL-13 augment MT3 expression that increases the intracellular free Zn^{2+} pool and may dampen superoxide defenses. Moreover, intracellular pathogens may exploit this mechanism for acquisition of the metal ion for survival

because such an increase in the free Zn^{2+} reservoir places the intracellular pathogen at an advantage: it assimilates a pool of Zn^{2+} that was once a part of the host. The finding establishes a link between MT-Zn^{2+} metabolism and the permissive nature of M(IL-4) macrophages to intracellular microbes. Figure 4.2 schematically outlines the distinct functions of MT1, MT2, and MT3 in macrophage defenses.

4.6.3 *Viral and Parasitic Infections*

Our knowledge of how the MT-Zn^{2+} axis controls immune responses to viruses and parasites is limited. Infection with viruses including coxsackievirus B type 3 and influenza A/PR8 upregulates MTs in the liver, lung, kidney, and spleen. The mechanism of induction involves MTF-1, STAT3 signaling, and glucocorticoids (Ghoshal et al. 2001; Ilbäck et al. 2004). In contrast, infection with the hepatitis C virus is associated with a reduction in MT expression in the liver. Increasing MT levels through Zn^{2+} supplementation decreases viral load pointing at a protective function of the MT-Zn^{2+} axis in viral clearance (Carrera et al. 2003; Read et al. 2018). Further studies are required to elucidate the precise mechanisms by which MTs influence the immune systems' ability to curtail viral uptake, replication, and shedding.

The parasite *T. cruzi*, the causative agent of Chagas disease, leads to cardiomyopathy and gastrointestinal inflammation. Infection with this pathogen reduces the expression of MT1 in the liver, while augmenting NO levels and oxidative stress.

Whether the benefits of reducing NO are conferred by restoration of MTs is unknown, but chemically scavenging NO in animals infected with *T. cruzi* restores MT1 expression and arrests the growth of this parasite (Gonzalez-Mejia et al. 2014).

4.7 Survival Edge: Microbes (Aim to) Get the Upper Hand

Several pathogens have developed counter-defense mechanisms to thrive within the host (Fig. 4.3). For example, to circumvent Zn intoxication by macrophages, *M. tuberculosis* induces heavy metal efflux P-type ATPases. CtpC, a P-type ATPase, is upregulated rapidly to expel Zn^{2+} from the microbe. A lack of CtpC causes Zn^{2+} retention within the mycobacterial cytoplasm, thereby poisoning it. Therefore, P1-type ATPases contribute to the defense armor of *M. tuberculosis* by dampening the toxic effects of Zn^{2+} (Botella et al. 2011). Group A *Streptococcus* growth is restricted by Zn^{2+} limitation caused by neutrophil-derived calprotectin. *Streptococcus pyogenes* encodes the Zn^{2+} importer AdcA and a Zn^{2+} sensor AdcR to compete with Zn^{2+} sequestration by calprotectin (Makthal et al. 2017). ZrfC, a plasma membrane Zn^{2+} transporter of *Aspergillus fumigatus*, has the ability to scavenge Zn^{2+} efficiently from lungs enabling it to grow even in the presence of calprotectin (Amich et al. 2014). *Neisseria meningitides* uses ZnuD, a high-affinity Zn^{2+} transporter, to circumvent Zn^{2+} deprivation (Lappann et al. 2013). This pathogen also responds to low

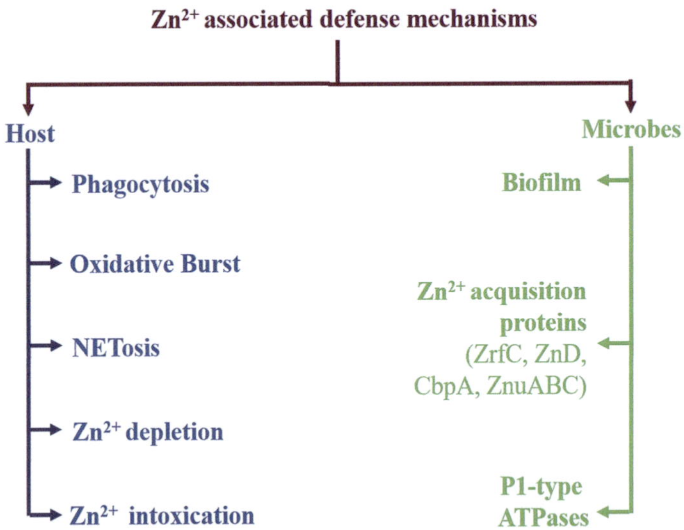

Fig. 4.3 Zn^{2+}-associated defense mechanisms. Immune cells employ various pathways (phagocytosis, oxidative burst, NETosis, Zn^{2+} depletion, and Zn^{2+} intoxication) to defend microbial invaders, whereas microbes also utilize counter-defense mechanisms including P1-type ATPases, Zn^{2+} acquisition transporters/proteins, and biofilm formation

Zn^{2+} by expression of curved DNA binding protein A (CbpA), which is a calprotectin receptor, on its outer membrane. This molecule facilitates the acquisition of Zn^{2+} bound to calprotectin by *N. meningitides*. Thus, the microbe defies a vital host defense mechanism and exploits it for its benefit (Stork et al. 2013). *Yersinia pestis* utilizes a zincophore, yersiniabactin (Ybt) synthetase, and the high-affinity Zn^{2+} transporter, ZnuABC, to obtain Zn^{2+}. These are crucial in the progression of lethal septicemic plague in mice (Bobrov et al. 2014). *S. typhimurium* can also express ZnuABC and thrives by subduing the host's Zn^{2+} deprivation strategies. These studies indicate that Zn^{2+} acquisition may be a "virulence determinant" in some pathogens (Liu et al. 2012).

Biofilms contain microbial communities associated with a polymeric matrix structure composed of factors such as extracellular DNA, polysaccharides, and proteins. Micromolar (100–250 μmol l^{-1}) concentrations of Zn^{2+} block biofilm formation by *Actinobacillus pleuropneumoniae*, *Haemophilus parasuis*, *S. typhimurium*, *Escherichia coli*, *S. aureus*, *Streptococcus suis*, and *Klebsiella pneumoniae* strains. Mechanistically, Zn^{2+} may interfere with the stability of extracellular DNA and polymers contained within the biofilm matrix and impede critical microbial processes such as iron homeostasis and energy metabolism of associated microbes (Hancock et al. 2010; Polyudova et al. 2018; Chan Wu et al. 2013a, b). Of note, in the context of biofilms, an excess of Zn^{2+} or a deficiency of it may exert the same effect: inhibition of biofilm formation. For example, biofilm formation by *Staphylococcus epidermidis*, *S. aureus*, and *S. pneumoniae* is inhibited by Zn^{2+} chelation or Zn^{2+} excess (Brown et al. 2017; Conrady et al. 2008; Formosa-Dague et al. 2016). It is plausible that Zn^{2+} concentrations within a narrow range are necessary to maintain structural integrity of the biofilm, while intoxicating amounts of the metal ion adversely impact the growth of microbes that facilitate biofilm development.

4.8 Concluding Remarks

The value of dietary Zn^{2+} intake to maintain immunological robustness has long been appreciated. The highly conserved nature of MTs and their ability to bind Zn^{2+} across prokaryotes and eukaryotes has prompted scientists to query their importance in cellular functions. Immune cells are no exception. As it turns out, this class of proteins fiercely guards the Zn^{2+} reservoir in immune cells and can dictate the ions' spatiotemporal presence both intracellularly and in the extracellular milieu. This attribute of MTs is notable, because the versatility of Zn^{2+} ions in biochemical processes demands that adequate amounts of Zn^{2+} be available for immune cells when and where they need it. Recent years have illuminated our knowledge of how the host taps into the MT-Zn^{2+} landscape to challenge microbial intrusion and overcome inflammatory damage caused by pathogen insult. These findings have opened newer apertures to explore the extent to which MTs orchestrate immunological responses. An important possibility to consider is that MT may function independent of Zn^{2+}, perhaps in its apo-form or by binding to another metal ion. Of note, the protein also interacts with Cu^{2+} ions to regulate Cu^{2+} homeostasis. Whether an MT-Cu^{2+} axis impacts immunological performance or the triad (MT- Zn^{2+}-Cu^{2+} axis) prevails over the two is unanswered. Nonetheless, MTs have surfaced prominently in the host-pathogen realm and will pave the path to a galvanizing story that continues to be told.

Acknowledgments This work was supported by 19CDA34770022 American Heart Association grant to KSV and NIH grant AI106269 to GSD. The authors declare no conflicts of interest.

Author Contributions DC and KSV wrote the manuscript. GD reviewed it and KSV supervised the work.

References

Agnello D, Lankford CS, Bream J, Morinobu A, Gadina M, O'shea JJ, Frucht DM (2003) Cytokines and transcription factors that regulate T helper cell differentiation: new players and new insights. J Clin Immunol 23(3):147–161

Amich J, Vicentefranqueira R, Mellado E, Ruiz-Carmuega A, Leal F, Calera JA (2014) The ZrfC alkaline zinc transporter is required for A spergillus fumigatus virulence and its growth in the presence of the Zn/Mn-chelating protein calprotectin. Cell Microbiol 16(4):548–564

Andreini C, Banci L, Bertini I, Rosato A (2006) Counting the zinc-proteins encoded in the human genome. J Proteome Res 5(1):196–201

Andrews GK (2001) Cellular zinc sensors: MTF-1 regulation of gene expression. In: Zinc Biochemistry, Physiology, and Homeostasis. Springer, Tokyo, pp 37–51

Aratake T, Higashi Y, Ueba Y, Hamada T, Shimizu T, Shimizu S et al (2018) The inhibitory role of intracellular free zinc in the regulation of Arg-1 expression in interleukin-4-induced activation of M2 microglia. Metallomics 10(10):1501–1509

Arizono K, Kagawa S-I, Hamada H, Ariyoshi T (1995) Nitric oxide mediated metallothionein induction by lipopolysaccharide. Res Commun Mol Pathol Pharmacol 90(1):49–58

Aydemir TB, Blanchard RK, Cousins RJ (2006) Zinc supplementation of young men alters metallothionein, zinc transporter, and cytokine gene expression in leukocyte populations. Proc Natl Acad Sci 103(6):1699–1704

Bao S, Liu M-J, Lee B, Besecker BY, Lai J-P, Guttridge DC, Knoell DL (2010) Zinc modulates the innate immune response *in vivo* to polymicrobial sepsis through regulation of NF-κB. Am J Phys Lung Cell Mol Phys 298(6):L744–L754

Bao B, Prasad AS, Beck FW, Bao GW, Singh T, Ali S, Sarkar FH (2011) Intracellular free zinc up-regulates IFN-γ and T-bet essential for Th1 differentiation in Con-A stimulated HUT-78 cells. Biochem Biophys Res Commun 407(4):703–707

Bates C, Bates P, Dardenne M, Prentice A, Lunn P, Northrop-Clewes C et al (1993) A trial of zinc supplementation in young rural Gambian children. Br J Nutr 69(1):243–255

Baum MK, Lai S, Sales S, Page JB, Campa A (2010) Randomized, controlled clinical trial of zinc supplementation to prevent immunological failure in HIV-infected adults. Clin Infect Dis 50(12):1653–1660

Beck FW, Li Y, Bao B, Prasad AS, Sarkar FH (2006) Evidence for reprogramming global gene expression during zinc deficiency in the HUT-78 cell line. Nutrition 22(10):1045–1056

Besecker BY, Exline MC, Hollyfield J, Phillips G, Disilvestro RA, Wewers MD, Knoell DL (2011) A comparison of zinc metabolism, inflammation, and disease severity in critically ill infected and noninfected adults early after intensive care unit admission. Am J Clin Nutr

93(6):1356–1364. Retrieved from https://www.ncbi.nlm.nih.gov/pubmed/21525204. https://doi.org/10.3945/ajcn.110.008417
Besold AN, Gilston BA, Radin JN, Ramsoomair C, Culbertson EM, Li CX et al (2018) Role of calprotectin in withholding zinc and copper from Candida albicans. Infect Immun 86(2):e00779–e00717
Bobrov AG, Kirillina O, Fetherston JD, Miller MC, Burlison JA, Perry RD (2014) The Y ersinia pestis siderophore, yersiniabactin, and the ZnuABC system both contribute to zinc acquisition and the development of lethal septicaemic plague in mice. Mol Microbiol 93(4):759–775
Bogden JD, Baker H, Frank O, Perez G, Kemp F, Bruening K, Louria D (1990) Micronutrient status and human immunodeficiency virus (HIV) infection. Ann N Y Acad Sci 587(1):189–195
Bonaventura P, Benedetti G, Albarède F, Miossec P (2015) Zinc and its role in immunity and inflammation. Autoimmun Rev 14(4):277–285
Botella H, Peyron P, Levillain F, Poincloux R, Poquet Y, Brandli I et al (2011) Mycobacterial P1-type ATPases mediate resistance to zinc poisoning in human macrophages. Cell Host Microbe 10(3):248–259
Brinkmann V, Reichard U, Goosmann C, Fauler B, Uhlemann Y, Weiss DS et al (2004) Neutrophil extracellular traps kill bacteria. Science 303(5663):1532–1535
Brody I (1977) Treatment of recurrent furunculosis with oral zinc. Lancet 310(8052):1358
Brown LR, Caulkins RC, Schartel TE, Rosch JW, Honsa ES, Schultz-Cherry S et al (2017) Increased zinc availability enhances initial aggregation and biofilm formation of streptococcus pneumoniae. Front Cell Infect Microbiol 7:233
Carlomagno M, Coghlan LG, McMurray D (1986) Chronic zinc deficiency and listeriosis in rats: acquired cellular resistance and response to vaccination. Med Microbiol Immunol 175(5):271–280
Carrera G, Paternain JL, Carrere N, Folch J, Courtade-Saïdi M, Orfila C et al (2003) Hepatic metallothionein in patients with chronic hepatitis C: relationship with severity of liver disease and response to treatment. Am J Gastroenterol 98(5):1142
Ciampo IRLD, Sawamura R, Ciampo LAD, Fernandes MIM (2018) Acrodermatitis enteropathica: clinical manifestations and pediatric diagnosis. Revista Paulista de Pediatria 36(2):238–241
Coghlan LG, Carlomagno MA, McMurray DN (1988) Effect of protein and zinc deficiencies on vaccine efficacy in guinea pigs following pulmonary infection with Listeria. Med Microbiol Immunol 177(5):255–263
Conrady DG, Brescia CC, Horii K, Weiss AA, Hassett DJ, Herr AB (2008) A zinc-dependent adhesion module is responsible for intercellular adhesion in staphylococcal biofilms. Proc Natl Acad Sci 105(49):19456–19461
Corbin BD, Seeley EH, Raab A, Feldmann J, Miller MR, Torres VJ et al (2008) Metal chelation and inhibition of bacterial growth in tissue abscesses. Science 319(5865):962–965
Cousins RJ (1998) A role of zinc in the regulation of gene expression. Proc Nutr Soc 57(2):307–311
Cousins RJ, Leinart AS (1988) Tissue-specific regulation of zinc metabolism and metallothionein genes by interleukin 1. FASEB J 2(13):2884–2890
Cousins RJ, Blanchard RK, Moore JB, Cui L, Green CL, Liuzzi JP et al (2003) Regulation of zinc metabolism and genomic outcomes. J Nutr 133(5):1521S–1526S
Crawford AC, Lehtovirta-Morley LE, Alamir O, Niemiec MJ, Alawfi B, Alsarraf M et al (2018) Biphasic zinc compartmentalisation in a human fungal pathogen. PLoS Pathog 14(5):e1007013
Crowthers KC, Kline V, Giardina C, Lynes MA (2000) Augmented humoral immune function in metallothionein-null mice. Toxicol Appl Pharmacol 166(3):161–172
Dalton T, Palmiter RD, Andrews GK (1994) Transcriptional induction of the mouse metallothionein-I gene in hydrogen peroxide-treated Hepa cells involves a composite major late transcription factor/antioxidant response element and metal response promoter elements. Nucleic Acids Res 22(23):5016–5023
Decoursey TE (2012) Voltage-gated proton channels. Compr Physiol 2(2):1355–1385. Retrieved from https://www.ncbi.nlm.nih.gov/pubmed/23798303. https://doi.org/10.1002/cphy.c100071

Dierichs L, Kloubert V, Rink L (2018) Cellular zinc homeostasis modulates polarization of THP-1-derived macrophages. Eur J Nutr 57(6):2161–2169
Driessen C, Hirv K, Kirchner H, Rink L (1995) Divergent effects of zinc on different bacterial pathogenic agents. J Infect Dis 171(2):486–489
Dubben S, Hönscheid A, Winkler K, Rink L, Haase H (2010) Cellular zinc homeostasis is a regulator in monocyte differentiation of HL-60 cells by 1α, 25-dihydroxyvitamin D3. J Leukoc Biol 87(5):833–844
Eide DJ (2004) The SLC39 family of metal ion transporters. Pflugers Arch 447(5):796–800
Falutz J, Tsoukas C, Gold P (1988) Zinc as a cofactor in human immunodeficiency virus—induced immunosuppression. JAMA 259(19):2850–2851
Feiler L, Smolin G, Okumoto M, Condon D (1982) Herpetic keratitis in zinc-deficient rabbits. Invest Ophthalmol Vis Sci 22(6):788–795
Fenwick P, Aggett P, Macdonald D, Huber C, Wakelin D (1990a) Zinc deficiency and zinc repletion: effect on the response of rats to infection with Trichinella spiralis. Am J Clin Nutr 52(1):166–172
Fenwick P, Aggett P, Macdonald DC, Huber C, Wakelin D (1990b) Zinc deprivation and zinc repletion: effect on the response of rats to infection with Strongyloides ratti. Am J Clin Nutr 52(1):173–177
Flagstad T, Andersen S, Nielsen K (1972) The course of experimental *Fasciola hepatica* infection in calves with a deficient cellular immunity. Res Vet Sci 13(5):468–477
Formosa-Dague C, Speziale P, Foster TJ, Geoghegan JA, Dufrêne YF (2016) Zinc-dependent mechanical properties of Staphylococcus aureus biofilm-forming surface protein SasG. Proc Natl Acad Sci 113(2):410–415
Fraker PJ, Caruso R, Kierszenbaum F (1982) Alteration of the immune and nutritional status of mice by synergy between zinc deficiency and infection with Trypanosoma cruzi. J Nutr 112(6):1224–1229
Friis H, Ndhlovu P, Mduluza T, Kaondera K, Sandström B, Michaelsen KF et al (1997) The impact of zinc supplementation on *Schistosoma mansoni* reinfection rate and intensities: a randomized, controlled trial among rural Zimbabwean schoolchildren. Eur J Clin Nutr 51(1):33
Ghoshal K, Majumder S, Zhu Q, Hunzeker J, Datta J, Shah M et al (2001) Influenza virus infection induces metallothionein gene expression in the mouse liver and lung by overlapping but distinct molecular mechanisms. Mol Cell Biol 21(24):8301–8317
Gläser R, Harder J, Lange H, Bartels J, Christophers E, Schröder J-M (2005) Antimicrobial psoriasin (S100A7) protects human skin from *Escherichia coli* infection. Nat Immunol 6(1):57
Gonzalez-Mejia ME, Torres-Rasgado E, Porchia LM, Salgado HR, Totolhua J-L, Ortega A et al (2014) Metallothionein-1 and nitric oxide expression are inversely correlated in a murine model of Chagas disease. Mem Inst Oswaldo Cruz 109(2):174–181
Haase H, Mazzatti DJ, White A, Ibs KH, Engelhardt G, Hebel S, … Rink L (2007) Differential gene expression after zinc supplementation and deprivation in human leukocyte subsets. Paper presented at the molecular medicine
Haerslev T, Jacobsen K, Nedergaard L, Zedeler K (1994) Immunohistochemical detection of metallothionein in primary breast carcinomas and their axillary lymph node metastases. Pathol Res Pract 190(7):675–681
Hambidge K, Walravens P, Neldner K (1977) The role of zinc in the pathogenesis and treatment of acrodermatitis enteropathica. Prog Clin Biol Res 14:329–342
Hancock V, Dahl M, Klemm P (2010) Abolition of biofilm formation in urinary tract Escherichia coli and Klebsiella isolates by metal interference through competition for fur. Appl Environ Microbiol 76(12):3836–3841
Hasan R, Rink L, Haase H (2013) Zinc signals in neutrophil granulocytes are required for the formation of neutrophil extracellular traps. Innate Immun 19(3):253–264
Hasan R, Rink L, Haase H (2016) Chelation of free Zn^{2+} impairs chemotaxis, phagocytosis, oxidative burst, degranulation, and cytokine production by neutrophil granulocytes. Biol Trace Elem Res 171(1):79–88

Hasegawa H, Suzuki K, Suzuki K, Nakaji S, Sugawara K (2000) Effects of zinc on the reactive oxygen species generating capacity of human neutrophils and on the serum opsonic activity *in vitro*. Lumin J Biol Chem Lumin 15(5):321–327

Hennigar SR, McClung JP (2016) Nutritional immunity: starving pathogens of trace minerals. Am J Lifestyle Med 10(3):170–173

Huang X, Feng Y, Fan W, Duan J, Duan Y, Xiong G et al (2019) Potential ability for metallothionein and vitamin E protection against cadmium immunotoxicity in head kidney and spleen of grass carp (Ctenopharyngodon idellus). Ecotoxicol Environ Saf 170:246–252

Huber KL, Cousins RJ (1993) Zinc metabolism and metallothionein expression in bone marrow during erythropoiesis. Am J Physiol Endocrinol Metab 264(5):E770–E775

Hujanen ES, Seppä ST, Virtanen K (1995) Polymorphonuclear leukocyte chemotaxis induced by zinc, copper and nickel *in vitro*. Biochimica et Biophysica Acta (BBA)-Gen Subjects 1245(2):145–152

Ilbäck N-G, Glynn AW, Wikberg L, Netzel E, Lindh U (2004) Metallothionein is induced and trace element balance changed in target organs of a common viral infection. Toxicology 199(2–3):241–250

Inoue K, O'Bryant Z, Xiong Z-G (2015) Zinc-permeable ion channels: effects on intracellular zinc dynamics and potential physiological/pathophysiological significance. Curr Med Chem 22(10):1248–1257

Jacob C, Maret W, Vallee BL (1998) Control of zinc transfer between thionein, metallothionein, and zinc proteins. Proc Natl Acad Sci 95(7):3489–3494

Karzakova L (2005) Laboratory manifestations of asymptomatic zinc deficiency. Klin Lab Diagn 12:39–41

Kehl-Fie TE, Skaar EP (2010) Nutritional immunity beyond iron: a role for manganese and zinc. Curr Opin Chem Biol 14(2):218–224

Khalili H, Soudbakhsh A, Hajiabdolbaghi M, Dashti-Khavidaki S, Poorzare A, Saeedi A, Sharififar R (2008) Nutritional status and serum zinc and selenium levels in Iranian HIV infected individuals. BMC Infect Dis 8(1):165

Kidd M, Qureshi M, Ferket P, Thomas LN (1994) Dietary zinc-methionine enhances mononuclear-phagocytic function in young turkeys. Biol Trace Elem Res 42(3):217–229

Kido T, Ishiwata K, Suka M, Yanagisawa H (2019) Inflammatory response under zinc deficiency is exacerbated by dysfunction of the T helper type 2 lymphocyte–M2 macrophage pathway. Immunology 156(4):356–372

Kim CH, Kim JH, Lee J, Ahn YS (2003) Zinc-induced NF-κB inhibition can be modulated by changes in the intracellular metallothionein level. Toxicol Appl Pharmacol 190(2):189–196

Langrish CL, McKenzie BS, Wilson NJ, de Waal Malefyt R, Kastelein RA, Cua DJ (2004) IL-12 and IL-23: master regulators of innate and adaptive immunity. Immunol Rev 202(1):96–105

Lappann M, Danhof S, Guenther F, Olivares-Florez S, Mordhorst IL, Vogel U (2013) In vitro resistance mechanisms of N eisseria meningitidis against neutrophil extracellular traps. Mol Microbiol 89(3):433–449

Laubach HE (1990) Effect of dietary zinc on larval burdens, tissue eosinophil numbers, and lysophospholipase activity of Ascaris suum infected mice. Acta Trop 47(4):205–211

Lee SR (2018) Critical role of zinc as either an antioxidant or a Prooxidant in cellular systems. Oxidative Med Cell Longev 2018: 9156285, 11. https://doi.org/10.1155/2018/9156285

Lee CM, Humphrey PA, Aboko-Cole GF (1983) Interaction of nutrition and infection: effect of zinc deficiency on resistance to Trypanosoma musculi. Int J Biochem 15(6):841–847

Leibbrandt ME, Koropatnick J (1994) Activation of human monocytes with lipopolysaccharide induces metallothionein expression and is diminished by zinc. Toxicol Appl Pharmacol 124(1):72–81

Li X, Chen H, Epstein PN (2004) Metallothionein protects islets from hypoxia and extends islet graft survival by scavenging most kinds of reactive oxygen species. J Biol Chem 279(1):765–771

Liu AL, Zhang ZM, Zhu BF, Liao ZH, Liu Z (2004) Metallothionein protects bone marrow stromal cells against hydrogen peroxide-induced inhibition of osteoblastic differentiation. Cell Biol Int 28(12):905–911

Liu JZ, Jellbauer S, Poe AJ, Ton V, Pesciaroli M, Kehl-Fie TE et al (2012) Zinc sequestration by the neutrophil protein calprotectin enhances Salmonella growth in the inflamed gut. Cell Host Microbe 11(3):227–239

Liu M-J, Bao S, Gálvez-Peralta M, Pyle CJ, Rudawsky AC, Pavlovicz RE et al (2013) ZIP8 regulates host defense through zinc-mediated inhibition of NF-κB. Cell Rep 3(2):386–400

Lynes MA, Borghesi LA, Youn J, Olson EA (1993) Immunomodulatory activities of extracellular metallothionein I. Metallothionein effects on antibody production. Toxicology 85(2–3):161–177

Makhijani EJYPR (1998) Flow cytometric determination of metallothionein levels in human peripheral blood lymphocytes: utility in environmental exposure assessment. J Toxicol Environ Health A 54(6):445–457

Makthal N, Nguyen K, Do H, Gavagan M, Chandrangsu P, Helmann JD et al (2017) A critical role of zinc importer AdcABC in group A streptococcus-host interactions during infection and its implications for vaccine development. EBioMedicine 21:131–141

Maret W (1994) Oxidative metal release from metallothionein via zinc-thiol/disulfide interchange. Proc Natl Acad Sci 91(1):237–241

Maret W (2000) The function of zinc metallothionein: a link between cellular zinc and redox state. J Nutr 130(5):1455S–1458S

Maret W (2006) Zinc coordination environments in proteins as redox sensors and signal transducers. Antioxid Redox Signal 8(9–10):1419–1441

Margoshes M, Vallee BL (1957) A cadmium protein from equine kidney cortex. J Am Chem Soc 79(17):4813–4814

Marone G, Columbo M, De Paulis A, Cirillo R, Giugliano R, Condorelli M (1986) Physiological concentrations of zinc inhibit the release of histamine from human basophils and lung mast cells. Agents Actions 18(1–2):103–106

Mayer LS, Uciechowski P, Meyer S, Schwerdtle T, Rink L, Haase H (2014) Differential impact of zinc deficiency on phagocytosis, oxidative burst, and production of pro-inflammatory cytokines by human monocytes. Metallomics 6(7):1288–1295

Maywald M, Wessels I, Rink L (2017) Zinc signals and immunity. Int J Mol Sci 18(10):2222

McMurray DN, Bartow RA, Mintzer CL, Hernandez-Frontera E (1990) Micronutrient Status and Immune Function in Tuberculosisa. Ann N Y Acad Sci 587(1):59–69

Minkus TM, Koski KG, Scott ME (1992) Marginal zinc deficiency has no effect on primary or challenge infections in mice with Heligmosomoides polygyrus (Nematoda). J Nutr 122(3):570–579

Mita M, Imura N, Kumazawa Y, Himeno S (2002) Suppressed proliferative response of spleen T cells from metallothionein null mice. Microbiol Immunol 46(2):101–107

Mita M, Satoh M, Shimada A, Okajima M, Azuma S, Suzuki JS et al (2008) Metallothionein is a crucial protective factor against Helicobacter pylori-induced gastric erosive lesions in a mouse model. Am J Physiol Gastrointest Liver Physiol *294*(4):G877–G884

Mossad SB, Macknin ML, Mendendorp SV, Mason P (1996) Zinc gluconate lozenges for treating the common cold: a randomized, double-blind, placebo-controlled study. Ann Intern Med 125(2):81–88

Nawar O, Akridge RE, Hassan E, El Gazar R, Doughty BL, Kemp W (1992) The effect of zinc deficiency on granuloma formation, liver fibrosis, and antibody responses in experimental schistosomiasis. Am J Trop Med Hyg 47(3):383–389

Nourani MR, Ebrahimi M, Roudkenar MH, Vahedi E, Ghanei M, Fooladi AAI (2011) Sulfur mustard induces expression of metallothionein-1A in human airway epithelial cells. Int J Gen Med 4:413

Novick S, Godfrey J, Godfrey N, Wilder H (1996) How does zinc modify the common cold? Clinical observations and implications regarding mechanisms of action. Med Hypotheses 46(3):295–302

Overbeck S, Rink L, Haase H (2008) Modulating the immune response by oral zinc supplementation: a single approach for multiple diseases. Arch Immunol Ther Exp 56(1):15–30

Palmiter RD (2004) Protection against zinc toxicity by metallothionein and zinc transporter 1. Proc Natl Acad Sci 101(14):4918–4923

Palmiter RD, Huang L (2004) Efflux and compartmentalization of zinc by members of the SLC30 family of solute carriers. Pflugers Arch 447(5):744–751

Pekarek R, Hoagland A, Powanda M (1977) Humoral and cellular immune responses in zinc deficient rats. Nutr Rep Int 16:267–276

Polyudova TV, Eroshenko DV, Korobov VP (2018) Plasma, serum, albumin, and divalent metal ions inhibit the adhesion and the biofilm formation of Cutibacterium (Propionibacterium) acnes. Aims Microbiol 4(1):165–172

Prasad AS (2000) Effects of zinc deficiency on Th1 and Th2 cytokine shifts. J Infect Dis 182(Supplement_1):S62–S68

Prasad AS (2014) Zinc is an antioxidant and anti-inflammatory agent: its role in human health. Front Nutr 1:14

Prasad AS, Schulert AR, Miale JR, Farid Z, Sandstead HH (1963) Zinc and iron deficiencies in male subjects with dwarfism and hypogonadism but without ancylostomiasis, schistosomiasis or severe anemia. Am J Clin Nutr 12(6):437–444

Qiu F, Chamberlin A, Watkins BM, Ionescu A, Perez ME, Barro-Soria R et al (2016) Molecular mechanism of Zn2+ inhibition of a voltage-gated proton channel. Proc Natl Acad Sci 113(40):E5962–E5971

Quaife CJ, Findley SD, Erickson JC, Froelick GJ, Kelly EJ, Zambrowicz BP, Palmiter RD (1994) Induction of a new metallothionein isoform (MT-IV) occurs during differentiation of stratified squamous epithelia. Biochemistry 33(23):7250–7259

Read SA, Parnell G, Booth D, Douglas MW, George J, Ahlenstiel G (2018) The antiviral role of zinc and metallothioneins in hepatitis C infection. J Viral Hepat 25(5):491–501

Richter M, Bonneau R, Girard M-A, Beaulieu C, Larivée P (2003) Zinc status modulates bronchopulmonary eosinophil infiltration in a murine model of allergic inflammation. Chest 123(3):446S

Ruel MT, Rivera JA, Santizo M-C, Lönnerdal B, Brown KH (1997) Impact of zinc supplementation on morbidity from diarrhea and respiratory infections among rural Guatemalan children. Pediatrics 99(6):808–813

Salvin S, Horecker B, Pan L-X, Rabin B (1987) The effect of dietary zinc and prothymosin α on cellular immune responses of RFJ mice. Clin Immunol Immunopathol 43(3):281–288

Savino W, Huang P, Corrigan A, Berrih S, Dardenne M (1984) Thymic hormone-containing cells. V. Immunohistological detection of metallothionein within the cells bearing thymulin (a zinc-containing hormone) in human and mouse thymuses. J Histochem Cytochem 32(9):942–946

Sazawal S, Black RE, Bhan MK, Bhandari N, Sinha A, Jalla S (1995) Zinc supplementation in young children with acute diarrhea in India. N Engl J Med 333(13):839–844

Sazawal S, Black RE, Bhan MK, Jalla S, Bhandari N, Sinha A, Majumdar S (1996) Zinc supplementation reduces the incidence of persistent diarrhea and dysentery among low socioeconomic children in India. J Nutr 126(2):443–450

Sazawal S, Black RE, Jalla S, Mazumdar S, Sinha A, Bhan MK (1998) Zinc supplementation reduces the incidence of acute lower respiratory infections in infants and preschool children: a double-blind, controlled trial. Pediatrics 102(1):1–5

Schroeder JJ, Cousins RJ (1990) Interleukin 6 regulates metallothionein gene expression and zinc metabolism in hepatocyte monolayer cultures. Proc Natl Acad Sci 87(8):3137–3141

Sciavolino PJ, Vilček J (1995) Regulation of metallothionein gene expression by TNF-α and IFN-β in human fibroblasts. Cytokine 7(3):242–250

Scott ME, Koski KG (2000) Zinc deficiency impairs immune responses against parasitic nematode infections at intestinal and systemic sites. J Nutr 130(5):1412S–1420S

Shankar AH, Prasad AS (1998) Zinc and immune function: the biological basis of altered resistance to infection. Am J Clin Nutr 68(2):447S–463S

Shankar A, Kumar N, Scott A (1995) Zinc-deficiency exacerbates experimental malaria infection in mice. Paper presented at the FASEB Journal
Shankar AH, Genton B, Baisor M, Paino J, Tamja S, Adiguma T et al (2000) The influence of zinc supplementation on morbidity due to Plasmodium falciparum: a randomized trial in preschool children in Papua New Guinea. Am J Trop Med Hygiene 62(6):663–669
Shi HN, Scott E, Stevenson M, Koski G (1994) Zinc deficiency impairs T cell function in mice with primary infection of Heligmosomoides polygyrus (Nematoda). Parasite Immunol 16(7):339–350
Singh K, Zaidi S, Raisuddin S, Saxena A, Murthy R, Ray P (1992) Effect of zinc on immune functions and host resistance against infection and tumor challenge. Immunopharmacol Immunotoxicol 14(4):813–840
Someya Y, Tanihata J, Sato S, Kawano F, Shirato K, Sugiyama M et al (2009) Zinc-deficiency induced changes in the distribution of rat white blood cells. J Nutr Sci Vitaminol 55(2):162–169
Stork M, Grijpstra J, Bos MP, Torres CM, Devos N, Poolman JT et al (2013) Zinc piracy as a mechanism of Neisseria meningitidis for evasion of nutritional immunity. PLoS Pathog 9(10):e1003733
Subramanian Vignesh K, Figueroa JAL, Porollo A, Caruso JA, Deepe GS Jr (2013) Granulocyte macrophage-colony stimulating factor induced Zn sequestration enhances macrophage superoxide and limits intracellular pathogen survival. Immunity 39(4):697–710
Subramanian Vignesh K, Figueroa JAL, Porollo A, Divanovic S, Caruso JA, Deepe GS Jr (2016) IL-4 induces metallothionein 3-and SLC30A4-dependent increase in intracellular Zn2+ that promotes pathogen persistence in macrophages. Cell Rep 16(12):3232–3246
Subramanian Vignesh K, Deepe G Jr (2017) Metallothioneins: emerging modulators in immunity and infection. Int J Mol Sci 18(10):2197
Supasai S, Aimo L, Adamo A, Mackenzie G, Oteiza P (2017) Zinc deficiency affects the STAT1/3 signaling pathways in part through redox-mediated mechanisms. Redox Biol 11:469–481
Tang AM, Graham NM, Saah AJ (1996) Effects of micronutrient intake on survival in human immunodeficiency virus type 1 infection. Am J Epidemiol 143(12):1244–1256
Tasci S, Sengil AZ, Altindis M, Arisoy K (1995) The effect of zinc supplementation in experimentally induced Toxoplasma gondii infection. J Egypt Soc Parasitol 25(3):745–751
Tate D, Miceli MV, Newsome DA (1995) Phagocytosis and H2O2 induce catalase and metallothionein gene expression in human retinal pigment epithelial cells. Invest Ophthalmol Vis Sci 36(7):1271–1279
Thorvaldsen JL, Mehra RK, Yu W, Sewell AK, Winge DR (1995) Analysis of copper-induced metallothionein expression using autonomously replicating plasmids in Candida glabrata. Yeast 11(15):1501–1511
Uchida Y, Takio K, Titani K, Ihara Y, Tomonaga M (1991) The growth inhibitory factor that is deficient in the Alzheimer's disease brain is a 68 amino acid metallothionein-like protein. Neuron 7(2):337–347
Ugajin T, Nishida K, Yamasaki S, Suzuki J, Mita M, Kubo M et al (2015) Zinc-binding metallothioneins are key modulators of IL-4 production by basophils. Mol Immunol 66(2):180–188
Ullio C, Brunk UT, Urani C, Melchioretto P, Bonelli G, Baccino FM, Autelli R (2015) Autophagy of metallothioneins prevents TNF-induced oxidative stress and toxicity in hepatoma cells. Autophagy 11(12):2184–2198
Uttra CKM, Devrajani BR, Shaikh HM, Shah SZA, Devrajani T, Shaikh N, Shah SFA (2011) Serum Zinc level in patients with pulmonary tuberculosis. Adv Biol Res 5(3):174–178
Vallee BL (1995) The function of metallothionein. Neurochem Int 27(1):23–33
Vruwink KG, Fletcher MP, Keen CL, Golub MS, Hendrickx AG, Gershwin ME (1991) Moderate zinc deficiency in rhesus monkeys. An intrinsic defect of neutrophil chemotaxis corrected by zinc repletion. J Immunol 146(1):244–249
Wellinghausen N, Driessen C, Rink L (1996) Stimulation of human peripheral blood mononuclear cells by zinc and related cations. Cytokine 8(10):767–771
Winqvist I, Olofsson T, Persson E (1985) Effect of zinc and other cations on the release of the eosinophil cationic protein. Scand J Clin Lab Invest 45(8):671–677

Wu C, Labrie J, Tremblay Y, Haine D, Mourez M, Jacques M (2013a) Zinc as an agent for the prevention of biofilm formation by pathogenic bacteria. J Appl Microbiol 115(1):30–40
Wu C, Pot C, Apetoh L, Thalhamer T, Zhu B, Murugaiyan G et al (2013b) Metallothioneins negatively regulate IL-27–induced type 1 regulatory T-cell differentiation. Proc Natl Acad Sci 110(19):7802–7807
Wu A, Tymoszuk P, Haschka D, Heeke S, Dichtl S, Petzer V et al (2017) Salmonella utilizes zinc to subvert antimicrobial host defense of macrophages via modulation of NF-κB signaling. Infect Immun 85(12):e00418–e00417
Zackular JP, Moore JL, Jordan AT, Juttukonda LJ, Noto MJ, Nicholson MR et al (2016) Dietary zinc alters the microbiota and decreases resistance to Clostridium difficile infection. Nat Med 22(11):1330

Chapter 5
Role of Zinc Signaling in Mast Cell, Basophil, T Cell, and B Cell

Keigo Nishida, Michiko Kato, and Ryota Uchida

Abstract Zinc is an essential heavy metal that plays an important role in nucleic acid metabolism, cell replication, and tissue growth and repair, and its deficiency is associated with a variety of disorders. Zinc directly affects the activity of several signaling molecules and of transcription factors that have a zinc-binding motif. However, it has not been clearly shown whether zinc itself acts as an intracellular signaling molecule like calcium. Several groups have presented evidence that zinc does indeed act as an intracellular signaling molecule, converting extracellular stimuli to intracellular signals and controlling various cell functions. This chapter summarizes our current understanding of zinc signaling, especially with regard to Zinc wave and the role of zinc signaling in Fc receptors and T- and B-cell antigen receptors, and discusses how these processes contribute to cytokine production.

Keywords Zinc · Zinc wave · L-type calcium channel · Cytokines · Zinc transporter · Signal transduction · Mast cells · Basophils · T cells · B cells

5.1 Introduction

Zinc is a trace element that is essential for metabolism in all organisms and is an important structural component of many enzymes and proteins including zinc finger proteins and zinc metalloenzymes (Prasad 1995; Vallee and Auld 1993). Zinc plays

K. Nishida (✉)
Laboratory of Immune Regulation, Graduate School of Pharmaceutical Sciences, Suzuka University of Medical Science, Suzuka, Mie, Japan

Laboratory of Immune Regulation, Faculty of Pharmaceutical Sciences, Suzuka University of Medical Science, Suzuka, Mie, Japan
e-mail: knishida@suzuka-u.ac.jp

M. Kato · R. Uchida
Laboratory of Immune Regulation, Graduate School of Pharmaceutical Sciences, Suzuka University of Medical Science, Suzuka, Mie, Japan

T. Fukada, T. Kambe (eds.), *Zinc Signaling*,
https://doi.org/10.1007/978-981-15-0557-7_5

an important role in the proper function of the immune system and the nervous system as well as growth. Zinc deficiency has been known to influence various pathological conditions including growth retardation, disorders in brain development, impaired immune function, and delayed wound healing (Hambidge 2000; Lansdown et al. 2007; Prasad 2008; Sensi et al. 2009).

The wide-ranging effects of zinc in the immune system and the nervous system have been demonstrated both in vivo and in vitro, and these effects mainly depend on the concentration of zinc (Frederickson et al. 2005; Rink and Gabriel 2000). Many studies have shown that zinc depletion disturbs immune function. For example, natural killer cell-mediated cytotoxic activity, antibody-mediated responses, and host defenses against pathogens and tumors have all been shown to be reduced in zinc-deficient mice (Fernandes et al. 1979; Fraker et al. 1982; Keen and Gershwin 1990). On the other hand, high concentrations of zinc can be cytotoxic and can induce apoptosis in T cells and B cells (Telford and Fraker 1995; Ibs and Rink 2003).

Accumulating evidence suggests the importance of special and temporal regulation of intracellular zinc concentration, and intricate regulation of zinc homeostasis has been discussed in many reports.

A phenomenon termed "Zinc wave" has been discovered in which a surge of free intracellular zinc occurs as a result of Fc epsilon receptor I (FcεRI) stimulation (Nishida and Yamasaki 2014; Yamasaki et al. 2007). Zinc wave can be observed within several minutes after FcεRI stimulation, and it originates from the endoplasmic reticulum (ER). Thus, extracellular stimuli can affect intracellular zinc levels through zinc transporters and zinc-permeable channels, and zinc acts as an intracellular signaling molecule (Haase et al. 2008; Inoue et al. 2015; Taylor et al. 2012; Yamasaki et al. 2012). Zinc signaling can be classified into at least two categories: late zinc signaling, which depends on transcriptional changes in zinc transporter expression, and early zinc signaling that involves Zinc wave, in which zinc is directly released from the cellular organelles by an extracellular stimulus such as FcεRI (Fukada et al. 2011; Hirano et al. 2008; Murakami and Hirano 2008; Nishida and Uchida 2018).

While many studies have shown that zinc has an important role in the immune system and that imbalances in zinc homeostasis lead to various disorders, questions remain as to how zinc homeostasis and signaling are regulated in mast and other immune cells and whether zinc transporters are involved in immune cell function. This chapter discusses the role of zinc signaling in cytokine production.

5.2 Zinc Homeostasis

Computational analysis by Andreini et al. has shown that approximately 10% of the human proteome represents zinc-binding proteins (Andreini et al. 2006), indicating the involvement of zinc and zinc-binding proteins in many cellular and physiological functions. The regulation of zinc homeostasis mainly involves zinc transporters and metallothioneins (Fig. 5.1).

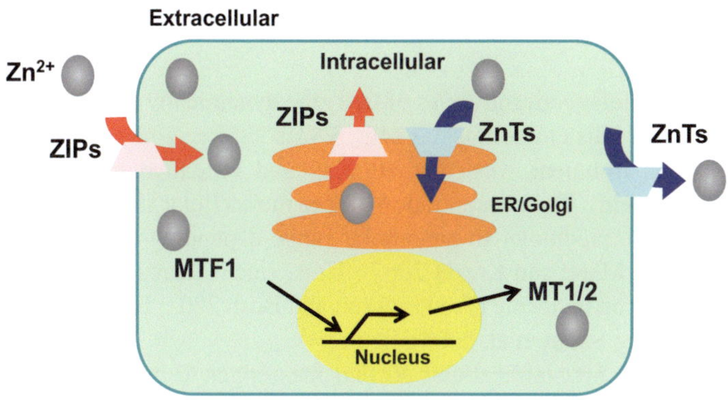

Fig. 5.1 Subcellular localization of zinc transporters and metallothioneins. An illustration of the subcellular localization and potential functions of ZIP and ZnT family members, based on currently available information. Arrows indicate the predicted direction of zinc mobilization. *Zn* Zinc, *ZIPs* Zrt- and Irt-like proteins, *ZnTs* Zinc transporters, *MTF1* Metal regulatory transcription factor 1, *MTs* Metallothionein 1/ metallothionein 2, *ER* Endoplasmic reticulum

Zinc transporters include Slc39/ZIP family and Slc30/ZnT family proteins that mediate the efflux and influx of zinc, respectively. The human SLC39/ZIP family consists of 14 members. They are responsible for the transport of zinc from the extracellular environment or from the intracellular compartments to increase intracellular zinc levels. The human SLC30/ZnT family consists of ten members that facilitate the efflux of zinc to reduce the cytoplasmic zinc by transporting it from the cytosol to the extracellular space or the intracellular compartments. Controlling the influx and efflux of zinc by the coordinated action of these proteins is essential for zinc homeostasis and its physiological function (Bafaro et al. 2017; Cousins et al. 2006; Hojyo and Fukada 2016; Kambe et al. 2017).

Metallothioneins (MTs), a family of small cysteine-rich proteins that bind zinc and other metal ions, are thought to be involved in the regulation of intracellular zinc concentration and the detoxification of nonessential heavy metals. Metallothioneins are considered as "zinc storage," maintaining intracellular free zinc concentration by binding to free intracellular zinc through cysteine (Thirumoorthy et al. 2011). When intracellular free zinc reaches a threshold concentration, zinc sensor MTF-1 (metal-responsive element-binding transcription factor 1) is activated and induces the expression of MTs, which sequester zinc ions (Andrews 2001). Thus, MTs and MTF-1 are considered key coordinators of zinc homeostasis.

5.3 Role of Zinc as a Second Messenger

Cytosolic and cellular concentrations of free zinc dynamically change in many cells in response to various environmental cues. Cytosolic free zinc is derived from the extracellular environment, vesicular storage, and oxidative release from zinc-sequestering proteins. The concentration of free intracellular zinc has been reported to fluctuate in the picomolar to nanomolar range depending on the physiological state of the cells (Maret and Krezel 2007). Thus, zinc is increasingly recognized as a potential signaling molecule (Frederickson and Bush 2001; Haase and Rink 2009; Hirano et al. 2008; Smart et al. 2004).

cAMP was the first intracellular second messenger to be discovered (Berthet et al. 1957), followed by calcium. Interestingly, zinc has also been recognized to play an essential role as a second messenger. Zinc itself affects a variety of signaling molecules including PKC, Ca/calmodulin-dependent protein kinase II (CaMKII), and Erk1/Erk 2 (Hubbard et al. 1991; Lengyel et al. 2000) (Ho et al. 2008). In addition, zinc activates ion channels such as the transient receptor potential ankyrin 1 (TRPA1) (Andersson et al. 2009; Hu et al. 2009), ATP-sensitive K^+ (Prost et al. 2004), and large-conductance Ca-activated K^+ (Hou et al. 2010) channels.

Involvement of zinc transporters in zinc signaling has been supported by many studies. For example, ZIP6/SLC39A6 has been reported to be required for the epithelial-mesenchymal transition (EMT) in the zebrafish gastrula organizer (Yamashita et al. 2004). Since ZIP6/SLC39A6 expression in the zebrafish organizer depends on STAT3 activation, any extracellular stimulus that regulates STAT3 activation could alter the intracellular zinc level by inducing changes in the expression of ZIP6/SLC39A6 or other zinc transporters. It has also been shown that TLR4 reduces the level of intracellular free zinc by inducing changes in the expression of zinc transporters in dendritic (Kitamura et al. 2006) and pulmonary endothelial cells (Thambiayya et al. 2011). These observations provide evidence that zinc functions as an intracellular second messenger and that it is capable of transducing extracellular stimuli into intracellular signaling events.

Function of zinc as a signaling molecule in immune cells has been implicated by a number of studies. In mast cells, FcεRI stimulation-induced PKC activation is dependent on ZnT5/Slc30a5 (Nishida et al. 2009). Furthermore, it has been reported that Slc39a10/ZIP10 is critically involved in the early development and functional expression of B cells important for antibody production (Hojyo et al. 2014; Miyai et al. 2014). Thus, extracellular stimuli can, by changing the expression of zinc transporters, affect intracellular signaling pathways by changing intracellular zinc status.

Zinc wave phenomenon, which causes a transient, transcription-independent increase of intracellular zinc, is also considered an important factor in zinc signaling (Yamasaki et al. 2007). It has been reported that intracellular zinc at the ER membrane is regulated by ZIP7/SLC39A7 (Taylor et al. 2008). It is also possible that transporters and channels other than ZIP or ZnT family members are involved in different cells and activated by different stimuli. The precise molecular mechanisms generating Zinc wave have not been fully elucidated. However, since Zinc wave does not involve extracellular zinc and is induced within several minutes after FcεRI stimulation, zinc signaling elicited by this phenomenon may be different from signaling that is initiated as a result of extracellular zinc mobilization.

5.4 Zinc Signaling Mediated by Antigen Receptors

Stimulation of antigen receptors such as FcεRI, B-cell receptors (BCRs), and T-cell receptors (TCRs) has been reported to induce rapid increase in free intracellular zinc such as the Zinc wave that occurs in mast cells and lymphocytes (Taniguchi et al. 2013; Yamasaki et al. 2012; Yu et al. 2011). In addition, it has been demonstrated that zinc-permeable channels and zinc transporters are responsible for the regulation of dynamic changes in cytosolic free zinc levels. In this section, we briefly describe the regulation of zinc signaling in different immune cells and discuss the targets of zinc signaling.

5.4.1 FcεRI-Mediated Zinc Signaling

L-type calcium channels (LTCCs) conduct zinc and can act as zinc-permeable channels on the plasma membrane of neurons and pancreatic β cells (Atar et al. 1995; Gyulkhandanyan et al. 2006; Sensi et al. 1997). However, it is not clear whether LTCCs are involved in the release of zinc from intracellular organs. LTCCs have been indicated to act as a gatekeeper of Zinc wave that occurs as a result of FcεRI stimulation in mast cells (Yamasaki et al. 2012).

It has been well established that LTCCs function as voltage-gated calcium channels on the plasma membrane. LTCCs are multi-subunit complexes composed of α_1, β, and α_2/δ subunits. The α_1 subunit functions as the voltage sensor, selective filter, and ion-conducting pore (Catterall 2000). The α_1 subunit on the cell surface is thought to require an association with the β subunit, which masks one or more ER retention signals (Bichet et al. 2000; Cornet et al. 2002).

In mast cells, LTCC containing the α1 subunit (α1D) Cav1.3 is predominantly expressed, while the expression of the LTCC β subunit, which is required for the localization of the α1 subunit to the plasma membrane, is very low. Furthermore, ZnT-1/Slc30a1 that interacts with the LTCC β subunit on the plasma membrane is expressed at high levels in mast cells. It has been shown that this interaction reduces the availability of free β subunit, thereby leading to the inhibition of α1D translocation to the plasma membrane (Levy et al. 2009). Therefore, high-level expression of ZnT-1 suggests that the majority of α1D are localized intracellularly in mast cells. α1D colocalizes partially with the ER marker calnexin but not with F-actin, which accumulates beneath the plasma membrane. These observations indicate that α1D localizes preferentially to intracellular organelle membranes, such as the ER membrane, rather than to the plasma membrane. This subcellular localization of α1D in mast cells suggests that it plays a different role in mast cells than in other cells, where it is located on the plasma membrane and acts as a calcium channel.

Consistent with this idea, an LTCC inhibitor verapamil attenuates Zinc wave in mast cells while cell survival, FcεRI expression, and FcεRI-mediated calcium elevation are not affected. Diltiazem, another type of LTCC antagonist, also inhibits Zinc wave without disturbing calcium elevation. Verapamil affects neither FcεRI-mediated calcium elevation in mast cells nor calcium-mediated signaling such as nuclear translocation of NFAT2 in response to calcium elevation (Yamasaki et al. 2012). On the other hand, treating mast cells with an LTCC agonist (s)-(-)-BayK8644 without antigen stimulation results in the elevation of intracellular zinc level but not calcium (Yamasaki et al. 2012). These findings support that the function of LTCC in mast cells differs from its function as a calcium channel in other cell types, such as neurons and pancreatic β cells. Since calcium influx primarily occurs via store-operated calcium entry (SOCE) in mast cells, LTCC in mast cells appears to have little effect on the regulation of intracellular calcium (Vig and Kinet 2009). LTCC agonists induce an increase in intracellular zinc even in the absence of calcium, and this increase is inhibited by verapamil. Similarly, stimulation of FcεRI induces zinc increase in mast cells in the presence or absence of an LTCC agonist, indicating that the mechanisms responsible for FcεRI-induced Zinc wave and LTCC agonist-induced elevation of zinc are probably similar.

In mast cells, suppression of α1D significantly reduces FcεRI-induced Zinc wave without affecting calcium elevation in mast cells, as in the case of verapamil treatment. Moreover, ectopic expression of wild-type α1D rescues the inhibition of Zinc wave by siRNA knockdown of α1D (Yamasaki et al. 2012). These observations suggest that the LTCC α1D subunit is a gatekeeper of Zinc wave.

This perspective raises questions of how zinc release from the ER into the cytoplasm is regulated by LTCC. The α1 subunit of LTCC contains a voltage sensor domain, and the channel activity is enhanced after membrane depolarization. The plasma membrane potential is hyperpolarized in mast cells after FcεRI stimulation (Shumilina et al. 2008; Vennekens et al. 2007). However, inhibiting the FcεRI-mediated plasma membrane hyperpolarization by high KCl treatment does not impair the induction of Zinc wave. Treating cells with bongkrekic acid, an adenine nucleotide translocase inhibitor, suppresses the depolarization of FcεRI-mediated intracellular membrane, but does not inhibit the induction of Zinc wave (Yamasaki et al. 2012). These observations suggest that the regulation of the membrane potential might not affect Zinc wave.

Phosphorylation of the pore-forming α1 subunit of LTCC has an additional effect on channel activity; in fact, cAMP-mediated channel activity is reduced by site-directed mutagenesis of the PKA consensus sites of α1D (Ramadan et al. 2009). However, a PKA inhibitor did not inhibit Zinc wave. Therefore, PKA might not assist in regulating the Zinc wave, at least not in mast cells. This event may be controlled by as yet unidentified regulatory proteins and mechanisms on the ER membrane.

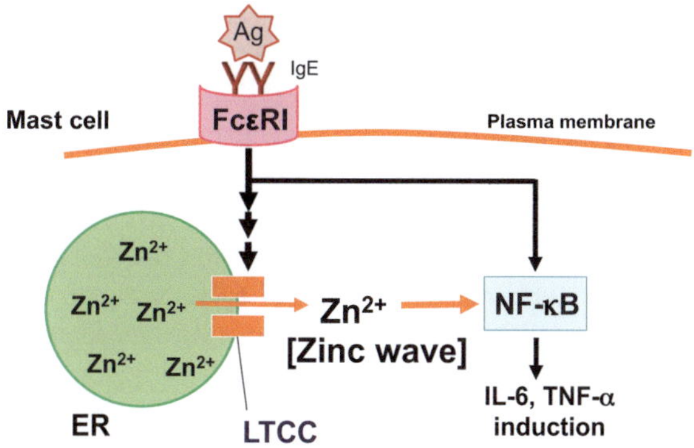

Fig. 5.2 Zinc wave participates in FcεRI-mediated cytokine production. Stimulation of FcεRI induces rapid elevation of intracellular zinc from the perinuclear region around the ER, and we named this phenomenon "Zinc wave." The α1D subunit of the LTCC is expressed on the ER membrane and acts as the gatekeeper of Zinc wave in mast cells. In addition, LTCC-mediated Zinc wave is a positive regulator of the DNA-binding activity of NF-κB and is involved in the regulation of cytokine production. *Ag* Antigen, *IgE* Immunoglobulin E, *FcεRI* High-affinity receptor for IgE, *Zn* Zinc, *ER* Endoplasmic reticulum, *LTCC* L-type calcium channel, *NF-κB* Nuclear factor-kappa B, *IL-6* Interleukin-6, *TNF-α* Tumor necrosis factor-alpha

These results show that LTCC's pore-forming α1D subunit, when expressed on the ER membrane, functions as a gatekeeper for the Zinc wave in mast cells and that this LTCC-mediated Zinc wave may function as an intracellular Zinc signal that can positively modify signal transduction to produce inflammatory cytokines (Fig. 5.2).

5.4.2 *BCR-Mediated Zinc Signaling*

The BCR-signaling pathway is critical for many cellular events, including cell growth, cell proliferation, and apoptosis (Dal Porto et al. 2004; Harwood and Batista 2008, 2010). BCR activation transduces the signal to several cascades, including the PI-3 K/Akt, PLCγ2/PKC, and Ras/Raf/ERK cascades (Brazil and Hemmings 2001; Hashimoto et al. 1998; Kurosaki 2011). These cascades are important for the differentiation of antibody-producing cells and memory B cells.

A study by Taniguchi et al. has provided a new insight that the activation of BCR signaling requires influx of zinc into the cytoplasm. Experiments using the DT40 chicken B cell line as a model have revealed that the zinc transporter ZIP9/SLC39A9 has a role in zinc signaling in B cells. Unlike Zinc wave, which originates from the ER, zinc release in this B cells occurs from the Golgi, and ZIP9 is considered to facilitate zinc transport from the Golgi to the cytoplasm (Taniguchi et al. 2013).

ZIP9-mediated zinc signaling affects the phosphorylation and activation of Akt and ERK that is initiated via BCR. In cells lacking Zip9, BCR-induced Akt and ERK phosphorylation is significantly attenuated while the enzymatic activity of protein tyrosine phosphatase (PTPase) is enhanced. Consistent with this observation, overexpressing hZIP9 decreased the PTPase activity in ZincPy-treated cZip9KO cells (Taniguchi et al. 2013). How ZIP9-mediated zinc signaling is regulated by BCR stimulation is currently unclear. It has been reported that ZIP7 is activated by protein kinase CK2 in a human breast cancer cell line (Taylor et al. 2012). This raises the possibility that ZIP9 and ZIP7 have similar function in the release of zinc. These findings support that ZIP9 on the Golgi body membrane regulates cytosolic zinc, enhancing the Akt and ERK phosphorylations in B cells (Fig. 5.3). Thus, zinc signaling in B cells may occur via a mechanism similar to that of Zinc wave in mast cells.

5.4.3 *TCR-Mediated Zinc Signaling*

Zinc signaling that resembles Zinc wave has also been described in T cells. Elevation of cytosolic zinc levels has been observed within 1 min after TCR stimulation. This increase is dependent on the concentration of extracellular zinc, and Zip6 has been shown to be responsible for the influx of zinc from the extracellular environment, as depletion of zinc from the culture medium or downregulation of Zip6 inhibited the increase in cytosolic zinc (Yu et al. 2011). Early events in TCR signaling include the phosphorylation of several signaling molecules immediately downstream of the

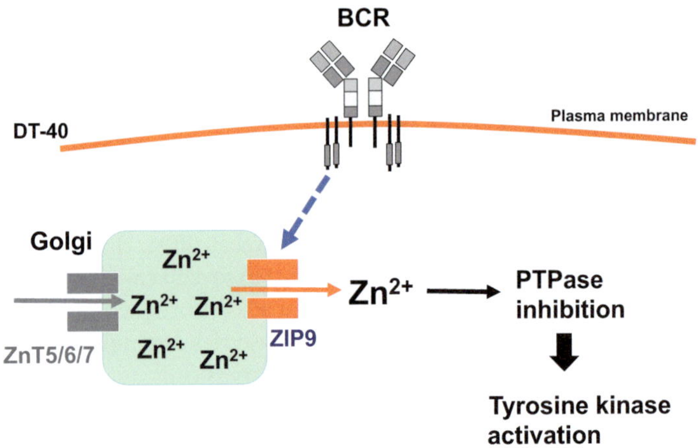

Fig. 5.3 Proposed action sites of intracellular zinc release by ZIP9 in DT40 cells for the activation of B cell receptor signaling. A proposed mechanism of zinc-induced PTPase inhibition by ZIP9, which leads to the activation of B cell receptor signaling in DT40 cells. Intracellular zinc is incorporated into the Golgi by ZnT5/ZnT6/ZnT7. Zn released into the cytosol from the Golgi by ZIP9 induction inhibits PTPase activity and induces the phosphorylation of Akt and ERK probably indirectly, by regulating upstream components of signal transduction. *BCR* B-cell receptor, *Zn* Zinc, *ZnT5/ZnT6/ZnT7* Zinc transporter 5/zinc transporter 6/zinc transporter7, *ZIP9* Zrt- and Irt-like protein9, *PTPase* Protein tyrosine phosphatases

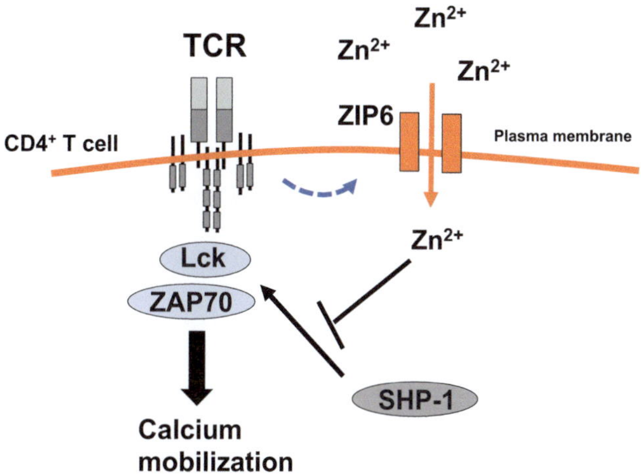

Fig. 5.4 Zinc functions as signaling molecule after T-cell activation. Cytoplasmic zinc concentrations increase within 1 min after TCR triggering, due to zinc influx via the transporter ZIP6. This increase, which is most pronounced in the immediate subsynaptic area, enhances TCR signaling, at least in part, by inhibiting the recruitment of SHP-1. *TCR* T-cell receptor, *Zn* Zinc, *ZIP6* Zrt- and Irt-like protein 6, *Lck* Lymphocyte-specific protein tyrosine kinase, *ZAP70* Zeta-chain-associated protein of 70 kDa, *SHP-1* SH2 domain-containing protein tyrosine phosphatase 1

TCR. The Src protein kinase Lck is primarily responsible for the phosphorylation of CD3 and ζ of the TCR complex and subsequent phosphorylation of ZAP70 which becomes associated with the TCR (Palacios and Weiss 2004). Negative feedback in TCR signaling is exerted by a tyrosine phosphatase SHP-1 which dephosphorylates ZAP70 and other molecules after it is recruited to the TCR complex (Altan-Bonnet and Germain 2005). Increased influx of zinc inhibits negative regulatory feedback loops, resulting in increased phosphorylation of ZAP70 and sustained calcium influx. These observations support that the influx of extracellular zinc upon TCR stimulation influences the early signaling response, and the role of zinc in tuning the TCR activation threshold when antigenic stimulation is suboptimal has been suggested (Yu et al. 2011) (Fig. 5.4).

5.5 Role of Zinc and Zinc Signaling in Cytokine Production

It has been reported in many studies that immune function is weakened by zinc depletion, indicating the role of zinc as a positive regulator in immune responses such as cytokine production. Here, we describe the effects of zinc and zinc homeostasis on biological events, particularly with regard to FcεRI-mediated production of cytokines.

5.5.1 Role of Zinc in Mast Cell-Mediated Cytokine Production

Immune cells such as mast cells, eosinophils, and basophils are involved in a variety of allergic reactions such as anaphylaxis, asthma, and atopic dermatitis (Galli et al. 2008; Kawakami et al. 2009; Metz et al. 2007). These allergic responses involve inflammatory reactions leading to the secretion of a variety of chemical mediators and cytokines.

Activation of mast cells results in the secretion of two classes of mediators. The first consists of preformed mediators that are stored in intracellular granules and rapidly secreted into activated cells. The second class of mediators are cytokines and chemokines, which are synthesized de novo and therefore secreted more slowly. These molecules play vital roles in inflammatory responses and are deeply involved in allergic diseases.

Requirement of zinc in degranulation and cytokine production in mast cells has been demonstrated in several studies. Artificially depleting intracellular zinc with the zinc chelator TPEN [*N,N,N,N*-tetrakis (2-pyridylmethyl) ethylenediamine] inhibits histamine release, cytokine production, and lipid mediator secretion. The inhibitory effects of TPEN were rescued by zinc supplement, and chelators of other metal ions did not affect mast cell functions (Kabu et al. 2006). These observations indicate the involvement of zinc in degranulation and cytokine production in mast cells. Similarly, it has been reported that depletion of intracellular zinc by TPEN or the clinically used heavy metal chelator DMPS (Torres-Alanis et al. 1995) results in the inhibition of the transcription of eotaxin and other chemokines in several human lung cell lines (Richter et al. 2003).

TPEN suppresses FcεRI-mediated cytokine production and the transcription of IL-6 and TNF-α mRNAs (Kabu et al. 2006). PKC is activated upon FcεRI stimulation and is involved in cytokine production through NF-κB activation (Klemm et al. 2006; Nechushtan et al. 2000), and TPEN inhibits FcεRI-mediated translocation of PKC to plasma membrane (Kabu et al. 2006). These results suggest PKC is a target of TPEN in the regulation of cytokine production. In fact, the zinc-binding domain of PKC is essential for plasma membrane translocation in response to FcεRI stimulation (Oancea et al. 1998).

ZnT5 was shown to be crucial in mast cell activation and the production of inflammatory cytokines. ZnT5 is abundantly expressed in mast cells and is upregulated by FcεRI stimulation, suggesting that ZnT5 is involved in mast cell-mediated allergic responses. Knocking ZnT5 out in mice affected delayed-type hypersensitivity but not immediate-type reactions such as anaphylaxis (Nishida et al. 2009). Consistent with this in vivo observation, it was shown that ZnT5 is required for FcεRI-mediated cytokine production, but not for degranulation, in mast cells. In *ZnT5-KO* mast cells, FcεRI-induced transcription of IL-6 and TNF-α was reduced. In addition, zinc and ZnT5 were shown to be required for plasma membrane translocation of PKC and subsequent nuclear translocation of NF-κB which controls the transcription of inflammatory cytokines (Nishida et al. 2009).

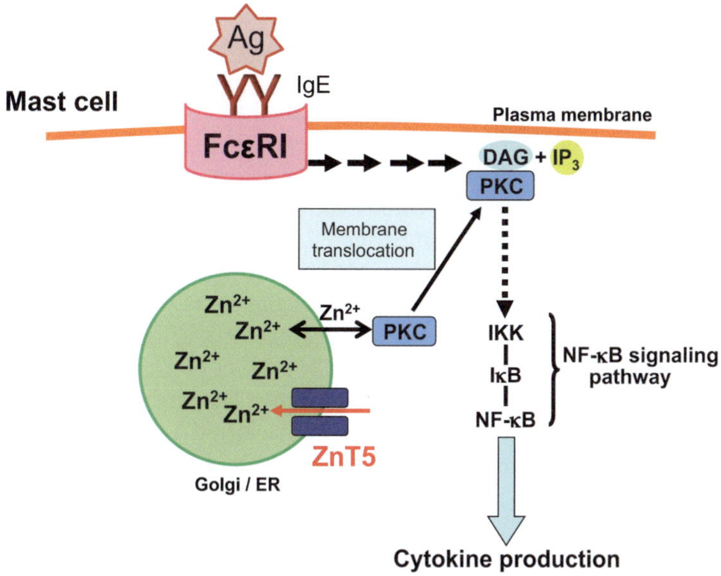

Fig. 5.5 Zinc transporters are involved in FcεRI-mediated mast cell activation. Zn is required in multiple steps of FcεRI-induced mast cell activation, including cytokine production. ZnT5 is required for the translocation of PKC to the plasma membrane and subsequent nuclear translocation of NF-κB, leading to the production of cytokines such as IL-6 and TNF-α. *Ag* Antigen, *IgE* Immunoglobulin E, *FcεRI* High-affinity receptor for IgE, *Zn* Zinc, *ER* Endoplasmic reticulum, *ZnT5* Zinc transporters5, *PKC* Protein kinase C, *DAG* Diacylglycerol, *IP3* Inositol triphosphate, *IKK* IκB kinase, *IκB* NF-κB inhibitor, *NF-κB* Nuclear factor-kappa B

Together, these findings suggest that zinc and zinc transporters are involved in mast cell-mediated allergic responses by regulating degranulation and cytokine production. Zinc and zinc transporters are also considered to modulate the PKC/NF-κB signaling pathway which regulates the expression of cytokine and chemokines (Fig. 5.5). From these findings, control of zinc can be viewed as a promising therapeutic strategy to alleviate allergies.

5.5.2 *Role of Zinc in Basophil-Mediated Cytokine Production*

Basophils represent less than 1% of peripheral blood leukocytes. Like mast cells, basophils express FcεRI on their cell surface and release cytokines and chemical mediators in response to FcεRI activation (Galli 2000; Karasuyama et al. 2017). Under physiological conditions, basophils circulate in the blood while mast cells reside in peripheral tissues. Infiltration of basophils into peripheral tissues is often observed in allergic inflammatory diseases such as atopic dermatitis and bronchial asthma (Karasuyama et al. 2018). The significance of basophils has long been overlooked in immunological studies due to their low abundance and morphological similarity to mast cells, but they are now recognized as a key player in the development of Th2-mediated allergic inflammation (Schroeder et al. 2001; Seder et al. 1991; Sokol et al. 2008; Voehringer et al. 2004).

It has been reported that zinc-binding metallothioneins (MTs) are required for FcεRI-induced IL-4 production in human and mouse basophils (Ugajin et al. 2015; Ugajin et al. 2016). FcεRI stimulation led to elevated transcription of Mt-1 and Mt-2 genes in primary mouse basophils, while it did not affect the expression of Zip and ZnT genes or Mt-3 and Mt-4. Furthermore, FcεRI-induced IL-4 production in basophils was inhibited in the absence of Mt-1 and Mt-2, suggesting the importance of MTs and zinc homeostasis in cytokine production in basophils. Involvement of MTs in FcεRI-induced calcineurin (CN)/nuclear factor of activated T cells (NFAT) signaling has also been described (Fig. 5.6). CN consists of a catalytic subunit CnA and a regulatory subunit CnB, forming a heterodimer. The catalytic domain contains a Fe^{3+}-Zn^{2+} dinuclear center; therefore zinc is considered a key catalytic cofactor (King and Huang 1984; Rusnak and Mertz 2000). On the other hand, it has been

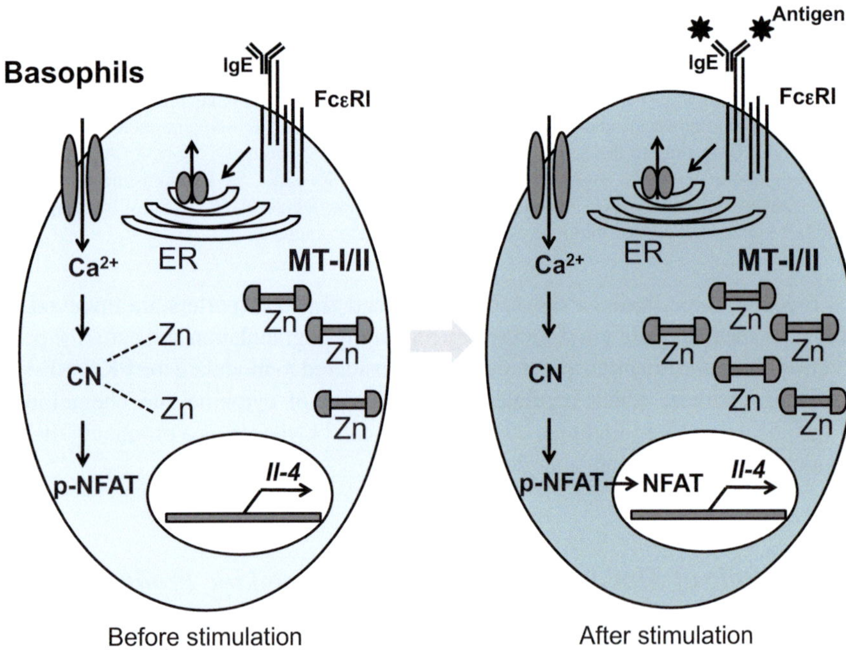

Fig. 5.6 Metallothioneins control in FcεRI-mediated Il-4 production in basophils. FcεRI stimulation activates basophils via signaling pathways that increase IL-4 production. Before stimulation, cytoplasmic free zinc ion inhibits CN. FcεRI stimulation induces the expression of metallothionein 1/ metallothionein 2, and metallothioneins can bind to cytoplasmic free zinc ion. As a result, CN/NFAT signaling pathway is activated and leads to increased IL-4-expression. *IgE* Immunoglobulin E, *FcεRI* High-affinity receptor for IgE, *ER* Endoplasmic reticulum, *Zn* Zinc, *MT-1/MT-2* Metallothionein 1/metallothionein 2, *Ca* Calcium, *CN* Calcineurin, *NFAT* Nuclear factor of activated T cells, *IL-4* Interleukin-4

reported that zinc inhibits the activity of CN in vitro (Aydemir et al. 2009; Huang et al. 2008; Takahashi et al. 2003). In basophils, deleting Mt1/Mt2 genes led to higher intracellular concentration of free zinc, and artificially increasing intracellular zinc levels attenuated CN activity. These findings suggest that MTs play an important role in the regulation of CN/NFAT signaling in basophils by regulating zinc homeostasis. A possible mechanism by which zinc regulates CN/NFAT signaling in basophils is that activated MTs sequester intracellular free zinc, thereby derepressing CN which leads to the activation of the signaling pathway. Zinc-dependent inhibition of cytokine expression has been described in T cells as well (Bao et al. 2003; Hayashi et al. 2008; Zhou et al. 2004). Thus, intracellular free zinc is suggested to function as a suppressor of cytokine production in immune cells. In contrast, zinc positively regulates the expression of cytokines. It has been shown that zinc supplementation enhances cytokine production in immune cells (Driessen et al. 1995; Rink and Kirchner 2000; Wellinghausen et al. 1997). These conflicting behaviors of zinc may be due to its concentration-dependent effect on different signaling molecules involved in cytokine expression.

5.6 Zinc Wave Is a Key Regulator of Cytokine Production in Immune Cells

Our recent study has shown that Zinc wave is also triggered by stimuli that do not mediate FcεRI, such as lipopolysaccharides (LPS) and cytokines. We investigated whether LTCC-mediated Zinc wave also occurs by activating TLR4 receptor which recognizes LPS and IL-33 receptor which recognizes the cytokine IL-33 in mast cells. A rapid rise in intracellular zinc was observed by stimulating TLR4 or IL-33 receptor, and this effect was inhibited by blocking LTCC, indicating that Zinc wave can be triggered by a variety of stimuli in mast cells. Furthermore, Zinc wave was found to be involved in the regulation of cytokine production triggered by the activation of TLR and cytokine receptors (Uchida et al. 2019).

Involvement of Zinc wave in LPS-mediated cytokine production led to the hypothesis that Zinc wave, originally described in mast cells, also occurs in dendritic cells. Zinc wave was indeed observed in dendritic cells in response to LPS stimulation, which was blocked by the LTCC antagonist nicardipine (Uchida et al. 2019). Correspondingly, TLR4-triggered production of IL-6 was also inhibited by chelating zinc or blocking LTCC in these cells. Collectively, these data suggest that LTCC-mediated Zinc wave is not a specific phenomenon in mast cells and it also functions to regulate cytokine production in dendritic cells.

5.7 Perspective

This chapter mainly discussed the regulatory mechanism of intracellular zinc signaling and its role in immune cells such as mast cells, basophils, T cells, and B cells. The observation of LTCC-induced Zinc wave not only in mast cells but also in dendritic cells is interesting as this suggests Zinc wave plays an important role in innate immune response as well. In addition, we reported that Zinc wave plays an important role in the regulation of cytokine production mediated by receptors other than the antigen receptor (Fig. 5.7). These new findings support that zinc signaling including Zinc wave is deeply involved in general defense responses such as innate immunity and acquired immunity. To understand how zinc signaling is involved in immune responses in vivo, further studies are needed including various in vivo experiments such as the evaluation of LTCC and zinc transporter knockouts in mouse models.

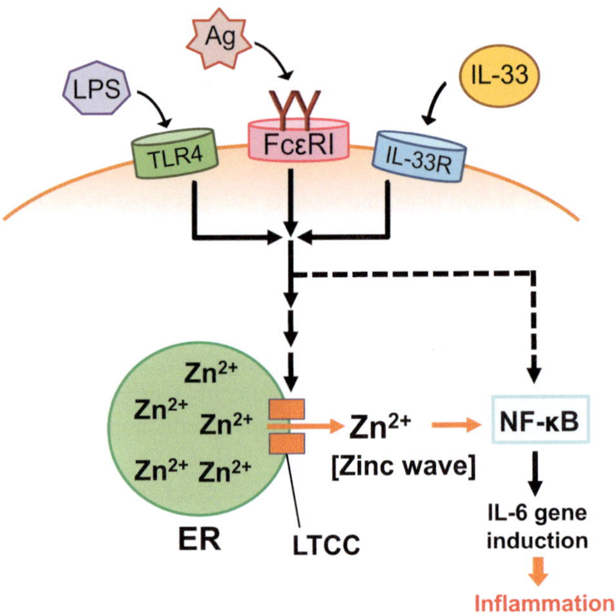

Fig. 5.7 LTCC-mediated Zinc wave in mast cells is caused by non-IgE-mediated stimulation. IgE-, LPS-, and IL-33-induced Zinc wave was observed in mast cells. Zinc wave controls IgE-, LPS-, and IL-33-induced cytokine gene expression. Thus, Zinc wave is activated by various stimuli and is linked to cytokine induction in immune cells. *LPS* Lipopolysaccharides, *Ag* Antigen, *IL-33* Interleukin-33, *FcεRI* High-affinity receptor for IgE, *ER* Endoplasmic reticulum, *Zn* Zinc, *LTCC* L-type calcium channel, *NF-κB* Nuclear factor-kappa B, *IL-6* Interleukin-6

References

Altan-Bonnet G, Germain RN (2005) Modeling T cell antigen discrimination based on feedback control of digital ERK responses. PLoS Biol 3:e356

Andersson DA, Gentry C, Moss S, Bevan S (2009) Clioquinol and pyrithione activate TRPA1 by increasing intracellular Zn2+. Proc Natl Acad Sci USA 106:8374–8379

Andreini C, Banci L, Bertini I, Rosato A (2006) Counting the zinc-proteins encoded in the human genome. J Proteome Res 5:196–201

Andrews GK (2001) Cellular zinc sensors: MTF-1 regulation of gene expression. Biometals 14:223–237

Atar D, Backx PH, Appel MM, Gao WD, Marban E (1995) Excitation-transcription coupling mediated by zinc influx through voltage-dependent calcium channels. J Biol Chem 270:2473–2477

Aydemir TB, Liuzzi JP, McClellan S, Cousins RJ (2009) Zinc transporter ZIP8 (SLC39A8) and zinc influence IFN-gamma expression in activated human T cells. J Leukoc Biol 86:337–348

Bafaro E, Liu Y, Xu Y, Dempski RE (2017) The emerging role of zinc transporters in cellular homeostasis and cancer. Signal Transduct Target Ther 2:e17029

Bao B, Prasad AS, Beck FW, Godmere M (2003) Zinc modulates mRNA levels of cytokines. Am J Physiol Endocrinol Metab 285:E1095–E1102

Berthet J, Rall TW, Sutherland EW (1957) The relationship of epinephrine and glucagon to liver phosphorylase. IV. Effect of epinephrine and glucagon on the reactivation of phosphorylase in liver homogenates. J Biol Chem 224:463–475

Bichet D, Cornet V, Geib S, Carlier E, Volsen S, Hoshi T, Mori Y, De Waard M (2000) The I-II loop of the Ca2+ channel alpha1 subunit contains an endoplasmic reticulum retention signal antagonized by the beta subunit. Neuron 25:177–190

Brazil DP, Hemmings BA (2001) Ten years of protein kinase B signalling: a hard Akt to follow. Trends Biochem Sci 26:657–664

Catterall WA (2000) Structure and regulation of voltage-gated Ca2+ channels. Annu Rev Cell Dev Biol 16:521–555

Cornet V, Bichet D, Sandoz G, Marty I, Brocard J, Bourinet E, Mori Y, Villaz M, De Waard M (2002) Multiple determinants in voltage-dependent P/Q calcium channels control their retention in the endoplasmic reticulum. Eur J Neurosci 16:883–895

Cousins RJ, Liuzzi JP, Lichten LA (2006) Mammalian zinc transport, trafficking, and signals. J Biol Chem 281:24085–24089

Dal Porto JM, Gauld SB, Merrell KT, Mills D, Pugh-Bernard AE, Cambier J (2004) B cell antigen receptor signaling 101. Mol Immunol 41:599–613

Driessen C, Hirv K, Wellinghausen N, Kirchner H, Rink L (1995) Influence of serum on zinc, toxic shock syndrome toxin-1, and lipopolysaccharide-induced production of IFN-gamma and IL-1 beta by human mononuclear cells. J Leukoc Biol 57:904–908

Fernandes G, Nair M, Onoe K, Tanaka T, Floyd R, Good RA (1979) Impairment of cell-mediated immunity functions by dietary zinc deficiency in mice. Proc Natl Acad Sci USA 76:457–461

Fraker PJ, Caruso R, Kierszenbaum F (1982) Alteration of the immune and nutritional status of mice by synergy between zinc deficiency and infection with Trypanosoma cruzi. J Nutr 112:1224–1229

Frederickson CJ, Bush AI (2001) Synaptically released zinc: physiological functions and pathological effects. Biometals 14:353–366

Frederickson CJ, Koh JY, Bush AI (2005) The neurobiology of zinc in health and disease. Nat Rev Neurosci 6:449–462

Fukada T, Yamasaki S, Nishida K, Murakami M, Hirano T (2011) Zinc homeostasis and signaling in health and diseases: zinc signaling. J Biol Inorg Chem 16:1123–1134

Galli SJ (2000) Mast cells and basophils. Curr Opin Hematol 7:32–39

Galli SJ, Tsai M, Piliponsky AM (2008) The development of allergic inflammation. Nature 454:445–454

Gyulkhandanyan AV, Lee SC, Bikopoulos G, Dai F, Wheeler MB (2006) The Zn2+-transporting pathways in pancreatic beta-cells: a role for the L-type voltage-gated Ca2+ channel. J Biol Chem 281:9361–9372

Haase H, Rink L (2009) Functional significance of zinc-related signaling pathways in immune cells. Annu Rev Nutr 29:133–152

Haase H, Ober-Blobaum JL, Engelhardt G, Hebel S, Heit A, Heine H, Rink L (2008) Zinc signals are essential for lipopolysaccharide-induced signal transduction in monocytes. J Immunol 181:6491–6502

Hambidge M (2000) Human zinc deficiency. J Nutr 130:1344S–1349S

Harwood NE, Batista FD (2008) New insights into the early molecular events underlying B cell activation. Immunity 28:609–619

Harwood NE, Batista FD (2010) Early events in B cell activation. Annu Rev Immunol 28:185–210

Hashimoto A, Okada H, Jiang A, Kurosaki M, Greenberg S, Clark EA, Kurosaki T (1998) Involvement of guanosine triphosphatases and phospholipase C-gamma2 in extracellular signal-regulated kinase, c-Jun NH2-terminal kinase, and p38 mitogen-activated protein kinase activation by the B cell antigen receptor. J Exp Med 188:1287–1295

Hayashi K, Ishizuka S, Yokoyama C, Hatae T (2008) Attenuation of interferon-gamma mRNA expression in activated Jurkat T cells by exogenous zinc via down-regulation of the calcium-independent PKC-AP-1 signaling pathway. Life Sci 83:6–11

Hirano T, Murakami M, Fukada T, Nishida K, Yamasaki S, Suzuki T (2008) Roles of zinc and zinc signaling in immunity: zinc as an intracellular signaling molecule. Adv Immunol 97(97):149–176

Ho Y, Samarasinghe R, Knoch ME, Lewis M, Aizenman E, DeFranco DB (2008) Selective inhibition of mitogen-activated protein kinase phosphatases by zinc accounts for extracellular signal-regulated kinase 1/2-dependent oxidative neuronal cell death. Mol Pharmacol 74:1141–1151

Hojyo S, Fukada T (2016) Zinc transporters and signaling in physiology and pathogenesis. Arch Biochem Biophys 611:43–50

Hojyo S, Miyai T, Fujishiro H, Kawamura M, Yasuda T, Hijikata A, Bin BH, Irie T, Tanaka J, Atsumi T et al (2014) Zinc transporter SLC39A10/ZIP10 controls humoral immunity by modulating B-cell receptor signal strength. Proc Natl Acad Sci USA 111:11786–11791

Hou S, Vigeland LE, Zhang G, Xu R, Li M, Heinemann SH, Hoshi T (2010) Zn^{2+} activates large conductance Ca^{2+}-activated K+ channel via an intracellular domain. J Biol Chem 285:6434–6442

Hu H, Bandell M, Petrus MJ, Zhu MX, Patapoutian A (2009) Zinc activates damage-sensing TRPA1 ion channels. Nat Chem Biol 5:183–190

Huang J, Zhang D, Xing W, Ma X, Yin Y, Wei Q, Li G (2008) An approach to assay calcineurin activity and the inhibitory effect of zinc ion. Anal Biochem 375:385–387

Hubbard SR, Bishop WR, Kirschmeier P, George SJ, Cramer SP, Hendrickson WA (1991) Identification and characterization of zinc binding sites in protein kinase C. Science 254:1776–1779

Ibs KH, Rink L (2003) Zinc-altered immune function. J Nutr 133:1452S–1456S

Inoue K, O'Bryant Z, Xiong ZG (2015) Zinc-permeable ion channels: effects on intracellular zinc dynamics and potential physiological/pathophysiological significance. Curr Med Chem 22:1248–1257

Kabu K, Yamasaki S, Kamimura D, Ito Y, Hasegawa A, Sato E, Kitamura H, Nishida K, Hirano T (2006) Zinc is required for FcepsilonRI-mediated mast cell activation. J Immunol 177:1296–1305

Kambe T, Matsunaga M, Takeda TA (2017) Understanding the contribution of zinc transporters in the function of the early secretory pathway. Int J Mol Sci 18:2179

Karasuyama H, Miyake K, Yoshikawa S, Yamanishi Y (2017) Multifaceted roles of basophils in health and disease. J Allergy Clin Immunol 142:370–380

Karasuyama H, Miyake K, Yoshikawa S, Kawano Y, Yamanishi Y (2018) How do basophils contribute to Th2 cell differentiation and allergic responses? Int Immunol 30:391–396

Kawakami T, Ando T, Kimura M, Wilson BS, Kawakami Y (2009) Mast cells in atopic dermatitis. Curr Opin Immunol 21:666–678
Keen CL, Gershwin ME (1990) Zinc deficiency and immune function. Annu Rev Nutr 10:415–431
King MM, Huang CY (1984) The calmodulin-dependent activation and deactivation of the phosphoprotein phosphatase, calcineurin, and the effect of nucleotides, pyrophosphate, and divalent metal ions. Identification of calcineurin as a Zn and Fe metalloenzyme. J Biol Chem 259:8847–8856
Kitamura H, Morikawa H, Kamon H, Iguchi M, Hojyo S, Fukada T, Yamashita S, Kaisho T, Akira S, Murakami M, Hirano T (2006) Toll-like receptor-mediated regulation of zinc homeostasis influences dendritic cell function. Nat Immunol 7:971–977
Klemm S, Gutermuth J, Hultner L, Sparwasser T, Behrendt H, Peschel C, Mak TW, Jakob T, Ruland J (2006) The Bcl10-Malt1 complex segregates Fc epsilon RI-mediated nuclear factor kappa B activation and cytokine production from mast cell degranulation. J Exp Med 203:337–347
Kurosaki T (2011) Regulation of BCR signaling. Mol Immunol 48:1287–1291
Lansdown AB, Mirastschijski U, Stubbs N, Scanlon E, Agren MS (2007) Zinc in wound healing: theoretical, experimental, and clinical aspects. Wound Repair Regen 15:2–16
Lengyel I, Fieuw-Makaroff S, Hall AL, Sim AT, Rostas JA, Dunkley PR (2000) Modulation of the phosphorylation and activity of calcium/calmodulin-dependent protein kinase II by zinc. J Neurochem 75:594–605
Levy S, Beharier O, Etzion Y, Mor M, Buzaglo L, Shaltiel L, Gheber LA, Kahn J, Muslin AJ, Katz A et al (2009) Molecular basis for zinc transporter 1 action as an endogenous inhibitor of L-type calcium channels. J Biol Chem 284:32434–32443
Maret W, Krezel A (2007) Cellular zinc and redox buffering capacity of metallothionein/thionein in health and disease. Mol Med 13:371–375
Metz M, Grimbaldeston MA, Nakae S, Piliponsky AM, Tsai M, Galli SJ (2007) Mast cells in the promotion and limitation of chronic inflammation. Immunol Rev 217:304–328
Miyai T, Hojyo S, Ikawa T, Kawamura M, Irie T, Ogura H, Hijikata A, Bin BH, Yasuda T, Kitamura H et al (2014) Zinc transporter SLC39A10/ZIP10 facilitates antiapoptotic signaling during early B-cell development. Proc Natl Acad Sci USA 111:11780–11785
Murakami M, Hirano T (2008) Intracellular zinc homeostasis and zinc signaling. Cancer Sci 99:1515–1522
Nechushtan H, Leitges M, Cohen C, Kay G, Razin E (2000) Inhibition of degranulation and interleukin-6 production in mast cells derived from mice deficient in protein kinase Cbeta. Blood 95:1752–1757
Nishida K, Uchida R (2018) Role of zinc signaling in the regulation of mast cell-, basophil-, and T cell-mediated allergic responses. J Immunol Res 2018:5749120
Nishida K, Yamasaki S (2014) Zinc signals in cellular functions and disorders. Springer, Tokyo
Nishida K, Hasegawa A, Nakae S, Oboki K, Saito H, Yamasaki S, Hirano T (2009) Zinc transporter Znt5/Slc30a5 is required for the mast cell-mediated delayed-type allergic reaction but not the immediate-type reaction. J Exp Med 206:1351–1364
Oancea E, Teruel MN, Quest AF, Meyer T (1998) Green fluorescent protein (GFP)-tagged cysteine-rich domains from protein kinase C as fluorescent indicators for diacylglycerol signaling in living cells. J Cell Biol 140:485–498
Palacios EH, Weiss A (2004) Function of the Src-family kinases, Lck and Fyn, in T-cell development and activation. Oncogene 23:7990–8000
Prasad AS (1995) Zinc: an overview. Nutrition 11:93–99
Prasad AS (2008) Zinc in human health: effect of zinc on immune cells. Mol Med 14:353–357
Prost AL, Bloc A, Hussy N, Derand R, Vivaudou M (2004) Zinc is both an intracellular and extracellular regulator of KATP channel function. J Physiol 559:157–167
Ramadan O, Qu Y, Wadgaonkar R, Baroudi G, Karnabi E, Chahine M, Boutjdir M (2009) Phosphorylation of the consensus sites of protein kinase A on alpha1D L-type calcium channel. J Biol Chem 284:5042–5049

Richter M, Cantin AM, Beaulieu C, Cloutier A, Larivee P (2003) Zinc chelators inhibit eotaxin, RANTES, and MCP-1 production in stimulated human airway epithelium and fibroblasts. Am J Physiol Lung Cell Mol Physiol 285:L719–L729
Rink L, Gabriel P (2000) Zinc and the immune system. Proc Nutr Soc 59:541–552
Rink L, Kirchner H (2000) Zinc-altered immune function and cytokine production. J Nutr 130:1407S–1411S
Rusnak F, Mertz P (2000) Calcineurin: form and function. Physiol Rev 80:1483–1521
Schroeder JT, MacGlashan DW Jr, Lichtenstein LM (2001) Human basophils: mediator release and cytokine production. Adv Immunol 77:93–122
Seder RA, Paul WE, Dvorak AM, Sharkis SJ, Kagey-Sobotka A, Niv Y, Finkelman FD, Barbieri SA, Galli SJ, Plaut M (1991) Mouse splenic and bone marrow cell populations that express high-affinity Fc epsilon receptors and produce interleukin 4 are highly enriched in basophils. Proc Natl Acad Sci USA 88:2835–2839
Sensi SL, Canzoniero LM, Yu SP, Ying HS, Koh JY, Kerchner GA, Choi DW (1997) Measurement of intracellular free zinc in living cortical neurons: routes of entry. J Neurosci 17:9554–9564
Sensi SL, Paoletti P, Bush AI, Sekler I (2009) Zinc in the physiology and pathology of the CNS. Nat Rev Neurosci 10:780–791
Shumilina E, Lam RS, Wolbing F, Matzner N, Zemtsova IM, Sobiesiak M, Mahmud H, Sausbier U, Biedermann T, Ruth P et al (2008) Blunted IgE-mediated activation of mast cells in mice lacking the Ca2+-activated K+ channel KCa3.1. J Immunol 180:8040–8047
Smart TG, Hosie AM, Miller PS (2004) Zn^{2+} ions: modulators of excitatory and inhibitory synaptic activity. Neuroscientist 10:432–442
Sokol CL, Barton GM, Farr AG, Medzhitov R (2008) A mechanism for the initiation of allergen-induced T helper type 2 responses. Nat Immunol 9:310–318
Takahashi K, Akaishi E, Abe Y, Ishikawa R, Tanaka S, Hosaka K, Kubohara Y (2003) Zinc inhibits calcineurin activity in vitro by competing with nickel. Biochem Biophys Res Commun 307:64–68
Taniguchi M, Fukunaka A, Hagihara M, Watanabe K, Kamino S, Kambe T, Enomoto S, Hiromura M (2013) Essential role of the zinc transporter ZIP9/SLC39A9 in regulating the activations of Akt and Erk in B-cell receptor signaling pathway in DT40 cells. PLoS One 8:e58022
Taylor K, Vichova P, Jordan N, Hiscox S, Hendley R, Nicholson R (2008) ZIP7-mediated intracellular zinc transport contributes to aberrant growth factor signaling in anti-hormone resistant breast cancer cells. Endocrinology 149:4912–4920
Taylor KM, Hiscox S, Nicholson RI, Hogstrand C, Kille P (2012) Protein kinase CK2 triggers cytosolic zinc signaling pathways by phosphorylation of zinc channel ZIP7. Sci Signal 5:ra11
Telford WG, Fraker PJ (1995) Preferential induction of apoptosis in mouse CD4+CD8+ alpha beta TCRloCD3 epsilon lo thymocytes by zinc. J Cell Physiol 164:259–270
Thambiayya K, Wasserloos KJ, Huang Z, Kagan VE, St Croix CM, Pitt BR (2011) LPS-induced decrease in intracellular labile zinc, [Zn]i, contributes to apoptosis in cultured sheep pulmonary artery endothelial cells. Am J Physiol Lung Cell Mol Physiol 300:L624–L632
Thirumoorthy N, Shyam Sunder A, Manisenthil Kumar K, Senthil Kumar M, Ganesh G, Chatterjee M (2011) A review of metallothionein isoforms and their role in pathophysiology. World J Surg Oncol 9:54
Torres-Alanis O, Garza-Ocanas L, Pineyro-Lopez A (1995) Evaluation of urinary mercury excretion after administration of 2,3-dimercapto-1-propane sulfonic acid to occupationally exposed men. J Toxicol Clin Toxicol 33:717–720
Uchida R, Xiang H, Arai H, Kitamura H, Nishida K (2019) L-type calcium channel-mediated zinc wave is involved in the regulation of IL-6 by stimulating non-IgE with LPS and IL-33 in mast cells and dendritic cells. Biol Pharm Bull 42:87–93
Ugajin T, Nishida K, Yamasaki S, Suzuki J, Mita M, Kubo M, Yokozeki H, Hirano T (2015) Zinc-binding metallothioneins are key modulators of IL-4 production by basophils. Mol Immunol 66:180–188

Ugajin T, Shibama S, Nishida K, Yokozeki H (2016) Metallothioneins are required for human basophil interleukin-4 gene induction via FcvarepsilonRIota stimulation. Allergol Int 65:466–468

Vallee BL, Auld DS (1993) Cocatalytic zinc motifs in enzyme catalysis. Proc Natl Acad Sci USA 90:2715–2718

Vennekens R, Olausson J, Meissner M, Bloch W, Mathar I, Philipp SE, Schmitz F, Weissgerber P, Nilius B, Flockerzi V, Freichel M (2007) Increased IgE-dependent mast cell activation and anaphylactic responses in mice lacking the calcium-activated nonselective cation channel TRPM4. Nat Immunol 8:312–320

Vig M, Kinet JP (2009) Calcium signaling in immune cells. Nat Immunol 10:21–27

Voehringer D, Shinkai K, Locksley RM (2004) Type 2 immunity reflects orchestrated recruitment of cells committed to IL-4 production. Immunity 20:267–277

Wellinghausen N, Martin M, Rink L (1997) Zinc inhibits interleukin-1-dependent T cell stimulation. Eur J Immunol 27:2529–2535

Yamasaki S, Sakata-Sogawa K, Hasegawa A, Suzuki T, Kabu K, Sato E, Kurosaki T, Yamashita S, Tokunaga M, Nishida K, Hirano T (2007) Zinc is a novel intracellular second messenger. J Cell Biol 177:637–645

Yamasaki S, Hasegawa A, Hojyo S, Ohashi W, Fukada T, Nishida K, Hirano T (2012) A novel role of the L-type calcium channel alpha1D subunit as a gatekeeper for intracellular zinc signaling: zinc wave. PLoS One 7:e39654

Yamashita S, Miyagi C, Fukada T, Kagara N, Che YS, Hirano T (2004) Zinc transporter LIVI controls epithelial-mesenchymal transition in zebrafish gastrula organizer. Nature 429:298–302

Yu M, Lee WW, Tomar D, Pryshchep S, Czesnikiewicz-Guzik M, Lamar DL, Li G, Singh K, Tian L, Weyand CM, Goronzy JJ (2011) Regulation of T cell receptor signaling by activation-induced zinc influx. J Exp Med 208:775–785

Zhou Z, Wang L, Song Z, Saari JT, McClain CJ, Kang YJ (2004) Abrogation of nuclear factor-kappaB activation is involved in zinc inhibition of lipopolysaccharide-induced tumor necrosis factor-alpha production and liver injury. Am J Pathol 164:1547–1556

Chapter 6
Review: The Role of Zinc Signaling in Reproduction

Atsuko Kageyama, Takafumi Namiki, Junya Ito, and Naomi Kashiwazaki

Abstract Human infertility has become a serious medical and social issue worldwide, especially in developed countries. Numerous assisted reproductive technologies (ARTs) (e.g., artificial insemination [AI], in vitro fertilization [IVF], and intracytoplasmic sperm injection [ICSI]) are widely used to treat infertility, but when embryos produced using these techniques are transferred, the pregnancy outcome is only approx. 20–50%, necessitating further improvement in ARTs. Many factors (including zinc) affect successful fertilization, further embryonic development, and the establishment and maintenance of pregnancy. Zinc signaling has many physiological functions in various types of cells, and it is thus speculated that zinc signaling has an important role in the abovementioned reproductive events, even in germ cells and reproductive tissues. Prior studies demonstrated that zinc deficiency results in a number of anomalies in the formation and maturation of spermatozoa, ovulation, and fertilization. During pregnancy, zinc deficiency causes spontaneous abortion, pregnancy-related toxemia, extended pregnancy or preterm birth, malformations, and retarded growth. The molecular mechanisms underlying zinc signaling in reproduction are not yet clear. In this chapter, we summarize the current knowledge about zinc signaling in reproduction—mainly in mammals—and we discuss recent advances in this field.

Keywords Zinc · Reproduction · Fertilization · Oocytes · Transporters

Authors Atsuko Kageyama, Takafumi Namiki, and Junya Ito have contributed equally to this chapter.

A. Kageyama · T. Namiki
Laboratory of Animal Reproduction, Graduate School of Veterinary Science, Azabu University, Sagamihara, Japan

J. Ito (✉) · N. Kashiwazaki
Laboratory of Animal Reproduction, Graduate School of Veterinary Science, Azabu University, Sagamihara, Japan

School of Veterinary Medicine, Azabu University, Sagamihara, Japan
e-mail: itoj@azabu-u.ac.jp

T. Fukada, T. Kambe (eds.), *Zinc Signaling*,
https://doi.org/10.1007/978-981-15-0557-7_6

6.1 Zinc Signaling in Female Reproduction

6.1.1 Gamete Genesis

Regarding gamete genesis and fertilization, it is well established that a divalent ion, i.e., Ca^{2+}, has the most important role in these events. Most mammalian oocytes are arrested at the prophase stage of the first meiosis, which is also called the germinal vesicle (GV) stage because these oocytes possess a GV. After a surge of luteinizing hormone (LH), GV oocytes resume meiosis and progress to the metaphase stage of the second meiosis (MII) through GV breakdown (GVBD). Oocytes that reach the MII stage are arrested again just before fertilization. The arrest at the first meiosis is regulated by the Mos/mitogen-activated protein kinase (MAPK) pathway (Choi et al. 1996) (Fig. 6.1).

Several research groups revealed that an optimal concentration of zinc in oocytes is required for the arrest through Mos/MAPK pathway called as “cytostatic factor (CSF)” (Hashimoto et al. 1994; Kim et al. 2010; Kong et al. 2012). Treatment with a zinc chelator, i.e., *N, N, N', N'*-tetrakis-(2-pyridylmethyl)-ethylenediamine (TPEN), can induce meiotic resumption from the GV stage (Kong et al. 2012; Wiersma et al. 1998; Taki et al. 2004). Kong et al. (2012) also reported that meiotic resumption induced by TPEN can be inhibited by an injection of Mos short interfering siRNA or treatment with cycloheximide, a potent inhibitor of protein synthesis.

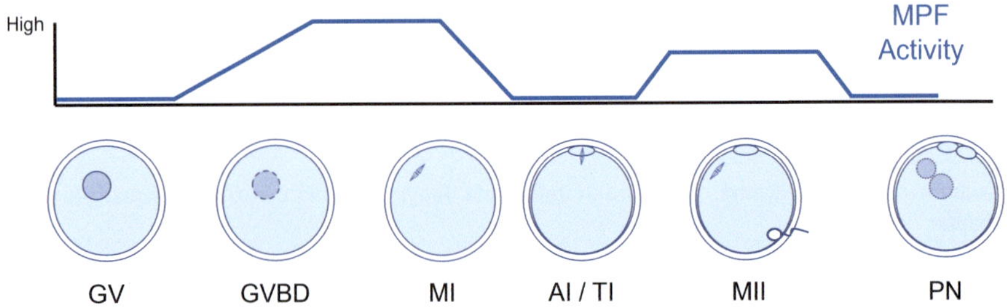

Fig. 6.1 The kinetics of zinc and meiosis-associated molecules during oocyte maturation in mammals. The top panel shows approximate relative zinc levels in GV oocytes, MII, and early embryos. Subsequent panels represent dynamics of important cellular activities over the course of oocyte maturation and fertilization. In this model, when intracellular zinc increases, MPF activity also increases and induces meiotic resumption. In AI/TI stage, when intracellular zinc is above a constant level, the CSF is activated, leading to CSF-mediated MII arrest. After maturation, zinc sparks cause reduction in cellular zinc and could lead to reduced CFS and MPF activity in fertilization. *MI* metaphase I; *AI/TI* anaphase I/telophase I

Taking the above findings together, it appears that at least in immature (GV) oocytes, zinc signaling has an inhibitory role in meiotic resumption via a suppression of CSF activity.

It has been shown that after the resumption of meiosis from the GV stage, large fluxes of zinc occur (Kim et al. 2010). In addition, TPEN treatment throughout the first meiosis resulted in a failure of progression to asymmetrical division (the extrusion of the first polar body) and in arrest at the telophase (Kim et al. 2010). An injection of nondegradable cyclin B1, which is a component of maturation promoting factor (MPF), partially rescued the arrest of zinc-insufficient oocytes, enabling them to enter MII (Bernhardt et al. 2011). MPF is also thought to be another component of CSF, suggesting that zinc may have different roles in each meiotic stage via the regulation of CSF activity.

The results of an in vivo study revealed that oocytes from mice fed a zinc-deficient diet exhibit defective maturation (Tian and Diaz 2012). Abnormal spindle formation was observed in these oocytes, and they were arrested at both MI and telophase I (Tian and Diaz 2012). These observations correspond to the results from oocytes treated with TPEN (Kim et al. 2010; Kong et al. 2012; Bernhardt et al. 2011, 2012). Together these findings indicate that zinc signaling has an essential role in reproduction (female reproduction in particular) and in the meiotic progression of oocytes both in vitro and in vivo.

6.1.2 *Oocyte Activation and Fertilization*

In most mammalian species, oocytes are arrested at the MII stage, prior to ovulation (Jones 2005; Madgwick and Jones 2007; Ducibella and Fissore 2008). "Oocyte activation" entails the progressive initiation of several events, i.e., cortical granule exocytosis, the inactivation of CSF, the resumption of meiosis and exit from MII arrest, the extrusion of the second polar body, and pronucleus (PN) formation (Ducibella and Fissore 2008; Schultz and Kopf 1995; Ito et al. 2011). The orderly completion of these events ensures the initiation of early embryonic development (Ducibella et al. 2002). In all species studied to date, oocyte activation requires a fertilization-associated increase in the intracellular concentration of Ca^{2+} (Stricker 1999). At fertilization, the cytoplasmic Ca^{2+} concentration in mammalian eggs rises and falls repeatedly in a phenomenon known as "Ca^{2+} oscillations" (Kline and Kline 1992; Miyazaki et al. 1993; Jones 1998a; Swann and Parrington 1999; Swann and Yu 2008). It is now acknowledged that upon the release of phospholipase C zeta (PLCζ) into the ooplasm, this enzyme hydrolyzes phosphatidyl inositol 4, 5-biphosphate (PIP_2) to produce inositol 1, 4, 5-trisphosphate (IP_3) and diacyl glycerol (Rebecchi and Pentyala 2000). IP_3 binds IP_3 receptor (IP_3R), which is distributed along the endoplasmic reticulum (ER, the main Ca^{2+} store of the cell), gating the receptor and inducing periodic intracellular Ca^{2+} increases (Berridge 2002; Lee et al. 2006; Ito et al. 2008a, 2010) (Fig. 6.2a).

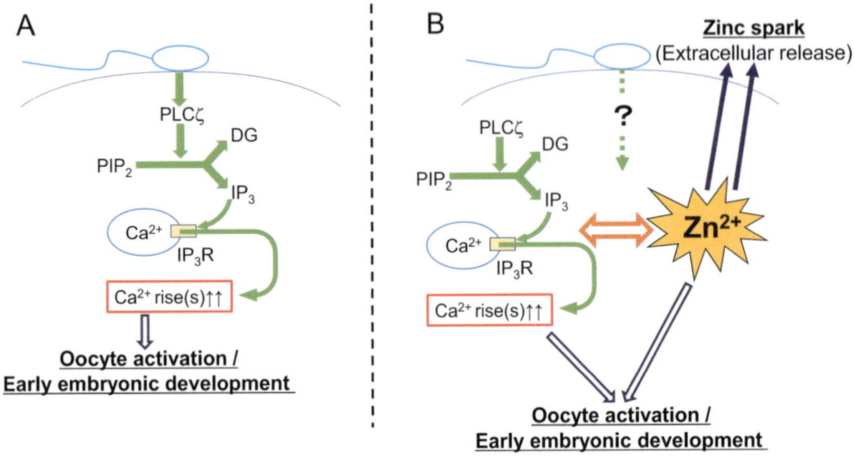

Fig. 6.2 The hypothesis of molecular mechanism during fertilization in mammals. (**a**) Calcium-dependent mechanism. PLCζ is released from the sperm into ooplasm. This enzyme hydrolyzes PIP_2 to produce IP_3 and DG. IP_3 binds IP_3R on the ER, stimulating Ca^{2+} release. (b) Calcium- and zinc-dependent mechanism. Zinc into the oocyte is released to the extracellular environment following fertilization and egg activation. This phenomenon is termed zinc spark. Ca^{2+} oscillation is necessary for mammalian fertilization, and zinc signal is also important for fertilization

Reproductive engineering techniques such as ICSI and somatic cell nuclear transfer (SCNT) have been developed in many mammalian species. ICSI is an important technique for not only experimental and domestic animals but also humans. The SCNT technique can be applied for "rejuvenation of the ovum" and for "avoiding of mitochondrial disease," which will become very important technologies in the future. Without the use of sperm, artificial stimuli are essential for oocyte activation when current assisted reproductive technologies (ARTs)—particularly SCNT and round spermatid injection (ROSI)—are applied (Kishigami and Wakayama 2007; Ogura et al. 1998; Wakayama et al. 1998).

To date, many artificial activation methods, including ethanol, electrical stimulation, and calcium ionophore, have been developed and widely applied for SCNT or ROSI-derived oocyte activation (Marcus 1990). However, most of these agents merely cause a transient increase in Ca^{2+}, and they do not induce repetitive intracellular Ca^{2+} oscillations similar to those that occur during normal fertilization. The oocyte activation of many invertebrate and vertebrate animals is induced by a transient increase in the number of calcium ions (Whitaker 2006), but the oocyte activation of mammalian species causes a repetitive Ca^{2+} increase (i.e., calcium oscillation) (Miyazaki et al. 1993; Swann and Parrington 1999; Kline 2000; Jones 1998b). These Ca^{2+} oscillations play a key role in the triggering of important events of oocyte activation such as cortical granule exocytosis, the resumption of meiosis, and exit from MII arrest. The low production of offspring with the use of SCNT or ROSI may be caused by calcium ion dynamics that differ from those of fertilization. It is necessary to clarify the fertilization mechanisms at the molecular level in order to solve this problem.

Although it has been known for several decades that Ca^{2+} is the essential trigger of egg activation, the mechanism by which a sperm causes the intracellular Ca^{2+} release remains to be established. In addition, the detailed mechanism underlying the Ca^{2+} release at fertilization in vertebrates remains to be fully clarified, especially regarding the role of the interaction between the sperm and oocyte in this process (Parrington 2001; Parrington et al. 2007). Three major hypotheses have been advanced to explain how sperm induce the Ca^{2+} release at fertilization (Ito et al. 2011; Whitaker 2006; Runft et al. 2002; Kurokawa et al. 2004; Swann et al. 2006). The membrane receptor hypothesis proposed that an oocyte surface receptor is engaged by a sperm ligand, triggering the production of a signaling pathway that triggers the Ca^{2+} release via an oocyte PLC (Swann and Parrington 1999) (Fig. 6.3a). The Ca^{2+} conduit hypothesis proposed that extracellular Ca^{2+} flows into the oocyte via the sperm during gamete fusion (Fig. 6.3b). Since both of these hypotheses rely on a sperm-oocyte interaction for the induction of Ca^{2+} responses, their plausibility was weakened by the finding that ICSI, which bypasses the sperm-oocyte interaction, is capable of inducing fertilization-like responses.

The sperm factor hypothesis can explain this finding, and there is direct experimental support for the hypothesis: an injection of mammalian sperm extracts was shown to induce Ca^{2+} responses (Fig. 6.3c) and oocyte activation events akin to those induced by natural fertilization. Therefore, the sperm factor hypothesis is widely supported at the present time.

Investigation of the "sperm factor" activity in the abovementioned mammalian sperm extracts revealed that this activity appeared to be based on a sperm-specific phospholipase C (PLC) with distinctive properties such as enhanced Ca^{2+} sensitivity compared to known PLC isoforms (Jones 1998a; Parrington et al. 2000; Rice et al. 2000; Wu et al. 2001). In 2002, the sperm-specific PLC named phospholipase C zeta (PLCζ) was discovered, and it appeared to have all the expected properties of an endogenous agent of oocyte activation (Saunders et al. 2002). An important

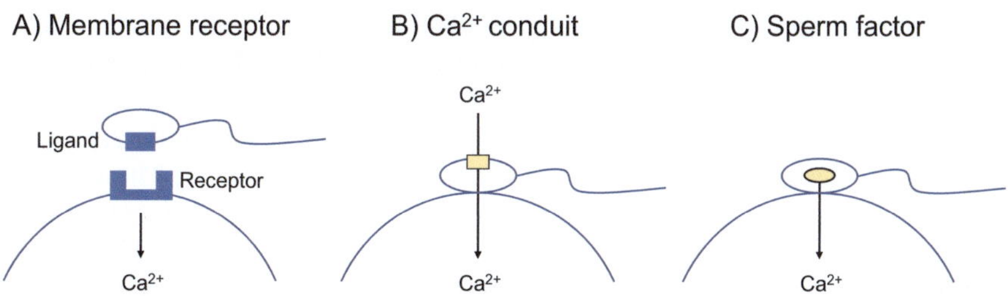

Fig. 6.3 Mechanism of intracellular Ca^{2+} increase in mammalian fertilization. (**a**) Membrane receptor hypothesis. A sperm ligand binds an oocyte surface receptor and triggers the Ca^{2+} release. (**b**) Ca^{2+} conduit hypothesis. The extracellular Ca^{2+} flows into the oocyte via the sperm during gamete fusion. (**c**) Sperm factor hypothesis. Sperm extracts are injected into the oocyte and induced Ca^{2+} response

structural difference between PLCζ and other PLC isoforms is that PLCζ is smaller than all of the previously identified isoforms and lacks a PH domain.

PLCζ was subsequently identified in many mammalian species (Ito et al. 2008b; Yoneda et al. 2006; Cox et al. 2002; Ross et al. 2008) and other animal species (Coward et al. 2005; Mizushima et al. 2009; Coward et al. 2011), and it was demonstrated that a microinjection of PLCζ complementary cRNA elicits Ca^{2+} oscillations in mouse oocytes that are equivalent to those seen during natural fertilization in mice. PLCζ is localized to acrosomal and post-acrosomal regions (Young et al. 2009). Nakai et al. showed that in pig sperm, PLCζ is also present in the sperm tail in addition to its expected localization in the sperm head (Nakai et al. 2011). The localization of PLCζ in sperm may differ according to the animal species. Taking the above-cited observations together, the general consensus is that PLCζ is probably the trigger of egg activation and embryo development, at least in mammals and probably in many other vertebrate species.

6.1.3 *The Role of Zinc Signaling in Fertilization: The Discovery of the "Zinc Sparks"*

Although the discussions of the mechanisms of fertilization have focused on calcium ion, it has been shown that meiotic resumption can be triggered artificially with the use of TPEN, and the meiotic resumption can be inhibited by overloading the oocyte with zinc ionophores (Bernhardt et al. 2012; Suzuki et al. 2010). In mice, TPEN treatment was also shown to be sufficient to activate MII-arrested oocytes injected with "inactivated" sperm heads in the absence of intracellular calcium oscillations, resulting in live births after embryo transfer (Suzuki et al. 2010). Thus, full-term development seems not to be dependent on Ca^{2+} release during MII exit, and Ca^{2+} oscillations do not assure an altered rate of full-term development compared to a single Ca^{2+} rise.

Whether the removal of zinc is a physiological phenomenon in fertilization remains unknown. The fertilization of a mature, zinc-enriched mouse egg triggers the ejection of zinc into the extracellular milieu in a series of coordinated events termed "zinc sparks" (Kim et al. 2011; Que et al. 2015) (Fig. 6.4). Zinc sparks were also observed in human eggs and bovine eggs following fertilization and egg activation (Duncan et al. 2016; Que et al. 2019). These reports establish that the zinc spark is a highly conserved event, at least in several mammalian species. These results established zinc as a crucial regulator of meiosis throughout the entirety of oocyte maturation (Kong et al. 2012; Bernhardt et al. 2011). Bernhardt et al. (2012) proposed a model in which zinc exerts a concentration-dependent regulation of meiosis through the CSF component Emi2, a zinc-binding protein, and zinc sparks could ensure the rapid and efficient inactivation of Emi2.

However, it is known that Emi2 inactivation results from Ca^{2+}-dependent mechanism (Hansen et al. 2006), and zinc sparking does not occur when Ca^{2+} is chelated

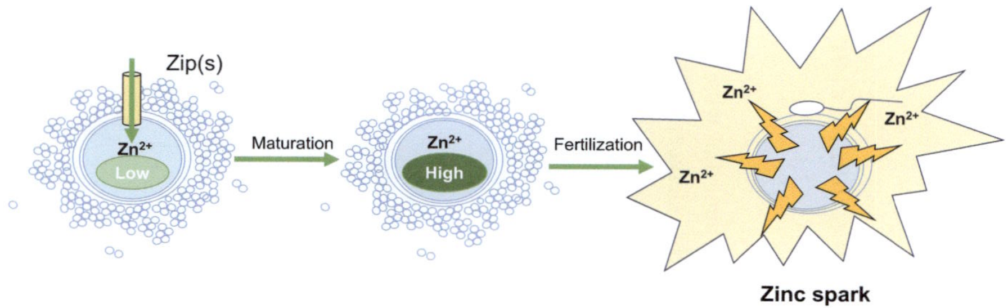

Fig. 6.4 Model of zinc influx and efflux during oocyte maturation and fertilization in mammals. This schematic diagram shows the zinc fluxes regulated by Zip(s). Zinc flow into low zinc level oocyte, and the oocytes resumed meiosis. Moreover, zinc level becomes higher in oocyte. Following fertilization, zinc-enriched oocyte induced zinc spark

(Duncan et al. 2016). The significance of the zinc spark is not clear. Zinc ions must accumulate in the oocyte in order to spark. The above-cited investigations showed that zinc sparks could not be induced in immature oocytes (such as the GV) that exist in the mouse ovary, and the acute accumulation of zinc during meiotic maturation is important (Kim et al. 2010). Zinc spark profiles revealed that zygotes that developed into blastocysts released more zinc than those that failed to develop, and the rate of embryo development and total cell number were higher than those of blastocysts (Zhang et al. 2016). Therefore, it is suggested that the amount of zinc ions at the zinc sparks may serve as an early biomarker of zygote quality in mouse models.

6.1.4 The Expression and Function of Zinc Transporters in Mammalian Oocytes

In earlier studies, mice fed a zinc-deficient diet were used (Matsuda and Watanabe 2003; Kim et al. 2009; Tian et al. 2014; Ueda et al. 2014). The influx and efflux of zinc ions are regulated by the zinc transporters ZIP and ZnT, respectively (Gaither and Eide 2001; Palmiter and Huang 2004; Eide 2006) (Fig. 6.5a). In mammals, a total of 23 zinc transporters (i.e., Zip1–14 and ZnT1–9) have been identified (Dempski 2012; Kambe 2012; Kambe et al. 2006). These zinc transporters are

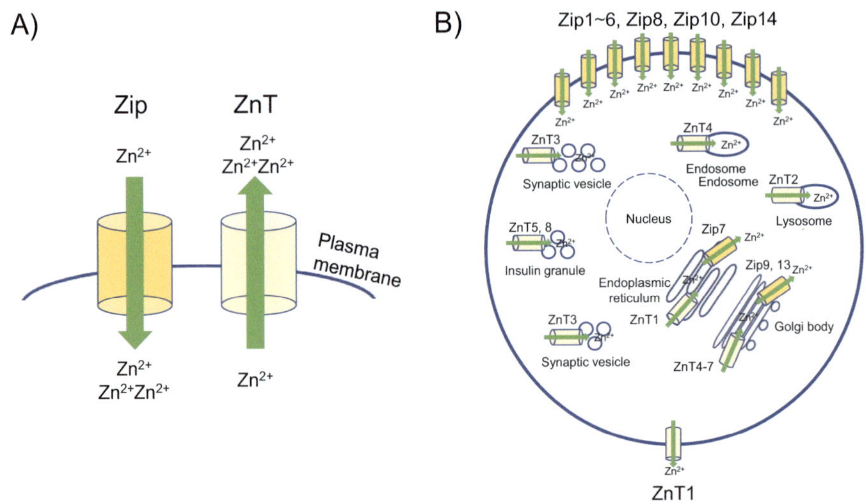

Fig. 6.5 Direction of zinc transport and subcellular localization of Zip and ZnT transporters. (**a**) Zinc transporters are localized to various intracellular compartments and the plasma membrane. Green arrows indicate the direction of zinc by Zip and ZnT, respectively. Zip function in zinc effluxes into the cytoplasm, while ZnT moves zinc in the opposite direction. (**b**) The localization of Zip and ZnT transporters is shown referring to the information available to many reports

expressed at various locations in the cells and tissues (Kambe 2011; Kambe et al. 2015) (Fig. 6.5b). In oocytes, it has been proposed that the influx of zinc is regulated by ZIPs (Fig. 6.4). Kong et al. reported that Zip6 and Zip10 were highly expressed in mouse oocytes (Kong et al. 2014) (Fig. 6.4b). We also confirmed the expression of both Zip6 and Zip10 mRNA by reverse transcription polymerase chain reaction (RT-PCR) and the presence of Zip10 protein by immunofluorescent staining (Kageyama et al. unpublished data).

Consequently, a determination of the functions of Zip6 and Zip10 in mouse oocytes could clarify the mechanisms of zinc signaling, including zinc sparks. Mice with a specific gene deletion (i.e., knockout [KO]) are useful for evaluating the function of specific genes, and we thus attempted to produce oocyte-specific Zip10 KO mice using a Cre/loxP strategy. Oocyte-specific Zip10 KO mice ($Zp3^{Cre/+}Zip10^{flox/flox}$: $Zip10^{d/d}$) were generated and used for further studies. We observed that the fertility of the $Zip10^{d/d}$ mice was significantly lower than that of the controls ($Zp3^{+/+}Zip10^{flox/flox}$: $Zip10^{f/f}$) (Kageyama et al. unpubl. data), suggesting that Zip10 has an essential role in the fertilization of mouse oocytes.

When oocytes derived from $Zip10^{d/d}$ mice were used for IVF, the fertilization (i.e., pronucleus formation) rate was significantly decreased compared to that of $Zip10^{f/f}$ mice (Kageyama et al. unpubl.). It remains unclear whether zinc sparks can occur in these Zip10-deficient oocytes. However, our findings demonstrated for the first time the importance of ZIP in mammalian fertilization. Further studies are required to clarify the molecular mechanisms underlying zinc signaling.

Reproductive engineering techniques such as ICSI, ROSI, and SCNT have been established in many mammalian species. ICSI and ROSI have been commonly used as ARTs. Artificial oocyte activation is necessary when using ICSI or ROSI because round spermatids have low oocyte activation ability. The methods that use Ca^{2+} ionophore (Funahashi et al. 1994; Ito and Shimada 2005), IP_3 injection (Amano et al. 2004), electrical pulse (Wang et al. 1998), ethanol (Kishikawa et al. 1999), etc. are considered artificial activation treatments (Fig. 6.6). As noted above, these treatments induce only a transient Ca^{2+} increase. Treatment with strontium chloride ($SrCl_2$) is an effective method for MII oocyte activation in mouse and rat models, as these oocytes treated by $SrCl_2$ show a repetitive Ca^{2+} increase (Kishigami and Wakayama 2007; Ogura et al. 1998; Wakayama et al. 1998; Sano et al. 2009). A reduction in the Ca^{2+} oscillation frequency causes abnormal gene expression and affects the weight variation in the offspring when these mouse embryos are transferred (Ozil et al. 2006).

It has not been established whether Ca^{2+} oscillation is essential for mammalian fertilization, and the observation of zinc sparks revealed that a zinc signal is also important for fertilization (Fig. 6.1b). Since it is still unclear how PLCζ regulates zinc sparking via calcium oscillations during fertilization in mammals, further experiments are required. The elucidation of the fertilization mechanisms of both Ca^{2+} oscillation and zinc spark can be expected to not only show the biological significance of these phenomena but to also be applied to reproductive technologies.

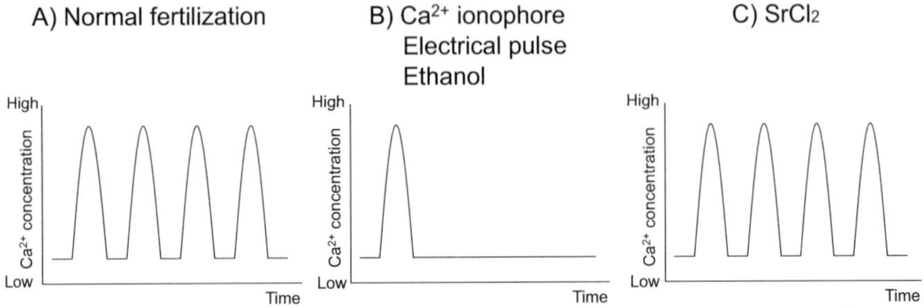

Fig. 6.6 Changes in intracellular Ca^{2+} in mammalian oocytes following sperm or artificial activation treatments show that the frequency of sperm- and artificial activation-induced Ca^{2+} oscillations appear very different. (**a**) Ca^{2+} oscillation induced by sperm, namely, normal fertilization, shows a repetitive increase. (**b**) Ca^{2+} oscillation induced by Ca^{2+} ionophore, electrical pulse, or ethanol. These treatments show only a transient Ca^{2+} increase. (**c**) Ca^{2+} oscillation induced by $SrCl_2$. These oocytes treated by $SrCl_2$ show repetitive intracellular Ca^{2+} oscillations similar to those that occur during normal fertilization

6.1.5 Pregnancy: Embryo Implantation, Decidualization, and Placentation

After fertilization and subsequent embryonic development, implantation of the embryo and the maintenance of pregnancy are regulated by two steroid hormones: estrogen (E2) and progesterone (P4) in most mammalian species. Although the menstrual cycles of humans and mice differ (28–30 days and 4 days, respectively), similar phenomena occur in these species. In humans, the pregestational phase lasts for the first 7 days after ovulation, and then endometrial receptivity for the embryos is achieved approx. 7–10 days after ovulation. The uterus then proceeds in the refractory phase for the rest of the cycle until menstruation continues (Namiki et al. 2018).

In humans, the 28–30-daymenstrual cycle begins with menstruation. The proliferative phase is under the influence of elevated E2 levels from the growing ovarian follicles, leading to the proliferation of epithelial, stromal, and vascular endothelium for the regeneration of the endometrium. Many developmental glands take on a serpentine form during the late growth phase. Elevated pituitary gonadotropin levels at this stage lead to follicle formation and the selection of dominant follicles. During the middle cycle, surges of gonadotropins (i.e., follicle-stimulating hormone [FSH] and LH) lead to ovulation on day 14. The luteal phase is characterized by endometrial thickening and the formation of the corpus luteum (CL) from ruptured follicles and subsequent P4 secretion in preparation for embryo implantation. Glandular cells are secreted with stromal cell differentiation with endometrial edema in preparation for implantation (Cha et al. 2012). Elevated E2 levels superimposed on P4 define a window of sensitivity during the mid-luteal phase that is conducive to embryo implantation and the subsequent pregnancy. In the absence of viable embryos, the receptive window spontaneously shifts to the refractory period and the CL is degraded, leading to hormone withdrawal and menstruation, thus resetting the cycle. Conversely, a transplanted blastocyst secretes chorionic gonadotropin to maintain the CL, thereby supporting a pregnancy.

In mice, uterine cell types respond differently to changing E2 and P4 levels. The day when the female mouse's vaginal plug is confirmed is the 1st day of pregnancy (D1). On D1, uterine epithelial cells proliferate under the influence of E2 before ovulation. On D4, when the elevated P4 secretion from the newly formed CL overlaps with the preimplantation ovarian E2 secretion, stromal cells show extensive proliferation with epithelial cell differentiation and glandular secretion. P4 is an absolute requirement for implantation in all species studied to date. Interestingly, ovarian estrogens are not important for implantation in many species—including hamsters, guinea pigs, rabbits, and pigs—in which embryonic estrogens are thought to play a role (Dey et al. 2004). E2 levels are elevated before the receptive phase (Thomas et al. 1973; Ghosh et al. 1994), but it remains unclear whether this midluteal estrogen is required for implantation in humans or nonhuman primates (Smitz et al. 1993; Rao et al. 2007; Wang and Dey 2006). Nevertheless, the abovementioned findings appear to support the estrogen requirement even in primates.

In mice, uterine sensitivity to implantation is divided into three phases: prereceptive (D1–D3), receptive (D4–D5), and refractory (beyond the afternoon of D5) (Fig. 6.7). Mouse embryos can implant in the uterus only during the receptive phase. The transition to the receptive phase requires a transient rise in E2 around noon on D4 under a high concentration of P4. When the receptive phase is complete, the uterus spontaneously shifts to the refractory period when the uterine environment becomes hostile to blastocyst survival. The progression through these phases is terminated and requires P4 withdrawal to reset the cycle from the refractory state (Namiki et al. 2018; Hirota 2019). The mouse uterus consists mainly of three types of cells, stromal cells (SCs), luminal epithelium (LE), and glandular epithelium (GE), and their physiological and functional interaction enables embryo implantation and pregnancy maintenance (Hantak et al. 2014). Embryo implantation (D4.5), decidualization (D5–), and placentation (D8) occur sequentially in the mouse uterus, and understanding the morphological and physiological changes in utero related to these steps is important for improving the implantation and pregnancy rate. When these steps' morphological and physiological changes occur in the uterine tissue, the embryo's development is normal, but when the changes do not occur, embryo implantation failure, abortion, delayed fetal development, and/or premature delivery may occur (Namiki et al. 2018).

It is thus important to elucidate the molecular mechanisms of phenomena that occur in the embryo and in utero following E2 and P4 signals in the uterus in order to improve the developmental rate of embryos. The expressions of the cytokine leukemia inhibitory factor (LIF) and its receptor (LIFR) in the endometrium are significantly reduced in the epithelial cells of infertile women (Margioula-Siarkou et al. 2016, 2017). In addition, E2 injection into the ovariectomized mice at D5 induced LIF expression and enabled the implantation of embryos into E2-primed uterus (Chen et al. 2000). In contrast, female mice with knockout of LIF gene ($Lif^{-/-}$) showed an infertility phenotype due to embryo implantation failure (Stewart et al. 1992). Thus, in addition to human clinical data, the results of experiments using knockout mice may contribute to our understanding of additional causes of infertility.

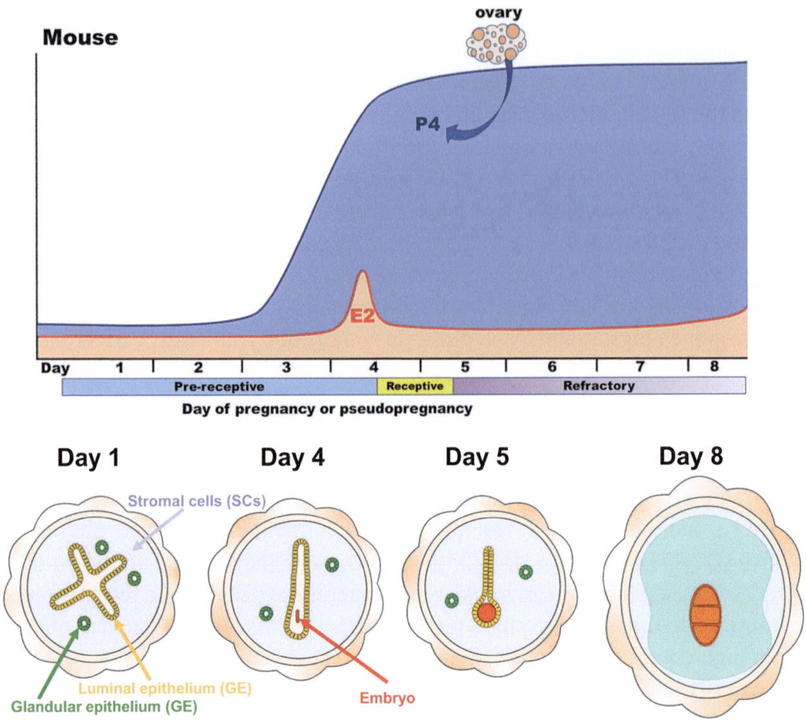

Fig. 6.7 Changes of hormonal levels and endometrium during early pregnancy in mice. (**a**) Estrogen (E2) and progesterone (P4) regulate the implantation window in mice, in which uterine sensitivity to accept the embryo is composed of "perceptive" (Days 1–3, the day we confirmed the vaginal plug was defined as Day 1.), "receptive" (Day 4), and "refractory" (Day 5 afternoon). On Day 4, an increase in the estrogen level is observed prior to the receptive stage. (**b**) Morphological changes of the uterus from Days 1 to 8 during early pregnancy in mice

6.1.6 Zinc and Pregnancy

The requirements for zinc during pregnancy and during lactation have been estimated from the zinc content of tissues produced during pregnancy and the zinc content of milk secreted during lactation (Wilson et al. 2016). The required amount of additional zinc for human pregnancy, estimated from the zinc concentration and weight of the tissue obtained, is approx. 100 mg (1540 mmol). The additional daily requirement during the second half of pregnancy, when fetal growth is fastest, is approx. 0.6 mg/day (9.2 mmol/day) (Chaffee and King 2012). In animal models of pregnancy, zinc deficiency was shown to limit fetal growth and, in severe cases, to cause teratogenic abnormalities (King 2000). Swanson and King (Chaffee and King 2012) revealed that low maternal serum Zn levels were associated with pregnancy hypertension and birth defects. Zinc deficiency during pregnancy can cause several abnormalities, including spontaneous abortion, pregnancy-related poisoning, long-term pregnancy, premature infants, malformations, and delayed growth (Favier 1992a). Decreased levels of zinc are also associated with blood dilution, decreased zinc binding protein, hormonal changes during pregnancy (Jameson 1976), and a decreased active transport of zinc from the mother to the fetus (Tamura et al. 1996). Pregnant and lactating women need an increased zinc intake because of normal embryogenesis, fetal growth, and lactation (Donangelo et al. 2005). In humans, the maternal serum zinc concentration during pregnancy decreases to approx. 50% of the normal level. In addition, the maternal zinc status (i.e., the maternal levels of serum zinc, plasma zinc, and placental zinc) is correlated with birth weight, and zinc transfer to the fetus is also dependent on the maternal serum zinc concentration (Henkel et al. 2003; Aitken and Clarkson 1987; Plante et al. 1994). The concentration of zinc in human amniotic fluid was reported to be proportional to newborns' weight (Irvine 1996). Low plasma zinc levels have been reported to be associated with pregnancy-related diseases such as prolonged delivery, high blood pressure, postpartum hemorrhage, spontaneous abortion, and congenital malformations (Gibson 1994). Interestingly, pregnant women taking 0.3 mg zinc/kg/day as zinc sulfate during the last trimester of pregnancy did not develop any reproductive effects (maternal weight gain, blood pressure, postpartum hemorrhage, or infection) (Mahomed et al. 1989).

It is clear that in mammals, the effects of zinc deficiency during pregnancy differ based on the degree, duration, and timing of the deficiency (Table 6.1). In rats, zinc

Table 6.1 Maternal and fetal effects due to zinc deficiency during pregnancy in different mammalian species

Species	Period of zinc deficiency		References
Mouse	**Zinc-deficient diet during 4–5 days before ovulation**		Tian et al. (2014), Beach et al. (1980), and Wilson et al. (2017)
	• High incidence of pregnancy loss (small litter size)		
	• Decrease pups weights and length		
	• Aberrant placentation (fetal placenta has become smaller)		
Rat	**Preweaning-maturation period**	**During pregnancy**	Hurley and Swenerton (1966), Gallaher and Hurley (1980), Caldwell et al. (1970), Oberleas et al. (1971), and Apgar (1970)
	• Growth suppression	• Premature fetus	
	• Aberrant estrus cycle	• Congenital abnormality	
	• Behavior disorder (no mating)	• Postpartum hemorrhage	
Human	**Decreased concentration of zinc in plasma during pregnancy**		Donangelo et al. (2005), Gibson (1994), and Mahomed et al. (1989)
	• Prolonged labor		
	• Hypertension		
	• Postpartum hemorrhage		
	• Spontaneous abortion		
	• Congenital abnormality		

deficiency during the period from weaning to maturation suppresses growth and results in extreme abnormalities in the estrus cycle and incomplete infertility (Hurley and Swenerton 1966). During the postnatal mating pregnancy period of rats, half of the fetuses are immature even at pregnancy term, and most exhibit extreme congenital malformation. Among these malformations are relatively mild ones such as cleft palate, mandibular defects, and abnormalities of the limbs; the serious malformations include those of the brain. The most common is tail malformation (72%), followed by syndactyly (64%), embryo malformation (54%), and brain malformations (47%) such as hydrocephalus, anencephaly, and extra-encephalopathy.

The incidence of malformations due to zinc deficiency during pregnancy is quite high, reaching its highest point when zinc deficiency lasts throughout the entire gestational period (0–21 days), but zinc deficiency at a more limited gestational age results in a more severe malformation in the middle period (6–12 days) than the early period (0–6 days) (Beach et al. 1980; Gallaher and Hurley 1980; Mutch and Hurley 1980; Keen and Hurley 1980). In addition, fetal absorption and malformations at the implantation site of the placenta are observed in most zinc-deficient female rats after pregnancy. These abnormalities are also the main cause of zinc deficiency in fetal tissues. Moreover, it is extremely difficult to deliver the offspring of zinc-deficient mother rats, and the births are associated with massive bleeding; there is also a significant influence of zinc deficiency on the aspects of birthing such as the birthing speed (Caldwell et al. 1970; Oberleas et al. 1971).

In zinc-deficient female rats, disordered endocrine function (e.g., impaired release of pituitary gonadotropin) is often observed, as are menstrual cycle abnormalities, amenorrhea, and follicle developmental disorder (Apgar 1970). In mice, it has been revealed that zinc deficiency during pregnancy has various effects on the pregnancy process and the fetus. Female mice fed a zinc-deficient diet (ZDD) for 4–5 days before ovulation showed an increased rate of fetal loss on D10.5 compared to the controls. In the ZDD group, there was a 46% increase in the proportion of implantation sites containing either no embryo or an embryo without a heartbeat. Female mice fed a ZDD for 4–5 days before ovulation showed decreased embryo length and weight values and a high incidence of pregnancy loss compared to the controls. The placental weight of the mice fed the ZDD was reduced by 37% compared to control mice (Tian et al. 2014). The decrease in placental weight was due to a 31% decrease in the area of the labyrinth and spongiotrophoblast layers (Tian et al. 2014; Wilson et al. 2017). These results strongly suggest the importance of zinc-dependent signal during pregnancy in mammals although molecular mechanism still remains unknown.

In humans, an insufficient intake of zinc not only results in poor genital development but can also delay men's growth. Zinc deficiency in women during pregnancy may result in an infant whose growth has been delayed (Chaffee and King 2012). Zinc deficiency in women may be associated with impaired synthesis/secretion of FSH and LH, abnormal ovarian development, disruption of the estrus cycle, prolonged gestation, teratogenicity, stillbirth, labor complexity, preeclampsia, toxemia, and more; it can lead to complications and can affect the newborn's birth weight (Bedwal and Bahuguna 1994).

Goldenberg et al. reported that daily zinc supplementation in women with very low plasma zinc levels in early pregnancy is associated with increased birth weight and head circumference in infants (Goldenberg and Tamura 1996). The zinc levels during prenatal care and parturition and the infant birth weights in the intrauterine growth-retarded group in that study were significantly lower than in those of a group of mothers with children showing normal growth. Goldenberg et al. (Roungsipragarn et al. 1999) further proposed that the measurement of maternal plasma zinc levels during late pregnancy could be used to identify mothers at higher risk of having a fetus with intrauterine growth retardation, indicating the importance of zinc in fetal development.

In their study conducted in India, Garg et al. (1993) reported that zinc supplementation significantly improved fetal growth. The effect on birth weight was better whenever the supplement was started in the first trimester than when it was started in the third trimester. In addition, the gestational age of infants increased more as the duration of zinc supplementation increased. The gestation period of the infants whose mothers received supplements from the first trimester averaged 39.4 weeks, and that of the infants whose mothers who received supplements only from the third trimester was 38.8 weeks. The femoral stem length is better among fetuses whose mothers have taken zinc supplements, and the increase tends to rise with gestational age, suggesting the importance of maternal zinc status for human fetal bone growth (Merialdi et al. 2004). There have been few comparative studies of the effects of zinc deficiency on female reproduction in other animals, but it was reported that zinc-deficient female chickens had low rates of hatching of fertilized eggs and high rates of severe growth defects and egg embryo mortality (Blamberg et al. 1960). The size of fetuses of zinc-deficient pigs was smaller than normal, and in rabbits, it was reported that during pregnancy, the zinc concentration in endometrium becomes extremely high (Lutwak-Mann and McIntosh 1969) and the transfer rate of zinc to the placenta becomes extremely high (McIntosh and Lutwak-Mann 1972).

In recent years, a relationship between zinc deficiency during pregnancy and various diseases has been reported. Alcoholism is one of the diseases that cause disorders of zinc metabolism. Alcohol-dependent changes in maternal zinc metabolism are related to the teratogenicity of alcohol. Flynn et al. were the first to report that plasma zinc levels are lower in pregnant alcoholics than in control pregnant women, and they revealed an inverse relationship between plasma zinc levels and blood zinc levels in these women (Flynn et al. 1981). In a rat model, when the pregnant dams were given diets that were marginal in zinc and alcohol, the teratogenicity was amplified (Taubeneck et al. 1994). Excessive alcohol intake may be associated with an increased risk of low maternal plasma zinc levels (Keen et al. 2010). However, supplementation with a large amount of zinc alone presents a risk of affecting the zinc-copper interaction for absorption at the intestinal level. Studies of laboratory animals have indicated that the consumption of a high-zinc diet may be associated with the induction of fetal copper deficiency (Uriu-Adams et al. 2010). In particular, in pregnant women, supplementation studies with single nutrients such as zinc may be inadequate because multiple essential micronutrients are required (Mello-Neto et al. 2013).

Based on the above findings, it is clear that zinc is an essential micronutrient during pregnancy, but its appropriate amount and precise functions are still unclear. The influx and efflux of zinc are regulated by zinc transporters, suggesting that ZIP and ZnT have essential roles in pregnancy in mammals. Future collaborations in clinical and basic research that expand our understanding of the function of zinc during pregnancy may help improve the pregnancy rate among humans as well as experimental animals and livestock.

6.2 Zinc Signaling in Male Reproduction

It has been believed that the optimal concentration of zinc is important for sex differentiation, spermatogenesis, and sperm functions in mammals. Older studies in rodents and humans showed that milder zinc deficiency causes oligospermic sterility (Al-Bader et al. 1999; Madding et al. 1986). Oligospermia developed in four of five volunteers on a restricted 30-week diet providing only 4 mg of zinc per day (Abbasi et al. 1980). This phenotype was rescued by zinc supplementation.

Male hypogonadism may stem from at least two mechanisms, including an impairment in the action of the Mullerian inhibiting factor that enables testicle differentiation (Favier 1992b). This regeneration was inhibited by zinc (Budzik et al. 1982). The other zinc-dependent mechanism is the synthesis of testosterone and its activity. In rodents and humans (Castro-Magana et al. 1981; Joven et al. 1985; Prasad 1979), zinc deficiency has been found to be associated with low testosterone levels (McClain et al. 1984). In addition, Favier et al. (Budzik et al. 1982) suggested that testosterone can modify zinc metabolism: in rats, testosterone treatment led to an increase in the intestinal uptake of zinc; indeed, the methyltestosterone treatment of small-sized children increases their zincemia (Castro-Magana et al. 1981). Hesketh (1982) used electron microscopy and observed that Leydig cells (which are a source of testosterone) in zinc-deficient rat testicles were smaller and showed endoplasmic reticulum abnormalities. It remains unknown whether zinc transporters have an essential role in these functions.

However, Chu et al. (2016) reported that zinc transporter 7 (ZnT7) may play an important role in the regulation of testosterone synthesis by modulating steroidogenic enzymes, and it may thus represent a therapeutic target in testosterone deficiency. Very recently, it was reported that the zinc transporter ZIPT-7.1, which is conserved as ZIP7 in mammals, regulates sperm activation in nematodes (Zhao et al. 2018). Another study showed that the expressions of ZIP5 and ZIP14 were increased in pachytene spermatocytes (Downey et al. 2016). Our unpublished data also suggest that the expression of some ZIPs is dramatically changed in a stage-dependent manner in male germ cells. These results strongly suggest that zinc transporters in male germ cells and other connected cells (Leydig cells and Sertoli cells) have essential roles in sperm function. Further studies using zinc transporter-deficient mice will help clarify the molecular mechanisms of zinc signaling in male reproduction.

Acknowledgments We apologize for many researchers whose work cannot be cited due to space limitation.

Funding Information This study was partially supported by Grants-in-Aid for Scientific Research from the Japan Society for the Promotion of Science (JSPS) (KAKENHI, 15H04584 to J.I.) and by International Exchange Committee of Azabu University to J.I. This study was also supported by Ministry of Education, Culture, Sports, Science and Technology (MEXT)-Supported Program for the Private University Research Branding Project, (2016–2019) to J.I.

References

Abbasi AA, Prasad AS, Rabbani P et al (1980) Experimental zinc deficiency in man. Effect on testicular function. J Lab Clin Med 96(3):544–550. doi:0022-2143(80)90159-6 [pii]

Aitken RJ, Clarkson JS (1987) Cellular basis of defective sperm function and its association with the genesis of reactive oxygen species by human spermatozoa. J Reprod Fertil 81(2):459–469

Al-Bader A, Omu AE, Dashti H (1999) Chronic cadmium toxicity to sperm of heavy cigarette smokers: immunomodulation by zinc. Arch Androl 43(2):135–140

Amano T, Mori T, Watanabe T (2004) Activation and development of porcine oocytes matured in vitro following injection of inositol 1,4,5-trisphosphate. Anim Reprod Sci 80(1–2):101–112. https://doi.org/10.1016/S0378-4320(03)00115-5

Apgar J (1970) Effect of zinc deficiency on maintenance of pregnancy in the rat. J Nutr 100(4):470–476. https://doi.org/10.1093/jn/100.4.470

Beach RS, Gershwin ME, Hurley LS (1980) Growth and development in postnatally zinc-deprived mice. J Nutr 110(2):201–211. https://doi.org/10.1093/jn/110.2.201

Bedwal RS, Bahuguna A (1994) Zinc, copper and selenium in reproduction. Experientia 50(7):626–640

Bernhardt ML, Kim AM, O'Halloran TV et al (2011) Zinc requirement during meiosis I-meiosis II transition in mouse oocytes is independent of the MOS-MAPK pathway. Biol Reprod 84(3):526–536. https://doi.org/10.1095/biolreprod.110.086488

Bernhardt ML, Kong BY, Kim AM et al (2012) A zinc-dependent mechanism regulates meiotic progression in mammalian oocytes. Biol Reprod 86(4):114. https://doi.org/10.1095/biolreprod.111.097253

Berridge MJ (2002) The endoplasmic reticulum: a multifunctional signaling organelle. Cell Calcium 32(5–6):235–249. doi:S0143416002001823 [pii]

Blamberg DL, Blackwood UB, Supplee WC et al (1960) Effect of zinc deficiency in hens on hatchability and embryonic development. Proc Soc Exp Biol Med 104:217–220

Budzik GP, Hutson JM, Ikawa H et al (1982) The role of zinc in mullerian duct regression. Endocrinology 110(5):1521–1525. https://doi.org/10.1210/endo-110-5-1521

Caldwell DF, Oberleas D, Clancy JJ et al (1970) Behavioral impairment in adult rats following acute zinc deficiency. Proc Soc Exp Biol Med 133(4):1417–1421

Castro-Magana M, Collipp PJ, Chen SY et al (1981) Zinc nutritional status, androgens, and growth retardation. Am J Dis Child 135(4):322–325

Cha J, Sun X, Dey SK (2012) Mechanisms of implantation: strategies for successful pregnancy. Nat Med 18(12):1754–1767. https://doi.org/10.1038/nm.3012

Chaffee BW, King JC (2012) Effect of zinc supplementation on pregnancy and infant outcomes: a systematic review. Paediatr Perinat Epidemiol 26(Suppl 1):118–137. https://doi.org/10.1111/j.1365-3016.2012.01289.x

Chen JR, Cheng JG, Shatzer T et al (2000) Leukemia inhibitory factor can substitute for nidatory estrogen and is essential to inducing a receptive uterus for implantation but is not essential for subsequent embryogenesis. Endocrinology 141(12):4365–4372. https://doi.org/10.1210/endo.141.12.7855

Choi T, Fukasawa K, Zhou R et al (1996) The Mos/mitogen-activated protein kinase (MAPK) pathway regulates the size and degradation of the first polar body in maturing mouse oocytes. Proc Natl Acad Sci USA 93(14):7032–7035

Chu Q, Chi ZH, Zhang X et al (2016) A potential role for zinc transporter 7 in testosterone synthesis in mouse Leydig tumor cells. Int J Mol Med 37(6):1619–1626. https://doi.org/10.3892/ijmm.2016.2576

Coward K, Ponting CP, Chang HY et al (2005) Phospholipase Czeta, the trigger of egg activation in mammals, is present in a non-mammalian species. Reproduction 130(2):157–163. doi:130/2/157 [pii]

Coward K, Ponting CP, Zhang N et al (2011) Identification and functional analysis of an ovarian form of the egg activation factor phospholipase C zeta (PLCzeta) in pufferfish. Mol Reprod Dev 78(1):48–56. https://doi.org/10.1002/mrd.21262

Cox LJ, Larman MG, Saunders CM et al (2002) Sperm phospholipase Czeta from humans and cynomolgus monkeys triggers Ca2+ oscillations, activation and development of mouse oocytes. Reproduction 124(5):611–623

Dempski RE (2012) The cation selectivity of the ZIP transporters. Curr Top Membr 69:221–245. https://doi.org/10.1016/B978-0-12-394390-3.00009-4

Dey SK, Lim H, Das SK et al (2004) Molecular cues to implantation. Endocr Rev 25(3):341–373. https://doi.org/10.1210/er.2003-0020

Donangelo CM, Zapata CL, Woodhouse LR et al (2005) Zinc absorption and kinetics during pregnancy and lactation in Brazilian women. Am J Clin Nutr 82(1):118–124. doi:82/1/118 [pii]

Downey AM, Hales BF, Robaire B (2016) Zinc transport differs in rat spermatogenic cell types and is affected by treatment with cyclophosphamide. Biol Reprod 95(1):22. https://doi.org/10.1095/biolreprod.116.140558

Ducibella T, Fissore R (2008) The roles of Ca2+, downstream protein kinases, and oscillatory signaling in regulating fertilization and the activation of development. Dev Biol 315(2):257–279. https://doi.org/10.1016/j.ydbio.2007.12.012

Ducibella T, Huneau D, Angelichio E et al (2002) Egg-to-embryo transition is driven by differential responses to Ca(2+) oscillation number. Dev Biol 250(2):280–291. doi:S0012160602907888 [pii]

Duncan FE, Que EL, Zhang N et al (2016) The zinc spark is an inorganic signature of human egg activation. Sci Rep 6:24737. https://doi.org/10.1038/srep24737

Eide DJ (2006) Zinc transporters and the cellular trafficking of zinc. Biochim Biophys Acta 1763(7):711–722. doi:S0167-4889(06)00059-0 [pii]

Favier AE (1992a) Hormonal effects of zinc on growth in children. Biol Trace Elem Res 32:383–398

Favier AE (1992b) The role of zinc in reproduction. Hormonal mechanisms. Biol Trace Elem Res 32:363–382

Flynn A, Miller SI, Martier SS et al (1981) Zinc status of pregnant alcoholic women: a determinant of fetal outcome. Lancet 1(8220 Pt 1):572–551. doi:S0140-6736(81)92029-8 [pii]

Funahashi H, Cantley TC, Stumpf TT et al (1994) In vitro development of in vitro-matured porcine oocytes following chemical activation or in vitro fertilization. Biol Reprod 50(5):1072–1077

Gaither LA, Eide DJ (2001) Eukaryotic zinc transporters and their regulation. Biometals 14(3–4):251–270

Gallaher D, Hurley LS (1980) Low zinc concentration in rat uterine fluid after 4 days of dietary deficiency. J Nutr 110(3):591–593. https://doi.org/10.1093/jn/110.3.591

Garg HK, Singhal KC, Arshad Z (1993) A study of the effect of oral zinc supplementation during pregnancy on pregnancy outcome. Indian J Physiol Pharmacol 37(4):276–284

Ghosh D, De P, Sengupta J (1994) Luteal phase ovarian oestrogen is not essential for implantation and maintenance of pregnancy from surrogate embryo transfer in the rhesus monkey. Hum Reprod 9(4):629–637

Gibson RS (1994) Zinc nutrition in developing countries. Nutr Res Rev 7(1):151–173. https://doi.org/10.1079/NRR19940010

Goldenberg RL, Tamura T (1996) Prepregnancy weight and pregnancy outcome. JAMA 275(14):1127–1128
Hansen DV, Tung JJ, Jackson PK (2006) CaMKII and polo-like kinase 1 sequentially phosphorylate the cytostatic factor Emi2/XErp1 to trigger its destruction and meiotic exit. Proc Natl Acad Sci USA 103(3):608–613. doi:0509549102 [pii]
Hantak AM, Bagchi IC, Bagchi MK (2014) Role of uterine stromal-epithelial crosstalk in embryo implantation. Int J Dev Biol 58(2–4):139–146. https://doi.org/10.1387/ijdb.130348mb
Hashimoto N, Watanabe N, Furuta Y et al (1994) Parthenogenetic activation of oocytes in c-mos-deficient mice. Nature 370(6484):68–71. https://doi.org/10.1038/370068a0
Henkel R, Baldauf C, Schill WB (2003) Resorption of the element zinc from spermatozoa by the epididymal epithelium. Reprod Domest Anim 38(2):97–101. doi:417 [pii]
Hesketh JE (1982) Effects of dietary zinc deficiency on leydig cell ultrastructure in the boar. J Comp Pathol 92(2):239–247. doi:0021-9975(82)90082-2 [pii]
Hirota Y (2019) Progesterone governs endometrial proliferation-differentiation switching and blastocyst implantation. Endocr J 66(3):199–206. https://doi.org/10.1507/endocrj.EJ18-0431
Hurley LS, Swenerton H (1966) Congenital malformations resulting from zinc deficiency in rats. Proc Soc Exp Biol Med 123(3):692–696
Irvine DS (1996) Glutathione as a treatment for male infertility. Rev Reprod 1(1):6–12
Ito J, Shimada M (2005) Timing of MAP kinase inactivation effects on emission of polar body in porcine oocytes activated by Ca2+ ionophore. Mol Reprod Dev 70(1):64–69. https://doi.org/10.1002/mrd.20182
Ito J, Yoon SY, Lee B et al (2008a) Inositol 1,4,5-trisphosphate receptor 1, a widespread Ca2+ channel, is a novel substrate of polo-like kinase 1 in eggs. Dev Biol 320(2):402–413. https://doi.org/10.1016/j.ydbio.2008.05.548
Ito M, Shikano T, Oda S et al (2008b) Difference in Ca2+ oscillation-inducing activity and nuclear translocation ability of PLCZ1, an egg-activating sperm factor candidate, between mouse, rat, human, and medaka fish. Biol Reprod 78(6):1081–1090. https://doi.org/10.1095/biolreprod.108.067801
Ito J, Yoshida T, Kasai Y et al (2010) Phosphorylation of inositol 1,4,5-triphosphate receptor 1 during in vitro maturation of porcine oocytes. Anim Sci J 81(1):34–41. https://doi.org/10.1111/j.1740-0929.2009.00699.x
Ito J, Parrington J, Fissore RA (2011) PLCzeta and its role as a trigger of development in vertebrates. Mol Reprod Dev 78(10–11):846–853. https://doi.org/10.1002/mrd.21359
Jameson S (1976) Zinc and copper in pregnancy, correlations to fetal and maternal complications. Acta Med Scand Suppl 593:5–20
Jones KT (1998a) Ca2+ oscillations in the activation of the egg and development of the embryo in mammals. Int J Dev Biol 42(1):1–10
Jones KT (1998b) Protein kinase C action at fertilization: overstated or undervalued? Rev Reprod 3(1):7–12
Jones KT (2005) Mammalian egg activation: from Ca2+ spiking to cell cycle progression. Reproduction 130(6):813–823. doi:130/6/813 [pii]
Joven J, Villabona C, Rubies-Prat J et al (1985) Hormonal profile and serum zinc levels in uraemic men with gonadal dysfunction undergoing haemodialysis. Clin Chim Acta 148(3):239–245. doi:0009-8981(85)90150-0 [pii]
Kambe T (2011) An overview of a wide range of functions of ZnT and Zip zinc transporters in the secretory pathway. Biosci Biotechnol Biochem 75(6):1036–1043. doi:JST.JSTAGE/bbb/110056 [pii]
Kambe T (2012) Molecular architecture and function of ZnT transporters. Curr Top Membr 69:199–220. https://doi.org/10.1016/B978-0-12-394390-3.00008-2
Kambe T, Suzuki T, Nagao M et al (2006) Sequence similarity and functional relationship among eukaryotic ZIP and CDF transporters. Genomics Proteomics Bioinformatics 4(1):1–9. https://doi.org/10.1016/S1672-0229(06)60010-7

Kambe T, Tsuji T, Hashimoto A et al (2015) The physiological, biochemical, and molecular roles of zinc transporters in zinc homeostasis and metabolism. Physiol Rev 95(3):749–784. https://doi.org/10.1152/physrev.00035.2014

Keen CL, Hurley LS (1980) Developmental changes in concentrations of iron, copper, and zinc in mouse tissues. Mech Ageing Dev 13(2):161–176. doi:0047-6374(80)90059-7 [pii]

Keen CL, Uriu-Adams JY, Skalny A et al (2010) The plausibility of maternal nutritional status being a contributing factor to the risk for fetal alcohol spectrum disorders: the potential influence of zinc status as an example. Biofactors 36(2):125–135. https://doi.org/10.1002/biof.89

Kim JT, Baek SH, Lee SH et al (2009) Zinc-deficient diet decreases fetal long bone growth through decreased bone matrix formation in mice. J Med Food 12(1):118–123. https://doi.org/10.1089/jmf.2007.0647

Kim AM, Vogt S, O'Halloran TV et al (2010) Zinc availability regulates exit from meiosis in maturing mammalian oocytes. Nat Chem Biol 6(9):674–681. https://doi.org/10.1038/nchembio.419

Kim AM, Bernhardt ML, Kong BY et al (2011) Zinc sparks are triggered by fertilization and facilitate cell cycle resumption in mammalian eggs. ACS Chem Biol 6(7):716–723. https://doi.org/10.1021/cb200084y

King JC (2000) Determinants of maternal zinc status during pregnancy. Am J Clin Nutr 71(5 Suppl):1334S–1343S. https://doi.org/10.1093/ajcn/71.5.1334s

Kishigami S, Wakayama T (2007) Efficient strontium-induced activation of mouse oocytes in standard culture media by chelating calcium. J Reprod Dev 53(6):1207–1215. doi:JST.JSTAGE/jrd/19067 [pii]

Kishikawa H, Wakayama T, Yanagimachi R (1999) Comparison of oocyte-activating agents for mouse cloning. Cloning 1(3):153–159. https://doi.org/10.1089/15204559950019915

Kline D (2000) Attributes and dynamics of the endoplasmic reticulum in mammalian eggs. Curr Top Dev Biol 50:125–154. doi:S0070-2153(00)50007-6 [pii]

Kline D, Kline JT (1992) Repetitive calcium transients and the role of calcium in exocytosis and cell cycle activation in the mouse egg. Dev Biol 149(1):80–89. doi:0012-1606(92)90265-I [pii]

Kong BY, Bernhardt ML, Kim AM et al (2012) Zinc maintains prophase I arrest in mouse oocytes through regulation of the MOS-MAPK pathway. Biol Reprod 87(1):11., 1-12. https://doi.org/10.1095/biolreprod.112.099390

Kong BY, Duncan FE, Que EL et al (2014) Maternally-derived zinc transporters ZIP6 and ZIP10 drive the mammalian oocyte-to-egg transition. Mol Hum Reprod 20(11):1077–1089. https://doi.org/10.1093/molehr/gau066

Kurokawa M, Sato K, Fissore RA (2004) Mammalian fertilization: from sperm factor to phospholipase Czeta. Biol Cell 96(1):37–45. https://doi.org/10.1016/j.biolcel.2003.11.003

Lee B, Yoon SY, Fissore RA (2006) Regulation of fertilization-initiated [Ca2+]i oscillations in mammalian eggs: a multi-pronged approach. Semin Cell Dev Biol 17(2):274–284. doi:S1084-9521(06)00022-X [pii]

Lutwak-Mann C, McIntosh JE (1969) Zinc and carbonic anhydrase in the rabbit uterus. Nature 221(5186):1111–1114

Madding CI, Jacob M, Ramsay VP et al (1986) Serum and semen zinc levels in normozoospermic and oligozoospermic men. Ann Nutr Metab 30(4):213–218. https://doi.org/10.1159/000177196

Madgwick S, Jones KT (2007) How eggs arrest at metaphase II: MPF stabilisation plus APC/C inhibition equals cytostatic factor. Cell Div 2:4. -1028-2-4. doi:1747-1028-2-4 [pii]

Mahomed K, James DK, Golding J et al (1989) Zinc supplementation during pregnancy: a double blind randomised controlled trial. BMJ 299(6703):826–830

Marcus GJ (1990) Activation of cumulus-free mouse oocytes. Mol Reprod Dev 26(2):159–162. https://doi.org/10.1002/mrd.1080260210

Margioula-Siarkou C, Prapas Y, Petousis S et al (2016) LIF and LIFR expression in the endometrium of fertile and infertile women: a prospective observational case control study. Mol Med Rep 13(6):4721–4728. https://doi.org/10.3892/mmr.2016.5142

Margioula-Siarkou C, Prapas Y, Petousis S et al (2017) LIF endometrial expression is impaired in women with unexplained infertility while LIF-R expression in all infertility sub-groups. Cytokine 96:166–172. doi:S1043-4666(17)30094-7 [pii]

Matsuda Y, Watanabe T (2003) Effects of oyster extract on the reproductive function of zinc-deficient mice: bioavailability of zinc contained in oyster extract. Congenit Anomalies (Kyoto) 43(4):271–279

McClain CJ, Gavaler JS, Van Thiel DH (1984) Hypogonadism in the zinc-deficient rat: localization of the functional abnormalities. J Lab Clin Med 104(6):1007–1015. doi:0022-2143(84)90057-X [pii]

McIntosh JE, Lutwak-Mann C (1972) Zinc transport in rabbit tissues. Some hormonal aspects of the turnover of zinc in female reproductive organs, liver and body fluids. Biochem J 126(4):869–876

Mello-Neto J, Rondo PH, Oshiiwa M et al (2013) Iron supplementation in pregnancy and breast-feeding and iron, copper and zinc status of lactating women from a human milk bank. J Trop Pediatr 59(2):140–144. https://doi.org/10.1093/tropej/fms055

Merialdi M, Caulfield LE, Zavaleta N et al (2004) Randomized controlled trial of prenatal zinc supplementation and fetal bone growth. Am J Clin Nutr 79(5):826–830. https://doi.org/10.1093/ajcn/79.5.826

Miyazaki S, Shirakawa H, Nakada K et al (1993) Essential role of the inositol 1,4,5-trisphosphate receptor/Ca2+ release channel in Ca2+ waves and Ca2+ oscillations at fertilization of mammalian eggs. Dev Biol 158(1):62–78. https://doi.org/10.1006/dbio.1993.1168

Mizushima S, Takagi S, Ono T et al (2009) Phospholipase Czeta mRNA expression and its potency during spermatogenesis for activation of quail oocyte as a sperm factor. Mol Reprod Dev 76(12):1200–1207. https://doi.org/10.1002/mrd.21097

Mutch PB, Hurley LS (1980) Mammary gland function and development: effect of zinc deficiency in rat. Am J Phys 238(1):E26–E31. https://doi.org/10.1152/ajpendo.1980.238.1.E26

Nakai M, Ito J, Sato K et al (2011) Pre-treatment of sperm reduces success of ICSI in the pig. Reproduction 142(2):285–293. https://doi.org/10.1530/REP-11-0073

Namiki T, Ito J, Kashiwazaki N (2018) Molecular mechanisms of embryonic implantation in mammals: lessons from the gene manipulation of mice. Reprod Med Biol 17(4):331–342. https://doi.org/10.1002/rmb2.12103

Oberleas D, Caldwell DF, Prasad AS (1971) Behavioral deficit with zinc deficiency. Psychopharmacol Bull 7(3):35

Ogura A, Suzuki O, Tanemura K et al (1998) Development of normal mice from metaphase I oocytes fertilized with primary spermatocytes. Proc Natl Acad Sci USA 95(10):5611–5615

Ozil JP, Banrezes B, Toth S et al (2006) Ca2+ oscillatory pattern in fertilized mouse eggs affects gene expression and development to term. Dev Biol 300(2):534–544. doi:S0012-1606(06)01131-6 [pii]

Palmiter RD, Huang L (2004) Efflux and compartmentalization of zinc by members of the SLC30 family of solute carriers. Pflugers Arch 447(5):744–751. https://doi.org/10.1007/s00424-003-1070-7

Parrington J (2001) Does a soluble sperm factor trigger calcium release in the egg at fertilization? J Androl 22(1):1–11

Parrington J, Lai FA, Swann K (2000) The soluble mammalian sperm factor protein that triggers Ca2+ oscillations in eggs: evidence for expression of mRNA(s) coding for sperm factor protein(s) in spermatogenic cells. Biol Cell 92(3–4):267–275

Parrington J, Davis LC, Galione A et al (2007) Flipping the switch: how a sperm activates the egg at fertilization. Dev Dyn 236(8):2027–2038. https://doi.org/10.1002/dvdy.21255

Plante M, de Lamirande E, Gagnon C (1994) Reactive oxygen species released by activated neutrophils, but not by deficient spermatozoa, are sufficient to affect normal sperm motility. Fertil Steril 62(2):387–393. doi:S0015-0282(16)56895-2 [pii]

Prasad AS (1979) Trace elements: biochemical and clinical effects of zinc and copper. Am J Hematol 6(1):77–87

Que EL, Bleher R, Duncan FE et al (2015) Quantitative mapping of zinc fluxes in the mammalian egg reveals the origin of fertilization-induced zinc sparks. Nat Chem 7(2):130–139. https://doi.org/10.1038/nchem.2133

Que EL, Duncan FE, Lee HC et al (2019) Bovine eggs release zinc in response to parthenogenetic and sperm-induced egg activation. Theriogenology 127:41–48. doi:S0093-691X(18)30911-7 [pii]

Rao AJ, Ramachandra SG, Ramesh V et al (2007) Establishment of the need for oestrogen during implantation in non-human primates. Reprod Biomed Online 14(5):563–571. doi:S1472-6483(10)61047-4 [pii]

Rebecchi MJ, Pentyala SN (2000) Structure, function, and control of phosphoinositide-specific phospholipase C. Physiol Rev 80(4):1291–1335. https://doi.org/10.1152/physrev.2000.80.4.1291

Rice A, Parrington J, Jones KT et al (2000) Mammalian sperm contain a Ca(2+)-sensitive phospholipase C activity that can generate InsP(3) from PIP(2) associated with intracellular organelles. Dev Biol 228(1):125–135. https://doi.org/10.1006/dbio.2000.9929

Ross PJ, Beyhan Z, Iager AE et al (2008) Parthenogenetic activation of bovine oocytes using bovine and murine phospholipase C zeta. BMC Dev Biol 8:16-213X-8-16. https://doi.org/10.1186/1471-213X-8-16

Roungsipragarn R, Borirug S, Herabutya Y (1999) Plasma zinc level and intrauterine growth retardation: a study in pregnant women in Ramathibodi hospital. J Med Assoc Thail 82(2):178–181

Runft LL, Jaffe LA, Mehlmann LM (2002) Egg activation at fertilization: where it all begins. Dev Biol 245(2):237–254. https://doi.org/10.1006/dbio.2002.0600

Sano D, Yamamoto Y, Samejima T et al (2009) A combined treatment with ethanol and 6-dimethylaminopurine is effective for the activation and further embryonic development of oocytes from Sprague-Dawley and Wistar rats. Zygote 17(1):29–36. https://doi.org/10.1017/S0967199408004875

Saunders CM, Larman MG, Parrington J et al (2002) PLC zeta: a sperm-specific trigger of Ca(2+) oscillations in eggs and embryo development. Development 129(15):3533–3544

Schultz RM, Kopf GS (1995) Molecular basis of mammalian egg activation. Curr Top Dev Biol 30:21–62. doi:S0070-2153(08)60563-3 [pii]

Smitz J, Bourgain C, Van Waesberghe L et al (1993) A prospective randomized study on oestradiol valerate supplementation in addition to intravaginal micronized progesterone in buserelin and HMG induced superovulation. Hum Reprod 8(1):40–45

Stewart CL, Kaspar P, Brunet LJ et al (1992) Blastocyst implantation depends on maternal expression of leukaemia inhibitory factor. Nature 359(6390):76–79. https://doi.org/10.1038/359076a0

Stricker SA (1999) Comparative biology of calcium signaling during fertilization and egg activation in animals. Dev Biol 211(2):157–176. https://doi.org/10.1006/dbio.1999.9340

Suzuki T, Yoshida N, Suzuki E et al (2010) Full-term mouse development by abolishing Zn2+–dependent metaphase II arrest without Ca2+ release. Development 137(16):2659–2669. https://doi.org/10.1242/dev.049791

Swann K, Parrington J (1999) Mechanism of Ca2+ release at fertilization in mammals. J Exp Zool 285(3):267–275. https://doi.org/10.1002/(SICI)1097-010X(19991015)285:3<267::AID-JEZ10>3.0.CO;2-P. [pii]

Swann K, Yu Y (2008) The dynamics of calcium oscillations that activate mammalian eggs. Int J Dev Biol 52(5–6):585–594. https://doi.org/10.1387/ijdb.072530ks

Swann K, Saunders CM, Rogers NT et al (2006) PLCzeta(zeta): a sperm protein that triggers Ca2+ oscillations and egg activation in mammals. Semin Cell Dev Biol 17(2):264–273. doi:S1084-9521(06)00031-0 [pii]

Taki M, Wolford JL, O'Halloran TV (2004) Emission ratiometric imaging of intracellular zinc: design of a benzoxazole fluorescent sensor and its application in two-photon microscopy. J Am Chem Soc 126(3):712–713. https://doi.org/10.1021/ja039073j

Tamura T, Goldenberg RL, Johnston KE et al (1996) Serum ferritin: a predictor of early spontaneous preterm delivery. Obstet Gynecol 87(3):360–365. doi:0029784495004378 [pii]

Taubeneck MW, Daston GP, Rogers JM et al (1994) Altered maternal zinc metabolism following exposure to diverse developmental toxicants. Reprod Toxicol 8(1):25–40

Thomas K, De Hertogh R, Pizarro M et al (1973) Plasma LH-HCG, 17 -estradiol, estrone and progesterone monitoring around ovulation and subsequent nidation. Int J Fertil 18(2):65–73

Tian X, Diaz FJ (2012) Zinc depletion causes multiple defects in ovarian function during the periovulatory period in mice. Endocrinology 153(2):873–886. https://doi.org/10.1210/en.2011-1599

Tian X, Anthony K, Neuberger T et al (2014) Preconception zinc deficiency disrupts postimplantation fetal and placental development in mice. Biol Reprod 90(4):83. https://doi.org/10.1095/biolreprod.113.113910

Ueda H, Nakai T, Konishi T et al (2014) Effects of zinc deficiency and supplementation on leptin and leptin receptor expression in pregnant mice. Biol Pharm Bull 37(4):581–587. doi:DN/JST.JSTAGE/bpb/b13-00813 [pii]

Uriu-Adams JY, Scherr RE, Lanoue L et al (2010) Influence of copper on early development: prenatal and postnatal considerations. Biofactors 36(2):136–152. https://doi.org/10.1002/biof.85

Wakayama T, Perry AC, Zuccotti M et al (1998) Full-term development of mice from enucleated oocytes injected with cumulus cell nuclei. Nature 394(6691):369–374. https://doi.org/10.1038/28615

Wang H, Dey SK (2006) Roadmap to embryo implantation: clues from mouse models. Nat Rev Genet 7(3):185–199. doi:nrg1808 [pii]

Wang WH, Abeydeera LR, Prather RS et al (1998) Functional analysis of activation of porcine oocytes by spermatozoa, calcium ionophore, and electrical pulse. Mol Reprod Dev 51(3):346–353. https://doi.org/10.1002/(SICI)1098-2795(199811)51:3<346::AID-MRD15>3.0.CO;2-0. [pii]

Whitaker M (2006) Calcium at fertilization and in early development. Physiol Rev 86(1):25–88. doi:86/1/25 [pii]

Wiersma A, Hirsch B, Tsafriri A et al (1998) Phosphodiesterase 3 inhibitors suppress oocyte maturation and consequent pregnancy without affecting ovulation and cyclicity in rodents. J Clin Invest 102(3):532–537. https://doi.org/10.1172/JCI2566

Wilson RL, Grieger JA, Bianco-Miotto T et al (2016) Association between maternal zinc status, dietary zinc intake and pregnancy complications: a systematic review. Nutrients 8(10):E641. 10.3390/nu8100641. doi:E641 [pii]

Wilson RL, Leemaqz SY, Goh Z et al (2017) Zinc is a critical regulator of placental morphogenesis and maternal hemodynamics during pregnancy in mice. Sci Rep 7(1):15137-017-15085-2. https://doi.org/10.1038/s41598-017-15085-2

Wu H, Smyth J, Luzzi V et al (2001) Sperm factor induces intracellular free calcium oscillations by stimulating the phosphoinositide pathway. Biol Reprod 64(5):1338–1349

Yoneda A, Kashima M, Yoshida S et al (2006) Molecular cloning, testicular postnatal expression, and oocyte-activating potential of porcine phospholipase Czeta. Reproduction 132(3):393–401. doi:132/3/393 [pii]

Young C, Grasa P, Coward K et al (2009) Phospholipase C zeta undergoes dynamic changes in its pattern of localization in sperm during capacitation and the acrosome reaction. Fertil Steril 91(5 Suppl):2230–2242. https://doi.org/10.1016/j.fertnstert.2008.05.021

Zhang N, Duncan FE, Que EL et al (2016) The fertilization-induced zinc spark is a novel biomarker of mouse embryo quality and early development. Sci Rep 6:22772. https://doi.org/10.1038/srep22772

Zhao Y, Tan CH, Krauchunas A et al (2018) The zinc transporter ZIPT-7.1 regulates sperm activation in nematodes. PLoS Biol 16(6):e2005069. https://doi.org/10.1371/journal.pbio.2005069

Chapter 7
Zinc Signaling in Skeletal Muscle

Kristyn Gumpper and Jianjie Ma

Abstract This chapter focuses on how zinc regulates several major functions of skeletal muscles. A common theme throughout this chapter will be how zinc concentration and localization mediated by ZIPs, ZnTs, and metallothioneins affect muscle contraction via zinc-binding proteins and modulation of the phosphatidylinositol 3-kinase/Akt/mammalian target of rapamycin (PI3K/Akt/mTOR) pathway to maintain homeostasis and how TRIM family proteins modulate the integrity of muscle fibers and their regeneration following injuries. Additionally, we examine how disruption in zinc homeostasis results in reduced function in some muscle-specific proteins and produces myopathies such as cachexia and impaired glucose metabolism.

Keywords Excitation-contraction coupling · Zinc-finger proteins · Zinc transporters · MuRF-1 · MG53 · Skeletal muscle repair · Aging and cachexia · RyR · Calcium channels

7.1 Introduction

Zinc is necessary for nearly all aspects of cellular function and regulation as a catalytic cofactor or structural stabilizer. Approximately 10% of proteins in the Protein Data Bank have zinc in their structure indexes (Chasapis et al. 2012; Wang et al. 2010). In the late 1970s and early 1980s, researchers assessed zinc metabolism to gain a clearer understanding of where zinc travels in the body and how fast it is excreted (Foster et al. 1979; Wastney et al. 1986; Wastney and Henkin 1988). The models from both short- and long-term studies suggest 90% of zinc in the body resides in tissues with slow zinc metabolism, such as skeletal muscle and bone. The authors of those studies suggested the low zinc kinetics in skeletal muscle may be a result from tight regulation of zinc within this tissue type.

K. Gumpper · J. Ma (✉)
Department of Surgery, Davis Heart and Lung Research Institute,
The Ohio State University Wexner Medical Center, Columbus, OH, USA
e-mail: Kristyn.Gumpper@osumc.edu; jianjie.ma@osumc.edu

T. Fukada, T. Kambe (eds.), *Zinc Signaling*,
https://doi.org/10.1007/978-981-15-0557-7_7

Zinc homeostasis is strongly regulated within the cell by multiple types of proteins. Transport is tightly controlled mainly through two protein families: zinc influx ZIP/Slc39 proteins and zinc efflux/sequestration ZnT/Slc30 proteins. ZIPs transport zinc into the cell or out of the cellular organelles when the cytoplasmic zinc concentration is low, whereas ZnTs transport zinc out of the cell or into the organelles for storage when the cytoplasmic zinc concentration is high. Mutations or deletions of ZIPs or ZnTs result in the dysregulation of zinc and a reduction in cell survival (Shusterman et al. 2017). For a more comprehensive summary of zinc transporters, see the review by T. Kimura and T. Kambe for the localization of ZIPs and ZnTs in the cell (Kimura and Kambe 2016). In addition to regulating movement of zinc into and around the cell, metallothioneins (MTs) are a small family of proteins that bind and sequester zinc, further regulating free zinc inside the cell. Together, ZnTs, ZIPs, and MTs are responsible for the tight regulation of zinc known as buffering and muffling. Excellent reviews of this phenomena were written by R.A. Colvin (Colvin et al. 2010), T. Kimura, and T. Kambe (Kimura and Kambe 2016). Zinc-binding proteins with zinc-finger domains often require zinc to perform functions, such as ubiquitination for many E3 ubiquitin ligases and transcription for zinc-binding transcription factors.

Tripartite motif (TRIM) proteins belong to a large family of E3 ubiquitin ligases with a conserved zinc-binding n-terminal RING-B-box-Coiled-coil (RBCC) domain and divergent c-terminal domains. In skeletal muscle, there are two subgroups of TRIM proteins essential for skeletal muscle survival and repair: (1) C-II containing a c-terminal subgroup one signature (COS) domain and (2) C-IV containing a PRY/SPRY c-terminal domain. Subgroup C-II contains TRIMs 63, 55, and 54 which are also known as Muscle RING Fingers, or MuRF-1, MuRF-2, or MuRF-3, respectively. Subgroup C-IV is the largest subgroup of TRIM proteins and contains the muscle-specific TRIM72 (Mitsugumin 53, or MG53) protein. These zinc-binding TRIM proteins are essential for maintaining skeletal muscle homeostasis as deletions or mutations in these proteins lead to multiple functional and metabolic disorders.

7.2 Muscle Contraction

Signaling between cells is integral for coordinating tissue function in multicellular organisms. For skeletal muscles, coordination between cells is required for functions like contraction, also known as excitation-contraction coupling. To contract, an action potential, the source of excitation, is emitted from a local neuron, causing depolarization within the T-tubules of skeletal muscle fibers through an influx of sodium ions. This depolarization causes calcium release from the sarcoplasmic reticulum (SR), allowing calcium to bind to troponin, revealing myosin binding sites on actin whose motion results in contraction. Skeletal muscle contractions are defined in two categories: twitch and tetanus. Twitch is defined as a single contraction/relaxation period produced by a single action potential from within the fibers. Tetanus is defined as a sustained muscle contraction caused by motor neurons

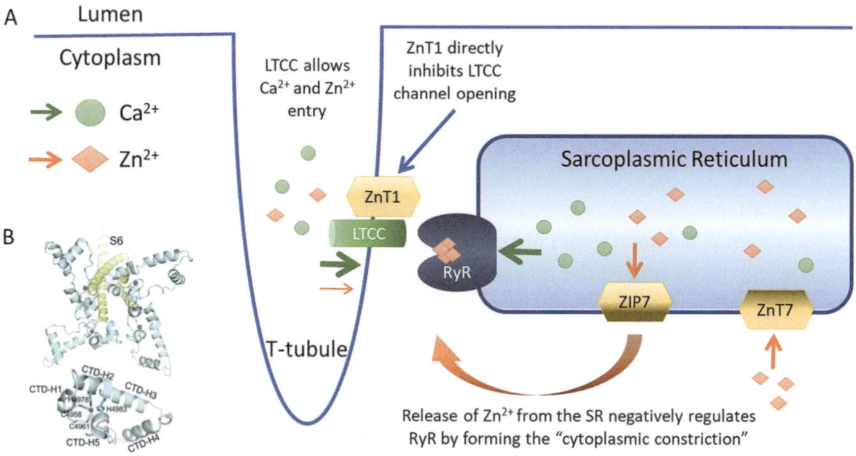

Fig. 7.1 Zinc signaling in skeletal muscle contraction. (**a**) ZnT1/L-type Ca^{2+} channel (LTCC) interaction modulates entry of zinc from extracellular space into the cytosol, which contributes to Ca^{2+} signaling associated with excitation-contraction coupling in skeletal muscle. ZIP7 and Znt7 regulate zinc transport and release across the sarcoplasmic reticulum (SR) in skeletal muscle. Intracellular zinc can also interfere with ryanodine receptor (RyR)-mediated Ca^{2+} release from the SR. (**b**) Near-atomic resolution of the 3-D structure of RyR1 reveals the binding motif for zinc that regulates open-close transition of the homotetrameric RyR1 channel (top) homotetramer with four zinc ions (bottom) single RyR peptide with one zinc ion. Zinc ions are indicated by purple spheres. (Adapted with permission from Yan et al. 2015)

releasing several action potentials in rapid succession. In other words, tetanus can be thought of as multiple twitches in quick succession without any relaxation in between. In 1963, Allen Isaacson and Alexander Sandow studied muscle twitch and tetanus in frogs and found that additional zinc increases the tension generated by a muscle after a single stimulating shock with peak twitch potentiation around 50 μM zinc (Isaacson and Sandow 1963). Tetanus, however, remains unchanged. Later studies further characterized zinc's effect on muscle twitch, noting 50 μM zinc not only increased twitch potentiation but also increased action potential and duration of the contraction and relaxation periods of the twitch (Taylor et al. 1972; Mashima and Washio 1964; Edman et al. 1966). These early studies point to zinc's involvement in regulating calcium release for interfiber communication for concerted contraction; however, at the time, the exact mechanism was unknown (see Fig. 7.1).

The ryanodine receptor (RyR) is a large calcium channel made up of four large RyR subunits, four smaller FKBP subunits, and other associated proteins that sit between the SR and plasma membrane of cells. There are 3 isoforms that share about 70% sequence identity and are differentially expressed throughout the body. RyR1 is primarily expressed in skeletal muscle, RyR2 is primarily expressed in the cardiac muscle, and RyR3 is expressed in other tissues. In the 1980s, the sequence and general structure of the RyR channel was elucidated, providing insight into how depolarization of T-tubules causes release of calcium from the SR (Takeshima et al. 1989). More recently, near-atomic resolution structural analysis of the RyR channel indicates there is a carboxy-terminal zinc-finger domain located in the cytosolic portion of the channel (Yan et al. 2015). When the calcium channel is formed as a homotetramer, the zinc-binding c-terminal domains of RyR contact each other to form a "cytoplasmic constriction" stabilized by 4 zinc ions. On top of the zinc cytoplasmic constriction site, other studies have suggested that zinc may have other binding sites on RyR or its ancillary proteins to modulate channel function (Xia et al. 2000).

Interactions between RyR channel proteins and ions like calcium and magnesium regulate the activation of the channel in a well-defined fashion. RyR1 is inactive at low Ca^{2+} concentrations; activated, or open, between 10 and 100 uM Ca^{2+}; and inactivated above 1 mM Ca^{2+} (Wei et al. 2016). This biphasic control of RyR is derived from two separate Ca^{2+} binding sites with the activating site having a higher affinity for Ca^{2+} than the inactivating site (Ching et al. 2000). Ca^{2+} binding can be inhibited by magnesium which binds to RyR at the same site as Ca^{2+}, but with a lower affinity than $Ca2^{+}$, inhibiting RyR channel function. Recent studies on cardiac RyR2 channel gating show the addition of a physiologic concentration of zinc (100 pM) to RyR2 incorporated into a phospholipid bilayer greatly increases the activation of the channel (Woodier et al. 2015). Furthermore, addition of zinc to the cytoplasmic face of the RyR2 channel indicated a biphasic regulation of the channel, peaking around 1 nM while inhibiting the channel at concentrations higher than 10 nM, just like Ca^{2+} regulation of RyR. Similar activation and inhibition was seen in studies on RyR1 binding to [^{3}H]ryanodine; however, the concentrations of activation and inhibition were higher, peaking between 100 nM and 1 uM with near-complete inhibition at 10 μM zinc (Xia et al. 2000). The biphasic control of RyR by zinc suggests it also has activation and inactivation sites that are separate from the calcium sites and are, as yet, undetermined (Wang et al. 2003). In addition to zinc directly interacting with calcium channels, zinc transporters can directly regulate calcium channels to mediate calcium signaling and subsequent muscle contraction.

The L-type calcium channel (LTCC), also known as the dihydropyridine channel, is a voltage-dependent calcium channel with 4 subunits and is responsible for excitation-contraction coupling by passing an inward calcium current into the SR, causing calcium-induced calcium release through the RyR channel into the cytoplasm. Not only can zinc enter the cell through the L-type calcium channel, but ZnT-1 directly inhibits the activity of the L-type calcium channel through the β-subunit by preventing α-subunit localization to the cell membrane, as indicated in Fig. 7.1 (Levy et al. 2009). ZnT inhibition of the L-type calcium channel appears to be zinc independent as expression of only the cytoplasmic c-terminal tail of ZnT-1 is enough to inhibit the channel, indicating an unusual dual function of the zinc transporter without any zinc channel function (Shusterman et al. 2017). Interestingly, in mast cells, the α-subunit of the L-type calcium channel is implicated in the release of zinc from the ER in a phenomena called "zinc waves" (Yamasaki et al. 2012). Moving forward, it will be essential to tease apart the apparent interplay between these two proteins and their roles in calcium and zinc regulation of muscle contraction.

7.3 Muscle Growth

When muscle cells experience injury or overload, skeletal muscle fibers are often injured and need to be repaired or replaced. To replace muscle fibers, satellite cells are activated to proliferate and differentiate into myotubes which mature into functional muscle fibers. Several growth factors like insulin and insulin-like growth factor (IGF) are involved in signaling for satellite cell differentiation into myocytes. Insulin and IGF activate proliferation and differentiation by binding to the insulin receptor or insulin-like growth factor receptor (IGFR) and activating the downstream phosphoinositide3-kinase (PI3K)/Akt/mTOR pathway. Zinc is a stabilizer of insulin and is a known insulin mimetic, which can have a synergetic effect on insulin stimulation of muscle cells (Ohashi et al. 2015; Miranda and Dey 2004). Nutritional zinc deficiency results in an inhibition of skeletal muscle growth, repair, and myoblast differentiation (Petrie et al. 1991; Jinno et al. 2014). Indeed, supplemental zinc in mouse diets increases phosphorylation of AKT, activating mTOR and its downstream effectors, $p70^{S6k}$ and 4E-BP1, for protein synthesis, promoting muscle growth and preventing muscle atrophy (Lynch et al. 2001; McClung et al. 2007).

Extracellular zinc alters proliferation and differentiation. In C2C12 cells (an immortalized myoblast cell line derived from mice), 20–50 μM zinc enhanced proliferation and differentiation into multinucleated myotubes (Ohashi et al. 2015; Mnatsakanyan et al. 2018). With more than 60 μM zinc, cell viability is greatly decreased, and neither proliferation nor differentiation occurs. Zinc promotes proliferation and differentiation through activation of the Akt and ERK signaling pathways. Through this, zinc can activate myogenic reserve cells which are mitotically quiescent, mono-nucleated myotube cells that arise during myotube formation from C2C12 cells (Ohashi et al. 2015; Mnatsakanyan et al. 2018). Differentiated myotubes absorbed more zinc from extracellular media than undifferentiated cells. Increased absorption of zinc can be correlated to increased expression of many of the 2 zinc transporter family proteins.

Differentiation of satellite cells or myoblasts to myotubes results in changes in the expression of most zinc transporters (Paskavitz et al. 2018). ZIP7 is the most well characterized of the ZIPs in skeletal muscle differentiation. The localization of zinc was correlated to the changing localization of ZIP7 in the myoblasts and differentiated myotubes (Mnatsakanyan et al. 2018). Indeed, silencing ZIP7 greatly reduces intracellular zinc content, pAkt, and the number of differentiated cells in the presence of extracellular zinc (Nimmanon et al. 2017). ZIP8, a plasma membrane ZIP, is also essential to skeletal muscle differentiation as siRNA knockdown myoblasts do not fuse to form myotubes (Gordon et al. 2018). In these models, knocking down ZIP8 results in a strong reduction in cellular manganese, iron, zinc, and calcium, which is concurrent with decreased differentiation and proliferation of myoblasts. Since there are many ions that pass through ZIP8, it will be essential to tease out how each ion affects proliferation and differentiation.

Several groups have noted internal zinc concentration increases and localization changes as myoblasts differentiate into myotubes (Mnatsakanyan et al. 2018;

Paskavitz et al. 2018). Zinc localizes throughout the cytoplasm in differentiated myotubes, but around the nucleus in undifferentiated myoblasts, most likely sequestered in the ER through ZIP7 action (Mnatsakanyan et al. 2018). Zinc localization near the nucleus in myoblasts may be related to zinc-finger domain containing transcription factors such as Zbtb4, Zbtb20, and Egr3, which require zinc for structural stability for the transcription of genes. One example is NF-κB, which reduces muscle atrophy and increases muscle mass (Kurosaka et al. 2017; Alonso-Martin et al. 2016). Although zinc localization clearly changes during myogenesis, there are very few publications assessing individual zinc transport proteins in myogenesis other than ZIP7 and ZIP8.

7.4 Skeletal Muscle Repair from Injury, Atrophy, and Cachexia

Skeletal muscle atrophy occurs when there is an imbalance in protein expression where degradation exceeds synthesis, resulting in a loss of muscle mass and function (Bodine and Baehr 2014). Cachexia is an advanced-stage cancer-related loss of muscle due to increased atrophy unrelated to nutritional deficits. Current therapies aim to increase protein synthesis and reduce protein degradation. In cachexia, the concentration of zinc in muscle is nearly double normal concentrations, while dropping in the serum leading to severe dyshomeostasis (Siren and Siren 2010).

Cachexia and aging-related muscle atrophy, also known as sarcopenia, are both associated with increased intracellular zinc and expression ZIP14 (Siren and Siren 2010; Larsson et al. 1987). ZIP14 is a plasma membrane zinc transporter that is overexpressed in skeletal muscle cells either in response to cancer-related secretion of cytokines TNF-α and TGF-β or aging-related increase in serum IL-6 (Wang et al. 2018; Aydemir et al. 2016). This results in an excess of zinc in the cytoplasm which is detrimental to cell survival (Khamzina et al. 2005). Unfortunately, blocking TNF-α or TGF-β is not enough to abolish the overexpression of ZIP14, and overexpression of ZIP14 in non-cancer models is not enough to induce muscle wasting; therefore; this suggests there must be other factors secreted from tumors inciting muscle wasting (Wang et al. 2018). Although blocking expression of ZIP14 may not be useful to prevent muscle wasting, TGF-β also contributes to wound healing and may signal differentiation of satellite cells into new muscle, potentially by upregulating ZIP14 expression (Wang et al. 2018; Leask and Abraham 2004; Hinz 2007).

Metallothioneins function in concert with ZIPs and ZnTs by regulating zinc to maintain homeostasis and restore homeostasis after injury, hypertrophy, or atrophy. Atrophying muscle from fasting, cachexia, uremia, or diabetes exhibits a significant increase in metallothionein expression (Wang et al. 2018; Latres et al. 2005; Urso et al. 2006; Lecker et al. 2004). These maladies are often associated with increased cytoplasmic zinc. Indeed, muscles undergoing sarcopenia, or aging-related muscle loss, have higher cytoplasmic zinc concentrations and increased expression of metallothioneins (Summermatter et al. 2017). Hypertrophy in skeletal muscle is induced through the inhibition of MT-1 and MT-2 together, causing an increase in pAkt, particularly in muscles with a high percentage of type IIb, fast-twitch fibers. This is consistent with early studies showing the addition of zinc increased force generated per twitch compared to fibers incubated without extracellular zinc (Isaacson and Sandow 1963; Taylor et al. 1972). It is important to note that zinc concentration in the cell does not appear to rely on MTs, as shown in MT null mice with sarcopenia (Summermatter et al. 2017).

Titin, one of the largest proteins in skeletal muscle, is like a molecular spring with a kinase domain that regulates protein expression in response to changes in mechanical load (Voelkel and Linke 2011). Zinc-finger protein nbr1 is a titin kinase ligand whose phosphorylation results in the activation of proteasomal degradation of proteins via zinc-finger proteins MuRF-1 and MuRF-2. MuRF-1 expression is increased in atrophic and cachectic skeletal muscle (Wang et al. 2018; Lecker et al. 2004; Lange et al. 2005). The importance of MuRF proteins is highlighted in a study of MuRF-2 and MuRF-3 double knockout mice. These mice exhibited protein aggregates, reduced force generation, and altered calcium handling (Lodka et al. 2016). The ubiquitin ligase function of MuRFs comes from their zinc-binding RING and B-box domains. It is possible that the mishandling of zinc during autophagy or other muscle wasting diseases like muscular dystrophy may alter MuRF's E3 ligase functions, reducing titin's ability to respond to sensed changes in mechanical load.

MG53 is a zinc-finger-containing protein highly expressed in and secreted from skeletal muscle and is integral in maintaining cellular homeostasis (Cai et al. 2009). MG53 was discovered in 2009 while elucidating what proteins are involved in maintaining the structure of the skeletal muscle T-tubule and calcium signaling. It turns out that MG53 localizes not just to the T-tubule but to all plasma membrane upon cell injury. Together with annexin V and caveolin, MG53 binds to the membrane lipid phosphatidylserine to form a membrane patch, repairing the cell. Cell membrane repair function requires dimerization of MG53 proteins through a disulfide bond near the coiled-coil section of the TRIM domain (Cai et al. 2009; Weisleder et al. 2012). As mentioned previously, MG53 as a TRIM protein has 2 zinc-finger domains in the RING-B-box region of the TRIM domain. MG53 is incapable of migrating to the plasma membrane upon injury without intact zinc-binding domains or if zinc is chelated, as indicated in Fig. 7.2 (Cai et al. 2015). Interestingly, though perhaps not surprisingly, MG53 has been implicated in regulating the PI3K/Akt/mTOR pathway during oxidative stress injuries by inducing phosphorylation of Akt, glycogen synthase kinase-3β (GSK-3β), and ERK1/ERK2 (Cao et al. 2010). Additionally, MG53's protection of cell membrane integrity can reverse muscle degeneration in several muscular dystrophy models via activation of the cell survival Akt, ERK1/ERK2, and GSK-3β pathway (Weisleder et al. 2012; He et al. 2012; Gushchina et al. 2017).

As an E3 ubiquitin ligase, MG53 has two known substrates: IRS-1 and focal adhesion kinase (FAK) (Nguyen et al. 2014). During myogenesis, total FAK is

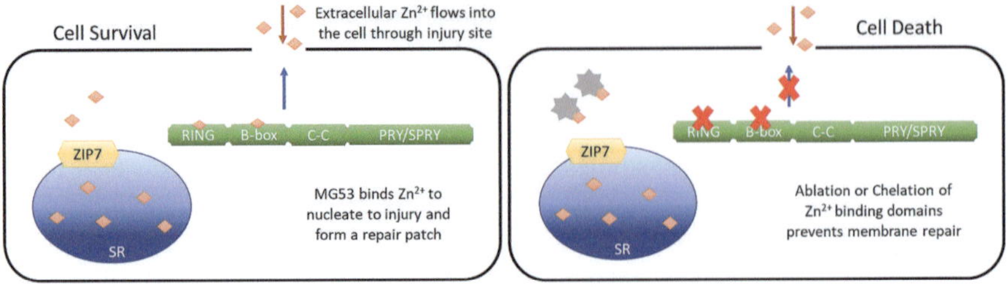

Fig. 7.2 MG53 binds zinc for membrane repair. Zinc binding to the RING and B-Box of MG53 contributes to repair of injury to skeletal muscle, through facilitation of intracellular vesicle trafficking to the injury site. ZIP-7-mediated movement of zinc across the SR membrane and injury response of zinc transport across the sarcolemmal membrane modulate the changes in zinc homeostasis associated with muscle injury

transiently reduced as it is targeted by MG53 for ubiquitination, a critical step after cell commitment to differentiation. After the temporary reduction in FAK, expression is increased again to promote the fusion of myoblasts into myotubes (Graham et al. 2015). Although MG53 requires zinc for function in plasma membrane repair and ubiquitinating FAK for myogenesis, it is unknown whether MG53 is also involved in zinc regulation in the cytoplasm to activate the cell survival pathways or if changes in zinc concentration or localization inside the cell can modulate MG53's membrane repair function.

7.5 Skeletal Muscle-Based Glucose Metabolism

Zinc is an essential nutrient in glucose metabolism. As mentioned before, zinc mediates the creation of insulin, stabilizes proteins, and regulates insulin sensitivity (Miranda and Dey 2004; Fujitani et al. 2014). A deficiency in dietary Zn^{2+} decreases the whole-body response to insulin, a condition that is exacerbated in diabetic mouse models and in humans (Myers et al. 2013). Many studies and reviews have described zinc's control of insulin in pancreatic β-islet cells, particularly focusing on the formation of insulin and the transport of zinc through ZIPs and ZnTs (Li 2014; Wijesekara et al. 2009; Huang 2014). ZnT8 in β-islet cells is the most characterized zinc transporter involved in insulin formation and secretion (Myers et al. 2012). Skeletal muscle is another major component of glucose metabolism, accounting for approximately 80% glucose uptake, and is involved in diabetic disease processes (Thiebaud et al. 1982; Abdul-Ghani and DeFronzo 2010).

Normally, when glucose concentration increases in the blood stream, β-islet cells secrete insulin. Elevation of insulin in the blood activates skeletal muscle to absorb glucose for storage and metabolism by binding to insulin receptors on the skeletal muscle plasma membrane. Insulin receptors are phosphorylated, leading to a signaling cascade that activates the transportation of GLUT-4 to the plasma membrane which facilitates glucose transport into the cell for further storage and use (Goodyear et al. 1991; Kristiansen et al. 1996, 1997). As it turns out, zinc has several significant roles in skeletal muscle-based glucose metabolism, insulin receptor activation, glucose transport, and glucose metabolism.

In 2004, Edward Miranda and Chinmoy Dey defined zinc's role in binding to the insulin receptor and modulating its downstream signaling in differentiated C2C12 mouse skeletal muscle cells (Miranda and Dey 2004). In their study, zinc alone activated the insulin receptor, though not as potently as insulin alone. Zinc and insulin together have an additive, or synergynisctic, effect on the phosphorylation status of the insulin receptor, increasing the receptor's sensitivity to insulin in the blood stream. Later studies in both human and mouse skeletal muscle-derived cells built upon the studies by Miranda and Dey, using human-derived skeletal muscle cells to follow the downstream signaling of the insulin receptor. Clear activation of phospho-Akt, phospho-ERK, phospho-SHP, and phospho-tyrosine occurred in cells with the addition of zinc with no activation in the presence of an insulin receptor inhibitor (Norouzi et al. 2018; Wu et al. 2016). Additionally, zinc restores glucose consumption in insulin-resistant L6 (a rat-derived immortalized cell line) myotubes, reestablishing GLUT-4 translocation in a dose-dependent manner (Wu et al. 2016).

Zinc is directly involved in glucose metabolism through GSK-3β, a constitutively expressed protein that phosphorylates glycogen synthase, an enzyme that catalyzes the conversion of glucose to glycogen. In vitro studies showed zinc can directly interact with GSK-3β, preventing its ability to phosphorylate glycogen synthase (Ilouz et al. 2002). In addition to binding to the IRs of skeletal muscle, zinc transport into and throughout skeletal muscle cells is necessary for maintaining homeostasis in glucose metabolism.

Zinc transporters ZIP7 (Slc39a7) and ZnT7 are necessary for zinc transport in muscle cells to regulate glucose metabolism genes. ZIP7 locates to the ER and Golgi apparatus and is known as the zinc "gatekeeper," regulating zinc signaling in the cell in response to changes in its phosphorylation state (Taylor et al. 2012). Studies knocking out ZIP7 in cell lines exhibited a reduction in the expression of genes implicated in glucose metabolism (Myers et al. 2013). Moreover, knocking down ZIP7 and treating with insulin results in significant reduction of GLUT4, pAkt, and insulin receptor substrates 1 and 2 (Irs1, Irs2) expression and reduction in glycogen synthesis. Overexpression of ZIP7 induces the expression of Irs1, Irs2, and pAkt, with a slight increase in GLUT4 expression (Myers et al. 2013). ZnT7 is also localized to the Golgi apparatus to sequester zinc (Huang et al. 2018). ZnT7 KO mice have a higher concentration of blood glucose and lower expression of pAkt, Irs1, Irs2, and GLUT4 without any changes in insulin secretion (Huang et al. 2007, 2012, 2018). Additionally, ZnT overexpression increases glucose uptake. Together ZIP7 and ZnT7 are necessary for the tight regulation of zinc concentrations inside skeletal muscle cells required for insulin-induced glucose uptake and metabolism. Although studies on ZIP7 and ZnT7 have not directly shown a coordinated effort between the two zinc transporters in skeletal muscle, studies in cardiomyocytes directly show ZIP7 and ZnT7 work in concert to regulate zinc concentrations in the cytoplasm to regulate glucose homeostasis (Tuncay et al.

2017). The coordination of ZIP7 and ZnT7 regulating glucose homeostasis may be integral to all glucose metabolizing-intensive cell types.

Since its initial discovery in 2009, many groups have worked toward understanding MG53 in physiology and how to target MG53 for tissue repair in regenerative medicine. As it turns out, MG53 may be a mediator in metabolic syndrome and insulin signaling in skeletal muscles; however this function is hotly debated (Cai et al. 2015; Ma et al. 2015; Hu and Xiao 2018). Several studies have shown that insulin receptor substrate 1 (IRS-1) and focal adhesion kinase (FAK) are E3 ligase substrates for MG53-mediated ubiquitination and degradation. Song et al. reported MG53 expression was markedly elevated in animal models of insulin resistance (Song et al. 2013). Adverse effects of MG53 as an E3 ligase in harnessing IRS1-mediated insulin signaling and metabolic function in muscle have been reported by a group from Peking University (Hu and Xiao 2018; Liu et al. 2015; Yi et al. 2013). Additionally, they suggest muscle samples derived from the db/db mice and human patients show an elevation of MG53; however this has not been reproducible by any other groups studying MG53. IRS-1 is ubiquitinated by MG53, preventing insulin signaling; however, without its zinc-binding motifs, there is disruption of its ubiquitination function (Cai et al. 2015; Song et al. 2013). There are three other homologous proteins in the IRS family: IRS-2, IRS-3, and IRS-4, all of which contribute to insulin signal transduction in skeletal muscle. An absence of IRS-1 or IRS-3 is not enough to induce type II diabetes. Only through a combination of IRS-1 and IRS-3 deficiency does a manifestation of diabetic phenotypes result, indicating that IRS-1 and IRS-3 serve overlapping physiological functions in insulin signaling Thus, MG53-mediated IRS-1 downregulation cannot solely induce type II diabetes. Indeed, we show MG53 expression remains unchanged in skeletal muscle after a high fat diet that resulted in metabolic syndrome, and less MG53 is secreted into the blood stream (Ma et al. 2015). Additional toxicological studies on MG53 by a third party indicate that repeated i.v. administration of MG53 results in no adverse effects with no changes to blood glucose. In the future, it will be essential to clarify whether MG53 is involved in causing diabetes or if diabetes alters MG53 function.

Zinc-α2-glycoprotein (ZAG), an adipokine, is another zinc-binding protein that is involved with glucose metabolism in skeletal muscles. ZAG improves insulin responsiveness and increases lean muscle mass of gastric but not soleus skeletal muscles by increasing phosphorylation of AkT, mTOR, and IRS-1. Increased phosphorylation in this signaling pathway increases the expression of GLUT1 and GLUT4, improving glucose uptake (Russell and Tisdale 2010a, b; Gao et al. 2018). Indeed, reduced serum ZAG is strongly correlated with obesity from increased circulating serum glucose (Liu et al. 2018). ZAG may be a possible zinc-based therapeutic to diabetes and obesity; however caution must be taken with this approach as it was initially described in relation to causing muscle loss in cancer patients (Todorov et al. 1998). Indeed, ZAG was originally characterized in relation to cachexia and has more recently been associated with sarcopenia. Just as increased ZAG secretion from tumors results in cachexia, increased plasma ZAG strongly correlates with increased frailty, particularly in older women (Lee et al. 2016).

7.6 Future Perspectives

Zinc's functions, transport, and zinc-binding proteins discussed herein are by no means exhaustive. Rather, this chapter is intended to highlight some of the more recent translational works being done in the field of zinc signaling in skeletal muscle while touching on some seminal studies that have led the field to where it is today. Many review articles already exist on zinc signaling; some even address aspects of zinc in skeletal muscle. This just exemplifies the importance of zinc research to the field of medicine. There are, however, still questions to be answered about the regulation of zinc and zinc's regulation of homeostasis.

Even though zinc is clearly an integral part of nutrition to aid in maintaining healthy skeletal muscle, there are still many unknowns that remain. ZIPs and ZnTs are differentially expressed in satellite cells during differentiation into mature muscle fibers; however, very little is still known about the specific roles of most of these transporters during and after that process. For example, although we know insulin incites ZIP7 to release zinc from the Golgi apparatus, resulting in further insulin-mediated signaling, knocking out ZIP7 does not result in embryonic lethality, suggesting another compensatory mechanisms for metabolism in skeletal muscle. Elucidating this mechanism will be essential for future diabetic research.

Atrophy is a large field of research itself, and there are still transcription factors that regulate and are regulated by zinc whose functions are still being elucidated. There is still the possibility of creating a muscle-specific zinc-chelating therapeutic to counter atrophy, particularly to prevent or treat cachexia or muscle wasting due to aging. Targeting ZIP14 or other zinc transporters' expression in satellite cells to incite differentiation may be a candidate for muscle wasting diseases like muscular dystrophy to promote muscle growth. As suggested by the study performed by Summermatter et al. (Summermatter et al. 2017), finding some therapeutic agent to block metallothioneins may be one of the several therapies related to countering skeletal muscle cachexia if targeting ZIP14 does not work. These proposed therapeutics would need to be carefully studied since manipulating zinc handling for improved muscle contraction may result in obesity and vice versa.

MG53 secretion from skeletal muscle can be a physiological myokine for tissue crosstalk in regenerative medicine. MG53 contains binding sites for zinc. Dissecting the cellular and molecular function of Zn by modulating the myokine function of MG53 will be an important endeavor for future research.

References

Abdul-Ghani MA, DeFronzo RA (2010) Pathogenesis of insulin resistance in skeletal muscle. J Biomed Biotechnol 2010:476279

Alonso-Martin S, Rochat A, Mademtzoglou D, Morais J, de Reyniès A, Auradé F, Chang TH-T, Zammit PS, Relaix F (2016) Gene expression profiling of muscle stem cells identifies novel regulators of postnatal myogenesis. Front Cell Dev Biol 4:58

Aydemir TB, Troche C, Kim J, Kim M-H, Teran OY, Leeuwenburgh C, Cousins RJ (2016) Aging amplifies multiple phenotypic defects in mice with zinc transporter Zip14 (Slc39a14) deletion. Exp Gerontol 85:88–94
Bodine SC, Baehr LM (2014) Skeletal muscle atrophy and the E3 ubiquitin ligases MuRF1 and MAFbx/atrogin-1. Am J Physiol Metab 307:E469–E484
Cai C, Masumiya H, Weisleder N et al (2009) MG53 nucleates assembly of cell membrane repair machinery. Nat Cell Biol 11:56–64
Cai C, Lin P, Zhu H, Ko J-K, Hwang M, Tan T, Pan Z, Korichneva I, Ma J (2015) Zinc binding to MG53 protein facilitates repair of injury to cell membranes. J Biol Chem 290:13830–13839
Cao C-M, Zhang Y, Weisleder N et al (2010) MG53 constitutes a primary determinant of cardiac ischemic preconditioning. Circulation 121:2565–2574
Chasapis CT, Loutsidou AC, Spiliopoulou CA, Stefanidou ME (2012) Zinc and human health: an update. Arch Toxicol 86:521–534
Ching LL, Williams AJ, Sitsapesan R (2000) Evidence for Ca(2+) activation and inactivation sites on the luminal side of the cardiac ryanodine receptor complex. Circ Res 87:201–206
Colvin RA, Holmes WR, Fontaine CP, Maret W (2010) Cytosolic zinc buffering and muffling: their role in intracellular zinc homeostasis. Metallomics 2:306
Edman KAP, Grieve DW, Nilsson E (1966) Studies of the excitation-contraction mechanism in the skeletal muscle and the myocardium. Pflugers Arch Gesamte Physiol Menschen Tiere 290:320–334
Foster DM, Aamodt RL, Henkin RI, Berman M (1979) Zinc metabolism in humans: a kinetic model. Am J Physiol Integr Comp Physiol 237:R340–R349
Fujitani Y, Tamaki M, Fukunaka A, Watada H (2014) Zinc and its role in the pathogenesis of type 2 diabetes. In: Zinc signals in cellular functions and disorders. Springer, Tokyo, pp 269–283
Gao S-X, Guo J, Fan G-Q, Qiao Y, Zhao R-Q, Yang X-J (2018) ZAG alleviates HFD-induced insulin resistance accompanied with decreased lipid depot in skeletal muscle in mice. J Lipid Res 59:2277–2286
Goodyear LJ, Hirshman MF, Horton ES (1991) Exercise-induced translocation of skeletal muscle glucose transporters. Am J Physiol Metab 261:E795–E799
Gordon SJV, Fenker DE, Vest KE, Padilla-Benavides T (2018) Manganese influx and expression of ZIP8 is essential in primary myoblasts and contributes to activation of SOD2. bioRxiv 494542
Graham ZA, Gallagher PM, Cardozo CP (2015) Focal adhesion kinase and its role in skeletal muscle. J Muscle Res Cell Motil 36:305–315
Gushchina LV, Bhattacharya S, McElhanon KE, Choi JH, Manring H, Beck EX, Alloush J, Weisleder N (2017) Treatment with recombinant human MG53 protein increases membrane integrity in a mouse model of limb girdle muscular dystrophy 2B. Mol Ther 25:2360–2371
He B, Tang R, Weisleder N et al (2012) Enhancing muscle membrane repair by gene delivery of MG53 ameliorates muscular dystrophy and heart failure in δ-Sarcoglycan-deficient hamsters. Mol Ther 20:727–735
Hinz B (2007) Formation and function of the myofibroblast during tissue repair. J Invest Dermatol 127:526–537
Hu X, Xiao R-P (2018) MG53 and disordered metabolism in striated muscle. Biochim Biophys Acta – Mol Basis Dis 1864:1984–1990
Huang L (2014) Zinc and its transporters, pancreatic β-cells, and insulin metabolism. Vitam Horm 95:365–390
Huang L, Yu YY, Kirschke CP, Gertz ER, Lloyd KKC (2007) Znt7(Slc30a7)-deficient mice display reduced body zinc status and body fat accumulation. J Biol Chem 282:37053–37063
Huang L, Kirschke CP, Lay Y-AE, Levy LB, Lamirande DE, Zhang PH (2012) Znt7-null mice are more susceptible to diet-induced glucose intolerance and insulin resistance. J Biol Chem 287:33883–33896
Huang L, Tepaamorndech S, Kirschke CP, Newman JW, Keyes WR, Pedersen TL, Dumnil J (2018) Aberrant fatty acid metabolism in skeletal muscle contributes to insulin resistance in zinc transporter 7 (znt7)-knockout mice. J Biol Chem 293:7549–7563

Ilouz R, Kaidanovich O, Gurwitz D, Eldar-Finkelman H (2002) Inhibition of glycogen synthase kinase-3beta by bivalent zinc ions: insight into the insulin-mimetic action of zinc. Biochem Biophys Res Commun 295:102–106

Isaacson A, Sandow A (1963) Effects of zinc on responses of skeletal muscle. J Gen Physiol 46:655–677

Jinno N, Nagata M, Takahashi T (2014) Marginal zinc deficiency negatively affects recovery from muscle injury in mice. Biol Trace Elem Res 158:65–72

Khamzina L, Veilleux A, Bergeron S, Marette A (2005) Increased activation of the mammalian target of rapamycin pathway in liver and skeletal muscle of obese rats: possible involvement in obesity-linked insulin resistance. Endocrinology 146:1473–1481

Kimura T, Kambe T (2016) The functions of metallothionein and ZIP and ZnT transporters: an overview and perspective. Int J Mol Sci 17:336

Kristiansen S, Hargreaves M, Richter EA (1996) Exercise-induced increase in glucose transport, GLUT-4, and VAMP-2 in plasma membrane from human muscle. Am J Physiol Metab 270:E197–E201

Kristiansen S, Hargreaves M, Richter EA (1997) Progressive increase in glucose transport and GLUT-4 in human sarcolemmal vesicles during moderate exercise. Am J Physiol Metab 272:E385–E389

Kurosaka M, Ogura Y, Funabashi T, Akema T (2017) Early growth response 3 (Egr3) contributes a maintenance of C2C12 myoblast proliferation. J Cell Physiol 232:1114–1122

Lange S, Xiang F, Yakovenko A et al (2005) The kinase domain of titin controls muscle gene expression and protein turnover. Science 308:1599–1603

Larsson S, Karlberg I, Selin E, Daneryd P, Peterson HI (1987) Trace element changes in serum and skeletal muscle compared to tumour tissue in sarcoma-bearing rats. In Vivo 1:131–140

Latres E, Amini AR, Amini AA, Griffiths J, Martin FJ, Wei Y, Lin HC, Yancopoulos GD, Glass DJ (2005) Insulin-like growth factor-1 (IGF-1) inversely regulates atrophy-induced genes via the phosphatidylinositol 3-kinase/Akt/mammalian target of rapamycin (PI3K/Akt/mTOR) pathway. J Biol Chem 280:2737–2744

Leask A, Abraham DJ (2004) TGF-β signaling and the fibrotic response. FASEB J 18:816–827

Lecker SH, Jagoe RT, Gilbert A, Gomes M, Baracos V, Bailey J, Price SR, Mitch WE, Goldberg AL (2004) Multiple types of skeletal muscle atrophy involve a common program of changes in gene expression. FASEB J 18:39–51

Lee Y-P, Chang C-H, Liu H-H, Chen C-Y, Chen C-Y, Hsu C-C, Chang C-I, Lin Y-T, Lee C-S, Tsai J-S (2016) Plasma zinc alpha2-glycoprotein levels correlate positively with frailty severity in female elders. https://doi.org/10.1097/MD.0000000000004753

Levy S, Beharier O, Etzion Y et al (2009) Molecular basis for zinc transporter 1 action as an endogenous inhibitor of L-type calcium channels. J Biol Chem 284:32434–32443

Li YV (2014) Zinc and insulin in pancreatic beta-cells. Endocrine 45:178–189

Liu F, Song R, Feng Y et al (2015) Upregulation of MG53 induces diabetic cardiomyopathy through transcriptional activation of peroxisome proliferation-activated receptor α. Circulation 131:795–804

Liu M, Zhu H, Dai Y, Pan H, Li N, Wang L, Yang H, Yan K, Gong F (2018) Zinc-α2-glycoprotein is associated with obesity in Chinese people and HFD-induced obese mice. Front Physiol 9:62

Lodka D, Pahuja A, Geers-Knörr C et al (2016) Muscle RING-finger 2 and 3 maintain striated-muscle structure and function. J Cachexia Sarcopenia Muscle 7:165–180

Lynch CJ, Patson BJ, Goodman SA, Trapolsi D, Kimball SR (2001) Zinc stimulates the activity of the insulin- and nutrient-regulated protein kinase mTOR. Am J Physiol Metab 281:E25–E34

Ma H, Liu J, Bian Z et al (2015) Effect of metabolic syndrome on mitsugumin 53 expression and function. PLoS One 10:e0124128

Mashima H, Washio H (1964) The effect of zinc on the electrical properties of membrane and the twitch tension in frog muscle fibres | Academic Article. Jpn J Physiol 14:538–550

McClung JP, Tarr TN, Barnes BR, Scrimgeour AG, Young AJ (2007) Effect of supplemental dietary zinc on the mammalian target of rapamycin (mTOR) signaling pathway in skeletal muscle and liver from post-absorptive mice. Biol Trace Elem Res 118:65–76

Miranda ER, Dey CS (2004) Effect of chromium and zinc on insulin signaling in skeletal muscle cells. Biol Trace Elem Res 101:19–36

Mnatsakanyan H, Serra RS i, Rico P, Salmerón-Sánchez M (2018) Zinc uptake promotes myoblast differentiation via Zip7 transporter and activation of Akt signalling transduction pathway. Sci Rep 8:13642

Myers SA, Nield A, Myers M (2012) Zinc transporters, mechanisms of action and therapeutic utility: implications for type 2 diabetes mellitus. J Nutr Metab 2012:173712

Myers SA, Nield A, Chew G-S, Myers MA (2013) The zinc transporter, Slc39a7 (Zip7) is implicated in glycaemic control in skeletal muscle cells. PLoS One 8:e79316

Nguyen N, Yi J-S, Park H, Lee J-S, Ko Y-G (2014) Mitsugumin 53 (MG53) ligase ubiquitinates focal adhesion kinase during skeletal myogenesis. J Biol Chem 289:3209–3216

Nimmanon T, Ziliotto S, Morris S, Flanagan L, Taylor KM (2017) Phosphorylation of zinc channel ZIP7 drives MAPK, PI3K and mTOR growth and proliferation signalling. Metallomics 9:471–481

Norouzi S, Adulcikas J, Sohal SS, Myers S (2018) Zinc stimulates glucose oxidation and glycemic control by modulating the insulin signaling pathway in human and mouse skeletal muscle cell lines. PLoS One 13:e0191727

Ohashi K, Nagata Y, Wada E, Zammit PS, Shiozuka M, Matsuda R (2015) Zinc promotes proliferation and activation of myogenic cells via the PI3K/Akt and ERK signaling cascade. Exp Cell Res 333:228–237

Paskavitz AL, Quintana J, Cangussu D, Tavera-Montañez C, Xiao Y, Ortiz-Miranda S, Navea JG, Padilla-Benavides T (2018) Differential expression of zinc transporters accompanies the differentiation of C2C12 myoblasts. J Trace Elem Med Biol 49:27–34

Petrie L, Chesters JK, Franklin M (1991) Inhibition of myoblast differentiation by lack of zinc. Biochem J 276(Pt 1):109–111

Russell ST, Tisdale MJ (2010a) Antidiabetic properties of zinc-α 2-glycoprotein in ob/ob mice. Endocrinology 151:948–957

Russell ST, Tisdale MJ (2010b) Mechanism of attenuation of skeletal muscle atrophy by zinc-α2-glycoprotein. Endocrinology 151:4696–4704

Shusterman E, Beharier O, Levy S et al (2017) Zinc transport and the inhibition of the L-type calcium channel are two separable functions of ZnT-1. Metallomics 9:228–238

Siren PMA, Siren MJ (2010) Systemic zinc redistribution and dyshomeostasis in cancer cachexia. J Cachexia Sarcopenia Muscle 1:23–33

Song R, Peng W, Zhang Y et al (2013) Central role of E3 ubiquitin ligase MG53 in insulin resistance and metabolic disorders. Nature 494:375–379

Summermatter S, Bouzan A, Pierrel E et al (2017) Blockade of metallothioneins 1 and 2 increases skeletal muscle mass and strength. Mol Cell Biol. https://doi.org/10.1128/MCB.00305-16

Takeshima H, Nishimura S, Matsumoto T et al (1989) Primary structure and expression from complementary DNA of skeletal muscle ryanodine receptor. Nature 339:439–445

Taylor SR, Preiser H, Sandow A (1972) Action potential parameters affecting excitation-contraction coupling. J Gen Physiol 59:421–436

Taylor KM, Hiscox S, Nicholson RI, Hogstrand C, Kille P (2012) Protein kinase CK2 triggers cytosolic zinc signaling pathways by phosphorylation of zinc channel ZIP7. Sci Signal 5:ra11–ra11

Thiebaud D, Jacot E, DeFronzo RA, Maeder E, Jequier E, Felber JP (1982) The effect of graded doses of insulin on total glucose uptake, glucose oxidation, and glucose storage in man. Diabetes 31:957–963

Todorov PT, McDevitt TM, Meyer DJ, Ueyama H, Ohkubo I, Tisdale MJ (1998) Purification and characterization of a tumor lipid-mobilizing factor. Cancer Res 58:2353–2358

Tuncay E, Bitirim VC, Durak A, Carrat GRJ, Taylor KM, Rutter GA, Turan B (2017) Hyperglycemia-induced changes in ZIP7 and ZnT7 expression cause Zn2+ release from the sarco(endo)plasmic reticulum and mediate ER stress in the heart. Diabetes 66:1346–1358

Urso ML, Scrimgeour AG, Chen Y-W, Thompson PD, Clarkson PM (2006) Analysis of human skeletal muscle after 48 h immobilization reveals alterations in mRNA and protein for extracellular matrix components. J Appl Physiol 101:1136–1148

Voelkel T, Linke WA (2011) Conformation-regulated mechanosensory control via titin domains in cardiac muscle. Pflugers Arch 462:143–154

Wang H, Chen K-Y, Wei Q-Q, Zhu P-H, Cheng X-Y (2003) Inhibition of ryanodine binding to sarcoplasmic reticulum vesicles of cardiac muscle by Zn2+ ions. Cell Physiol Biochem 11:83–92

Wang C, Vernon R, Lange O, Tyka M, Baker D (2010) Prediction of structures of zinc-binding proteins through explicit modeling of metal coordination geometry. Protein Sci 19:NA–NA

Wang G, Biswas AK, Ma W et al (2018) Metastatic cancers promote cachexia through ZIP14 upregulation in skeletal muscle. Nat Med 24:770–781

Wastney ME, Henkin RI (1988) Development and application of a model for zinc metabolism in humans. Prog Food Nutr Sci 12:243–254

Wastney ME, Aamodt RL, Rumble WF, Henkin RI (1986) Kinetic analysis of zinc metabolism and its regulation in normal humans. Am J Physiol 251:R398–R408

Wei R, Wang X, Zhang Y et al (2016) Structural insights into Ca(2+)-activated long-range allosteric channel gating of RyR1. Cell Res 26:977–994

Weisleder N, Takizawa N, Lin P et al (2012) Recombinant MG53 protein modulates therapeutic cell membrane repair in treatment of muscular dystrophy. Sci Transl Med 4:139ra85

Wijesekara N, Chimienti F, Wheeler MB (2009) Zinc, a regulator of islet function and glucose homeostasis. Diabetes Obes Metab 11:202–214

Woodier J, Rainbow RD, Stewart AJ, Pitt SJ (2015) Intracellular zinc modulates cardiac ryanodine receptor-mediated calcium release. J Biol Chem 290:17599–17610

Wu Y, Lu H, Yang H, Li C, Sang Q, Liu X, Liu Y, Wang Y, Sun Z (2016) Zinc stimulates glucose consumption by modulating the insulin signaling pathway in L6 myotubes: essential roles of Akt–GLUT4, GSK3β and mTOR–S6K1. J Nutr Biochem 34:126–135

Xia RH, Cheng XY, Wang H, Chen KY, Wei QQ, Zhang XH, Zhu PH (2000) Biphasic modulation of ryanodine binding to sarcoplasmic reticulum vesicles of skeletal muscle by Zn2+ ions. Biochem J 345(Pt 2):279–286

Yamasaki S, Hasegawa A, Hojyo S, Ohashi W, Fukada T, Nishida K, Hirano T (2012) A novel role of the L-type calcium channel α1D subunit as a gatekeeper for intracellular zinc signaling: zinc wave. PLoS One 7:e39654

Yan Z, Bai X, Yan C et al (2015) Structure of the rabbit ryanodine receptor RyR1 at near-atomic resolution. Nature 517:50–55

Yi J-S, Park JS, Ham Y-M et al (2013) MG53-induced IRS-1 ubiquitination negatively regulates skeletal myogenesis and insulin signalling. Nat Commun 4:2354

Chapter 8
Zinc Signaling in Aging Heart Function

Belma Turan, Deniz Billur, and Yusuf Olgar

Abstract Zinc is an important micronutrient for mammalians, whereas free/labile zinc ion (Zn^{2+}) level ($[Zn^{2+}]_i$) in cells can be detrimental if it is over the physiological range. The cellular $[Zn^{2+}]_i$ is regulated by proteins, responsible for its influx (ZIP family) into cells or efflux (ZnT-family) from cells. In addition, it has been shown that there is a close relationship between cellular $[Zn^{2+}]_i$ and increasing oxidative stress, and both are also responsible for the dysregulation of the excitation/contraction coupling in the heart. Although age-dependent changes in the structure and function of the left ventricle are closely related to fibrosis, cellular evidence showed the importance of mitochondria as a potential target for aging medicine. However, several studies have shown that zinc supplementation provides important improvements in the maintenance of cardiovascular disorders, and it is needed to know the exact role of $[Zn^{2+}]_i$ and its transporters in aging heart function. Although there is very little information related to the role of Zn^{2+} transporters in mammalian aging, some studies have shown their importance in cardiovascular disorders including failing human heart. Since few studies examined the role of ZIP14 as an inflammation responsive transporter and amplified with aging, we demonstrated an important data on the ZIP14 status in organelles such as the sarcoplasmic reticulum and mitochondria in aging rat's ventricular cardiomyocytes. As a summary, we discussed our data in the light of literature data as well.

Keywords Heart · Free zinc ion · Mitochondria · Aging · Zinc-transporters · ZIP14

B. Turan (✉) · Y. Olgar
Department of Biophysics, Faculty of Medicine, Ankara University, Ankara, Turkey
e-mail: belma.turan@medicine.ankara.edu.tr

D. Billur
Department of Histology-Embryology, Faculty of Medicine, Ankara University, Ankara, Turkey

T. Fukada, T. Kambe (eds.), *Zinc Signaling*,
https://doi.org/10.1007/978-981-15-0557-7_8

8.1 Introduction

Aging is a physiological process, genetically programmed and modified by environmental influences and lifestyle of individuals. Although the percentage of aging people among the populations is increasing worldwide, the rate of aging can vary widely among individuals (Strait and Lakatta 2012; Heidenreich et al. 2013). It is well known that the aging process is not a pathological condition, but it is a continuum progressing throughout the life span of humans. Understanding the basis of mammalian aging and producing innovative treatment approaches will surely improve the life of aging people.

In the general aspects of this progress, the gradual loss in many organ functions of humans is named as physiologic aging. Coronary artery disease (CAD) has also an increased risk factor between aging people, while there is also an increased incidence of cerebrovascular, renal, and pulmonary disease that can accelerate the loss of function, particularly leading to heart failure (HF). Therefore, it is an obvious reality why HF is a consequence of age-associated disease (Shioi and Inuzuka 2012; Veronica and Esther 2012; Quarles et al. 2015; Bradford et al. 2017). The aging process in humans can be defined as a combination of morphological and functional changes and can take place over time, in part, related to decreased elasticity in tissues (Kirkwood 2005). In aging humans, there exists increased resistance to cardiac pumping which, in turn, induces an increased workload against that increased resistance being the underlying cause of HF in aging individuals (Quarles et al. 2015). In the aging heart, there are marked increases in the wall thickness and mass of the left ventricle. At the cellular level, all these changes result in a significant increase in cardiomyocyte dimension with a marked decrease in the number of alive and functional cardiomyocytes (Olgar et al. 2018a).

Both experimental and clinical studies pointed out the association between aging and both whole body and cardiac insulin resistance (Paolisso et al. 1999; Kannel 1999; Schocken 2000; Wilson and Kannel 2002; Gottdiener et al. 2002; Barbieri et al. 2001; Boudina 2013). Insulin resistance is present in 30–50% of humans over the age of 65 years (Kannel 1999; Wilson and Kannel 2002; Senni et al. 1999), while it is about 60–70% in aging experimental animals (Olgar et al. 2018a). However, there are important controversies between results, particularly related to humans, whether insulin resistance is an absolute consequence of aging or results of increased body weight-related complications (Paolisso et al. 1999; Barbieri et al. 2001). As an answer to these controversies, we demonstrated that the aging male rats (24 months old) (about their 70) had marked insulin resistance compared to those of adult male rats (6 months old) even if they did have normal body weight and blood glucose level (Olgar et al. 2018a).

It has been also mentioned that mitochondrial dysfunction has an important role in aging-associated disorders in mammalians since the properly working mitochondria are vital for mammalian cells. However, there are many controversies in this field with several unresolved questions (Payne and Chinnery 2015). The main role of mitochondria is to provide the continuous ATP supply and modulate the cytosolic

Ca^{2+} signaling, influencing cellular reactive oxygen species (ROS) levels and regulating the redox state of cells (Beal 2002; Ristow 2012; Lee and Oh 2010). We, in our early studies, have demonstrated that both external and internal oxidants, including ROS, caused a 30-fold increase in cytosolic free Zn^{2+} level in cardiomyocytes ($[Zn^{2+}]_i$) parallel to only twofold increase in $[Ca^{2+}]_i$ (Turan et al. 1997). Furthermore, the cellular ATP level has significant effects on both steady-state K^+-channel currents and inward-rectifier K^+ channel currents, while there is no effect on transient K^+-channel currents (Degirmenci et al. 2018). Furthermore, the high $[Zn^{2+}]_i$ could induce marked activation in ATP-sensitive K^+ channel currents, depending on the cellular ATP levels (Degirmenci et al. 2018). These data point out that high $[Zn^{2+}]_i$ can be considered as much more biologically toxic than that of high $[Ca^{2+}]_i$. Moreover, the studies strongly pointed out the role of cellular high $[Zn^{2+}]_i$ on the function of mitochondria via movement of $[Zn^{2+}]_i$ between cytosolic and mitochondrial pools by affecting the opening of permeability transition pores together with mitochondrial membrane potential and release of some proapoptotic agents from the mitochondria in the heart preparations (Jiang et al. 2001; Sensi et al. 2003; Tuncay et al. 2013; Tuncay et al. 2019; Billur et al. 2016). Therefore, the present chapter focused on the documents of already presented data in this field, particularly associated with the role of $[Zn^{2+}]_i$ and Zn^{2+} transporters in aging heart function.

8.2 Role of Zinc in Mammalian Heart Function

Trace element zinc is a multipurpose element for the mammalian body and plays an important role for the regulation of several cellular signaling mechanisms such as zinc ion, Zn^{2+} (Stefanidou et al. 2006). Zn^{2+}, being a catalytic and structural component of proteins, contributes to cellular proliferation in mammalian cells, including cardiomyocytes (Berg and Shi 1996; Tatsumi and Fliss 1994). The Zn^{2+} dyshomeostasis in the body can lead to many diseases, including neurological and cardiovascular diseases (Prasad et al. 1990; Prasad 2013; Evans 1986; Jurowski et al. 2014). Zinc compounds are used as therapeutic agents in many diseases not only in those associated with zinc deficiency but also in acute diarrhea in children, Wilson's disease, and diseases associated with aging such as prevention of both blindness and infection-related disorders in the elderly (Prasad 2008, 2013; Jurowski et al. 2014; Prasad et al. 1993). Following new discoveries on the role of Zn^{2+} in biological systems, particularly related to its second messenger role in the immune system, it gets much more attention for the clinical studies (Yamasaki et al. 2007; Haase and Rink 2014; Hojyo and Fukada 2016). Both external and internal stimuli can induce the release of free Ca^{2+} from metalloproteins (Turan et al. 1997; Tuncay et al. 2011) and from intracellular Zn^{2+} stores such as in the perinuclear area, at most from the endoplasmic reticulum (ER) in mast cells (Yamasaki et al. 2007) as well as sarco(endo) plasmic reticulum, S(E)R, in cardiomyocytes, which is a phenomenon named as Zn^{2+} wave (Tuncay et al. 2011, 2017). Zn^{2+} release can occur during resting state of cardiomyocytes, similar to that of Ca^{2+} release, and is named as Zn^{2+} sparks as well

(Tuncay et al. 2011). It has been also shown that Zn^{2+} wave was dependent on Ca^{2+} influx besides the activation of some kinases (Tuncay et al. 2011; Hershfinkel et al. 2001), while that Zn^{2+} wave can contribute to various intracellular signaling pathways.

Besides studies about the relationship between zinc deficiency and associated diseases in mammalians, several clinical trials with randomized controls performed in patients with type 2 diabetes have shown that zinc supplementation provides important improvements in the process of glycemic control (Capdor et al. 2013) and lipidemia control (Foster et al. 2010), which are closely related to the maintenance of cardiovascular disorders (CVDs). There are many articles focused on the relationship between zinc status and CVDs in mammalians. For the demonstration of the relationship between zinc status and CVD in humans, the authors monitored zinc intake or serum zinc level in patients (Soinio et al. 2007; Mursu et al. 2011; de Oliveira Otto et al. 2011). Some studies demonstrated low serum zinc level in atherosclerosis, coronary artery disease, angina, and cardiac ischemia (Eby and Halcomb 2006; Giannoglou et al. 2010; Efeovbokhan et al. 2014). More importantly, a higher rate of HF in patients with zinc deficiency has been also emphasized. The systemic body level of zinc, although it participates in a number of biochemical signaling mechanisms in mammalian cells, is not a good biomarker for any pathological condition, particularly associated with many CVDs (Pan et al. 2017). The body mass zinc level of humans is about 2–3 g, while the heart contains 0.4% of body zinc (King et al. 2000; Jackson et al. 1982).

Although there are conflicting reports on the relationship between serum zinc levels and HF, when statistical adjustments are performed due to age and sex, most of the data reported no significant association between zinc intake and CVDs. Moreover, Otto et al. reported a greater risk of CVDs with increased dietary zinc intake (de Oliveira Otto et al. 2011). However, it has been also announced that CVD risk was not associated with total dietary zinc intake. On the other hand, some studies pointed out the importance of excessive zinc intake in CVDs due to its highly toxic effects (Turan et al. 1997; Brown et al. 2000; Fosmire 1990; Plum et al. 2010; Romanjuk et al. 2016). Interestingly, in a recent study, authors take into consideration the role of zinc in immunity, energy metabolism, and antioxidative defense system, and they determine the zinc status of trained athletes compared with control individuals. Their meta-analysis demonstrated that athletes have lower serum zinc level, although they have higher total dietary zinc intake (Chu et al. 2018).

The basal level of intracellular free Zn^{2+} in mammalian cells ($[Zn^{2+}]_i$) is generally very low since it is mostly found to be bound to proteins, while this level is increased under external and/or internal stimulations. The $[Zn^{2+}]_i$ in ventricular rabbit and rat cardiomyocytes was measured as less than 1 nmol under physiological condition, while this level increased markedly under electrical stimulation, similar to that of the cytosolic level of free Ca^{2+} ($[Ca^{2+}]_i$) (Turan et al. 1997; Tuncay et al. 2011; Chabosseau et al. 2014). We and others have demonstrated the significant modulation of $[Zn^{2+}]_i$ in cardiomyocytes with the redox state of the cells, including under both increases in ROS and reactive nitrogen species (RNS) (Turan et al. 1997; Tuncay et al. 2011; St Croix et al. 2002; Jang et al. 2007). In addition, it has been demonstrated that $[Zn^{2+}]_i$ in ventricular cardiomyocytes from rats could increase markedly under hyperglycemia, at most, due to its uncontrolled leaking from S(E)R through opposing changes in the expression of Zn^{2+} transporters, ZIP7 and ZnT7, with important contribution to cardiac dysfunction in diabetic mammalians (Tuncay et al. 2017). Consequently, these experimental studies clearly provided important information related to the detrimental contribution of excess $[Zn^{2+}]_i$ in ventricular cardiomyocytes via affecting excitation-contraction coupling.

8.3 Aging and Insufficient Heart Function

It is clear that not only humans but also all alive species in the natural are aging. For humans, aging is a physiological process and started with their birth. The physiologic changes are, therefore, accompanying their maturation process. All over the world, the percentage of elderly individuals is rising among the populations; however, proper maintenance of their lives is considered an essential clinical handicap (Mirzaei et al. 2016; Hammadah et al. 2017; Ferrucci et al. 2018; Zhang et al. 2018; Anderson et al. 2019). Many findings in aging individuals and/or elderly patients are accepted as abnormal events in younger patients, although they are not directly responsible for a special symptom since symptoms in older individuals often can arise due to multiple alterations in their body.

Aging in humans is one of the major risk factors for HF, while elderly individuals are more likely to develop the cardiac disease, and they cannot recover properly, following a heart attack. In spite of healthy lifestyle desires, with many proposals for this program, such as exercise and well-balanced daily diet, it seems impossible to avoid the ill effects of aging in humans. As mentioned in the introduction, the marked structural and functional distortions are common alterations in the heart of aging humans, which are the dominant risk factors for cardiovascular diseases, which are the leading cause of death among almost all humans with increasing elderly populations (Shioi and Inuzuka 2012; Veronica and Esther 2012; Quarles et al. 2015; Bradford et al. 2017; Chiao and Rabinovitch 2015). The prevalence rate of cardiovascular diseases for Americans 60–79 years of age is greater than 70%, while it is higher than 80% for Americans age greater than 80 years (Douaud et al. 2014).

Although the phenotypes of cardiac aging have been well characterized, several recent studies have emphasized the contribution of multiple molecular mechanisms underlying the development of cardiac aging, such as altered nutrients and related changes in growth signaling as well as changes in insulin-like growth factor-1 (IGF-1) signaling (Puche et al. 2008; Kennedy et al. 2007; de Rivera et al. 2006; Wessells et al. 2004). In addition, remodeling in the adverse extracellular matrix and chronic activation in neurohormonal signaling are involved in the development of aging insufficient heart function. Moreover, recent studies have shown that cellular impaired $[Ca^{2+}]_i$ homeostasis and overproduction of superoxides in the mitochondria and mitochondrial dysfunction are the contributors to cardiac aging in mammalians (Strait and Lakatta 2012; Shioi and Inuzuka 2012; Chiao and Rabinovitch 2015; Borlaug and Kass 2006; Dai et al. 2009; Judge et al. 2005; Zieman and Kass 2004; Dai et al. 2014).

8.4 Role of the Mitochondria in Aging Heart

It is well accepted that any disruption in the basal level $[Ca^{2+}]_i$ is closely associated with the impartment of mitochondria and the altered level of mitochondrial free Ca^{2+} ($[Ca^{2+}]_{mit}$) in cardiomyocytes, particularly under pathological condition (Shioi and Inuzuka 2012; Dai et al. 2014). Moreover, the altered level of $[Ca^{2+}]_{mit}$ can lead to important impairments in different intracellular signal transduction pathways, including cellular oxidative stress and apoptosis. In these regards, it can be a noticeable fact that the acceleration rate of most studies in the field of mitochondria and CVD is related with the demonstration of the importance of mitochondria as a potential target for mitochondrial medicine (Giorgi et al. 2012; Nazarewicz et al. 2013a, b; Chung et al. 1998). Taking these studies into consideration, one can point out why aging is associated with not only high $[Ca^{2+}]_i$ but also high $[Ca^{2+}]_{mit}$-related mitochondrial dyshomeostasis mechanisms in cardiomyocytes (Raza et al. 2007), in a similar manner to that of neurons (Thayer and Miller 1990; Herraiz-Martinez et al. 2015; Santulli et al. 2015). Even early studies emphasized the role of mitochondria in energy production and in the modulation of $[Ca^{2+}]_{mit}$ homeostasis in cardiomyocytes, being fundamental players in several processes for heart function (Carafoli and Lehninger 1971; Denton et al. 1980; Rizzuto et al. 1992). In other words, mitochondria in mammalian cells are critical organelle related to metabolic energy generation through oxidative phosphorylation, participating in the redox state of the cells, including $[Ca^{2+}]_i$ homeostasis (Stanley et al. 2005). Therefore, the loss of cellular $[Ca^{2+}]_i$ homeostasis leads to cellular dysfunction through the impaired mitochondrial function. Supporting this statement, recent studies have shown the importance and the molecular structure of the main proteins involved in the handling of $[Ca^{2+}]_{mit}$ for mitochondria-based pharmacological strategies, including mitochondrial fusion and fission proteins (Payne and Chinnery 2015; Giorgi et al. 2012; Dibb et al. 2004; Eisner et al. 2017). The fusion process involves mitofusin 1 and 2 (Mfn1 and Mfn2), and they mediate outer mitochondrial membrane fusion. In addition, optic atrophy protein 1 (OPA1) regulates inner mitochondrial membrane fusion in mammalian cells (Mishra and Chan 2016). On the other hand, the fission process is mediated by dynamin-related protein 1 (Drp1) and mitofission protein Fis1, which are closely related with a polarized and a depolarized mitochondrion (Hamacher-Brady and Brady 2016).

If one wants to give detailed information about the role of mitochondria in elderly heart function, the statement on the important contribution of mitochondria to aging due to the balancing key role in the cellular processes is very revealing. Age-associated cellular changes are affecting the mitochondrial membrane potential leading to mitochondrial depolarization via the opening of membrane transition pores (Payne and Chinnery,2015; Elmore et al. 2001). Therefore, it can be summarized that although aging is a physiological process, it is closely associated with not only decreased mitophagy but also with impaired mitochondrial functions (Moreira et al. 2017). If any of these essential processes are changed, the maintenance of a healthy mitochondrial function is disrupted, further leading to organ dysfunction (i.e., HF in mammalians). Considering the radius of muscle fibers in heart tissue (H&E staining) from aging rats with marked insulin resistance, there was a 60% significant increase in systolic and diastolic function compared to those of adult rats (Olgar et al. 2018a). Light microscopy analysis of those tissues stained with Mallory's Azan presented significant loss in the integrity of heart muscle myofibrils and increases in the connective tissue around the myofibrils together with marked fibrous increases in interstitial regions. Furthermore, marked irregularity in the mitochondria including their fractionation around the myofilaments and higher amount of lysosomes were detected with electron microscopy examinations in aging rat's left ventricular tissue (Olgar et al. 2018a). Moreover, there were an apparent irregular arrangement of myofilaments and Z-lines and enlarged T-tubules in this aging group. Indeed, previous studies also demonstrated that aging is associated with not only systemic insulin resistance but also myocardial insulin resistance, being independent of body weight. It is closely related to mitochondrial dysfunction associated with impairment in the insulin action of cells (Bhashyam et al. 2007).

As can be seen in Fig. 8.1, there are a number of changes in the morphology of left ventricular heart tissue from insulin-resistant aging rats (24-month-old Wistar male rats) compared to those of adults (6-month-old Wistar male rats). The light microscopy photomicrographs showed that there was a marked muscle hypertrophy due to enlarged diameter in age group samples stained with hematoxylin and eosin (B) with normal appearance in the adult group (A). The left ventricular heart tissue samples stained with Mallory's Azan also demonstrated important morphological changes including an increased amount of collagen fiber (blue stained) in aging rats (D) with normal muscle fibers (red stained) in the adult group (C). All these light microscopy data clearly indicate the existence of fibrosis in insulin-resistant aging heart tissue. Electron microscopy analysis performed in isolated left ventricular cardiomyocytes also demonstrated the important alterations in age group such as irregularly partitioned and clustered intermyofibrillar mitochondria (in part via increased fission process), enlargement in T-tubules, swelling and vesiculation of the sarcoplasmic reticulum (SR), and irregularly arranged Z-lines (Fig. 8.2b–d). However, there were regularly arranged intermyofibrillar mitochondria, T-tubules, SR, and the components of the sarcomere with normal Z-lines in the adult group (Fig. 8.2a). Therefore, both experimental and clinical data strongly demonstrate the essential role in the development of the aging-associated processes in cardiac insufficient function (Lesnefsky et al. 2016).

Indeed, early studies demonstrated that mitochondrial function is impaired in aging mammalian's myocardium at various levels including the activity of the electron transport chain as well as depressed antioxidant capacity of the mitochondria in aging heart samples (Corsetti et al. 2008; Tate and Herbener 1976; Schmucker and Sachs 1985). An increased ROS formation in cells, at most associated with increased

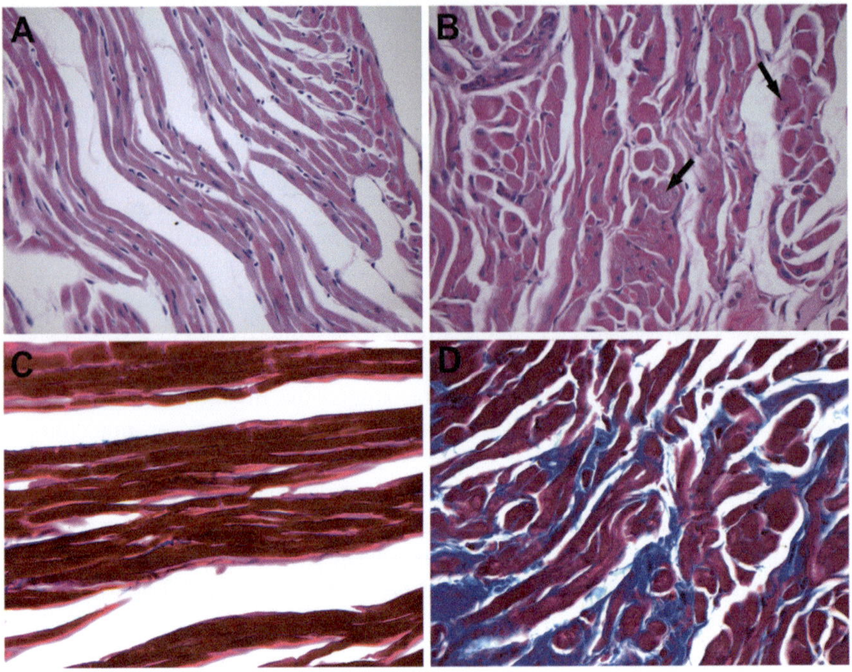

Fig. 8.1 Light microscopy photomicrographs of the left ventricular heart tissue. Heart tissues were prepared for light microscopy analysis, as described elsewhere (Billur et al. 2016). Light micrographs showed normal appearance in adult group samples (6-month-old male Wistar rats) stained with either hematoxylin and eosin or Mallory's Azan (**a** and **c**, respectively) while there was marked cardiac muscle hypertrophy (arrow) in hematoxylin and eosin-stained or Mallory's Azan-stained age group samples (24-month-old male Wistar rats) (**b** and **d**, respectively). The blue color indicated the collagen fiber (age samples), and the red indicated the muscle fiber (adult group) in tissue samples stained with Mallory's Azan. Magnification ×400

$[Ca^{2+}]_{mit}$ and depressed antioxidant capacity of mitochondria, is associated with an increased number of damaged and/or dead cells. Indeed, either mouse with a mitochondrial-targeted catalase overexpression or cardiomyocytes treated with direct mitochondria targeting antioxidant (MitoTEMPO) demonstrated marked cardioprotection against aging-associated alterations in the heart (Dai et al. 2014; Schriner et al. 2005). However, the exact mechanisms by which aged mitochondria affect cardiac function are not exactly shown, but it seems increased ROS generations and reduced antioxidant capacity of the mitochondria in aged hearts are important candidates responsible for cardiovascular dysfunction. Consistent with these statements, studies have shown the association between $[Ca^{2+}]_{mit}$ increase in mitochondria together with increases in mitochondrial superoxide production and decreases in ATP production (Mansouri et al. 2006). All of them are the main oxidative stress sources in HF, further leading to electrical instability, at most, responsible for sudden cardiac death (Dey et al. 2018; Owada et al. 2017).

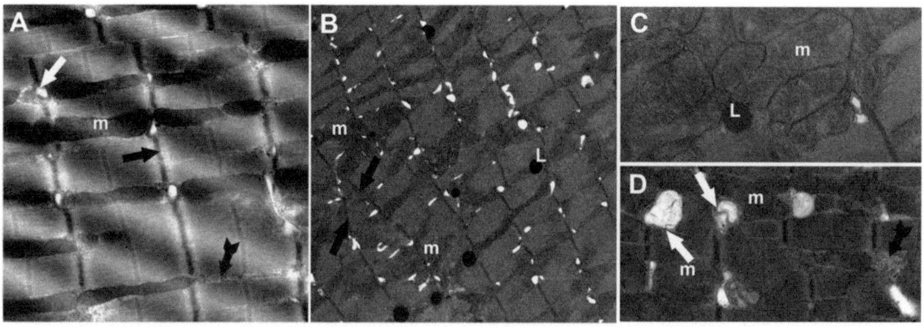

Fig. 8.2 Representative electron microscopy photomicrographs of cardiomyocytes isolated from the left ventricle of the heart. The electron microscopic examination of isolated cardiomyocytes in the adult group (6-month-old male Wistar rats) showed regularly arranged intermyofibrillar mitochondria, T-tubules, SR, and the components of sarcomere including Z-lines (**a**). Representative electron micrographs showed irregularly partitioned and clustered intermyofibrillar mitochondria, enlarged T-tubules, swelling and vesiculation of SR, and irregularly arranged Z-lines in isolated left ventricular cardiomyocytes from aged group (24-month-old male Wistar rats) (**b**–**d**). Shortened symbols: *m* mitochondrion, *L* lysosome, *arrow* Z line, *white arrow* T tubules, *tailed arrow* sarcoplasmic reticulum. Magnification for A,C,D ×21,560 and magnification for B ×10,000

All these studies indicate that mitochondria-targeting antioxidants, like MitoTEMPO, have important cardioprotective effect in age mammalian heart, besides other therapies (Boengler et al. 2017; Jeong et al. 2016). Consequently, increases in mitochondrial oxidative stress play an important role in the development of age-related functional alterations, and, therefore, any enhancement in mitochondrial antioxidant defense can provide important therapeutic benefits in elderly cardiac complications.

8.5 Role of Zinc in Mitochondrial Dysfunction in Aging Heart

The zinc level in heart tissues is about or less than 1 g and is shown to be positively correlated with ejection fraction in humans (Jackson et al. 1982; Oster et al. 1993). Furthermore, Turan et al., for the first time, demonstrated that its free concentration in mammalian cardiomyocytes loaded with specific fluorescence dye was less than 1 nM, while Palmer et al. showed Zn^{2+} localization into the sarcomere I-band (Turan et al. 1997; Palmer et al. 2006). Cellular free Zn^{2+} has an important role in cardiomyocytes affecting Ca^{2+} signaling for the regulation of cellular responses (Yi et al. 2013) as well as in excitation-contraction coupling to the organization of Ca^{2+} dynamics (Tuncay et al. 2011). Miyamoto et al. performed a work focusing on the role of Ca^{2+} dysregulation on the induction of mitochondrial depolarization and apoptosis in activated heterotrimeric G protein α subunit ($G\alpha_q$) that expressed cardiomyocytes with measuring cytosolic Ca^{2+} and mitochondrial membrane potential, simultaneously, by confocal microscopy (Miyamoto et al. 2005). In a later study, Raza et al. demonstrated the relationship between aging and elevated $[Ca^{2+}]_i$ and altered Ca^{2+} homeostasis-associated mechanisms in hippocampal neurons, while Giorgi et al. discussed widely the importance of mitochondrial Ca^{2+} homeostasis as potential target event for mitochondrial medicine (Giorgi et al. 2012; Raza et al. 2007). Furthermore, the interactive roles of Zn^{2+} and Ca^{2+} in mitochondrial dysfunction have been studied and demonstrated that depolarization-induced Zn^{2+} uptake induced high amount of cell death and enhanced accumulation of $[Ca^{2+}]_{min}$ in Ca^{2+}- and Zn^{2+}-containing media (Pivovarova et al. 2014), whereas Pitt and Stewart examined the role of $[Ca^{2+}]_i$ in the regulation of Ca^{2+} release in mammalian cardiomyocytes (Pitt and Stewart 2015). In addition, Joseph et al. demonstrated mitochondrial oxidative stress-associated increased SR Ca^{2+} leak via hyperphosphorylated RyR2 in cardiac lipid overload condition (Joseph et al. 2017).

However, the exact role of Zn^{2+} in heart function under physiological condition is not known very well yet. Since zinc, as an essential micronutrient, is important for the functioning of the immune system (Doets et al. 2012) and zinc deficiency is particularly very common in elderly humans (Joshi and Bakowska 2011), zinc-rich diet can be recommended to elderly population. Furthermore, the role of intracellular Zn^{2+} release and Zn^{2+} toxicity in the brain, Zn^{2+} dyshomeostasis, and neuronal injury and the role of Zn^{2+} in aging, oxidative stress, and Alzheimer's disease are widely documented (McCord and Aizenman 2014). Moreover, the role zinc in aging and immunosenescence is widely discussed by Cabrera and others focusing on documents related to zinc and oxidative inflammatory aging (Haase and Rink 2014; Cabrera 2015).

The importance of zinc in age-related conditions cannot be related only with cellular $[Zn^{2+}]_I$ deficiency. Studies demonstrated that any alteration in $[Zn^{2+}]_I$ homeostasis can arise due to alterations in Zn^{2+}-transporting proteins, particularly in CVDs (ZIP and ZnT families) (Tuncay et al. 2018; Olgar et al. 2018b; Mocchegiani et al. 2010). Supporting these statements, Giacconi et al. determined the effect of Zn^{2+} influx into the lymphocytes from young adults than in lymphocytes from elderly donors via the upregulation of the ZIP genes (Giacconi et al. 2012).

To develop new sensitive and specific biomarkers for the determination of free Zn^{2+} in cells with reliable techniques is a new research field. Another emerging aim in this field is to quantify and cellularly locate Zn^{2+} transporters. In these regards, by using FRET-based recombinant-targeted Zn^{2+} probes, authors demonstrated that the Zn^{2+} level in sarco(endo)plasmic reticulum ($[Zn^{2+}]_{ser}$) is much higher than the cytoplasm while it is less than the mitochondria, $[Zn^{2+}]_{mit}$ (Chabosseau et al. 2014). Both light and electron microscopy analyses also showed that increased $[Zn^{2+}]_i$ in cardiomyocytes could induce marked increases in the size of the cells, indicating hypertrophic cells, markedly increasing the mitochondrial matrix to the cristae area together with marked clustering and vacuolated mitochondrion (Billur et al. 2016). In confocal studies, authors showed the role of increased $[Zn^{2+}]_i$ on the opening of mitochondrial permeability transition pores which further leads to loss of mitochondrial membrane potential (Sensi et al. 2003; Sharpley and Hirst 2006; Kumari et al. 2017; Xi et al. 2009). However, Rajapaksed et al. demonstrated how externally zinc exposure-protected oxidative stress is associated with retinal pigment epithelium death, due to its reducing action on damaged mitochondria (Rajapakse et al. 2017).

Freshly isolated left ventricular cardiomyocytes, when exposed to zinc compounds such as zinc ionophore and zinc pyrithione (ZnPT), showed marked degenerative changes including damages (swelling and cytoplasmic vacuolizations) in light microscopic examinations of left ventricular cardiomyocytes isolated from adult (6-month-old) male rats in a concentration-dependent manner (Fig. 8.3b–d) compared to the controls (Fig. 8.3a). Electron microscopy examination of these isolated cardiomyocytes demonstrated gradual enlargement in T-tubules and fragmentation in the mitochondria with increasing concentrations of ZnPT-incubated cells (Fig. 8.4b–d). The cells incubated without ZnPT showed normal appearance of SR, regularly arranged intermyofibrillar mitochondria and T-tubule, and normal-sized Z-lines (Fig. 8.4a).

Despite the central role of the mitochondria in metabolism, the relationship between mitochondrial quality and insulin action remains unclear. In these regards, Jeong et al. performed a study in order to show the role of mitochondrial oxidative stress in the development of cardiac diastolic dysfunction in metabolic syndrome

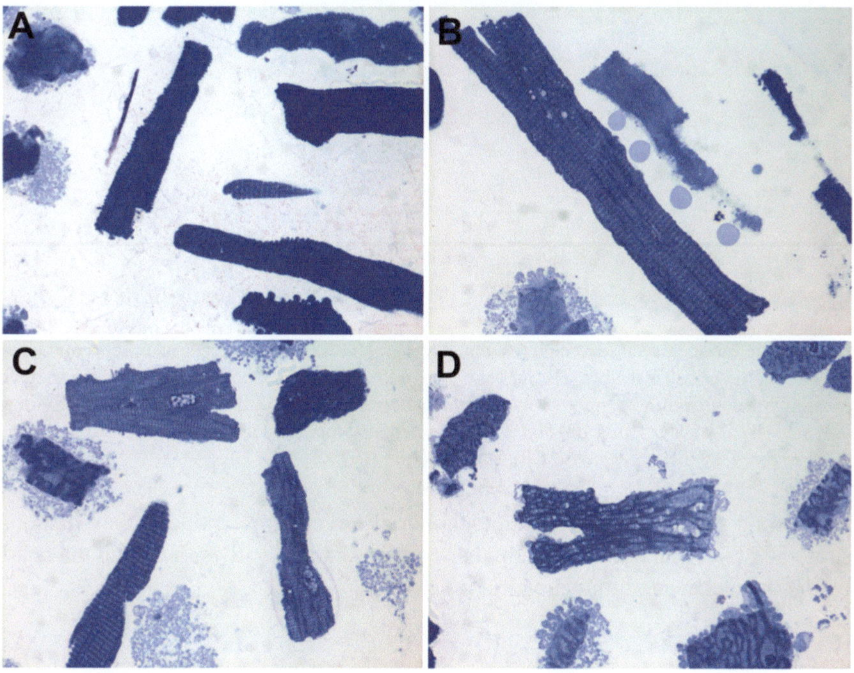

Fig. 8.3 Representative light micrograph images. Samples were prepared as described elsewhere (Billur et al. 2016). Images obtained from 1 μm semi-thin sections of isolated left ventricular cardiomyocytes isolated from 6-month-old male rat heart and stained with Toluidine blue in (**a**). The degenerative changes in cardiomyocytes incubated with zinc ionophore and zinc pyrithione (ZnPT) in a concentration-dependent manner. ZnPT incubations with either 0.01 μM (**b**), 0,1 μM (**c**), or 1 μM ZnPT (**d**). Magnification a–d ×1,000

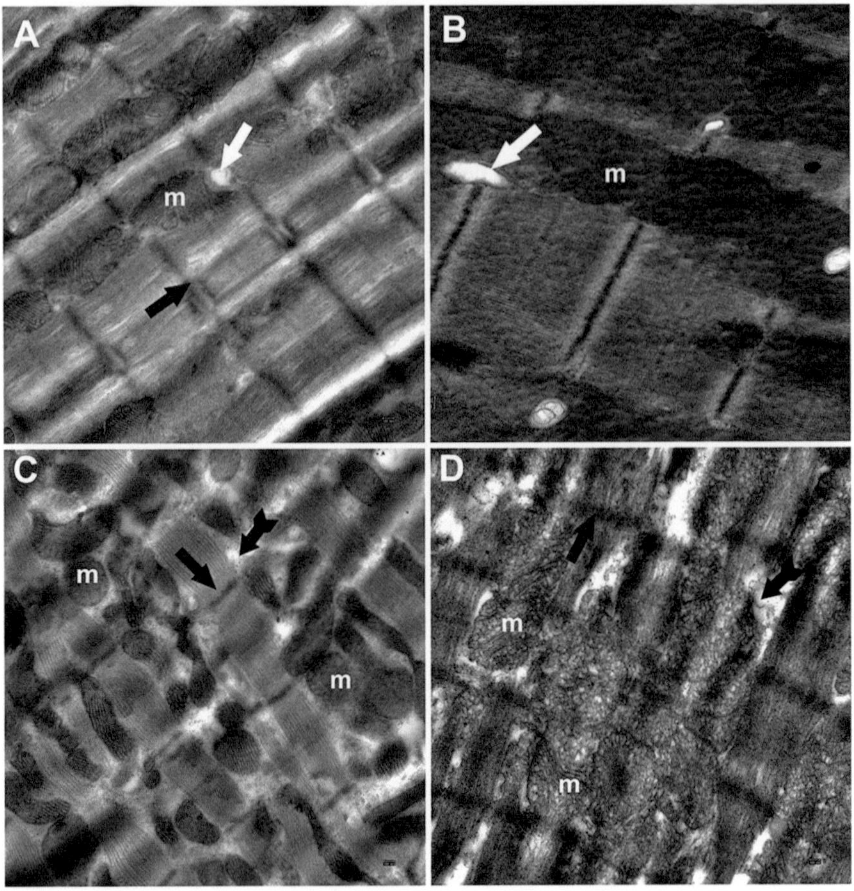

Fig. 8.4 Representative electron microscopy micrographs. Samples for electron microscopy were prepared as described elsewhere (Billur et al. 2016). Images obtained from isolated left ventricular cardiomyocytes from adult rats (6-month-old male rats) in (**a**). ZnPT incubations with either 0.01 μM (**b**), 0,1 μM (**c**), or 1 μM ZnPT (**d**). Shortened symbols: *m* mitochondrion, *arrow* Z-line, *white arrow* T tubule, *tailed arrow* SR. Magnification A-D ×21,560

rats (Jeong et al. 2016). Using mitochondrial targeting antioxidant, they demonstrated that one can prevent insulin resistance and diastolic dysfunction in the heart by the maintenance of mitochondria function.

8.6 Role of Zinc Transporters in Mammalian Heart Function

Several studies have shown that high $[Zn^{2+}]_i$ can induce severe detrimental effects in cardiomyocytes, including the increased basal level of $[Ca^{2+}]_i$, depolarization in mitochondrial membrane potential, and increases in oxidative stress (Turan et al. 1997;

Degirmenci et al. 2018; Tuncay et al. 2011, 2016, 2017; Turan and Tuncay 2017). $[Zn^{2+}]_i$ homeostasis in cells are maintained by different proteins including metalloproteins, and two types of Zn^{2+} transporter families, responsible from Zn^{2+} influx into the cytosol (ZIPs, solute-linked transporters, SLC39) and Zn^{2+} efflux from the cytosol (ZnTs, SLC30) (Kimura and Kambe 2016).

Although there are a number of studies on the important roles of Zn^{2+} status in cells, there are a limited number of studies related to the direct role of alterations associated with Zn^{2+} transporters for CVDs. In this concept, researchers demonstrated the important role of ZIP7 induction of ER stress in yeast through the release of Zn^{2+} from S(E)R (Ellis et al. 2004; Hogstrand et al. 2009; Taylor et al. 2012). Another study also showed that ZIP7 is localized to the Golgi inducing the Zn^{2+} release from its lumen into the cytosol (Huang et al. 2005). It has been also mentioned that some endogenous factors including protein kinase 2 (CK2) played an important role for the activity of ZIP7 (Taylor et al. 2012), which are closely associated with the contribution of $[Zn^{2+}]_i$ dyshomeostasis to pathological conditions in cells (Grubman et al. 2014; Groth et al. 2013). Latter studies provided important data related with the role of ZIP7 together with ZnT7 (which are localized to S(E)R) in the development of diabetic cardiomyopathy through increased due to the high amount of Zn^{2+} release from S(E)R (Tuncay et al. 2017). In further studies by Turan's team, it has been also shown that protein expression levels of some Zn^{2+} transporters, ZIP7 and ZIP14, were increased, while ZIP8 was significantly decreasing in MetS rat's cardiomyocytes. Furthermore, in the same study, the authors demonstrated that ZnT7 was decreased with no change in ZnT8. Moreover, these changes were found to be associated with the alterations in the protein expression levels of MMP-2 and MMP-9 in the same ventricular cardiomyocytes (Olgar and Turan 2018). In addition, Olgar and his coworkers examined the role of Zn^{2+} transporters in the development of heart failure (HF) in patients scheduled to undergo orthotopic heart transplantation for end-stage HF as well as in ventricular cardiomyocytes-induced HF via the induction of ER stress (Olgar et al. 2018b). They showed that ZIP8, ZIP14, and ZnT8 are localized to both sarcolemma and S(E)R in ventricular cardiomyocytes and ZIP14 and ZnT8 were significantly increased with decreased ZIP8 level in these HF cardiomyocytes. Those data also significantly demonstrated the correlation between the induction of ER stress, the elevation of $[Zn^{2+}]_i$ due to altered expression levels of these Zn^{2+} transporters, and increased oxidative stress, at least via the activated PKCα in HF cardiomyocytes. In another study, authors provided more information about these transporters via demonstrating the important role of ZIP7 and ZnT7 in cardiac dysfunction via affecting S(E)R-mitochondria interaction in hyperglycemic cardiomyocytes (Tuncay et al. 2018).

The above data are also supported by others. It has been shown that an intestinal epithelium-specific ZIP7 deletion in mice was shown to be related with increased apoptosis in the stem cell-derived cells, via a role with increased ER stress in these samples. They further showed that the upregulation of ZIP7 played an important role in the maintenance of $[Zn^{2+}]_i$ homeostasis under ER stress (Ohashi et al. 2016). In addition, Fukada's team well-documented the roles of Zn^{2+} transporters in health and diseases in their several early and recent review articles (Kimura and Kambe 2016; Fukada and Kambe 2018; Kambe et al. 2008, 2015). In their wide documentation, a number of genetic disorders caused by mutations in the genes encoding several Zn^{2+} transporters, such as ZIP4 in acrodermatitis enteropathica, ZIP13 in the spondylocheiro dysplastic form of Ehlers-Danlos syndrome, ZnT2 in transient neonatal zinc deficiency, ZnT8 in diabetes, and ZnT10 in Parkinsonism and dystonia, have been shown. In another review article, Hara et al. documented the expressions and modification levels of several Zn^{2+} transporters in mammalian cells/tissues, focusing on their physiological roles such as ZnT2, ZnT3, ZnT4, and ZnT8 as well as ZIP4, ZIP5, ZIP6, ZIP7, ZIP8, ZIP10, ZIP12, ZIP13, and ZIP14 (Hara et al. 2017).

As a summary, all the above studies strongly emphasized that a number of cellular actors can interact with cellular Zn^{2+} for their functions, acting not only as an accessory molecule for proteins but also as a signaling molecule, much similar to Ca^{2+} (Hara et al. 2017; Maret 2013). More importantly, known data imply that the Zn^{2+} transporter ZIP7, particularly, has an important role in the proper heart function in mammalians.

8.7 Role of Zinc Transporter ZIP14 in Aging Mammalian Heart Function

The named ZnT and ZIP proteins are classified as Zn^{2+} transporters, which are responsible either in Zn^{2+} influx into the cytosol or Zn^{2+} efflux from the cytosol in cells. However, some of them can transfer the other trace metal ions, including iron ion as well as other cations (Eide 2011). Among ZIPs, ZIP14 (SLC39A14) was first identified in uncharacterized human genes, and it is most highly expressed in the liver (Nomura et al. 1994) and in epithelial cells (Zhu et al. 1997), as well as associated with the estrogen-related gene LIV-1 (Taylor and Nicholson 2003). In a later study, Taylor et al. have demonstrated that ZIP14 located to the plasma membrane could function as Zn^{2+} influx transporter in CHO cells. Furthermore, into cell, demonstrate properly its role that the plasma membrane located in ZIP14 was regulated with interleukin-6 and is regulated in the liver, contributing to the hypozincemia of the acute-phase response to inflammation and infection (Liuzzi et al. 2005). Furthermore, it was also shown that ZIP14-mediated the Zn^{2+} uptake and non-transferrin-bound iron uptake into cells, indicating its role in zinc and iron metabolism in hepatocytes (Cousins et al. 2006; Jiang et al. 2018; Gartmann et al. 2018), in human proximal tubular epithelial cells (van Raaij et al. 2019), and in the promotion of cachexia in metastatic cancers through ZIP14 upregulation in skeletal muscle (Wang et al. 2018). The mediation role ZIP14 in the uptake of non-transferrin-bound iron was also given by the study of Zhao et al. in the latter period (Zhao et al. 2010).

Another important role of ZIP14 has been shown by Aydemir et al., who demonstrated that hepatic ZIP14-mediated Zn^{2+} transport could contribute to endosomal insulin receptor trafficking and glucose metabolism (Aydemir et al. 2016a).

Supporting this study, Turan's lab provided important information related to the possible role of ZIP14 in insulin-resistant MetS rats with high blood glucose. In those studies, the protein expression level of ZIP14 was found to be markedly increased being parallel to increased $[Zn^{2+}]_i$ in MetS-rat cardiomyocytes (Olgar and Turan 2018). More importantly, they demonstrated the similar increased protein level of ZIP14 in HF human heart with a correlation between the induction of ER stress, the elevation of $[Zn^{2+}]_i$, and increased oxidative stress (Olgar et al. 2018b).

Although there is very little information related with the role of Zn^{2+} transporters, particularly ZIP14 in mammalian aging, Aydemir et al. demonstrated the role of ZIP14 as an inflammation responsive transporter, and it has important phenotypic effects, amplified with aging (Aydemir et al. 2016b). In these regards, the ZIP14 situation in aging male rat (24 months old) in the left ventricular cardiomyocytes compared to those of adult male rats (6 months old) was examined by Western blot analysis. As can be seen in Fig. 8.5a, the total protein expression level of ZIP14 in aging rats' cardiomyocytes was not significantly different from that of adult value.

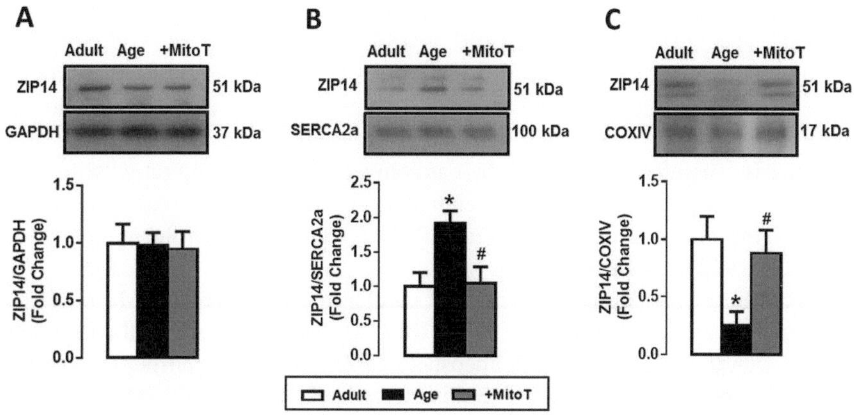

Fig. 8.5 Western blot analysis of subcellular protein expression levels of ZIP14 in isolated left ventricular cardiomyocyte homogenates. Cardiomyocytes were freshly isolated from 6-month-old (adult) and 24-month-old (aging) male Wistar rats (Turan et al. 1997). Age cardiomyocytes were incubated with mitochondria-targeted antioxidant MitoTEMPO (1 μM) for 3 h. Though whole cell homogenates were extracted with conventional NP-40 lysis buffer (250 mM NaCl, 1% NP-40, and 50 mM Tris-HCl; pH 8.0 and 1XPIC), isolation of mitochondrial fractions and sarcoplasmic reticulum fractions from freshly isolated cardiomyocytes were performed by different densiometric centrifugal forces using Mitochondria Isolation Kit for Cultured Cells (Thermo, 89,874) and Endoplasmic Reticulum Isolation Kit (Sigma-Aldrich, ER0100), respectively. Protein levels were measured by Bradford Assay (Pierce Biotechnology, USA) with bovine serum albumin as the standard method, and equal amount of protein preparations was run on SDS-polyacrylamide gels and blotted with a primary antibody against ZIP14 (Thermo, PA5–21077; 1:300), GAPDH (Santa Cruz, Sc365062; 1:10000), COXIV (Cell Signalling, ab1474; 1:10000), and SERCA2a (Santa Cruz, Sc-376235; 1:5000) to detect their protein levels. Immunoreactive bands were detected by a chemiluminescent reaction (ECL kit, Advansta). ZIP14 protein levels presented as whole-cell (A), sarcoplasmic reticulum-graded (B), and mitochondrial-graded lysates (C) for all groups. Data were presented as mean ± SEM. Significance level accepted as $*p < 0.05$ vs. Adult group, $^{\#}p < 0.05$ vs. Age group

However, its level in isolated mitochondria (Fig. 8.5b) was significantly depressed in age group compared with the adult group, while it was significantly high in the isolated S(E)R part (Fig. 8.5c) in the same cells. These data imply that the protein expression level of this transporter, ZIP14, has a muffling role between S(E)R and the mitochondria in aging heart cells.

8.8 Conclusion

Literature search shows that the functional significance of ZIP14-mediated transport of Zn^{2+} either into or out of cardiomyocytes is not known yet, although its role for Zn^{2+} transport in the liver, intestine, adipose tissue, and bones is documented (Cousins et al. 2006; Pinilla-Tenas et al. 2011; Guthrie et al. 2015; Tominaga et al. 2005; Hojyo et al. 2011) providing information about its marked tissue specificity of expression. ZIP14 can also transport other metal ions, such as iron ions, manganese ions, and cadmium ions (Zhao et al. 2010; Jenkitkasemwong et al. 2015; Fujishiro et al. 2011, 2012; Tuschl et al. 2016).

Most of the data have shown that increased oxidative stress plays an important role in the disruption of both cellular Zn^{2+} homeostasis and the responsible Zn^{2+} transporters. In this regard, our data given in Fig.8.3 can support clearly this hypothesis. Therefore, we treated the cardiomyocytes from aging rat with an antioxidant directly targeting the mitochondria, MitoTEMPO. As can be seen in Fig. 8.5, the altered level of depressed ZIP14 expression in either the mitochondria or S(E)R could be augmented fully.

Taken into consideration the marked cardioprotective effect of MitoTEMPO treatment of aging rats (Olgar et al. 2018a), we have summarized the evidence related with the role of cellular homeostasis of Zn^{2+} and Zn^{2+} transporters and the protective action of mitochondrial antioxidant MitoTEMPO in Fig. 8.6. Under the light of other documents, we propose that both oxidative and nitrosative stresses are increasing in aging heart, leading to alterations in both cellular Ca^{2+} and Zn^{2+} homeostasis, through alterations of both S(E)R and mitochondria. All these changes can underline the serious alterations in cardiac function (Olgar et al. 2018a, b; Tuncay et al. 2013; Tuncay et al. 2018). If one can follow a strategy to protect the mitochondria during the aging period, then it can help to have a proper heart function in age mammalians.

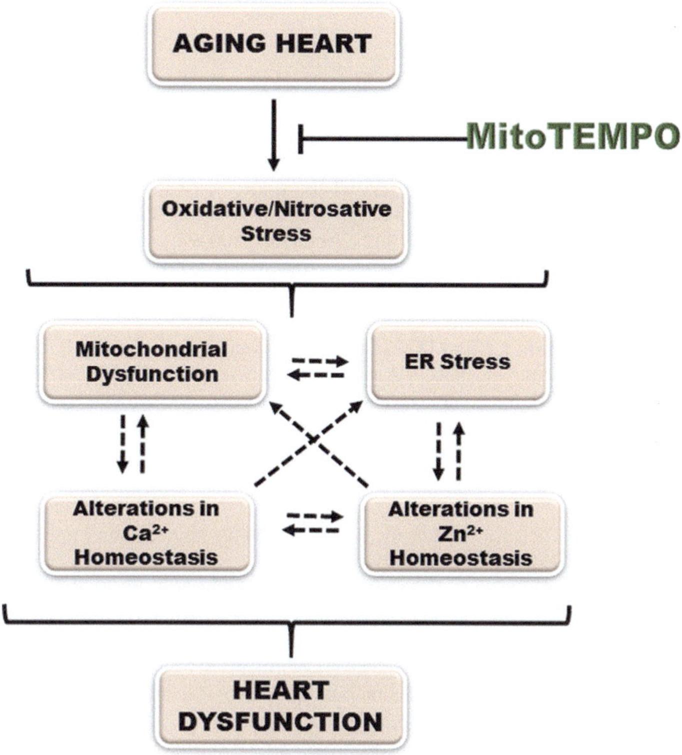

Fig. 8.6 A pathway giving possible mechanisms underlines the changes in the heart during the aging process and a mitochondria-targeting antioxidant MitoTEMPO providing cardioprotection/therapy via cellular Zn^{2+} homeostasis against aging-associated development of cardiovascular insufficiencies/dysfunction in insulin-resistant mammalian heart. Taken into consideration the marked cardioprotective effect of MitoTEMPO treatment of aging rats (Olgar et al. 2018a), we have summarized the evidence related with the role of cellular homeostasis of Zn^{2+} and the protective action of mitochondrial antioxidant MitoTEMPO

Acknowledgments This work has been supported by a grant from TUBITAK (SBAG-216S979) to Belma Turan.

Compliance with Ethical Standards Animal experimental protocols were performed in accordance with the standards of the European Community guidelines on the care and use of laboratory animals and approved by the Ankara University with a reference number of 2016-18-165.

Conflict of Interest No potential conflicts of interest relevant to this article were reported.

References

Anderson R, Lagnado A, Maggiorani D, Walaszczyk A, Dookun E, Chapman J, Birch J, Salmonowicz H, Ogrodnik M, Jurk D, Proctor C, Correia-Melo C, Victorelli S, Fielder E, Berlinguer-Palmini R, Owens A, Greaves LC, Kolsky KL, Parini A, Douin-Echinard V, LeBrasseur NK, Arthur HM, Tual-Chalot S, Schafer MJ, Roos CM, Miller JD, Robertson N, Mann J, Adams PD, Tchkonia T, Kirkland JL, Mialet-Perez J, Richardson GD, Passos JF (2019) Length-independent telomere damage drives post-mitotic cardiomyocyte senescence. The EMBO J 38 (5). https://doi.org/10.15252/embj.2018100492

Aydemir TB, Troche C, Kim MH, Cousins RJ (2016a) Hepatic ZIP14-mediated zinc transport contributes to endosomal insulin receptor trafficking and glucose metabolism. J Biol Chem 291(46):23939–23951. https://doi.org/10.1074/jbc.M116.748632

Aydemir TB, Troche C, Kim J, Kim MH, Teran OY, Leeuwenburgh C, Cousins RJ (2016b) Aging amplifies multiple phenotypic defects in mice with zinc transporter Zip14 (Slc39a14) deletion. Exp Gerontol 85:88–94. https://doi.org/10.1016/j.exger.2016.09.013

Barbieri M, Rizzo MR, Manzella D, Paolisso G (2001) Age-related insulin resistance: is it an obligatory finding? The lesson from healthy centenarians. Diabetes Metab Res Rev 17(1):19–26

Beal MF (2002) Oxidatively modified proteins in aging and disease. Free Radic Biol Med 32(9):797–803

Berg JM, Shi Y (1996) The galvanization of biology: a growing appreciation for the roles of zinc. Science (New York, NY) 271(5252):1081–1085

Bhashyam S, Parikh P, Bolukoglu H, Shannon AH, Porter JH, Shen YT, Shannon RP (2007) Aging is associated with myocardial insulin resistance and mitochondrial dysfunction. Am J Physiol Heart Circ Physiol 293(5):H3063–H3071. https://doi.org/10.1152/ajpheart.00163.2007

Billur D, Tuncay E, Okatan EN, Olgar Y, Durak AT, Degirmenci S, Can B, Turan B (2016) Interplay between cytosolic free Zn2+ and mitochondrion morphological changes in rat ventricular cardiomyocytes. Biol Trace Elem Res:1–12

Boengler K, Kosiol M, Mayr M, Schulz R, Rohrbach S (2017) Mitochondria and ageing: role in heart, skeletal muscle and adipose tissue. J Cachexia Sarcopenia Muscle 8(3):349–369. https://doi.org/10.1002/jcsm.12178

Borlaug BA, Kass DA (2006) Mechanisms of diastolic dysfunction in heart failure. Trends Cardiovasc Med 16(8):273–279. https://doi.org/10.1016/j.tcm.2006.05.003

Boudina S (2013) Cardiac aging and insulin resistance: could insulin/insulin-like growth factor (IGF) signaling be used as a therapeutic target? Curr Pharm Des 19(32):5684–5694

Bradford WH, Omens JH, Sheikh F (2017) Vinculin at the heart of aging. Ann Transl Med 5(3):62. https://doi.org/10.21037/atm.2017.01.65

Brown AM, Kristal BS, Effron MS, Shestopalov AI, Ullucci PA, Sheu KF, Blass JP, Cooper AJ (2000) Zn2+ inhibits alpha-ketoglutarate-stimulated mitochondrial respiration and the isolated alpha-ketoglutarate dehydrogenase complex. J Biol Chem 275(18):13441–13447

Cabrera AJ (2015) Zinc, aging, and immunosenescence: an overview. Pathobiol Aging Age Relat Dis 5:25592. https://doi.org/10.3402/pba.v5.25592

Capdor J, Foster M, Petocz P, Samman S (2013) Zinc and glycemic control: a meta-analysis of randomised placebo controlled supplementation trials in humans. J Trace Elem Med Biol 27(2):137–142. https://doi.org/10.1016/j.jtemb.2012.08.001

Carafoli E, Lehninger AL (1971) A survey of the interaction of calcium ions with mitochondria from different tissues and species. Biochem J 122(5):681–690

Chabosseau P, Tuncay E, Meur G, Bellomo EA, Hessels A, Hughes S, Johnson PR, Bugliani M, Marchetti P, Turan B, Lyon AR, Merkx M, Rutter GA (2014) Mitochondrial and ER-targeted eCALWY probes reveal high levels of free Zn2+. ACS Chem Biol 9(9):2111–2120. https://doi.org/10.1021/cb5004064

Chiao YA, Rabinovitch PS (2015) The aging heart. Cold Spring Harb Perspect Med 5(9):a025148. https://doi.org/10.1101/cshperspect.a025148

Chu A, Holdaway C, Varma T, Petocz P, Samman S (2018) Lower serum zinc concentration despite higher dietary zinc intake in athletes: a systematic review and meta-analysis. Sports Med 48(2):327–336. https://doi.org/10.1007/s40279-017-0818-8

Chung KT, Wong TY, Wei CI, Huang YW, Lin Y (1998) Tannins and human health: a review. Crit Rev Food Sci Nutr 38(6):421–464. https://doi.org/10.1080/10408699891274273

Corsetti G, Pasini E, D'Antona G, Nisoli E, Flati V, Assanelli D, Dioguardi FS, Bianchi R (2008) Morphometric changes induced by amino acid supplementation in skeletal and cardiac muscles of old mice. Am J Cardiol 101(11A):26E–34E. https://doi.org/10.1016/j.amjcard.2008.02.078

Cousins RJ, Liuzzi JP, Lichten LA (2006) Mammalian zinc transport, trafficking, and signals. J Biol Chem 281(34):24085–24089. https://doi.org/10.1074/jbc.R600011200

Dai DF, Santana LF, Vermulst M, Tomazela DM, Emond MJ, MacCoss MJ, Gollahon K, Martin GM, Loeb LA, Ladiges WC, Rabinovitch PS (2009) Overexpression of catalase targeted to mitochondria attenuates murine cardiac aging. Circulation 119(21):2789–2797. https://doi.org/10.1161/circulationaha.108.822403

Dai DF, Chiao YA, Marcinek DJ, Szeto HH, Rabinovitch PS (2014) Mitochondrial oxidative stress in aging and healthspan. Longev Healthspan 3:6. https://doi.org/10.1186/2046-2395-3-6

Degirmenci S, Olgar Y, Durak A, Tuncay E, Turan B (2018) Cytosolic increased labile Zn(2+) contributes to arrhythmogenic action potentials in left ventricular cardiomyocytes through protein thiol oxidation and cellular ATP depletion. J Trace Elements Med Biol 48:202–212. https://doi.org/10.1016/j.jtemb.2018.04.014

Denton RM, McCormack JG, Edgell NJ (1980) Role of calcium ions in the regulation of intramitochondrial metabolism. Effects of Na+, Mg2+ and ruthenium red on the Ca2+−stimulated oxidation of oxoglutarate and on pyruvate dehydrogenase activity in intact rat heart mitochondria. Biochem J 190(1):107–117

Dey S, DeMazumder D, Sidor A, Foster DB, O'Rourke B (2018) Mitochondrial ROS drive sudden cardiac death and chronic proteome remodeling in heart failure. Circ Res 123(3):356–371. https://doi.org/10.1161/circresaha.118.312708

Dibb KM, Rueckschloss U, Eisner DA, Isenberg G, Trafford AW (2004) Mechanisms underlying enhanced cardiac excitation contraction coupling observed in the senescent sheep myocardium. J Mol Cell Cardiol 37(6):1171–1181. https://doi.org/10.1016/j.yjmcc.2004.09.005

Doets EL, Cavelaars AE, Dhonukshe-Rutten RA, van't Veer P, de Groot LC (2012) Explaining the variability in recommended intakes of folate, vitamin B12, iron and zinc for adults and elderly people. Public Health Nutr 15(5):906–915. https://doi.org/10.1017/s1368980011002643

Douaud G, Groves AR, Tamnes CK, Westlye LT, Duff EP, Engvig A, Walhovd KB, James A, Gass A, Monsch AU, Matthews PM, Fjell AM, Smith SM, Johansen-Berg H (2014) A common brain network links development, aging, and vulnerability to disease. Proc Natl Acad Sci 111(49):17648. https://doi.org/10.1073/pnas.1410378111

Eby GA, Halcomb WW (2006) High-dose zinc to terminate angina pectoris: a review and hypothesis for action by ICAM inhibition. Med Hypotheses 66(1):169–172. https://doi.org/10.1016/j.mehy.2005.06.013

Efeovbokhan N, Bhattacharya SK, Ahokas RA, Sun Y, Guntaka RV, Gerling IC, Weber KT (2014) Zinc and the prooxidant heart failure phenotype. J Cardiovasc Pharmacol 64(4):393–400. https://doi.org/10.1097/fjc.0000000000000125

Eide DJ (2011) The oxidative stress of zinc deficiency. Metallomics 3(11):1124–1129. https://doi.org/10.1039/c1mt00064k

Eisner V, Cupo RR, Gao E, Csordas G, Slovinsky WS, Paillard M, Cheng L, Ibetti J, Chen SR, Chuprun JK, Hoek JB, Koch WJ, Hajnoczky G (2017) Mitochondrial fusion dynamics is robust in the heart and depends on calcium oscillations and contractile activity. Proc Natl Acad Sci U S A 114(5):E859–e868. https://doi.org/10.1073/pnas.1617288114

Ellis CD, Wang F, MacDiarmid CW, Clark S, Lyons T, Eide DJ (2004) Zinc and the Msc2 zinc transporter protein are required for endoplasmic reticulum function. J Cell Biol 166(3):325–335. https://doi.org/10.1083/jcb.200401157

Elmore SP, Qian T, Grissom SF, Lemasters JJ (2001) The mitochondrial permeability transition initiates autophagy in rat hepatocytes. FASEB J 15(12):2286–2287. https://doi.org/10.1096/fj.01-0206fje
Evans GW (1986) Zinc and its deficiency diseases. Clin Physiol Biochem 4(1):94–98
Ferrucci L, Levine ME, Kuo PL, Simonsick EM (2018) Time and the metrics of aging. Circ Res 123(7):740–744. https://doi.org/10.1161/circresaha.118.312816
Fosmire GJ (1990) Zinc toxicity. Am J Clin Nutr 51(2):225–227. https://doi.org/10.1093/ajcn/51.2.225
Foster M, Petocz P, Samman S (2010) Effects of zinc on plasma lipoprotein cholesterol concentrations in humans: a meta-analysis of randomised controlled trials. Atherosclerosis 210(2):344–352. https://doi.org/10.1016/j.atherosclerosis.2009.11.038
Fujishiro H, Doi M, Enomoto S, Himeno S (2011) High sensitivity of RBL-2H3 cells to cadmium and manganese: an implication of the role of ZIP8. Metallomics 3(7):710–718. https://doi.org/10.1039/c1mt00020a
Fujishiro H, Yano Y, Takada Y, Tanihara M, Himeno S (2012) Roles of ZIP8, ZIP14, and DMT1 in transport of cadmium and manganese in mouse kidney proximal tubule cells. Metallomics 4(7):700–708. https://doi.org/10.1039/c2mt20024d
Fukada T, Kambe T (2018) Welcome to the world of zinc signaling. Int J Mol Sci 19(3). https://doi.org/10.3390/ijms19030785
Gartmann L, Wex T, Grungreiff K, Reinhold D, Kalinski T, Malfertheiner P, Schutte K (2018) Expression of zinc transporters ZIP4, ZIP14 and ZnT9 in hepatic carcinogenesis-an immunohistochemical study. J Trace Elements Med Biol 49:35–42. https://doi.org/10.1016/j.jtemb.2018.04.034
Giacconi R, Malavolta M, Costarelli L, Busco F, Galeazzi R, Bernardini G, Gasparini N, Mocchegiani E (2012) Comparison of intracellular zinc signals in nonadherent lymphocytes from young-adult and elderly donors: role of zinc transporters (Zip family) and proinflammatory cytokines. J Nutr Biochem 23(10):1256–1263. https://doi.org/10.1016/j.jnutbio.2011.07.005
Giannoglou GD, Konstantinou DM, Kovatsi L, Chatzizisis YS, Mikhailidis DP (2010) Association of reduced zinc status with angiographically severe coronary atherosclerosis: a pilot study. Angiology 61(5):449–455. https://doi.org/10.1177/0003319710366702
Giorgi C, Baldassari F, Bononi A, Bonora M, De Marchi E, Marchi S, Missiroli S, Patergnani S, Rimessi A, Suski JM, Wieckowski MR, Pinton P (2012) Mitochondrial Ca(2+) and apoptosis. Cell Calcium 52(1):36–43. https://doi.org/10.1016/j.ceca.2012.02.008
Gottdiener JS, McClelland RL, Marshall R, Shemanski L, Furberg CD, Kitzman DW, Cushman M, Polak J, Gardin JM, Gersh BJ, Aurigemma GP, Manolio TA (2002) Outcome of congestive heart failure in elderly persons: influence of left ventricular systolic function. The Cardiovascular Health Study. Ann Intern Med 137(8):631–639
Groth C, Sasamura T, Khanna MR, Whitley M, Fortini ME (2013) Protein trafficking abnormalities in Drosophila tissues with impaired activity of the ZIP7 zinc transporter Catsup. Development 140(14):3018–3027. https://doi.org/10.1242/dev.088336
Grubman A, Lidgerwood GE, Duncan C, Bica L, Tan JL, Parker SJ, Caragounis A, Meyerowitz J, Volitakis I, Moujalled D, Liddell JR, Hickey JL, Horne M, Longmuir S, Koistinaho J, Donnelly PS, Crouch PJ, Tammen I, White AR, Kanninen KM (2014) Deregulation of subcellular biometal homeostasis through loss of the metal transporter, Zip7, in a childhood neurodegenerative disorder. Acta Neuropathol Commun 2:25. https://doi.org/10.1186/2051-5960-2-25
Guthrie GJ, Aydemir TB, Troche C, Martin AB, Chang SM, Cousins RJ (2015) Influence of ZIP14 (slc39A14) on intestinal zinc processing and barrier function. Am J Physiol Gastrointest Liver Physiol 308(3):G171–G178. https://doi.org/10.1152/ajpgi.00021.2014
Haase H, Rink L (2014) Zinc signals and immune function. Biofactors 40(1):27–40. https://doi.org/10.1002/biof.1114
Hamacher-Brady A, Brady NR (2016) Mitophagy programs: mechanisms and physiological implications of mitochondrial targeting by autophagy. Cell Mol Life Sci 73(4):775–795. https://doi.org/10.1007/s00018-015-2087-8

Hammadah M, Al Mheid I, Wilmot K, Ramadan R, Abdelhadi N, Alkhoder A, Obideen M, Pimple PM, Levantsevych O, Kelli HM, Shah A, Sun YV, Pearce B, Kutner M, Long Q, Ward L, Ko YA, Hosny Mohammed K, Lin J, Zhao J, Bremner JD, Kim J, Waller EK, Raggi P, Sheps D, Quyyumi AA, Vaccarino V (2017) Telomere shortening, regenerative capacity, and cardiovascular outcomes. Circ Res 120(7):1130–1138. https://doi.org/10.1161/circresaha.116.309421

Hara T, Takeda TA, Takagishi T, Fukue K, Kambe T, Fukada T (2017) Physiological roles of zinc transporters: molecular and genetic importance in zinc homeostasis. J Physiol Sci 67(2):283–301. https://doi.org/10.1007/s12576-017-0521-4

Heidenreich PA, Albert NM, Allen LA, Bluemke DA, Butler J, Fonarow GC, Ikonomidis JS, Khavjou O, Konstam MA, Maddox TM, Nichol G, Pham M, Pina IL, Trogdon JG (2013) Forecasting the impact of heart failure in the United States: a policy statement from the American Heart Association. Circ Heart Fail 6(3):606–619. https://doi.org/10.1161/HHF.0b013e318291329a

Herraiz-Martinez A, Alvarez-Garcia J, Llach A, Molina CE, Fernandes J, Ferrero-Gregori A, Rodriguez C, Vallmitjana A, Benitez R, Padro JM, Martinez-Gonzalez J, Cinca J, Hove-Madsen L (2015) Ageing is associated with deterioration of calcium homeostasis in isolated human right atrial myocytes. Cardiovasc Res 106(1):76–86. https://doi.org/10.1093/cvr/cvv046

Hershfinkel M, Moran A, Grossman N, Sekler I (2001) A zinc-sensing receptor triggers the release of intracellular Ca2+ and regulates ion transport. Proc Natl Acad Sci U S A 98(20):11749–11754. https://doi.org/10.1073/pnas.201193398

Hogstrand C, Kille P, Nicholson RI, Taylor KM (2009) Zinc transporters and cancer: a potential role for ZIP7 as a hub for tyrosine kinase activation. Trends Mol Med 15(3):101–111. https://doi.org/10.1016/j.molmed.2009.01.004

Hojyo S, Fukada T (2016) Zinc transporters and signaling in physiology and pathogenesis. Arch Biochem Biophys 611:43–50. https://doi.org/10.1016/j.abb.2016.06.020

Hojyo S, Fukada T, Shimoda S, Ohashi W, Bin BH, Koseki H, Hirano T (2011) The zinc transporter SLC39A14/ZIP14 controls G-protein coupled receptor-mediated signaling required for systemic growth. PLoS One 6(3):e18059. https://doi.org/10.1371/journal.pone.0018059

Huang L, Kirschke CP, Zhang Y, Yu YY (2005) The ZIP7 gene (Slc39a7) encodes a zinc transporter involved in zinc homeostasis of the Golgi apparatus. J Biol Chem 280(15):15456–15463. https://doi.org/10.1074/jbc.M412188200

Jackson MJ, Jones DA, Edwards RH (1982) Tissue zinc levels as an index of body zinc status. Clin Physiol 2(4):333–343

Jang Y, Wang H, Xi J, Mueller RA, Norfleet EA, Xu Z (2007) NO mobilizes intracellular Zn2+ via cGMP/PKG signaling pathway and prevents mitochondrial oxidant damage in cardiomyocytes. Cardiovasc Res 75(2):426–433. https://doi.org/10.1016/j.cardiores.2007.05.015

Jenkitkasemwong S, Wang CY, Coffey R, Zhang W, Chan A, Biel T, Kim JS, Hojyo S, Fukada T, Knutson MD (2015) SLC39A14 is required for the development of hepatocellular iron overload in murine models of hereditary hemochromatosis. Cell Metab 22(1):138–150. https://doi.org/10.1016/j.cmet.2015.05.002

Jeong EM, Chung J, Liu H, Go Y, Gladstein S, Farzaneh-Far A, Lewandowski ED, Dudley SC, Jr (2016) Role of mitochondrial oxidative stress in glucose tolerance, insulin resistance, and cardiac diastolic dysfunction. J Am Heart Assoc 5(5). https://doi.org/10.1161/jaha.115.003046

Jiang D, Sullivan PG, Sensi SL, Steward O, Weiss JH (2001) Zn(2+) induces permeability transition pore opening and release of pro-apoptotic peptides from neuronal mitochondria. J Biol Chem 276(50):47524–47529. https://doi.org/10.1074/jbc.M108834200

Jiang S, Yan K, Sun B, Gao S, Yang X, Ni Y, Ma W, Zhao R (2018) Long-term high-fat diet decreases hepatic Iron storage associated with suppressing TFR2 and ZIP14 expression in rats. J Agric Food Chem 66(44):11612–11621. https://doi.org/10.1021/acs.jafc.8b02974

Joseph LC, Barca E, Subramanyam P, Komrowski M, Pajvani U, Colecraft HM, Hirano M, Morrow JP (2017) Correction: inhibition of NAPDH oxidase 2 (NOX2) prevents oxidative stress and mitochondrial abnormalities caused by saturated fat in cardiomyocytes. PLoS One 12(3):e0174525. https://doi.org/10.1371/journal.pone.0174525

Joshi DC, Bakowska JC (2011) Determination of mitochondrial membrane potential and reactive oxygen species in live rat cortical neurons. J Visual Exp (51). https://doi.org/10.3791/2704

Judge S, Jang YM, Smith A, Hagen T, Leeuwenburgh C (2005) Age-associated increases in oxidative stress and antioxidant enzyme activities in cardiac interfibrillar mitochondria: implications for the mitochondrial theory of aging. FASEB J 19(3):419–421. https://doi.org/10.1096/fj.04-2622fje
Jurowski K, Szewczyk B, Nowak G, Piekoszewski W (2014) Biological consequences of zinc deficiency in the pathomechanisms of selected diseases. J Biol Inorg Chem 19(7):1069–1079. https://doi.org/10.1007/s00775-014-1139-0
Kambe T, Geiser J, Lahner B, Salt DE, Andrews GK (2008) Slc39a1 to 3 (subfamily II) Zip genes in mice have unique cell-specific functions during adaptation to zinc deficiency. Am J Physiol Regul Integr Comp Physiol 294(5):R1474–R1481. https://doi.org/10.1152/ajpregu.00130.2008
Kambe T, Fukue K, Ishida R, Miyazaki S (2015) Overview of inherited zinc deficiency in infants and children. J Nutr Sci Vitaminol 61(Suppl):S44–S46. https://doi.org/10.3177/jnsv.61.S44
Kannel WB (1999) Current status of the epidemiology of heart failure. Curr Cardiol Rep 1(1):11–19
Kennedy BK, Steffen KK, Kaeberlein M (2007) Ruminations on dietary restriction and aging. Cell Mol Life Sci 64(11):1323–1328. https://doi.org/10.1007/s00018-007-6470-y
Kimura T, Kambe T (2016) The functions of metallothionein and ZIP and ZnT transporters: an overview and perspective. Int J Mol Sci 17(3):336. https://doi.org/10.3390/ijms17030336
King JC, Shames DM, Woodhouse LR (2000) Zinc homeostasis in humans. J Nutr 130 (5S Suppl):1360S–1366S. https://doi.org/10.1093/jn/130.5.1360S
Kirkwood TB (2005) Understanding the odd science of aging. Cell 120(4):437–447. https://doi.org/10.1016/j.cell.2005.01.027
Kumari A, Singh KP, Mandal A, Paswan RK, Sinha P, Das P, Ali V, Bimal S, Lal CS (2017) Intracellular zinc flux causes reactive oxygen species mediated mitochondrial dysfunction leading to cell death in Leishmania donovani. PLoS One 12(6):e0178800. https://doi.org/10.1371/journal.pone.0178800
Lee HY, Oh BH (2010) Aging and arterial stiffness. Circ J 74(11):2257–2262
Lesnefsky EJ, Chen Q, Hoppel CL (2016) Mitochondrial metabolism in aging heart. Circ Res 118(10):1593–1611. https://doi.org/10.1161/circresaha.116.307505
Liuzzi JP, Lichten LA, Rivera S, Blanchard RK, Aydemir TB, Knutson MD, Ganz T, Cousins RJ (2005) Interleukin-6 regulates the zinc transporter Zip14 in liver and contributes to the hypozincemia of the acute-phase response. Proc Natl Acad Sci U S A 102(19):6843–6848. https://doi.org/10.1073/pnas.0502257102
Mansouri A, Muller FL, Liu Y, Ng R, Faulkner J, Hamilton M, Richardson A, Huang TT, Epstein CJ, Van Remmen H (2006) Alterations in mitochondrial function, hydrogen peroxide release and oxidative damage in mouse hind-limb skeletal muscle during aging. Mech Ageing Dev 127(3):298–306. https://doi.org/10.1016/j.mad.2005.11.004
Maret W (2013) Zinc and human disease. Met Ions Life Sci 13:389–414. https://doi.org/10.1007/978-94-007-7500-8_12
McCord MC, Aizenman E (2014) The role of intracellular zinc release in aging, oxidative stress, and Alzheimer's disease. Front Aging Neurosci 6:77. https://doi.org/10.3389/fnagi.2014.00077
Mirzaei H, Di Biase S, Longo VD (2016) Dietary interventions, cardiovascular aging, and disease: animal models and human studies. Circ Res 118(10):1612–1625. https://doi.org/10.1161/circresaha.116.307473
Mishra P, Chan DC (2016) Metabolic regulation of mitochondrial dynamics. J Cell Biol 212(4):379–387. https://doi.org/10.1083/jcb.201511036
Miyamoto S, Howes AL, Adams JW, Dorn GW 2nd, Brown JH (2005) Ca2+ dysregulation induces mitochondrial depolarization and apoptosis: role of Na+/Ca2+ exchanger and AKT. J Biol Chem 280(46):38505–38512. https://doi.org/10.1074/jbc.M505223200
Mocchegiani E, Malavolta M, Costarelli L, Giacconi R, Cipriano C, Piacenza F, Tesei S, Basso A, Pierpaoli S, Lattanzio F (2010) Zinc, metallothioneins and immunosenescence. Proc Nutr Soc 69(3):290–299. https://doi.org/10.1017/s0029665110001862
Moreira OC, Estebanez B, Martinez-Florez S, de Paz JA, Cuevas MJ, Gonzalez-Gallego J (2017) Mitochondrial function and mitophagy in the elderly: effects of exercise. Oxidative Med Cell Longev 2017:2012798. https://doi.org/10.1155/2017/2012798

Mursu J, Robien K, Harnack LJ, Park K, Jacobs DR Jr (2011) Dietary supplements and mortality rate in older women: the Iowa women's health study. Arch Intern Med 171(18):1625–1633. https://doi.org/10.1001/archinternmed.2011.445

Nazarewicz RR, Dikalova A, Bikineyeva A, Ivanov S, Kirilyuk IA, Grigor'ev IA, Dikalov SI (2013a) Does scavenging of mitochondrial superoxide attenuate cancer prosurvival signaling pathways? Antioxid Redox Signal 19(4):344–349. https://doi.org/10.1089/ars.2013.5185

Nazarewicz RR, Dikalova AE, Bikineyeva AT, Ivanov SV, Dikalov SI (2013b) Mitochondria-targeted antioxidant mitoTEMPO inhibits glycolysis and induces melanoma cell death by blocking ROSsensitive survival and metabolic pathways. 27 (1_supplement):793.797–793.797. https://doi.org/10.1096/fasebj.27.1_supplement.793.7

Nomura N, Nagase T, Miyajima N, Sazuka T, Tanaka A, Sato S, Seki N, Kawarabayasi Y, Ishikawa K, Tabata S (1994) Prediction of the coding sequences of unidentified human genes. II. The coding sequences of 40 new genes (KIAA0041-KIAA0080) deduced by analysis of cDNA clones from human cell line KG-1 (supplement). DNA Res 1(5):251–262

Ohashi W, Kimura S, Iwanaga T, Furusawa Y, Irie T, Izumi H, Watanabe T, Hijikata A, Hara T, Ohara O, Koseki H, Sato T, Robine S, Mori H, Hattori Y, Watarai H, Mishima K, Ohno H, Hase K, Fukada T (2016) Zinc transporter SLC39A7/ZIP7 promotes intestinal epithelial self-renewal by resolving ER stress. PLoS Genet 12(10):e1006349. https://doi.org/10.1371/journal.pgen.1006349

Olgar Y, Turan B (2018) A SGLT2 inhibitor dapagliflozin comparison with insulin exerts important effects on Zn2+–transporters in cardiomyocytes from insulin-resistant metabolic syndrome rats through inhibition of oxidative stress. Can J Physiol Pharmacol. https://doi.org/10.1139/cjpp-2018-0466

Olgar Y, Degirmenci S, Durak A, Billur D, Can B, Kayki-Mutlu G, Arioglu-Inan E, Turan B (2018a) Aging related functional and structural changes in the heart and aorta: MitoTEMPO improves aged-cardiovascular performance. Exp Gerontol 110:172–181. https://doi.org/10.1016/j.exger.2018.06.012

Olgar Y, Durak A, Tuncay E, Bitirim CV, Ozcinar E, Inan MB, Tokcaer-Keskin Z, Akcali KC, Akar AR, Turan B (2018b) Increased free Zn(2+) correlates induction of sarco(endo)plasmic reticulum stress via altered expression levels of Zn(2+) -transporters in heart failure. J Cell Mol Med 22(3):1944–1956. https://doi.org/10.1111/jcmm.13480

de Oliveira Otto MC, Alonso A, Lee DH, Delclos GL, Jenny NS, Jiang R, Lima JA, Symanski E, Jacobs DR Jr, Nettleton JA (2011) Dietary micronutrient intakes are associated with markers of inflammation but not with markers of subclinical atherosclerosis. J Nutr 141(8):1508–1515. https://doi.org/10.3945/jn.111.138115

Oster O, Dahm M, Oelert H (1993) Element concentrations (selenium, copper, zinc, iron, magnesium, potassium, phosphorous) in heart tissue of patients with coronary heart disease correlated with physiological parameters of the heart. Eur Heart J 14(6):770–774

Owada T, Yamauchi H, Saitoh SI, Miura S, Machii H, Takeishi Y (2017) Resolution of mitochondrial oxidant stress improves aged-cardiovascular performance. Coron Artery Dis 28(1):33–43. https://doi.org/10.1097/mca.0000000000000434

Palmer BM, Vogt S, Chen Z, Lachapelle RR, Lewinter MM (2006) Intracellular distributions of essential elements in cardiomyocytes. J Struct Biol 155(1):12–21. https://doi.org/10.1016/j.jsb.2005.11.017

Pan Z, Choi S, Ouadid-Ahidouch H, Yang JM, Beattie JH, Korichneva I (2017) Zinc transporters and dysregulated channels in cancers. Front Biosci (Landmark edition) 22:623–643

Paolisso G, Tagliamonte MR, Rizzo MR, Scheen A, D'Onofrio F, Lefebvre P (1999) [Is aging associated with a diminution of insulin sensitivity? Roles of IGF1 and dehydroepiandrosterone]. J Ann Diabetol Hotel Dieu :63–74

Payne BA, Chinnery PF (2015) Mitochondrial dysfunction in aging: much progress but many unresolved questions. Biochim Biophys Acta 1847(11):1347–1353. https://doi.org/10.1016/j.bbabio.2015.05.022

Pinilla-Tenas JJ, Sparkman BK, Shawki A, Illing AC, Mitchell CJ, Zhao N, Liuzzi JP, Cousins RJ, Knutson MD, Mackenzie B (2011) Zip14 is a complex broad-scope metal-ion transporter whose

functional properties support roles in the cellular uptake of zinc and nontransferrin-bound iron. Am J Physiol Cell Physiol 301(4):C862–C871. https://doi.org/10.1152/ajpcell.00479.2010
Pitt SJ, Stewart AJ (2015) Examining a new role for zinc in regulating calcium release in cardiac muscle. Biochem Soc Trans 43(3):359–363. https://doi.org/10.1042/bst20140285
Pivovarova NB, Stanika RI, Kazanina G, Villanueva I, Andrews SB (2014) The interactive roles of zinc and calcium in mitochondrial dysfunction and neurodegeneration. J Neurochem 128(4):592–602. https://doi.org/10.1111/jnc.12489
Plum LM, Rink L, Haase H (2010) The essential toxin: impact of zinc on human health. Int J Environ Res Public Health 7(4):1342–1365. https://doi.org/10.3390/ijerph7041342
Prasad AS (2008) Clinical, immunological, anti-inflammatory and antioxidant roles of zinc. Exp Gerontol 43(5):370–377. https://doi.org/10.1016/j.exger.2007.10.013
Prasad AS (2013) Discovery of human zinc deficiency: its impact on human health and disease. Adv Nutr (Bethesda, Md) 4(2):176–190. https://doi.org/10.3945/an.112.003210
Prasad AS, Miale A Jr, Farid Z, Sandstead HH, Schulert AR (1990) Clinical and experimental. Zinc metabolism in patients with the syndrome of iron deficiency anemia, hepatosplenomegaly, dwarfism, and hypogonadism. 1963. J Lab Clin Med 116(5):737–749
Prasad AS, Fitzgerald JT, Hess JW, Kaplan J, Pelen F, Dardenne M (1993) Zinc deficiency in elderly patients. Nutrition 9(3):218–224
Puche JE, Garcia-Fernandez M, Muntane J, Rioja J, Gonzalez-Baron S, Castilla Cortazar I (2008) Low doses of insulin-like growth factor-I induce mitochondrial protection in aging rats. Endocrinology 149(5):2620–2627. https://doi.org/10.1210/en.2007-1563
Quarles EK, Dai DF, Tocchi A, Basisty N, Gitari L, Rabinovitch PS (2015) Quality control systems in cardiac aging. Ageing Res Rev 23(Pt A):101–115. https://doi.org/10.1016/j.arr.2015.02.003
van Raaij SEG, Srai SKS, Swinkels DW, van Swelm RPL (2019) Iron uptake by ZIP8 and ZIP14 in human proximal tubular epithelial cells. Biometals 32(2):211–226. https://doi.org/10.1007/s10534-019-00183-7
Rajapakse D, Curtis T, Chen M, Xu H (2017) Zinc protects oxidative stress-induced RPE death by reducing mitochondrial damage and preventing lysosome rupture. Oxidative Med Cell Longev 2017:6926485. https://doi.org/10.1155/2017/6926485
Raza M, Deshpande LS, Blair RE, Carter DS, Sombati S, DeLorenzo RJ (2007) Aging is associated with elevated intracellular calcium levels and altered calcium homeostatic mechanisms in hippocampal neurons. Neurosci Lett 418(1):77–81. https://doi.org/10.1016/j.neulet.2007.03.005
Ristow M (2012) Interview with Michael Ristow. Aging 4(1):2. https://doi.org/10.18632/aging.100430
de Rivera C, Shukitt-Hale B, Joseph JA, Mendelson JR (2006) The effects of antioxidants in the senescent auditory cortex. Neurobiol Aging 27(7):1035–1044. https://doi.org/10.1016/j.neurobiolaging.2005.05.003
Rizzuto R, Simpson AW, Brini M, Pozzan T (1992) Rapid changes of mitochondrial Ca2+ revealed by specifically targeted recombinant aequorin. Nature 358(6384):325–327. https://doi.org/10.1038/358325a0
Romanjuk A, Lyndin M, Moskalenko R, Gortinskaya O, Lyndina Y (2016) The role of heavy metal salts in pathological biomineralization of breast cancer tissue. Adv Clin Exp Med 25(5):907–910. https://doi.org/10.17219/acem/34472
Santulli G, Xie W, Reiken SR, Marks AR (2015) Mitochondrial calcium overload is a key determinant in heart failure. Proc Natl Acad Sci U S A 112(36):11389–11394. https://doi.org/10.1073/pnas.1513047112
Schmucker DL, Sachs HG (1985) Age-dependent alterations in rat ventricular myocardium: a quantitative analysis. Mech Ageing Dev 31(1):89–101
Schocken DD (2000) Epidemiology and risk factors for heart failure in the elderly. Clin Geriatr Med 16(3):407–418
Schriner SE, Linford NJ, Martin GM, Treuting P, Ogburn CE, Emond M, Coskun PE, Ladiges W, Wolf N, Van Remmen H, Wallace DC, Rabinovitch PS (2005) Extension of murine life span by

overexpression of catalase targeted to mitochondria. Science (New York, NY) 308(5730):1909–1911. https://doi.org/10.1126/science.1106653
Senni M, Tribouilloy CM, Rodeheffer RJ, Jacobsen SJ, Evans JM, Bailey KR, Redfield MM (1999) Congestive heart failure in the community: trends in incidence and survival in a 10-year period. Arch Intern Med 159(1):29–34
Sensi SL, Ton-That D, Sullivan PG, Jonas EA, Gee KR, Kaczmarek LK, Weiss JH (2003) Modulation of mitochondrial function by endogenous Zn2+ pools. Proc Natl Acad Sci U S A 100(10):6157–6162. https://doi.org/10.1073/pnas.1031598100
Sharpley MS, Hirst J (2006) The inhibition of mitochondrial complex I (NADH:ubiquinone oxidoreductase) by Zn2+. J Biol Chem 281(46):34803–34809. https://doi.org/10.1074/jbc.M607389200
Shioi T, Inuzuka Y (2012) Aging as a substrate of heart failure. J Cardiol 60(6):423–428. https://doi.org/10.1016/j.jjcc.2012.07.015
Soinio M, Marniemi J, Laakso M, Pyorala K, Lehto S, Ronnemaa T (2007) Serum zinc level and coronary heart disease events in patients with type 2 diabetes. Diabetes Care 30(3):523–528. https://doi.org/10.2337/dc06-1682
St Croix CM, Wasserloos KJ, Dineley KE, Reynolds IJ, Levitan ES, Pitt BR (2002) Nitric oxide-induced changes in intracellular zinc homeostasis are mediated by metallothionein/thionein. Am J Physiol Lung Cell Mol Physiol 282(2):L185–L192. https://doi.org/10.1152/ajplung.00267.2001
Stanley WC, Recchia FA, Lopaschuk GD (2005) Myocardial substrate metabolism in the normal and failing heart. Physiol Rev 85(3):1093–1129. https://doi.org/10.1152/physrev.00006.2004
Stefanidou M, Maravelias C, Dona A, Spiliopoulou C (2006) Zinc: a multipurpose trace element. Arch Toxicol 80(1):1–9. https://doi.org/10.1007/s00204-005-0009-5
Strait JB, Lakatta EG (2012) Aging-associated cardiovascular changes and their relationship to heart failure. Heart Fail Clin 8(1):143–164. https://doi.org/10.1016/j.hfc.2011.08.011
Tate EL, Herbener GH (1976) A morphometric study of the density of mitochondrial cristae in heart and liver of aging mice. J Gerontol 31(2):129–134
Tatsumi T, Fliss H (1994) Hypochlorous acid mobilizes intracellular zinc in isolated rat heart myocytes. J Mol Cell Cardiol 26(4):471–479. https://doi.org/10.1006/jmcc.1994.1058
Taylor KM, Nicholson RI (2003) The LZT proteins; the LIV-1 subfamily of zinc transporters. Biochim Biophys Acta 1611(1–2):16–30
Taylor KM, Hiscox S, Nicholson RI, Hogstrand C, Kille P (2012) Protein kinase CK2 triggers cytosolic zinc signaling pathways by phosphorylation of zinc channel ZIP7. Sci Signal 5(210):ra11. https://doi.org/10.1126/scisignal.2002585
Thayer SA, Miller RJ (1990) Regulation of the intracellular free calcium concentration in single rat dorsal root ganglion neurones in vitro. J Physiol 425:85–115
Tominaga K, Kagata T, Johmura Y, Hishida T, Nishizuka M, Imagawa M (2005) SLC39A14, a LZT protein, is induced in adipogenesis and transports zinc. FEBS J 272(7):1590–1599. https://doi.org/10.1111/j.1742-4658.2005.04580.x
Tuncay E, Bilginoglu A, Sozmen NN, Zeydanli EN, Ugur M, Vassort G, Turan B (2011) Intracellular free zinc during cardiac excitation-contraction cycle: calcium and redox dependencies. Cardiovasc Res 89(3):634–642. https://doi.org/10.1093/cvr/cvq352
Tuncay E, Okatan EN, Vassort G, Turan B (2013) ss-blocker timolol prevents arrhythmogenic Ca(2)(+) release and normalizes Ca(2)(+) and Zn(2)(+) dyshomeostasis in hyperglycemic rat heart. PloS One 8(7):e71014. https://doi.org/10.1371/journal.pone.0071014
Tuncay E, Bitirim V, Durak A, Rutter GA, Turan B (2016) 362 – both reactive ROS and RNS contribute to intracellular free Zn2+ regulation in cardiomyocytes via zinc transporter ZIP7 under hyperglycemia. Free Radic Biol Med 100:S153. https://doi.org/10.1016/j.freeradbiomed.2016.10.403
Tuncay E, Bitirim VC, Durak A, Carrat GRJ, Taylor KM, Rutter GA, Turan B (2017) Hyperglycemia-induced changes in ZIP7 and ZnT7 expression cause Zn(2+) release from the Sarco(endo)plasmic reticulum and mediate ER stress in the heart. Diabetes 66(5):1346–1358. https://doi.org/10.2337/db16-1099

Tuncay E, Bitirim CV, Olgar Y, Durak A, Rutter GA, Turan B (2018) Zn(2+)-transporters ZIP7 and ZnT7 play important role in progression of cardiac dysfunction via affecting sarco(endo) plasmic reticulum-mitochondria coupling in hyperglycemic cardiomyocytes. Mitochondrion. https://doi.org/10.1016/j.mito.2017.12.011

Tuncay E, Olgar Y, Durak A, Degirmenci S, Bitirim CV, Turan B (2019) beta3 -adrenergic receptor activation plays an important role in the depressed myocardial contractility via both elevated levels of cellular free Zn(2+) and reactive nitrogen species. J Cell Physiol. https://doi.org/10.1002/jcp.28015

Turan B, Tuncay E (2017) Impact of labile zinc on heart function: from physiology to pathophysiology. Int J Mol Sci 18(11). https://doi.org/10.3390/ijms18112395

Turan B, Fliss H, Desilets M (1997) Oxidants increase intracellular free Zn2+ concentration in rabbit ventricular myocytes. Am J Phys 272(5 Pt 2):H2095–H2106

Tuschl K, Meyer E, Valdivia LE, Zhao N, Dadswell C, Abdul-Sada A, Hung CY, Simpson MA, Chong WK, Jacques TS, Woltjer RL, Eaton S, Gregory A, Sanford L, Kara E, Houlden H, Cuno SM, Prokisch H, Valletta L, Tiranti V, Younis R, Maher ER, Spencer J, Straatman-Iwanowska A, Gissen P, Selim LA, Pintos-Morell G, Coroleu-Lletget W, Mohammad SS, Yoganathan S, Dale RC, Thomas M, Rihel J, Bodamer OA, Enns CA, Hayflick SJ, Clayton PT, Mills PB, Kurian MA, Wilson SW (2016) Mutations in SLC39A14 disrupt manganese homeostasis and cause childhood-onset parkinsonism-dystonia. Nat Commun 7:11601. https://doi.org/10.1038/ncomms11601

Veronica G, Esther RR (2012) Aging, metabolic syndrome and the heart. Aging Dis 3(3):269–279

Wang G, Biswas AK, Ma W, Kandpal M, Coker C, Grandgenett PM, Hollingsworth MA, Jain R, Tanji K, Lomicronpez-Pintado S, Borczuk A, Hebert D, Jenkitkasemwong S, Hojyo S, Davuluri RV, Knutson MD, Fukada T, Acharyya S (2018) Metastatic cancers promote cachexia through ZIP14 upregulation in skeletal muscle. Nat Med 24(6):770–781. https://doi.org/10.1038/s41591-018-0054-2

Wessells RJ, Fitzgerald E, Cypser JR, Tatar M, Bodmer R (2004) Insulin regulation of heart function in aging fruit flies. Nat Genet 36(12):1275–1281. https://doi.org/10.1038/ng1476

Wilson PW, Kannel WB (2002) Obesity, diabetes, and risk of cardiovascular disease in the elderly. Am J Geriatr Cardiol 11(2):119–123,125

Xi J, Wang H, Mueller RA, Norfleet EA, Xu Z (2009) Mechanism for resveratrol-induced cardioprotection against reperfusion injury involves glycogen synthase kinase 3beta and mitochondrial permeability transition pore. Eur J Pharmacol 604(1–3):111–116. https://doi.org/10.1016/j.ejphar.2008.12.024

Yamasaki S, Sakata-Sogawa K, Hasegawa A, Suzuki T, Kabu K, Sato E, Kurosaki T, Yamashita S, Tokunaga M, Nishida K, Hirano T (2007) Zinc is a novel intracellular second messenger. J Cell Biol 177(4):637–645. https://doi.org/10.1083/jcb.200702081

Yi T, Vick JS, Vecchio MJ, Begin KJ, Bell SP, Delay RJ, Palmer BM (2013) Identifying cellular mechanisms of zinc-induced relaxation in isolated cardiomyocytes. Am J Physiol Heart Circ Physiol 305(5):H706–H715. https://doi.org/10.1152/ajpheart.00025.2013

Zhang W, Song M, Qu J, Liu GH (2018) Epigenetic modifications in cardiovascular aging and diseases. Circ Res 123(7):773–786. https://doi.org/10.1161/circresaha.118.312497

Zhao N, Gao J, Enns CA, Knutson MD (2010) ZRT/IRT-like protein 14 (ZIP14) promotes the cellular assimilation of iron from transferrin. J Biol Chem 285(42):32141–32150. https://doi.org/10.1074/jbc.M110.143248

Zhu H, Cong JP, Shenk T (1997) Use of differential display analysis to assess the effect of human cytomegalovirus infection on the accumulation of cellular RNAs: induction of interferon-responsive RNAs. Proc Natl Acad Sci U S A 94(25):13985–13990

Zieman S, Kass D (2004) Advanced glycation end product cross-linking: pathophysiologic role and therapeutic target in cardiovascular disease. Congest Heart Fail 10(3):144–149; quiz 150-141

Chapter 9
Zinc Signaling in the Life and Death of Neurons

Elias Aizenman

Abstract ZnT3 (SLC30a3), a zinc transporter, packages zinc into glutamatergic synaptic vesicles, imparting the metal with potential neurotransmitter-like properties. In fact, zinc is released from nerve terminals in an activity-dependent fashion and interacts with neurotransmitter receptors, influencing their function. This is most notably the case for the NMDA receptor, where synaptically released zinc can effectively regulate ion channel function. Critically, synaptically released zinc also directly activates a receptor (mZnR/GPR39) whose primary ligand is the metal itself, regulating neuronal excitation. In essence, then, zinc *is* a neurotransmitter. Zinc is also, in and of itself, a critical and ubiquitous second messenger. Intracellular zinc in neurons can be liberated from metal-binding proteins, such as metallothionein, via redox-regulated processes. The liberated zinc, in turn, can activate both pro-survival and pro-cell death signaling cascades, the latter of which are associated with the injury-mediated enhancement of functional Kv2.1 potassium channels in the cell membrane, leading to the cell death-promoting loss of cytoplasmic potassium. It is thus not altogether surprising that cellular zinc is tightly regulated through the presence of metal-binding proteins and multiple zinc transporters, many of which are covered in detail elsewhere in this book. This rich array of zinc signaling processes in neurons clearly point to the many vital and critical roles the metal plays in brain function and dysfunction.

Keywords Zinc · Neurotransmission · Neurodegeneration · Metallothionein · Oxidative stress · Potassium channel

---

E. Aizenman (✉)
Department of Neurobiology and Pittsburgh Institute for Neurodegenerative Diseases, University of Pittsburgh School of Medicine, Pittsburgh, PA, USA
e-mail: redox@pitt.edu

T. Fukada, T. Kambe (eds.), *Zinc Signaling*,
https://doi.org/10.1007/978-981-15-0557-7_9

9.1 Zinc Rocks!

Or rather, zinc comes from rocks, most commonly sphalerite (zinc sulfide). And if you ever find yourself in Pittsburgh, Pennsylvania, you may view many fine samples of this most widespread zinc-containing ore at the Hillman Hall of Minerals and Gems, within the Carnegie Museum of Natural History in the neighborhood of Oakland (Fig. 9.1). Industrial zinc is extracted from sphalerite by heated chemical reduction or "smelting." And if you then happen to find yourself Down Under atop a boat sailing along the Derwent river just north of Hobart, Tasmania, you will likely pass by the old, massive Lutana zinc works (now Nyrstar), one of the largest zinc smelting facilities in the world (Fig. 9.2). We, of course, must eat the zinc that flows through our bodies (Prasad 2013; Gibson et al. 2016), but it is interesting (at least to this author) to sometimes step back and take in the fact that zinc is, after all, a supernova product (Umeda and Nomoto 2002; Timmes et al. 1995) that somehow, through complex evolutionary processes (Berg and Shi 1996; Bini 2010; Hilgers and Ludwig 2001), has found its way to fill a critical signaling niche within our brain (Frederickson et al. 2005; Sensi et al. 2011). Yes, think about that.

9.2 Zinc Is a Neurotransmitter

Over the years, there have been various iterations of the criteria listed as necessary for a molecule to meet and thereby achieve the status of being classified as a neurotransmitter (Purves et al. 2001). Within this author's own interpretation of these criteria, herein are listed the requirements already met, or not, by the metal zinc:

Fig. 9.1 Zinc-containing rock. Sphalerite sample (with ankerite –lighter-colored mineral; from Kosovo), displayed at the Hillman Hall of Minerals and Gems at the Carnegie Museum of Natural History in Pittsburgh, PA. Many other sphalerite-containing samples can be found in the museum's collection. (Photograph by the author)

Fig. 9.2 A zinc smelter. The Nyrstar zinc works in Lutana, just north of Hobart, Tasmania. The body of water in front is the Derwent River, flowing from Lake St. Clair south to the Tasman Sea. (Photograph by the author). *"And so instead Mr. Hung dug out a small workshop beneath his old Zinc Company house in Lutana and set to work building fake antique chairs and carving ersatz whalebone to complement our more sordid fictions."* (From Richard Flanagan's *Gould's Book of Fish* (2001))

1. *Zinc must be present in the presynaptic cell (a synthetic process would be a bonus!).* Although our brains sometimes explode with great ideas, they are not even close to generating the conditions necessary to produce zinc, such as the energy present within the expanding gas that follows a supernova explosion (Timmes et al. 1995). As such, we must consume the zinc that neurons will eventually utilize (Prasad 2013; Gibson et al. 2016). Nonetheless, zinc is indeed present in the presynaptic cell, as it is present in *all* cells. And as such, a mechanism must be in place to put zinc, as a neurotransmitter, in the *right* place within the neuron, that is, in synaptic vesicles. This brings us to the next point.
2. *A transport process must be present to place zinc within synaptic vesicles.* ZnT3 (SLC30a3) is a zinc-selective transporter present mostly in a subpopulation of synaptic vesicles that normally also contain glutamate. In fact, this transporter is both necessary and sufficient for synaptic vesicles to sequester zinc (Cole et al. 1999; Wenzel et al. 1997; Palmiter et al. 1996). And why not all glutamatergic vesicles have ZnT3, and thereby contain zinc (Lavoie et al. 2011)? One possible explanation lies with the fact that the tail end of the zinc transporter interacts

with the adaptor-like protein complex AP-3, which may localize to only a subpopulation of vesicles, hence determining the vesicular sorting of ZnT3 (Salazar et al. 2004). Indeed, deletion of the gene for the delta subunit of AP-3 results in a reduction in vesicular zinc in the so-called *mocha* mouse (Kantheti et al. 1998). It must be noted, however, that not all of the vesicular zinc is absent in this mutant mouse (Stoltenberg et al. 2004). In contrast, deletion of ZnT3 gene results in a complete absence of vesicular, or "synaptic," zinc (Cole et al. 1999). And so, yes, zinc is sequestered into synaptic vesicles by the zinc transporter ZnT3, rendering the metal as a potential transmitter. Before we move on to the next point, it must be noted that an interesting co-dependent functional relationship exists between ZnT3 and the glutamate vesicular transporter Vglut1. That is, the transport properties of either one of these molecules are facilitated by the presence of the other within the same vesicle (Salazar et al. 2004).

3. *Zinc must be released from presynaptic terminals by an activity-dependent process where it would be free to diffuse across the synaptic cleft.* Believe it or not, this was a point of contention until recent years. In spite the fact that several classic papers strongly suggested synaptic zinc release did in fact occur (Assaf and Chung 1984; Howell et al. 1984), a lack of state-of-the-art fluorescent indicators or rapid zinc chelators led some authors to prematurely conclude that even though vesicular zinc could be "externalized" following synaptic activity, it did not diffuse across the cleft but somehow remained anchored to the presynaptic membrane to form what was termed a "zinc veneer" (Kay 2003). But as technology continued to improve, evidence for synaptic zinc release began, once again, to accumulate (Qian and Noebels 2005). The advent of highly selective and sensitive extracellular ratiometric zinc indicators, such as LZ9 (Anderson et al. 2015), and, especially, the development of high-affinity, *fast*, and selective zinc chelators, such as ZX1 (Pan et al. 2011; Anderson et al. 2015), allowed investigators to finally demonstrate, without a doubt, that zinc, indeed, was synaptically released following presynaptic neuronal activity. Moreover, synaptic zinc could diffuse across the synaptic cleft to modify postsynaptic function, the topic of the next point in this section.
4. *Zinc must act synaptically* via *zinc-sensing receptors.* In addition to the well-known neuromodulatory effects of zinc on classical neurotransmitter receptors, in particular the NMDA ligand-gated ion channel (Peters et al. 1987; Westbrook and Mayer 1987; Traynelis et al. 1998; Paoletti et al. 1997; Vergnano et al. 2014; Anderson et al. 2015), synaptically released zinc will specifically and directly activate the metabotropic zinc receptor (mZnR) GPR39 (Hershfinkel et al. 2001; Holst et al. 2007; Yasuda et al. 2007), first described by Michal Hershfinkel, Israel Sekler, and colleagues. The physiological effects of postsynaptic mZnR/GPR39 activation by synaptically released zinc are inhibitory in nature. In one example, mZnR/GPR39 activation produces an enhanced, syntaxin-dependent insertion of KCC2 in the postsynaptic membrane, leading to increased synaptic inhibition mediated by GABAA channels by shifting the chloride reversal potential in a hyperpolarizing direction (Chorin et al. 2011; Saadi et al. 2012). Of interest, *intracellular zinc blocks* KCC2 activity (Hershfinkel et al. 2009). In a

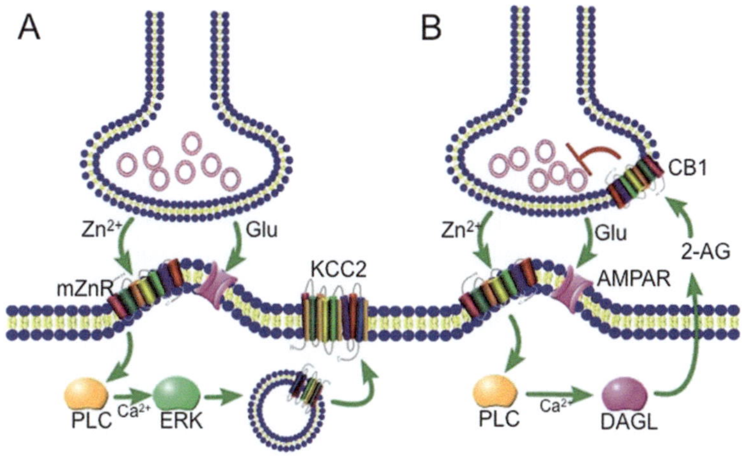

Fig. 9.3 Synaptic zinc activation of mZnR/GPR39. In (**a**), synaptic zinc leads to an ERK-dependent, SNARE-mediated insertion of KCC2 transporter molecules in the plasma membrane, driving the reversal potential for GABAA currents in the hyperpolarized direction and rendering synaptic inhibition more effective. In (**b**), activity-dependent synaptic zinc release triggers endocannabinoid synthesis and release from postsynaptic neurons, with subsequent inhibition of transmitter release via activation of presynaptic CB1 receptors

second example, synaptic zinc activation of mZnR/GPR39 induces postsynaptic production of the endocannabinoid 2-AG, which, in turn, diffuses to the presynaptic terminal to inhibit further excitatory glutamatergic neurotransmission via CB1 receptors (Perez-Rosello et al. 2013). Thus, in both cases (Fig. 9.3), activation of mZnR/GPR39 leads to decreased excitation. Not surprisingly, genetic deletion of mZnR/GPR39 produces hypersensitivity to seizure-inducing stimuli (Gilad et al. 2015), a phenotype not too dissimilar from that previously observed in ZnT3 null mice (Cole et al. 2000).

5. *Following activity-dependent release, zinc must be removed from the synaptic cleft to terminate its activity.* A number of potential ZIP (SLC39a) zinc transporters may serve as vehicles for zinc entry into a non-synaptic compartment (e.g., astrocytes, pre- or postsynaptic neurons) following its synaptic release. In fact, it has been observed that genetic deletion of ZIP1 and ZIP3 decreased zinc accumulation in CA1 neurons of the hippocampus following pathogenic zinc release stimulation (Qian et al. 2011). This does not necessarily mean that these transporters normally regulate cleft zinc concentrations, but they are the only ones thus far implicated in this process. It must be also mentioned that a non-synonymous single nucleotide polymorphism in ZIP8 was recently associated with schizophrenia in European populations, potentially implicating this transporter in the normal physiological function of synaptic zinc (Carrera et al. 2012; Li et al. 2016). In addition to ZIPs, zinc may gain entry into the postsynaptic cells (Li et al. 2001; Frederickson et al. 1989; Koh et al. 1996) via a number of conduits, including calcium channels (Kerchner et al. 2000; Dietz et al. 2008;

Sheline et al. 2002; Kim et al. 2000; Sensi et al. 1997) and calcium-permeable AMPA receptors (Yin and Weiss 1995; Weiss et al. 1993; Sensi et al. 1997), essentially removing it from the synaptic cleft. Finally, there is evidence that synaptically released zinc can diffuse to extra-synaptic sites (Anderson et al. 2015), suggesting that movement away from the cleft may also contribute to the termination of its neurotransmitter actions.

The information presented here strongly suggests that, by most criteria generally attributed to classical neuronal messengers, zinc is, indeed, a bona fide neurotransmitter. It is present in presynaptic vesicles, released in an activity-dependent manner, has specific receptors that, when activated by zinc, alter the excitability of the postsynaptic cells (in addition to serving as allosteric neuromodulator at other receptors), and has several putative mechanisms to terminate its actions. It must also be noted that, although not as common as in the glutamatergic system, some zinc-containing terminals release GABA as well, resulting in inhibition of GABAA receptors (Ruiz et al. 2004), a previously known effect of the metal (Smart and Constanti 1982). At this point, the reader is directed to a more comprehensive treatise on synaptic zinc, including its effects on plasticity, namely, an outstanding review recently published by Brendan McAllister and Richard Dyck (McAllister and Dyck 2017). Moreover, for elegant analyses of the actions of synaptic zinc on cortical circuit function and its role in behavior, the reader is referred to recent studies from Richard Dyck and Thanos Tzounopoulos' laboratories (Anderson et al. 2017; Kumar et al. 2019; McAllister et al. 2018; Patrick Wu and Dyck 2018).

9.3 Intracellular Zinc Signaling

9.3.1 Translocation of Synaptic Zinc

As mentioned in the prior section, synaptically released zinc has been observed to translocate from presynaptic terminals to postsynaptic neurons (Li et al. 2001; Frederickson et al. 1989). For the most part, this phenomenon has been closely associated with neuronal injury, as the stimuli utilized to observe the translocation have been generally injurious themselves (Qian et al. 2011; Koh et al. 1996; Frederickson et al. 1989; Yin et al. 2002). In an extremely interesting, recently described study by Paul Rosenberg, Larry Benowitz, and co-workers, retinal ganglion cell death observed following optic nerve crush was ascribed to zinc translocation, originating from amacrine cells (Li et al. 2017). In this system, expression of ZnT3 in amacrine cells was observed only after retinal ganglion cell axonal injury. Indeed, ZnT3 null mice showed attenuated injury, as were animals treated intravitreally with extracellular zinc chelators (Li et al. 2017). But not all cases of trans-synaptic zinc translocation have been associated with injurious processes. For example, mossy fiber-originating zinc translocation has been observed to activate TrkB in CA3 hippocampal neurons via Src-dependent, BDNF-independent mechanisms, at least in vitro, resulting in synaptic potentiation (Huang et al. 2008). Other investigators have similarly reported effects of synaptic zinc translocation on heterosynaptic facilitation of cortical inputs to CA3 hippocampal neurons (Eom et al. 2019). More recently, the zinc transporter ZnT1, which moves zinc from the cytoplasm to the extracellular space, was found to interact with the highly zinc-sensitive NMDA receptor subunit GluN2A (Mellone et al. 2015), suggesting the possibility that synaptic zinc translocation could then undertake a circuitous route to modulate glutamatergic neurotransmission mediated by NMDA receptors (Krall et al. 2018).

John Weiss and colleagues recently reported that the selective vulnerability of CA3 pyramidal hippocampal neurons to limbic seizures is likely due to zinc translocation-induced toxicity, gaining access to cells primarily via calcium-permeable AMPA-gated receptor channels (Medvedeva et al. 2017). CA1 neurons, which are selectively injured following global ischemia (instead of seizures), also perish as a result of an intracellular increase in zinc. However, in contrast to CA3 neurons, the source of the injurious zinc in CA1 cells is not synaptic, but, rather, hastily liberated metal that had been previously bound inside the postsynaptic cells by metal-binding proteins such as metallothionein (Medvedeva et al. 2017). This intracellular source of zinc, as a primary driver of neuronal injury, is the next topic of this chapter.

9.3.2 *Intracellular Liberation of Zinc*

As recently and brilliantly discussed by Wolfgang Maret, zinc itself is not a redox-active metal, but it both regulates and is highly regulated by redox processes (Maret 2019). In fact, the cytosolic zinc-binding protein metallothionein, which has a very high affinity for metal ($Kd = 1 \times 10^{-14}$ M at neutral pH), can readily release zinc upon mild oxidative conditions, given its very negative (−365 mV) redox potential (Maret and Vallee 1998). This property makes metallothionein ideally suited to transfer zinc to nascent zinc-binding proteins (Jacob et al. 1998), without altering the intracellular free concentrations of the metal, which can be highly deleterious to the cell (Koh et al. 1996; Aras and Aizenman 2011; Aras et al. 2009a). It is, in fact, this source of zinc that may be a critical component of a large number of neurodegenerative cell signaling pathways (Hara and Aizenman 2014; McCord and Aizenman 2014; Pal et al. 2004; Aras and Aizenman 2011).

A clear hint that zinc translocation from ZnT3-containing vesicles was not the only mechanism for unregulated intracellular free zinc elevation came from a set of studies by Jae Koh and colleagues. First, this group observed that zinc still accumulated in hippocampal neurons following seizures in ZnT3-deficient mice (Lee et al. 2000). In a follow-up study, these investigators showed that zinc accumulation in the hippocampus was significantly depleted in metallothionein 3 (MT3) knockout mice and even more so in double ZnT3/MT3 null animals (Lee et al. 2003). MT3 is constitutively expressed and is the primary isoform of metallothionein present in the brain (Vasak and Meloni 2017; Hidalgo et al. 2001). Nearly concurrent with Jae

Koh's group observations (Lee et al. 2000), our group, in collaboration with Ian Reynolds, stumbled upon intracellular zinc liberation following oxidative injury, as a primary cause of cell death in neurons (Aizenman et al. 2000).

And so enter 2,2′-dithiodipyridine or DTDP. Our group was in search of a cell-permeant oxidant that could target intracellular thiol groups. The idea at the time was to induce the release of calcium from intracellular storage sites, namely, the endoplasmic reticulum (ER), independently from calcium entry from extracellular sources, such as the NMDA receptor. We were fortunate that down the hallway from our laboratory, Guy Salama and colleagues had observed that DTDP and related thiol oxidants could effectively release calcium from the sarcoplasmic reticulum of skeletal and cardiac myocytes (Zaidi et al. 1989; Prabhu and Salama 1990). Salama and co-workers went on to show that oxidant-induced intracellular calcium release was due to the action of DTDP and related compounds on a redox-sensitive gate within the ryanodine receptor complex (Salama et al. 1992). Of note, this work led to subsequent observations that nitric oxide was also an effective trigger of intracellular calcium liberation from sarcoplasmic reticulum by targeting the same gate (Stoyanovsky et al. 1997; Kobzik et al. 1994).

Our group began using DTDP in neurons previously loaded with "traditional" calcium fluorescent indicators such as Fura-2, being well aware that these compounds were also highly effective in detecting fluctuating intracellular zinc levels (Grynkiewicz et al. 1985; Cheng and Reynolds 1998), although we had no reason to suspect we were looking at anything but calcium at the time. DTDP, as advertised, produced substantial "calcium" transients in neurons and dramatic neurotoxicity, while sparing the surrounding glia. Moreover, glutamate receptor antagonist could not protect from DTDP's toxic effects. We had essentially found what we were looking for, or so we thought. With NMDA receptor antagonists rapidly losing favor as potential neuroprotectants in humans (Olney 1994), we thought we could target toxic calcium loads in neurons via a different mechanistic pathway.

This author gives full credit to Ian Reynolds for righting our ship, which, admittedly, had gone astray. Ian had been exploring the toxic effects of zinc in neurons (Dineley et al. 2000) and had arrived to the now classic papers by Wolfgang Maret, Bert Valle, and colleagues investigating the mobility of zinc from seemingly stable zinc/sulfur complexes in metallothionein following the treatment with redox-active ligands (Jacob et al. 1998; Jiang et al. 1998; Maret et al. 1997; Maret and Vallee 1998). With this seminal work, Maret, Valle, and co-workers showed effective and complete liberation of zinc from metallothionein utilizing none other than the thiol oxidant DTDP (Maret and Vallee 1998). Upon this revelation, we naturally adjusted course and showed that intracellular liberation of zinc, not calcium, was primarily responsible for the toxic effects of DTDP on neurons (Aizenman et al. 2000). Later studies by our group (Hara and Aizenman 2004) and others confirmed without a doubt that oxidants such as DTDP induce intracellular zinc release from metal-binding stores in intact cells (Gibon et al. 2011a, b; Lien et al. 2018). We later showed that other, more physiological (or pathophysiological) stimuli, such as microglial-derived peroxynitrite (Knoch et al. 2008; Zhang et al. 2004), glutamate receptor stimulation (Aras et al. 2009a), and ischemic injury (Aras et al. 2009b), could also effectively trigger the liberation of intracellular zinc.

Our group and others had now arrived at the realization that intracellular liberation of zinc was a critical trigger in neuronal injury responses. The critical question to answer now was the obvious "how?" Although there were clearly many paths to take, this author's group embarked on a signaling discovery trail that was initially guided by earlier discoveries in the laboratories of John Cidlowski at NIEHS and Shan Ping Yu together with Dennis Choi, then at Washington University in St. Louis. These research teams had suggested that a critical component of cell injury pathways was the eventual loss of cytoplasmic potassium, which, in turn, created a permissive environment for cell death-inducing proteases and nucleases to be optimally activated (Bortner et al. 1997; Hughes et al. 1997; Hughes and Cidlowski 1999; Montague et al. 1999; Wei et al. 2003; Yu et al. 1997, 1998, 2001). We set out to investigate whether potassium efflux was necessary for intracellular zinc-mediated cell death, and, if so, what cellular components were harnessed to accomplish this cellular task. A brief summary of how our group pieced this signaling pathway together is described in the next section.

9.3.3 A Zinc/Potassium Continuum in Neuronal Cell Death

In our first study on this topic (McLaughlin et al. 2001), rat cortical neurons in culture briefly exposed to a lethal concentration of DTDP were monitored for the appearance of enhanced potassium currents under voltage-clamp conditions, in a manner very similar to that described by Yu and colleagues in serum-deprived and staurosporine treated cells (Yu et al. 1997). Delayed-rectifier, TEA (tetraethylammonium)-sensitive potassium currents were indeed dramatically enhanced following treatment, albeit it took approximately 3 h for this effect to be clearly measurable. This suggested to us that complex signaling cascades were in place to link the intracellular release of zinc with the loss of intracellular potassium. We observed that intracellular zinc chelation and p38 inhibitors could both prevent the DTDP-induced enhanced currents and inhibit cell death, while caspase inhibitors only attenuated the latter. That is, the enhanced currents, in and of themselves, were necessary, but not sufficient, to trigger cell death. The current enhancement was an indirect manifestation of the cellular loss of intracellular potassium, which created a cell death-*enabling* environment, rather than triggering a cell death-*inducing* event. Our study clearly showed that intracellular zinc liberation was an upstream trigger mediating the apoptotic surge of potassium currents, occurring in a p38-dependent fashion (McLaughlin et al. 2001).

Connecting the dots between zinc liberation and potassium current enhancement required the identification of the ion channel responsible for the augmented currents. Together with Ed Levitan and in collaboration with Jeanne Nerbonne, the author's team zeroed in on Kv2.1 as the channel responsible for the injury-mediated enhanced currents (Pal et al. 2003). The use of dominant-negative mutated or truncated Kv2.1 channels in neurons was sufficient to prove that this channel was responsible for the apoptotic current surge. Moreover, preventing the enhanced currents with the dominant-negative constructs was highly neuroprotective. Once Kv2.1 was identified as the channel to target, the pieces of the puzzle fell together quite nicely in a series of subsequent studies. Since we knew p38 activation was important, we readily identified its upstream kinase, downstream from the zinc-liberation event, as being the apoptosis signaling kinase 1 or ASK-1 (Aras and Aizenman 2005). Moreover, in collaboration with the Levitan laboratory, we discovered that the enhanced potassium currents during the cell death process was mediated by the syntaxin-dependent insertion of a reserve pool of Kv2.1 channels, rather than a modification of pre-existing functional channels at the cell membrane (Pal et al. 2006). Several years later we showed that the de novo syntaxin-dependent insertion of Kv2.1 channels occurred at pre-existing membrane channel clusters (Justice et al. 2017), hubs of seemingly silent Kv2.1 aggregates that aid in ion channel exocytosis (Deutsch et al. 2012). Channel declustering by non-lethal ischemic injury (Aras et al. 2009b; Schulien et al. 2016) may be, in fact, a neuroprotective strategy neurons employ in the phenomenon known as ischemic preconditioning. Importantly, sublethal, intracellular zinc liberation may be an important trigger for this neuroprotective cascade to occur (Aras et al. 2009a, b; Schulien et al. 2016; Shah et al. 2014; Lee et al. 2008).

Additional studies in our laboratory showed that p38's involvement in the zinc-triggered apoptotic Kv2.1-mediated current surge was due to a direct phosphorylation of the channel by the kinase at serine 800, in the intracellular C-terminal domain (Redman et al. 2007). We later went on to show that zinc also facilitated the phosphorylation of tyrosine 124 on the N-terminal domain of the channel via Src kinase, primarily by inhibiting a phosphatase (protein tyrosine phosphatase epsilon or PTPε), which also targets this site (Tiran et al. 2003; Redman et al. 2009). In fact, phosphorylation has to occur at both sites for the Kv2.1 exocytotic process to occur (Redman et al. 2009), with the Src-mediated step seemingly required to proceed first (He et al. 2015). We later went on to show that, not surprisingly, calcium, in addition to zinc, had also a role to play. We observed that either oxidant or zinc-mediated liberation of calcium from the endoplasmic reticulum (Schulien et al. 2016) led to the activation of CaMKII, the binding of this kinase to syntaxin. Syntaxin then, in turn, binds and facilitates the membrane insertion of the dual phosphorylated Kv2.1 (McCord and Aizenman 2013).

The interaction of Kv2.1 with syntaxin is a critical and perhaps unique step necessary for the completion of the neuronal cell death cascade. Removing the syntaxin binding site and interfering with the interaction of the channel with the SNARE protein are highly neuroprotective strategies both in vitro and in vivo (McCord et al. 2014; Yeh et al. 2017), importantly, without altering the endogenous levels of Kv2.1-mediated functional currents in neurons. This exciting development has led us to the targeted discovery of small molecules as novel neuroprotective agents (Yeh et al. 2018). An illustrated summary of the pathways described in this section is presented in Fig. 9.4. A potential key role of the mitochondria in

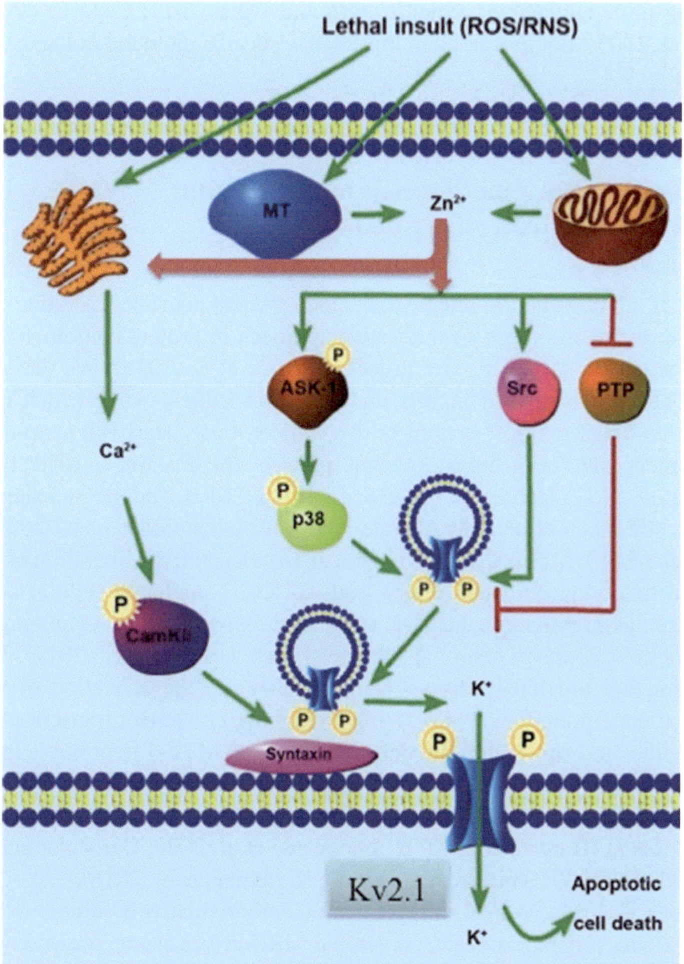

Fig. 9.4 A zinc/potassium continuum in neuronal cell death. Cell death-enabling pathway triggered by intracellular zinc, released from metallothionein (MT), with a possible interplay with mitochondrial zinc pools, following oxidative or nitrosative injury (Aizenman et al. 2000; Aras and Aizenman 2011). The liberated zinc, in turn, activates p38 via ASK-1 (Aras and Aizenman 2005; McLaughlin et al. 2001). Activation of p38 leads to phosphorylation of residue S800 in Kv2.1 (Redman et al. 2007). Zinc also inhibits protein tyrosine phosphatase epsilon (PTPε), enabling Src phosphorylation of residue Y124 of Kv2.1 (Redman et al. 2009). The dual phosphorylated channel interacts with the SNARE protein syntaxin at a C-terminal region (C1a) (McCord et al. 2014; Pal et al. 2006), requiring the activation and binding of CaMKII (McCord and Aizenman 2013). The latter had been activated via the oxidative release of calcium from the ER (McCord and Aizenman 2013). Zinc can also trigger calcium release from the ER via activation of ryanodine receptors (Schulien et al. 2016). The interaction of Kv2.1 with syntaxin leads to the insertion of the potassium channel in the cell membrane, producing large measurable delayed rectifier currents under voltage clamp, and, during the injurious process in dying cells, the loss of intracellular potassium from the cytoplasm (McLaughlin et al. 2001)

regulating intracellular zinc (Shuttleworth and Weiss 2011; Capasso et al. 2005; Sensi et al. 2003), not discussed in this chapter, is also included in the overall diagram in Fig. 9.4.

9.4 Tapping the Zinc/Potassium Continuum as a Neuroprotective Strategy

The lack of effective neuroprotective therapies to treat neurodegenerative disorders recently drove us to design an innovative approach to protect neurons from dying during the course of the disease process (Justice et al. 2018). We designed this potential new therapy based on three underlying principles, two of which have been discussed in detail in prior sections of this chapter. First, nearly all neurodegenerative disorders have been linked, at some point of the disease, to oxidative injury (Cicero et al. 2017; Cobley et al. 2018; Liu et al. 2017), and, as indicated earlier, oxidative stress can effectively liberate zinc from intracellular metal-binding proteins (Maret 2019; McCord and Aizenman 2014). Second, the liberation of intracellular zinc following oxidative injury leads to Kv2.1-mediated cellular potassium loss, which, as highlighted above, is a required element in the completion of cell death programs of many neuronal subtypes (Shah and Aizenman 2014). The third component, not previously discussed here, relates to the activation of the zinc-selective transcription factor MTF-1 (metal regulatory element transcription factor 1) by the liberated zinc. Zinc, and only zinc, binds to MTF-1 to promote its nuclear translocation and binding to the MRE (metal regulatory element) in the promoter region of various zinc-sensitive genes, including ZnT1 and inducible isoforms of metallothionein (Daniels et al. 2002; Langmade et al. 2000; Carpenter and Palmer 2017; Andrews 2000; Smirnova et al. 2000; Hardyman et al. 2016).

The idea was this. If a neuron were sufficiently oxidatively injured to trigger a substantial zinc-liberation event, an ectopic MRE-driven gene construct would be present, driving the expression of a protein or peptide designed to interfere with the apoptotic insertion of Kv2.1 (Fig. 9.5). The potential breakthrough concept was thus the design of a normally silent gene, to be activated as a neuroprotective agent only upon lethal injury. For our proof-of-concept study (Justice et al. 2018), we opted to drive the expression of the hepatitis C-derived protein NS5A with the MTF-1/MRE strategy. We had previously shown that NS5A is an effective inhibitor of the Src-mediated phosphorylation of Kv2.1, an earlier noted signaling step required for the interaction of the channel with syntaxin (Redman et al. 2009; Clemens et al. 2015). Hepatitis C evolved this property to retard or inhibit normal, ongoing hepatocyte cell death in order to increase the probability of the virus in establishing a chronic infection in that organ (Mankouri et al. 2009). For some unknown reason, hepatocytes also utilize Kv2.1 to die (Mankouri et al. 2009), although whether zinc is required to drive this process is not yet known. We were very pleased to observe that neurons previously transfected with and MRE-driven

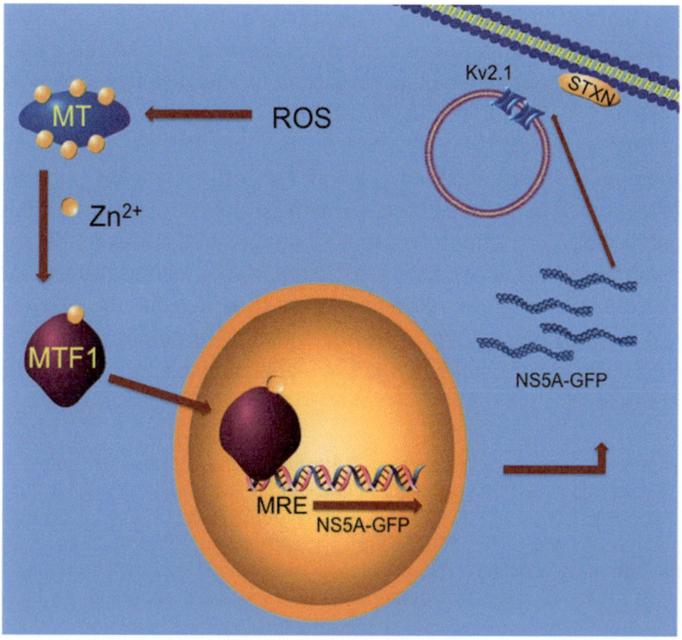

Fig. 9.5 Zinc-dependent expression of a neuroprotective protein. Oxidative injury induces the liberation of intracellular metal-binding sources, primarily metallothionein (MT). The liberated zinc, in turn, binds and promotes the nuclear translocation MTF-1, activating the expression of the hepatitis C-derived protein NS5A (tagged with GFP). NS5A prevent phosphorylation and syntaxin-dependent membrane insertion of Kv2.1, preventing the apoptotic potassium current surge and thereby blocking the cell death cascade mechanism

NS5A constructs where highly resistant to a normally lethal stimulus, also failing to observe the enhancement of potassium currents following injury (Justice et al. 2018). As predicted, unchallenged neurons had no detectable levels of the viral protein. Encouraged by these results, we plan to test this new method of neuroprotection in in vivo models of both acute and chronic neurodegeneration in the near future.

9.5 Concluding Remarks

Zinc has long known to be sequestered within synaptic vesicles in vast regions of the brain, including the cerebral cortex and the hippocampus. With the increasing development of a large range of highly specific zinc probes, a little mental elbow grease, and an unending curiosity on the part of old and new zinc neuro-aficionados, the metal has been finally yielding its secrets, revealing its many roles in neuronal function and dysfunction. In addition to the so-called synaptic zinc, non-vesicular intracellular pools of zinc are important contributors to a vast array of signaling networks that not only influence brain physiology but also are critical for determining whether neurons will live or die. With apologies to colleagues whose work was not described in detail, this review did not intend to cover the full gamut of zinc signaling components in the brain, as, in fact, two other chapters in this volume are also devoted to zinc function in the central nervous system. However, it is hoped that the examples provided here illustrate the fact that detailed, mechanistic studies of zinc signaling networks in neurons can not only offer a window into the complex biology of this metal in the brain but also point toward novel venues for therapies in complex brain disorders. It must be saliently pointed out, however, that the work is very far from over. Indeed, we have only begun to scratch the surface of the large number of cellular and molecular components that integrate the neurobiology of zinc. Hopefully, this brief glimpse into the signaling of neuronal zinc will beckon more young scientists to join us in this fascinating and rewarding branch of neurobiological research.

Acknowledgments This book chapter is dedicated to the memory of Sumon Pal, Ph.D. (1976–2018, pictured below), whose work in the author's laboratory revealed Kv2.1 as the ion channel responsible for the cell death surge of potassium currents following an injurious zinc signal in neurons (Pal et al. 2003, 2004, 2006). The author also wishes to thank all of the members of his laboratory who have contributed to zinc-related research during the past two decades, as well as his many collaborators. This work has been supported by NIH grant NS043277 to the author.

References

Aizenman E, Stout AK, Hartnett KA, Dineley KE, McLaughlin B, Reynolds IJ (2000) Induction of neuronal apoptosis by thiol oxidation: putative role of intracellular zinc release. J Neurochem 75(5):1878–1888

Anderson CT, Radford RJ, Zastrow ML, Zhang DY, Apfel UP, Lippard SJ, Tzounopoulos T (2015) Modulation of extrasynaptic NMDA receptors by synaptic and tonic zinc. Proc Natl Acad Sci USA 112(20):E2705–E2714. https://doi.org/10.1073/pnas.1503348112

Anderson CT, Kumar M, Xiong S, Tzounopoulos T (2017) Cell-specific gain modulation by synaptically released zinc in cortical circuits of audition. elife 6:e29893. https://doi.org/10.7554/eLife.29893

Andrews GK (2000) Regulation of metallothionein gene expression by oxidative stress and metal ions. Biochem Pharmacol 59(1):95–104

Aras MA, Aizenman E (2005) Obligatory role of ASK1 in the apoptotic surge of K+ currents. Neurosci Lett 387(3):136–140. https://doi.org/10.1016/j.neulet.2005.06.024

Aras MA, Aizenman E (2011) Redox regulation of intracellular zinc: molecular signaling in the life and death of neurons. Antioxid Redox Signal 15(8):2249–2263. https://doi.org/10.1089/ars.2010.3607

Aras MA, Hara H, Hartnett KA, Kandler K, Aizenman E (2009a) Protein kinase C regulation of neuronal zinc signaling mediates survival during preconditioning. J Neurochem 110(1):106–117. https://doi.org/10.1111/j.1471-4159.2009.06106.x

Aras MA, Saadi RA, Aizenman E (2009b) Zn2+ regulates Kv2.1 voltage-dependent gating and localization following ischemia. Eur J Neurosci 30(12):2250–2257. https://doi.org/10.1111/j.1460-9568.2009.07026.x

Assaf SY, Chung SH (1984) Release of endogenous Zn2+ from brain tissue during activity. Nature 308(5961):734–736

Berg JM, Shi Y (1996) The galvanization of biology: a growing appreciation for the roles of zinc. Science 271(5252):1081–1085

Bini E (2010) Archaeal transformation of metals in the environment. FEMS Microbiol Ecol 73:1):1–1)16. https://doi.org/10.1111/j.1574-6941.2010.00876.x

Bortner CD, Hughes FM Jr, Cidlowski JA (1997) A primary role for K+ and Na+ efflux in the activation of apoptosis. J Biol Chem 272(51):32436–32442

Capasso M, Jeng JM, Malavolta M, Mocchegiani E, Sensi SL (2005) Zinc dyshomeostasis: a key modulator of neuronal injury. J Alzheimers Dis 8(2):93–108; discussion 209-115

Carpenter MC, Palmer AE (2017) Native and engineered sensors for Ca(2+) and Zn(2+): lessons from calmodulin and MTF1. Essays Biochem 61(2):237–243. https://doi.org/10.1042/EBC20160069

Carrera N, Arrojo M, Sanjuan J, Ramos-Rios R, Paz E, Suarez-Rama JJ, Paramo M, Agra S, Brenlla J, Martinez S, Rivero O, Collier DA, Palotie A, Cichon S, Nothen MM, Rietschel M, Rujescu D, Stefansson H, Steinberg S, Sigurdsson E, St Clair D, Tosato S, Werge T, Stefansson K, Gonzalez JC, Valero J, Gutierrez-Zotes A, Labad A, Martorell L, Vilella E, Carracedo A, Costas J (2012) Association study of nonsynonymous single nucleotide polymorphisms in schizophrenia. Biol Psychiatry 71(2):169–177. https://doi.org/10.1016/j.biopsych.2011.09.032

Cheng C, Reynolds IJ (1998) Calcium-sensitive fluorescent dyes can report increases in intracellular free zinc concentration in cultured forebrain neurons. J Neurochem 71(6):2401–2410

Chorin E, Vinograd O, Fleidervish I, Gilad D, Herrmann S, Sekler I, Aizenman E, Hershfinkel M (2011) Upregulation of KCC2 activity by zinc-mediated neurotransmission via the mZnR/GPR39 receptor. J Neurosci 31(36):12916–12926. https://doi.org/10.1523/JNEUROSCI.2205-11.2011

Cicero CE, Mostile G, Vasta R, Rapisarda V, Signorelli SS, Ferrante M, Zappia M, Nicoletti A (2017) Metals and neurodegenerative diseases. A systematic review. Environ Res 159:82–94. https://doi.org/10.1016/j.envres.2017.07.048

Clemens K, Yeh CY, Aizenman E (2015) Critical role of casein kinase 2 in hepatitis C NS5A-mediated inhibition of Kv2.1 K(+) channel function. Neurosci Lett 609:48–52. https://doi.org/10.1016/j.neulet.2015.10.021

Cobley JN, Fiorello ML, Bailey DM (2018) 13 reasons why the brain is susceptible to oxidative stress. Redox Biol 15:490–503. https://doi.org/10.1016/j.redox.2018.01.008

Cole TB, Wenzel HJ, Kafer KE, Schwartzkroin PA, Palmiter RD (1999) Elimination of zinc from synaptic vesicles in the intact mouse brain by disruption of the ZnT3 gene. Proc Natl Acad Sci USA 96(4):1716–1721

Cole TB, Robbins CA, Wenzel HJ, Schwartzkroin PA, Palmiter RD (2000) Seizures and neuronal damage in mice lacking vesicular zinc. Epilepsy Res 39(2):153–169

Daniels PJ, Bittel D, Smirnova IV, Winge DR, Andrews GK (2002) Mammalian metal response element-binding transcription factor-1 functions as a zinc sensor in yeast, but not as a sensor of cadmium or oxidative stress. Nucleic Acids Res 30(14):3130–3140

Deutsch E, Weigel AV, Akin EJ, Fox P, Hansen G, Haberkorn CJ, Loftus R, Krapf D, Tamkun MM (2012) Kv2.1 cell surface clusters are insertion platforms for ion channel delivery to the plasma membrane. Mol Biol Cell 23(15):2917–2929. https://doi.org/10.1091/mbc.E12-01-0047

Dietz RM, Weiss JH, Shuttleworth CW (2008) Zn2+ influx is critical for some forms of spreading depression in brain slices. J Neurosci 28(32):8014–8024. https://doi.org/10.1523/JNEUROSCI.0765-08.2008

Dineley KE, Scanlon JM, Kress GJ, Stout AK, Reynolds IJ (2000) Astrocytes are more resistant than neurons to the cytotoxic effects of increased [Zn(2+)](i). Neurobiol Dis 7(4):310–320. https://doi.org/10.1006/nbdi.2000.0303

Eom K, Ho Hyun J, Lee DG, Kim S, Jeong HJ, Kang JS, Ho WK, Lee SH (2019) Intracellular Zn(2+) signaling facilitates mossy fiber input-induced heterosynaptic potentiation of direct cortical inputs in hippocampal CA3 pyramidal cells. J Neurosci. https://doi.org/10.1523/JNEUROSCI.2130-18.2019

Flanagan R (2001) Gould's book of fish. Pan Macmillan Australia, Sydney

Frederickson CJ, Hernandez MD, McGinty JF (1989) Translocation of zinc may contribute to seizure-induced death of neurons. Brain Res 480(1–2):317–321

Frederickson CJ, Koh JY, Bush AI (2005) The neurobiology of zinc in health and disease. Nat Rev Neurosci 6(6):449–462. https://doi.org/10.1038/nrn1671

Gibon J, Richaud P, Bouron A (2011a) Hyperforin changes the zinc-storage capacities of brain cells. Neuropharmacology 61(8):1321–1326. https://doi.org/10.1016/j.neuropharm.2011.08.002

Gibon J, Tu P, Bohic S, Richaud P, Arnaud J, Zhu M, Boulay G, Bouron A (2011b) The over-expression of TRPC6 channels in HEK-293 cells favours the intracellular accumulation of zinc. Biochim Biophys Acta 1808(12):2807–2818. https://doi.org/10.1016/j.bbamem.2011.08.013

Gibson RS, King JC, Lowe N (2016) A review of dietary zinc recommendations. Food Nutr Bull 37(4):443–460. https://doi.org/10.1177/0379572116652252

Gilad D, Shorer S, Ketzef M, Friedman A, Sekler I, Aizenman E, Hershfinkel M (2015) Homeostatic regulation of KCC2 activity by the zinc receptor mZnR/GPR39 during seizures. Neurobiol Dis 81:4–13. https://doi.org/10.1016/j.nbd.2014.12.020

Grynkiewicz G, Poenie M, Tsien RY (1985) A new generation of Ca2+ indicators with greatly improved fluorescence properties. J Biol Chem 260(6):3440–3450

Hara H, Aizenman E (2004) A molecular technique for detecting the liberation of intracellular zinc in cultured neurons. J Neurosci Methods 137(2):175–180. https://doi.org/10.1016/j.jneumeth.2004.02.018

Hara H, Aizenman E (2014) Oxidative stress and neuronal zinc signaling. In: Fukada T, Kambe T (eds) Zinc signals in cellular functions and disorders. Springer, Tokyo, pp 55–87

Hardyman JE, Tyson J, Jackson KA, Aldridge C, Cockell SJ, Wakeling LA, Valentine RA, Ford D (2016) Zinc sensing by metal-responsive transcription factor 1 (MTF1) controls metallothionein and ZnT1 expression to buffer the sensitivity of the transcriptome response to zinc. Metallomics 8(3):337–343. https://doi.org/10.1039/c5mt00305a

He K, McCord MC, Hartnett KA, Aizenman E (2015) Regulation of pro-apoptotic phosphorylation of Kv2.1 K+ channels. PLoS One 10(6):e0129498. https://doi.org/10.1371/journal.pone.0129498

Hershfinkel M, Moran A, Grossman N, Sekler I (2001) A zinc-sensing receptor triggers the release of intracellular Ca2+ and regulates ion transport. Proc Natl Acad Sci USA 98(20):11749–11754. https://doi.org/10.1073/pnas.201193398

Hershfinkel M, Kandler K, Knoch ME, Dagan-Rabin M, Aras MA, Abramovitch-Dahan C, Sekler I, Aizenman E (2009) Intracellular zinc inhibits KCC2 transporter activity. Nat Neurosci 12(6):725–727. https://doi.org/10.1038/nn.2316

Hidalgo J, Aschner M, Zatta P, Vasak M (2001) Roles of the metallothionein family of proteins in the central nervous system. Brain Res Bull 55(2):133–145

Hilgers MT, Ludwig ML (2001) Crystal structure of the quorum-sensing protein LuxS reveals a catalytic metal site. Proc Natl Acad Sci USA 98(20):11169–11174. https://doi.org/10.1073/pnas.191223098

Holst B, Egerod KL, Schild E, Vickers SP, Cheetham S, Gerlach LO, Storjohann L, Stidsen CE, Jones R, Beck-Sickinger AG, Schwartz TW (2007) GPR39 signaling is stimulated by zinc ions but not by obestatin. Endocrinology 148(1):13–20. https://doi.org/10.1210/en.2006-0933

Howell GA, Welch MG, Frederickson CJ (1984) Stimulation-induced uptake and release of zinc in hippocampal slices. Nature 308(5961):736–738

Huang YZ, Pan E, Xiong ZQ, McNamara JO (2008) Zinc-mediated transactivation of TrkB potentiates the hippocampal mossy fiber-CA3 pyramid synapse. Neuron 57(4):546–558. https://doi.org/10.1016/j.neuron.2007.11.026

Hughes FM Jr, Cidlowski JA (1999) Potassium is a critical regulator of apoptotic enzymes in vitro and in vivo. Adv Enzym Regul 39:157–171

Hughes FM Jr, Bortner CD, Purdy GD, Cidlowski JA (1997) Intracellular K+ suppresses the activation of apoptosis in lymphocytes. J Biol Chem 272(48):30567–30576

Jacob C, Maret W, Vallee BL (1998) Control of zinc transfer between thionein, metallothionein, and zinc proteins. Proc Natl Acad Sci USA 95(7):3489–3494. https://doi.org/10.1073/pnas.95.7.3489

Jiang LJ, Maret W, Vallee BL (1998) The glutathione redox couple modulates zinc transfer from metallothionein to zinc-depleted sorbitol dehydrogenase. Proc Natl Acad Sci USA 95(7):3483–3488. https://doi.org/10.1073/pnas.95.7.3483

Justice JA, Schulien AJ, He K, Hartnett KA, Aizenman E, Shah NH (2017) Disruption of KV2.1 somato-dendritic clusters prevents the apoptogenic increase of potassium currents. Neuroscience 354:158–167. https://doi.org/10.1016/j.neuroscience.2017.04.034

Justice JA, Manjooran DT, Yeh CY, Hartnett-Scott KA, Schulien AJ, Kosobucki GJ, Mammen S, Palladino MJ, Aizenman E (2018) Molecular neuroprotection induced by zinc-dependent expression of hepatitis C-derived protein NS5A targeting Kv2.1 potassium channels. J Pharmacol Exp Ther 367(2):348–355. https://doi.org/10.1124/jpet.118.252338

Kantheti P, Qiao X, Diaz ME, Peden AA, Meyer GE, Carskadon SL, Kapfhamer D, Sufalko D, Robinson MS, Noebels JL, Burmeister M (1998) Mutation in AP-3 delta in the mocha mouse links endosomal transport to storage deficiency in platelets, melanosomes, and synaptic vesicles. Neuron 21(1):111–122

Kay AR (2003) Evidence for chelatable zinc in the extracellular space of the hippocampus, but little evidence for synaptic release of Zn. J Neurosci 23(17):6847–6855

Kerchner GA, Canzoniero LM, Yu SP, Ling C, Choi DW (2000) Zn2+ current is mediated by voltage-gated Ca2+ channels and enhanced by extracellular acidity in mouse cortical neurones. J Physiol 528(Pt 1):39–52

Kim AH, Sheline CT, Tian M, Higashi T, McMahon RJ, Cousins RJ, Choi DW (2000) L-type Ca(2+) channel-mediated Zn(2+) toxicity and modulation by ZnT-1 in PC12 cells. Brain Res 886(1–2):99–107

Knoch ME, Hartnett KA, Hara H, Kandler K, Aizenman E (2008) Microglia induce neurotoxicity via intraneuronal Zn(2+) release and a K(+) current surge. Glia 56(1):89–96. https://doi.org/10.1002/glia.20592

Kobzik L, Reid MB, Bredt DS, Stamler JS (1994) Nitric oxide in skeletal muscle. Nature 372(6506):546–548. https://doi.org/10.1038/372546a0

Koh JY, Suh SW, Gwag BJ, He YY, Hsu CY, Choi DW (1996) The role of zinc in selective neuronal death after transient global cerebral ischemia. Science 272(5264):1013–1016

Krall RF, Moutal A, Khanna M, Hershfinkel M, Aizenman E, Tzounopoulos T (2018) Zinc transporter ZnT1 regulates zinc inhibition of NMDARs. Paper presented at the Society for Neuroscience, San Diego

Kumar M, Xiong S, Tzounopoulos T, Anderson CT (2019) Fine control of sound frequency tuning and frequency discrimination acuity by synaptic zinc signaling in mouse auditory cortex. J Neurosci 39(5):854–865. https://doi.org/10.1523/JNEUROSCI.1339-18.2018

Langmade SJ, Ravindra R, Daniels PJ, Andrews GK (2000) The transcription factor MTF-1 mediates metal regulation of the mouse ZnT1 gene. J Biol Chem 275(44):34803–34809. https://doi.org/10.1074/jbc.M007339200

Lavoie N, Jeyaraju DV, Peralta MR 3rd, Seress L, Pellegrini L, Toth K (2011) Vesicular zinc regulates the Ca2+ sensitivity of a subpopulation of presynaptic vesicles at hippocampal mossy fiber terminals. J Neurosci 31(50):18251–18265. https://doi.org/10.1523/JNEUROSCI.4164-11.2011

Lee JY, Cole TB, Palmiter RD, Koh JY (2000) Accumulation of zinc in degenerating hippocampal neurons of ZnT3-null mice after seizures: evidence against synaptic vesicle origin. J Neurosci 20(11):RC79

Lee JY, Kim JH, Palmiter RD, Koh JY (2003) Zinc released from metallothionein-iii may contribute to hippocampal CA1 and thalamic neuronal death following acute brain injury. Exp Neurol 184(1):337–347

Lee JY, Kim YJ, Kim TY, Koh JY, Kim YH (2008) Essential role for zinc-triggered p75NTR activation in preconditioning neuroprotection. J Neurosci 28(43):10919–10927. https://doi.org/10.1523/JNEUROSCI.3421-08.2008

Li Y, Hough CJ, Suh SW, Sarvey JM, Frederickson CJ (2001) Rapid translocation of Zn(2+) from presynaptic terminals into postsynaptic hippocampal neurons after physiological stimulation. J Neurophysiol 86(5):2597–2604. https://doi.org/10.1152/jn.2001.86.5.2597

Li M, Wu DD, Yao YG, Huo YX, Liu JW, Su B, Chasman DI, Chu AY, Huang T, Qi L, Zheng Y, Group CNW, DietGen C, Luo XJ (2016) Recent positive selection drives the expansion of a schizophrenia risk nonsynonymous variant at SLC39A8 in Europeans. Schizophr Bull 42(1):178–190. https://doi.org/10.1093/schbul/sbv070

Li Y, Andereggen L, Yuki K, Omura K, Yin Y, Gilbert HY, Erdogan B, Asdourian MS, Shrock C, de Lima S, Apfel UP, Zhuo Y, Hershfinkel M, Lippard SJ, Rosenberg PA, Benowitz L (2017) Mobile zinc increases rapidly in the retina after optic nerve injury and regulates ganglion cell survival and optic nerve regeneration. Proc Natl Acad Sci USA 114(2):E209–E218. https://doi.org/10.1073/pnas.1616811114

Lien CF, Lee WS, Wang IC, Chen TI, Chen TL, Yang KT (2018) Intermittent hypoxia-generated ROS contributes to intracellular zinc regulation that limits ischemia/reperfusion injury in adult rat cardiomyocyte. J Mol Cell Cardiol 118:122–132. https://doi.org/10.1016/j.yjmcc.2018.03.014

Liu Z, Zhou T, Ziegler AC, Dimitrion P, Zuo L (2017) Oxidative stress in neurodegenerative diseases: from molecular mechanisms to clinical applications. Oxidative Med Cell Longev 2017:2525967. https://doi.org/10.1155/2017/2525967

Mankouri J, Dallas ML, Hughes ME, Griffin SD, Macdonald A, Peers C, Harris M (2009) Suppression of a pro-apoptotic K+ channel as a mechanism for hepatitis C virus persistence. Proc Natl Acad Sci USA 106(37):15903–15908. https://doi.org/10.1073/pnas.0906798106

Maret W (2019) The redox biology of redox-inert zinc ions. Free Radic Biol Med 134:311–326. https://doi.org/10.1016/j.freeradbiomed.2019.01.006

Maret W, Vallee BL (1998) Thiolate ligands in metallothionein confer redox activity on zinc clusters. Proc Natl Acad Sci USA 95(7):3478–3482

Maret W, Larsen KS, Vallee BL (1997) Coordination dynamics of biological zinc "clusters" in metallothioneins and in the DNA-binding domain of the transcription factor Gal4. Proc Natl Acad Sci USA 94(6):2233–2237. https://doi.org/10.1073/pnas.94.6.2233

McAllister BB, Dyck RH (2017) Zinc transporter 3 (ZnT3) and vesicular zinc in central nervous system function. Neurosci Biobehav Rev 80:329–350. https://doi.org/10.1016/j.neubiorev.2017.06.006

McAllister BB, Wright DK, Wortman RC, Shultz SR, Dyck RH (2018) Elimination of vesicular zinc alters the behavioural and neuroanatomical effects of social defeat stress in mice. Neurobiol Stress 9:199–213. https://doi.org/10.1016/j.ynstr.2018.10.003

McCord MC, Aizenman E (2013) Convergent Ca2+ and Zn2+ signaling regulates apoptotic Kv2.1 K+ currents. Proc Natl Acad Sci USA 110(34):13988–13993. https://doi.org/10.1073/pnas.1306238110

McCord MC, Aizenman E (2014) The role of intracellular zinc release in aging, oxidative stress, and Alzheimer's disease. Front Aging Neurosci 6:77. https://doi.org/10.3389/fnagi.2014.00077

McCord MC, Kullmann PH, He K, Hartnett KA, Horn JP, Lotan I, Aizenman E (2014) Syntaxin-binding domain of Kv2.1 is essential for the expression of apoptotic K+ currents. J Physiol 592(16):3511–3521. https://doi.org/10.1113/jphysiol.2014.276964

McLaughlin B, Pal S, Tran MP, Parsons AA, Barone FC, Erhardt JA, Aizenman E (2001) p38 activation is required upstream of potassium current enhancement and caspase cleavage in thiol oxidant-induced neuronal apoptosis. J Neurosci 21(10):3303–3311

Medvedeva YV, Ji SG, Yin HZ, Weiss JH (2017) Differential vulnerability of CA1 versus CA3 pyramidal neurons after ischemia: possible relationship to sources of Zn2+ accumulation and its entry into and prolonged effects on mitochondria. J Neurosci 37(3):726–737. https://doi.org/10.1523/JNEUROSCI.3270-16.2016

Mellone M, Pelucchi S, Alberti L, Genazzani AA, Di Luca M, Gardoni F (2015) Zinc transporter-1: a novel NMDA receptor-binding protein at the postsynaptic density. J Neurochem 132(2):159–168. https://doi.org/10.1111/jnc.12968

Montague JW, Bortner CD, Hughes FM Jr, Cidlowski JA (1999) A necessary role for reduced intracellular potassium during the DNA degradation phase of apoptosis. Steroids 64(9):563–569

Olney JW (1994) Neurotoxicity of NMDA receptor antagonists: an overview. Psychopharmacol Bull 30(4):533–540

Pal S, Hartnett KA, Nerbonne JM, Levitan ES, Aizenman E (2003) Mediation of neuronal apoptosis by Kv2.1-encoded potassium channels. J Neurosci 23(12):4798–4802

Pal S, He K, Aizenman E (2004) Nitrosative stress and potassium channel-mediated neuronal apoptosis: is zinc the link? Pflugers Arch 448(3):296–303. https://doi.org/10.1007/s00424-004-1256-7

Pal SK, Takimoto K, Aizenman E, Levitan ES (2006) Apoptotic surface delivery of K+ channels. Cell Death Differ 13(4):661–667. https://doi.org/10.1038/sj.cdd.4401792

Palmiter RD, Cole TB, Quaife CJ, Findley SD (1996) ZnT-3, a putative transporter of zinc into synaptic vesicles. Proc Natl Acad Sci USA 93(25):14934–14939

Pan E, Zhang XA, Huang Z, Krezel A, Zhao M, Tinberg CE, Lippard SJ, McNamara JO (2011) Vesicular zinc promotes presynaptic and inhibits postsynaptic long-term potentiation of mossy fiber-CA3 synapse. Neuron 71(6):1116–1126. https://doi.org/10.1016/j.neuron.2011.07.019

Paoletti P, Ascher P, Neyton J (1997) High-affinity zinc inhibition of NMDA NR1-NR2A receptors. J Neurosci 17(15):5711–5725

Patrick Wu HP, Dyck RH (2018) Signaling by synaptic zinc is required for whisker-mediated, fine texture discrimination. Neuroscience 369:242–247. https://doi.org/10.1016/j.neuroscience.2017.11.020

Perez-Rosello T, Anderson CT, Schopfer FJ, Zhao Y, Gilad D, Salvatore SR, Freeman BA, Hershfinkel M, Aizenman E, Tzounopoulos T (2013) Synaptic Zn2+ inhibits neurotransmitter release by promoting endocannabinoid synthesis. J Neurosci 33(22):9259–9272. https://doi.org/10.1523/JNEUROSCI.0237-13.2013

Peters S, Koh J, Choi DW (1987) Zinc selectively blocks the action of N-methyl-D-aspartate on cortical neurons. Science 236(4801):589–593

Prabhu SD, Salama G (1990) Reactive disulfide compounds induce Ca2+ release from cardiac sarcoplasmic reticulum. Arch Biochem Biophys 282(2):275–283

Prasad AS (2013) Discovery of human zinc deficiency: its impact on human health and disease. Adv Nutr 4(2):176–190. https://doi.org/10.3945/an.112.003210

Purves D, Augustine GJ, Fitspatrick D, Katz LC, LaMantia A-S, McNamara JO, Williams SM (2001) Neuroscience, 2nd edn. Sinauer Associates, Sunderland

Qian J, Noebels JL (2005) Visualization of transmitter release with zinc fluorescence detection at the mouse hippocampal mossy fibre synapse. J Physiol 566(Pt 3):747–758. https://doi.org/10.1113/jphysiol.2005.089276

Qian J, Xu K, Yoo J, Chen TT, Andrews G, Noebels JL (2011) Knockout of Zn transporters zip-1 and zip-3 attenuates seizure-induced CA1 neurodegeneration. J Neurosci 31(1):97–104. https://doi.org/10.1523/JNEUROSCI.5162-10.2011

Redman PT, He K, Hartnett KA, Jefferson BS, Hu L, Rosenberg PA, Levitan ES, Aizenman E (2007) Apoptotic surge of potassium currents is mediated by p38 phosphorylation of Kv2.1. Proc Natl Acad Sci USA 104(9):3568–3573. https://doi.org/10.1073/pnas.0610159104

Redman PT, Hartnett KA, Aras MA, Levitan ES, Aizenman E (2009) Regulation of apoptotic potassium currents by coordinated zinc-dependent signalling. J Physiol 587(Pt 18):4393–4404. https://doi.org/10.1113/jphysiol.2009.176321

Ruiz A, Walker MC, Fabian-Fine R, Kullmann DM (2004) Endogenous zinc inhibits GABA(A) receptors in a hippocampal pathway. J Neurophysiol 91(2):1091–1096. https://doi.org/10.1152/jn.00755.2003

Saadi RA, He K, Hartnett KA, Kandler K, Hershfinkel M, Aizenman E (2012) SNARE-dependent upregulation of potassium chloride co-transporter 2 activity after metabotropic zinc receptor activation in rat cortical neurons in vitro. Neuroscience 210:38–46. https://doi.org/10.1016/j.neuroscience.2012.03.001

Salama G, Nigam M, Shome K, Finkel MS, Lagenaur C, Zaidi NF (1992) Ryanodine-affinity chromatography purifies 106 kD Ca2+ release channels from skeletal and cardiac sarcoplasmic reticulum. Cell Calcium 13(10):635–647

Salazar G, Love R, Werner E, Doucette MM, Cheng S, Levey A, Faundez V (2004) The zinc transporter ZnT3 interacts with AP-3 and it is preferentially targeted to a distinct synaptic vesicle subpopulation. Mol Biol Cell 15(2):575–587. https://doi.org/10.1091/mbc.e03-06-0401

Schulien AJ, Justice JA, Di Maio R, Wills ZP, Shah NH, Aizenman E (2016) Zn(2+) -induced Ca(2+) release via ryanodine receptors triggers calcineurin-dependent redistribution of cortical neuronal Kv2.1 K(+) channels. J Physiol 594(10):2647–2659. https://doi.org/10.1113/JP272117

Sensi SL, Canzoniero LM, Yu SP, Ying HS, Koh JY, Kerchner GA, Choi DW (1997) Measurement of intracellular free zinc in living cortical neurons: routes of entry. J Neurosci 17(24):9554–9564

Sensi SL, Ton-That D, Sullivan PG, Jonas EA, Gee KR, Kaczmarek LK, Weiss JH (2003) Modulation of mitochondrial function by endogenous Zn2+ pools. Proc Natl Acad Sci USA 100(10):6157–6162. https://doi.org/10.1073/pnas.1031598100

Sensi SL, Paoletti P, Koh JY, Aizenman E, Bush AI, Hershfinkel M (2011) The neurophysiology and pathology of brain zinc. J Neurosci 31(45):16076–16085. https://doi.org/10.1523/Jneurosci.3454-11.2011

Shah NH, Aizenman E (2014) Voltage-gated potassium channels at the crossroads of neuronal function, ischemic tolerance, and neurodegeneration. Transl Stroke Res 5(1):38–58. https://doi.org/10.1007/s12975-013-0297-7

Shah NH, Schulien AJ, Clemens K, Aizenman TD, Hageman TM, Wills ZP, Aizenman E (2014) Cyclin e1 regulates Kv2.1 channel phosphorylation and localization in neuronal ischemia. J Neurosci 34(12):4326–4331. https://doi.org/10.1523/JNEUROSCI.5184-13.2014

Sheline CT, Ying HS, Ling CS, Canzoniero LM, Choi DW (2002) Depolarization-induced 65zinc influx into cultured cortical neurons. Neurobiol Dis 10(1):41–53

Shuttleworth CW, Weiss JH (2011) Zinc: new clues to diverse roles in brain ischemia. Trends Pharmacol Sci 32(8):480–486. https://doi.org/10.1016/j.tips.2011.04.001

Smart TG, Constanti A (1982) A novel effect of zinc on the lobster muscle GABA receptor. Proc R Soc Lond B Biol Sci 215(1200):327–341. https://doi.org/10.1098/rspb.1982.0045

Smirnova IV, Bittel DC, Ravindra R, Jiang H, Andrews GK (2000) Zinc and cadmium can promote rapid nuclear translocation of metal response element-binding transcription factor-1. J Biol Chem 275(13):9377–9384

Stoltenberg M, Nejsum LN, Larsen A, Danscher G (2004) Abundance of zinc ions in synaptic terminals of mocha mutant mice: zinc transporter 3 immunohistochemistry and zinc sulphide autometallography. J Mol Histol 35(2):141–145

Stoyanovsky D, Murphy T, Anno PR, Kim YM, Salama G (1997) Nitric oxide activates skeletal and cardiac ryanodine receptors. Cell Calcium 21(1):19–29

Timmes FX, Woosley SE, Weaver TA (1995) Galactic chemical evolution – hydrogen through zinc. Astrophys J Suppl S 98(2):617–658. https://doi.org/10.1086/192172

Tiran Z, Peretz A, Attali B, Elson A (2003) Phosphorylation-dependent regulation of Kv2.1 channel activity at tyrosine 124 by Src and by protein-tyrosine phosphatase epsilon. J Biol Chem 278(19):17509–17514. https://doi.org/10.1074/jbc.M212766200

Traynelis SF, Burgess MF, Zheng F, Lyuboslavsky P, Powers JL (1998) Control of voltage-independent zinc inhibition of NMDA receptors by the NR1 subunit. J Neurosci 18(16):6163–6175

Umeda H, Nomoto K (2002) Nucleosynthesis of zinc and iron peak elements in population III type II supernovae: comparison with abundances of very metal poor halo stars. Astrophys J 565(1):385–404. https://doi.org/10.1086/323946

Vasak M, Meloni G (2017) Mammalian metallothionein-3: new functional and structural insights. Int J Mol Sci 18(6):E1117. https://doi.org/10.3390/ijms18061117

Vergnano AM, Rebola N, Savtchenko LP, Pinheiro PS, Casado M, Kieffer BL, Rusakov DA, Mulle C, Paoletti P (2014) Zinc dynamics and action at excitatory synapses. Neuron 82(5):1101–1114. https://doi.org/10.1016/j.neuron.2014.04.034

Wei L, Yu SP, Gottron F, Snider BJ, Zipfel GJ, Choi DW (2003) Potassium channel blockers attenuate hypoxia- and ischemia-induced neuronal death in vitro and in vivo. Stroke 34(5):1281–1286. https://doi.org/10.1161/01.STR.0000065828.18661.FE

Weiss JH, Hartley DM, Koh JY, Choi DW (1993) AMPA receptor activation potentiates zinc neurotoxicity. Neuron 10(1):43–49

Wenzel HJ, Cole TB, Born DE, Schwartzkroin PA, Palmiter RD (1997) Ultrastructural localization of zinc transporter-3 (ZnT-3) to synaptic vesicle membranes within mossy fiber boutons in the hippocampus of mouse and monkey. Proc Natl Acad Sci USA 94(23):12676–12681

Westbrook GL, Mayer ML (1987) Micromolar concentrations of Zn2+ antagonize NMDA and GABA responses of hippocampal neurons. Nature 328(6131):640–643. https://doi.org/10.1038/328640a0

Yasuda S, Miyazaki T, Munechika K, Yamashita M, Ikeda Y, Kamizono A (2007) Isolation of Zn2+ as an endogenous agonist of GPR39 from fetal bovine serum. J Recept Signal Transduct Res 27(4):235–246. https://doi.org/10.1080/10799890701506147

Yeh CY, Bulas AM, Moutal A, Saloman JL, Hartnett KA, Anderson CT, Tzounopoulos T, Sun D, Khanna R, Aizenman E (2017) Targeting a potassium channel/syntaxin interaction ameliorates cell death in ischemic stroke. J Neurosci 37(23):5648–5658. https://doi.org/10.1523/JNEUROSCI.3811-16.2017

Yeh CH, Ye Z, Gaur S, Moutal A, Henton AM, Kouvaros S, Saloman JL, Hartnett-Scott KA, Khanna M, Camacho CJ, Aizenman E (2018) Neuroprotection by small molecule inhibition of Kv2.1-syntaxin interaction. In: Societry for Neuroscience

Yin HZ, Weiss JH (1995) Zn(2+) permeates Ca(2+) permeable AMPA/kainate channels and triggers selective neural injury. Neuroreport 6(18):2553–2556

Yin HZ, Sensi SL, Ogoshi F, Weiss JH (2002) Blockade of Ca2+-permeable AMPA/kainate channels decreases oxygen-glucose deprivation-induced Zn2+ accumulation and neuronal loss in hippocampal pyramidal neurons. J Neurosci 22(4):1273–1279

Yu SP, Yeh CH, Sensi SL, Gwag BJ, Canzoniero LM, Farhangrazi ZS, Ying HS, Tian M, Dugan LL, Choi DW (1997) Mediation of neuronal apoptosis by enhancement of outward potassium current. Science 278(5335):114–117

Yu SP, Farhangrazi ZS, Ying HS, Yeh CH, Choi DW (1998) Enhancement of outward potassium current may participate in beta-amyloid peptide-induced cortical neuronal death. Neurobiol Dis 5(2):81–88. https://doi.org/10.1006/nbdi.1998.0186

Yu SP, Canzoniero LM, Choi DW (2001) Ion homeostasis and apoptosis. Curr Opin Cell Biol 13(4):405–411

Zaidi NF, Lagenaur CF, Abramson JJ, Pessah I, Salama G (1989) Reactive disulfides trigger Ca2+ release from sarcoplasmic reticulum via an oxidation reaction. J Biol Chem 264(36):21725–21736

Zhang Y, Wang H, Li J, Jimenez DA, Levitan ES, Aizenman E, Rosenberg PA (2004) Peroxynitrite-induced neuronal apoptosis is mediated by intracellular zinc release and 12-lipoxygenase activation. J Neurosci 24(47):10616–10627. https://doi.org/10.1523/JNEUROSCI.2469-04.2004

Chapter 10
Possible Therapeutic Roles of Metallothionein-3 and Zinc in Endosome-Autophagosome-Lysosome Pathway (EALP) Dysfunction in Astrocytes

Ha Na Kim, Bo-Ra Seo, Sook-Jeong Lee, and Jae-Young Koh

Abstract The accumulation of abnormal protein aggregates contributes to the pathological progression of diverse neurodegenerative diseases. An increasing body of evidence indicates that defects in the protein clearance system play a crucial role in this process. Cargoes delivered via endosomes, phagosomes, and autophagosomes converge on lysosomes for degradation, which process is collectively called "the endosome-autophagosome-lysosome pathway" or EALP. As such, dysfunction of lysosomes may result in the accumulation of all these upstream vesicles/cargoes and may play a key role in diverse neurodegenerative conditions.

Over the years, we found that Zn-metallothionein-3 (MT3), the brain-enriched form of metallothionein, regulates lysosomal functions in cortical astrocytes. Zn-MT3 appears to interact with β-actin and activate c-Abl kinase. As a result, Zn-MT3 plays a role in maintaining lysosomal acidity, a prerequisite for vesicle fusion as well as cargo degradation. The reported downregulation of MT3 in Alzheimer's disease (AD), hence, may contribute to lysosomal dysfunction in AD. Of interest, raising intracellular free zinc levels also caused lysosomal acidification and normalization of degradation, even in the context of arrested autophagy. Pending further research on the mechanisms of these effects, we propose that mea-

H. N. Kim · B.-R. Seo
Neural Injury Lab, Biomedical Research Center, Asan Institute for Life Sciences, Asan Medical Center, Seoul, Korea
e-mail: say7101@amc.seoul.kr

S.-J. Lee
Department of Bioactive Material Science, Chonbuk National University, Jeonju, South Korea
e-mail: sj@jbnu.ac.kr

J.-Y. Koh (✉)
Neural Injury Lab, Biomedical Research Center, Asan Institute for Life Sciences, Asan Medical Center, Seoul, Korea

Department of Neurology, University of Ulsan College of Medicine, Seoul, South Korea
e-mail: jkko@amc.seoul.kr

T. Fukada, T. Kambe (eds.), *Zinc Signaling*,
https://doi.org/10.1007/978-981-15-0557-7_10

sures increasing Zn-MT3 and/or intracellular/lysosomal free zinc may be useful in normalizing lysosomal functions in neurodegenerative conditions.

Keywords Lysosome · Zinc · MT3 · Autophagy · Alzheimer

10.1 Introduction

Eukaryotic cells continuously produce and degrade a great number of cellular macroconstituents, the balance of which is essential for their survival. Accumulation of certain waste proteins can lead to the formation of toxic protein oligomers and aggregates (Komatsu et al. 2006). In addition, accumulation of dysfunctional organelles such as mitochondria and peroxisomes may contribute to increases in oxidative stress (Jin 2006). Collectively, these events may cause severe cellular dysfunction, eventually leading to their demise, which may underlie diverse pathologies associated with aging.

Macromolecule degradation in eukaryotic cells takes place in diverse compartments, including proteasomes, peroxisomes, late endosomes, and lysosomes. While proteasomes and peroxysomes engage in degradation of specific components, lysosomes are specialized for the all-purpose, high-capacity degradation of large proteins, protein aggregates, and organelles (De Duve and Wattiaux 1996). Cargoes are delivered to lysosomes via diverse routes that include processes, such as macroautophagy, chaperone-mediated autophagy, microautophagy, endocytosis, and phagocytosis, collectively referred to as the endosome-autophagosome-lysosome pathway (EALP). As the final effector organelle in the EALP, lysosomes are equipped with more than 60 acidic hydrolases that are able to degrade almost all cellular macroconstituents (Settembre et al. 2013). Although the mode of activation of each enzyme may differ, autocatalytic conversion of a proform to an active form seems to be the main mechanism for many lysosomal enzymes such as cathepsins (Stoka et al. 2016).

The average life span of humans is long compared to those of most other mammals. While with aging other dividing cells may remove excess waste products and replace damaged cells by cell replacement, in postmitotic central neurons, clearance of waste materials may completely depend on adequate lysosomal function throughout life. Therefore, lysosomal dysfunction tends to affect the central nervous system (CNS) to a greater extent than other tissues or organs in humans. One example of such dysfunction is a group of disorders termed lysosomal storage disorders (LSDs). The LSDs include neuronal ceroid lipofuscinosis (NCL, also known as Batten's disease), Niemann–Pick type C (NPC), and mucolipidosis type IV (MLIV), which are caused by defects in *CLN1–CLN3*, *NPC1*, and *MCOLN1* (TRPML1) genes, respectively. All of these genetic defects are linked to lysosomal dysfunction (Nixon 2013). Other examples are aging-related neurodegenerative disorders that include Alzheimer's disease (AD), Parkinson's disease (PD), and amyotrophic lateral sclerosis (ALS).

10.2 Contributors to Lysosomal Dysfunction in Aging-Related Neurodegenerative Diseases and Their Mechanisms

In central neurons, inadequate lysosomal degradation would allow a variety of waste products to gradually accumulate in cells. Some of the waste products can be degraded by alternative paths such as proteasomes or dumped out of the cell as exosomes through the formation of multivesicular bodies (MVBs). However, since these processes may not be sufficient to remove all waste products, such waste products will gradually accumulate in organelles and the cytosol, and will interfere with various cell functions. It is not yet completely clear which waste products are toxic to neurons, but in neurodegenerative conditions, oligomers of peptides and proteins, such as amyloid-β, phospho-tau, α-synuclein, and transactive response (TAR) DNA-binding protein 43 (TDP-43), are regarded as the main culprits.

Aβ, the main component of senile plaques in AD, is produced mainly in late endosomes from the plasma membrane protein amyloid precursor protein (APP), by the action of β- secretase-1 (BACE1). BACE1 is a transmembrane aspartic protease responsible for most of the β-secretase activity, but there is no direct evidence to support a causative role for increased BACE1 activity in AD. Instead, a growing body of evidence indicates that Aβ is rather a normal product of APP metabolism that serves diverse physiological functions (Muller et al. 2017; Manucat-Tan et al. 2019). For instance, a recent study reported that Aβ acts as an antimicrobial peptide in the brain (Kumar et al. 2016). Since BACE1 works best at acidic pH (around pH 4.5) (Huse et al. 2000; Tesco et al. 2007), stagnation of APP in acidic late endosomes, as may occur under conditions of lysosomal dysfunction, could increase the Aβ production from APP. Combined with decreased lysosomal degradation, this late-endosomal retention of BACE1 and APP would result in the accumulation of Aβ. In fact, by yet unidentified mechanism, Aβ itself can cause lysosomal alkalinization and dysfunction (Guo et al. 2017), which would further promote Aβ accumulation. Accordingly, a small increase in Aβ production caused by APP mutations could induce mild lysosomal dysfunction, which, in turn, leads to further increases in Aβ levels. This type of a positive feedback loop would result in a gradual increase in Aβ accumulation. Intriguingly, mutant presenilin-1 (PSEN1), which has been suggested to increase Aβ levels by virtue of its role as a component of γ-secretase, also causes lysosomal alkalinization by inhibiting vacuolar ATPase (V-ATPase) assembly (Lee et al. 2010a; Coffey et al. 2014). Hence, a common denominator in the effects of both mutated APP and presenilin-1 (PS-1) may be lysosomal dysfunction.

Another potential contributor to aberrant lysosomal degradation in AD is hyperphosphorylated tau, which accumulates in the cytosol leading to the formation of neurofibrillary tangles, another hallmark of AD. The main physiological function of tau is to bind to microtubules in order to stabilize the structure. Dissociation of tau from microtubules disrupts retrograde transport of peripherally derived endosomes and autophagosomes to lysosomes localized in the vicinity of the nucleus, interfering with degradation of cargoes. Conversely, inhibition of lysosomal degradation further aggravates phospho-tau accumulation (Hamano et al. 2008). Although mutations in the tau gene cause tauopathies without producing conspicuous Aβ accumulation (Roberson et al. 2007), mutations in APP genes cause accumulation of Aβ, α-synuclein, and tau (Boutajangout et al. 2004; Samura et al. 2006). These findings indicate that aberrant APP processing and Aβ may have broader effects on the EALP than tau. Alternatively, lysosomal dysfunction alone may not be sufficient for Aβ accumulation. Perhaps, accompanied abnormality in the endocytosis of APP may be a prerequisite. In fact, proteins involved in endocytosis may be upregulated by aging (Alsaqati et al. 2018).

Recent studies have found that a substantial fraction of genes involved in PD are related to endosomal trafficking and/or lysosomal function. To name a few, *VPS35, GBA, ATP13A2, ATP6AP2, DNAJC13/RME-8, RAB7L1*, and GAK (cyclin G-associated kinase) are such genes (Perrett et al. 2015). The *ATP13A2/PARK9* gene encodes a lysosomal ATPase that transports cations, and the *ATP6AP2* gene encodes a transmembrane protein that is a component of V-ATPase. The resultant functional defects in lysosomes, in combination with environmental factors, likely contribute to accumulation of α-synuclein aggregates in midbrain dopaminergic neurons as well as in cortical neurons. Reciprocally, as in the case of Aβ in AD, thus-formed α-synuclein aggregates can further impair macroautophagy (Moors et al. 2017; Rahman and Rhim 2017), again giving rise to a vicious cycle.

Likewise, lysosomal dysfunction has been implicated in the pathogenesis of other neurodegenerative diseases, including ALS, Huntington's disease (HD), and other trinucleotide repeat disorders (Cipolat Mis et al. 2016; Croce and Yamamoto 2019; Yamada et al. 2002). Aggregates of SOD-1 (superoxide dismutase-1) or TDP-43, which are associated with ALS, disrupt the EALP (Chen et al. 2012; Han et al. 2015; Xia et al. 2016; Leibiger et al. 2018). Aggregates of mutant huntingtin, and likely those of other polyglutamine proteins, again inhibit the EALP (Croce and Yamamoto 2019). Hence, it is tempting to speculate that a common mechanism underlying neurodegenerative diseases, especially those accompanied by accumulation of aggregated proteins, may be lysosomal or EALP dysfunction. Hence, measures targeting normalization of the EALP function may prove to be preventive against or therapeutically effective in diverse neurodegenerative diseases.

10.3 Possible Measures to Restore Lysosomal Function

A number of kinases, membrane proteins, transport machinery, signaling membrane phospholipids, and cations, such as Ca^{2+}, Mg^{2+}, and Zn^{2+}, are reported to regulate the EALP (Shen et al. 2011; Liuzzi et al. 2014; Venkatachalam et al. 2015; De Craene et al. 2017; Nascimbeni et al. 2017). Hence, it may not be easy to elucidate which steps are critically affected in a particular neurodegenerative disease. To correct abnormalities in the EALP degradation in neurodegenerative disease models, a number of investigations have focused on finding measures to promote early steps such as autophagosome formation. However, if the main defect is the lysosomal dysfunction causing arrested autophagy or more broadly arrested EALP, activation of early steps of EALP alone may not be sufficient to restore the degradation. In fact, while it has been reported that levels of beclin-1, an upstream activator of autophagy, are reduced in AD brains (Pickford et al. 2008; Jaeger et al. 2010), little evidence supports that autophagosome formation in AD is defective. To the contrary, an increasing body of evidence indicates that upstream vesicles including autophagosomes and late endosomes accumulate likely as a result of inefficient fusion between these vesicles and lysosomes (Boland et al. 2008; Lee et al. 2011). Directly supporting this, a recent study presented evidence that autophagosome formation is not reduced, but is instead upregulated, in the early stage of AD (Sanchez-Varo et al. 2012). If so, therapies aiming at enhancing autophagosome formation per se would be futile or may even aggravate accumulation of vacuoles in AD.

In contrast, as discussed above, multiple lines of evidence indicate that lysosomal dysfunction occurs in and contributes to neurodegenerative disorders. Hence, measures that increase lysosomal function in the EALP may prove effective as therapy for proteinopathic neurodegenerative conditions. One possible way to increase lysosomal function is to upregulate lysosomes and lysosomal proteins, including enzymes. Although the level of transcription factor EB (TFEB), the master transcriptional activator of lysosomal proteins, is not reduced in AD brains, further increasing it by delivering viral TFEB constructs reduces Aβ and phospho-tau levels in AD mice (Xiao et al. 2015; Polito et al. 2014). Hence, TFEB may be a viable target for the development of drugs to enhance lysosomal degradation. There appear to be diverse ways to increase the level and activity of TFEB in addition to introducing the corresponding gene. For example, AKT, mTORC1 (mammalian target of rapamycin complex 1), and ERK-2 (extracellular signal-regulated kinase-2) phosphorylate TFEB to inhibit its translocation to nuclei; thus, inhibitors of these kinases may upregulate TFEB activity. Notably, the disaccharide sugar trehalose activates TFEB and induces lysosomal biogenesis (Rusmini et al. 2019), which may underlie the beneficial effect of trehalose on the accumulation of TDP-43 in a cell model of ALS (Wang et al. 2018).

Although TFEB activates lysosomes, it acts in the entire EALP. Are there measures that more specifically target the lysosome? There appear to be several such measures. For instance, GBA (encoding for the lysosomal enzyme glucocerebrosidase), mutations of which gene are linked to PD, may be one. Whereas homozygous mutations lead to Gaucher disease, heterozygosity is a risk factor for PD. GBA interacts with α-synuclein and its mutations disrupt functions of lysosomes, including lysosomal recycling. Whether this defect is caused by accumulation of glucosylceramide, the substrate of GBA, in lysosomes, or a deficiency in GBA's nonenzymatic functions is yet unknown. Regardless, treatment of GBA-deficient fibroblasts with imiglucerase, a recombinant human GBA, is effective in normalizing lysosomal functions. Another potential example is progranulin (PGRN), a gene involved in neuronal ceroid lipofuscinosis. Again, haploinsufficiency of PGRN results in frontotemporal lobar degeneration (FTLD) accompanied by TDP-43 accumulation (Baker et al. 2006; Cruts et al. 2006). Although it remains unknown why a PGRN deficiency induces neurodegenerative diseases, several lines of evidence implicate PRGN in lysosomal function. First, PGRN facilitates the acidification of lysosomes and maturation of cathepsin D (CTSD) (Tanaka et al. 2017). Second, PGRN may act through its C-terminal granulin E domain to function as a chaperone that regulates multiple lysosomal enzymes, including GBA and CTSD (Jian et al. 2017). PGRN also has links to TFEB. The promoter of *GRN* contains TFEB-binding sites, and PGRN expression is upregulated by TFEB overexpression (Sardiello et al. 2009). Furthermore, a reduction or complete deletion of PGRN changes the expression of genes associated with lysosomal function and lipid metabolism, indicative of lysosomal dysfunction (Evers et al. 2017). Hence, measures to increase levels of PRGC may help restore lysosomal dysfunction.

Lastly, in many cases, lysosomal dysfunction may be overcome by manipulating luminal pH. In fact, for optimal activity of lysosomal enzymes, the lysosomal pH should be kept around 4.5–5.5. For this, V-ATPase, the lysosomal proton pump, is required. In diverse cell models of proteinopathic neurodegenerative diseases, it is found that lysosomal pH is shifted toward the alkaline direction, a change that may be brought about by downregulating V-ATPase. The end results of lysosomal alkalinization are: (1) decreased fusion between cargo vesicles and lysosomes and (2) suboptimal enzyme activities. Regardless of the specific cause, re-acidification of lysosomes tends to normalize fusion as well as degradation. Hence, measures that help re-acidify lysosomes may prove useful in hindering the progression of proteinopathic neurodegenerative diseases. A recent study reported that acidic nanoparticles may be useful for this purpose (Bourdenx et al. 2016; Trudeau et al. 2016).

10.4 Emerging Role of Metallothionein-3 (MT3) in Regulation of Endocytosis and Lysosome Functions in Astrocytes

Metallothioneins (MTs) are small-sized proteins with seven metal-binding sites. While under normal conditions zinc is the main metal bound to MTs, under pathological conditions, their apo-forms may function as buffers for toxic heavy metals such as lead, cadmium, and copper. Recently, however, a growing body of evidence has indicated that MTs may have additional functions in cells. For instance, oxidative or nitrative stress can release zinc from MTs. The released zinc may then be involved in diverse signaling events such as protein kinase C (PKC) and Erk activation (Aras et al. 2009; Lin et al. 2011; He and Aizenman 2010; Noh et al. 1999). In addition, MT3, a CNS-enriched form, binds to β-actin and contributes to signaling such as c-Abl activation. The latter mechanism appears to affect clathrin-mediated endocytosis and lysosomal function, both of which are major constituents of the EALP (Lee et al. 2015).

One known finding about MT3 biology in neurodegenerative disease is that it is downregulated in Alzheimer's disease, especially in astrocytes. While the significance of this change has yet to be determined, based on results obtained from astrocyte cell culture experiments, it seems possible that MT3 downregulation may contribute to dysfunction in the EALP, especially in Aβ endocytosis as well as its lysosomal degradation in astrocytes. Hence, while the cascade leading to MT3 depletion in AD is unknown, it may play as a component in Aβ accumulation in the AD brain.

In AD, Aβ is mainly produced in neurons and secreted to the extracellular fluid at least partly via the exosome pathway. The accumulation of Aβ in the extracellular fluid (ECF) depends on its secretion rate and aggregation rate as well as its removal rate. For the latter, clearance via the glymphatic and vascular system and removal by microglial phagocytosis are well-known processes. In addition, a growing body of evidence indicates that Aβ endocytosis by astrocytes also plays a significant role in its clearance (Tarasoff-Conway et al. 2015; Dominguez-Prieto et al. 2018). In this regard, it seems significant that the absence of MT3 markedly reduces Aβ uptake via the clathrin-mediated endocytosis (Fig. 10.1a). While this effect may reduce Aβ burden in astrocytes, it may decrease the clearance of Aβ from the ECF. This effect may be linked to the MT3's interaction with β-actin. MT3 promotes polymerization of actin, which is a necessary step in clathrin-mediated endocytosis.

The other effect of MT3 in the EALP is on lysosomes. We found that the absence of MT3 in astrocytes resulted in the reduction of autophagy flux (Fig. 10.1b) (Lee et al. 2010b). One of the changes induced by MT3 deletion was lysosomal alkalinization. Combined with reduced lysosomal enzyme activities, MT3 deletion resulted in autophagosome accumulation and reduced lysosomal degradation. Again, the interaction between MT3 and β-actin, and the consequent activation of c-Abl kinase, is a necessary event for this effect. Of interest, the N-terminal portion of MT3 containing TCPCP motif, a unique component of MT3 among MTs, was required for the interaction of MT3 with actin, and thus to restore lysosomal function. Consistently, MT1 and MT2 did not interact with β-actin and had no effect on lysosomal biology. Finally, zinc binding to MT3 is also required, as depletion of zinc with TPEN abrogated the effect. As will be discussed in the next section, increasing intracellular free zinc alone also has the function-boosting effect on lysosomes, including the acidity. Hence, zinc released from MT3 might be an additional contributing mechanism to increasing lysosomal function.

Since MT3 expression is limited mainly to CNS cells such as astrocytes and neurons, the MT3-dependent regulation of endocytosis and lysosomal function may be unique to the CNS. As such, MT3 may be a good target for the development of lysosome-boosting drugs in the CNS. Further studies are needed to show whether MT3 or any of derived peptides, when given systemically or into the CSF, has desired effects on brain lysosomes.

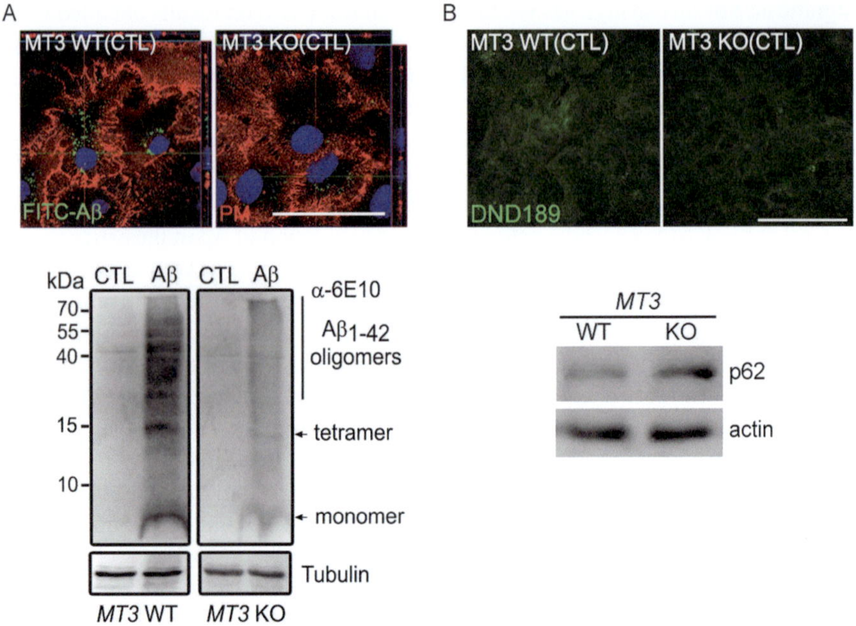

Fig. 10.1 MT3 controls lysosomal pH and clathrin-mediated endocytosis of Aβ. (**a**) MT3 KO (knockout) decreases Aβ endocytosis. Confocal fluorescence micrographs of MT3 WT (wild-type) and KO astrocytes. Cells were incubated with 1 μM fluorescein isothiocyanate (FITC)-Aβ (green dots) for 15 min at 37 °C. The plasma membrane was stained with Alexa Fluor 594-WGA (wheat germ agglutinin) (red). The FITC-Aβ uptake was noticeably reduced in MT3 KO cells compared with WT cells. Western blots (lower) for 6E10 and corresponding β-actin in MT3 WT and KO astrocytes. Astrocytes from MT3 WT and KO were incubated with 1 μM Aβ. After 24 h, the cells were lysed and immunoblotted with an anti-6E10 antibody. Compared to MT3 WT cells, MT3 KO cells reduced Aβ uptake (56). (http://creativecommons.org/licenses/by/4.0/). (**b**) Lysosomal pH in MT3 WT and KO astrocytes. Confocal fluorescence photomicrographs of LysoSensor Green DND-189-loaded MT3 WT (left) and KO (right) astrocytes. Western blots (lower) for p62, a marker of autophagy flux, and corresponding β-actin in MT3 WT and KO astrocytes. Compared to MT3 WT cells, MT3 KO cells exhibited more alkaline lysosomal pH and reduced autophagy flux

10.5 Measures Increasing Cytosolic and/or Lysosomal Free Zinc Levels May Help Overcome Lysosomal Dysfunction

Another potential therapeutic strategy for re-acidification of lysosomes is to raise intracellular or lysosomal free Zn^{2+} levels. Simple exposure of cultured cells to Zn^{2+}-enriched media or to a Zn^{2+} ionophore such as clioquinol is sufficient to achieve this effect. Clioquinol increases cytosolic and lysosomal Zn^{2+} levels and activates autophagy, resulting in degradation of mutant huntingtin aggregates (Park et al. 2011). Increasing intracellular or lysosomal Zn^{2+} levels by a zinc ionophore clioquinol reverse lysosomal pH changes and autophagy arrest (Fig. 10.2). Consistent with these changes, Zn-clioquinol reduces levels of Aβ or mutant huntingtin in the respective cell models (Seo et al. 2015).

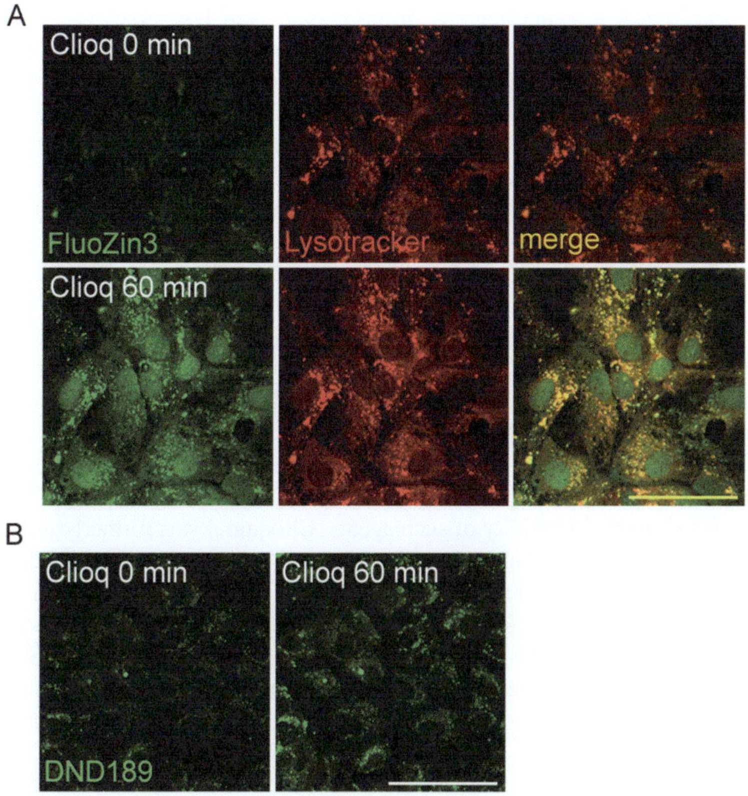

Fig. 10.2 A zinc ionophore clioquinol increases Zn^{2+} and proton levels in lysosomes. (**a**) FluoZin3-AM and LysoTracker-loaded astrocytes before (upper, Clioq 0 min) and after a 1 h treatment (lower, Clioq 60 min) with 5 μM clioquinol, a zinc ionophore. Addition of clioquinol increased FluoZin3 signals in lysosomes, resulting in an increase in yellow spots in the merged image. (**b**) For lysosomal pH evaluation, astrocytes were loaded with Lysosensor DND189 before (left, Clioq 0 min) and after a 1 h treatment (right, Clioq 60 min) with 5 μM clioquinol. Clioquinol increased lysosomal acidity

Lysosomes contain various potential Zn^{2+} transport routes, including ZnT2 (Zn^{2+} transporter-2) and ZnT4 ((Zn^{2+} transporter-4), as well as ATP13A2/PARK9. Zn^{2+} transporters (ZnTs) are Zn^{2+} –H^+ antiporters that, upon activation, transfer Zn^{2+} out of the cytosol, thereby reducing cytosolic Zn^{2+} levels (Kambe et al. 2015). ZnT1 ((Zn^{2+} transporter-1) moves Zn^{2+} from the cytosol to the extracellular space, and ZnT2 and ZnT4 transport Zn^{2+} into acidic organelles, such as endosomes, lysosomes, and secretory vesicles. It was recently reported that ZnT2 interacts with V-ATPase, and further that loss of ZnT2 disrupts V-ATPase assembly, impairing vesicle acidification (Lee et al. 2017). Another player may be ATP13A2/PARK9, a lysosomal type 5 P-type ATPase. Mutations in ATP13A2 are associated with early-onset Parkinsonism, known as Kufor–Rakeb syndrome (KRS). Studies using *ATP13A2*$^{-/-}$ cells from a KRS patient revealed that *ATP13A2* encodes a Zn^{2+} transporter that serves to sequester Zn^{2+} in endosomes and lysosomes (Tsunemi and Krainc 2014). These studies showed that mutation or knockdown of the corresponding gene results in reduced lysosomal Zn^{2+} levels, increased lysosomal pH, and reduced lysosomal degradation, a mechanism that may contribute to the pathogenesis of Parkinsonism. Hence, as is also true in this case, lysosomal Zn^{2+} levels seem to be linked to lysosomal acidification. In theory, the action of ZnTs as Zn^{2+}/H^+ antiporters is predicted to alkalinize lysosomes. One possible explanation for this apparent paradox is that re-acidification may be a physical property of high Zn^{2+} levels in lysosomes, reflecting the fact that Zn^{2+} in solution lowers the pH (Kiedrowski 2012). Another possibility is that cytosolic or lysosomal Zn^{2+} somehow activates V-ATPase or other indirect routes of H^+ influx. For instance, Zn^{2+} activates membrane protein kinase C (PKC), which is known to upregulate V-ATPase activity (Heming and Bidani 1995). Although further studies are required to elucidate the underlying mechanism, methods that raise lysosomal Zn^{2+} levels may be helpful for overcoming the lysosomal dysfunction that contributes to the pathogenesis of diverse neurodegenerative diseases. One caveat in using Zn^{2+} ionophores as therapeutics is the potential toxicity of such agents. Clioquinol was formerly used as an antimicrobial drug in Japan, but was withdrawn because of a serious side effect termed subacute myelo-optic neuropathy (SMON) (Konagaya 2015). In culture conditions, clioquinol can kill neurons and astrocytes by excessively increasing intracellular free Zn^{2+} levels (Park et al. 2011). In light of this progress, developing Zn^{2+} ionophores with optimal physical parameters that do not raise intracellular free Zn^{2+} levels above the cytotoxic threshold may prove to be critical for the clinical use of such agents.

10.6 Conclusions

In this paper, we discussed the possibility that abnormalities in the EALP, especially lysosomal dysfunction and the resultant arrested autophagy, may act as core pathogenic events in diverse proteinopathic neurodegenerative disorders. In addition, we discussed some possible measures that can be taken to normalize lysosomal functions under these conditions, and thereby restore normal flux through the EALP. In particular, we presented evidence showing that measures that raise Zn^{2+} levels as well as those that normalize Zn-MT3 functions may be effective in restoring lysosomal acidity and catabolic flux through the EALP.

Acknowledgments This research was supported by grants from the National Research Foundation of Korea (NRF) funded by the Ministry of Science, ICT (NRF-2016R1E1A1A01941212 and NRF-2017M3C7A1028949 for Jae-Young Koh) and the Korea Health Technology R&D Project through the Korea Health Industry Development Institute (KHIDI), funded by the Ministry of Health & Welfare, Republic of Korea (HI14C1913).

References

Alsaqati M, Thomas RS, Kidd EJ (2018) Proteins involved in endocytosis are upregulated by ageing in the Normal human brain: implications for the development of Alzheimer's disease. J Gerontol A Biol Sci Med Sci 73:289–298

Aras MA, Hara H, Hartnett KA, Kandler K, Aizenman E (2009) Protein kinase C regulation of neuronal zinc signaling mediates survival during preconditioning. J Neurochem 110:106–117

Baker M, Mackenzie IR, Pickering-Brown SM, Gass J, Rademakers R, Lindholm C, Snowden J, Adamson J, Sadovnick AD, Rollinson S, Cannon A, Dwosh E, Neary D, Melquist S, Richardson A, Dickson D, Berger Z, Eriksen J, Robinson T, Zehr C, Dickey CA, Crook R, McGowan E, Mann D, Boeve B, Feldman H, Hutton M (2006) Mutations in progranulin cause tau-negative frontotemporal dementia linked to chromosome 17. Nature 442:916–919

Boland B, Kumar A, Lee S, Platt FM, Wegiel J, Yu WH, Nixon RA (2008) Autophagy induction and autophagosome clearance in neurons: relationship to autophagic pathology in Alzheimer's disease. J Neurosci 28:6926–6937

Bourdenx M, Daniel J, Genin E, Soria FN, Blanchard-Desce M, Bezard E, Dehay B (2016) Nanoparticles restore lysosomal acidification defects: implications for Parkinson and other lysosomal-related diseases. Autophagy 12:472–483

Boutajangout A, Authelet M, Blanchard V, Touchet N, Tremp G, Pradier L, Brion JP (2004) Characterisation of cytoskeletal abnormalities in mice transgenic for wild-type human tau and familial Alzheimer's disease mutants of APP and presenilin-1. Neurobiol Dis 15:47–60

Chen S, Zhang X, Song L, Le W (2012) Autophagy dysregulation in amyotrophic lateral sclerosis. Brain Pathol 22:110–116

Cipolat Mis MS, Brajkovic S, Frattini E, Di Fonzo A, Corti S (2016) Autophagy in motor neuron disease: key pathogenetic mechanisms and therapeutic targets. Mol Cell Neurosci 72:84–90

Coffey EE, Beckel JM, Laties AM, Mitchell CH (2014) Lysosomal alkalization and dysfunction in human fibroblasts with the Alzheimer's disease-linked presenilin 1 A246E mutation can be reversed with cAMP. Neuroscience 263:111–124

Croce KR, Yamamoto A (2019) A role for autophagy in Huntington's disease. Neurobiol Dis 122:16–22

Cruts M, Gijselinck I, van der Zee J, Engelborghs S, Wils H, Pirici D, Rademakers R, Vandenberghe R, Dermaut B, Martin JJ, can Duijn C, Peeters K, Sciot R, Santens P, De Pooter T, Mattheijssens M, Van den Broeck M, Cuijt I, Vennekens K, De Deyn PP, Kumar-Singh S, Van Broeckhoven C (2006) Null mutations in progranulin cause ubiquitin-positive frontotemporal dementia linked to chromosome 17q21. Nature 442:920–924

De Craene JO, Bertazzi DL, Bar S, Friant S (2017) Phosphoinositides, major actors in membrane trafficking and lipid signaling pathways. Int J Mol Sci 18:634

De Duve C, Wattiaux R (1996) Functions of lysosomes. Annu Rev Physiol 28:435–492

Dominguez-Prieto M, Velasco A, Tabernero A, Medina JM (2018) Endocytosis and Transcytosis of amyloid-beta peptides by astrocytes: a possible mechanism for amyloid-beta clearance in Alzheimer's disease. J Alzheimers Dis 65:1109–1124

Evers BM, Rodriguez-Navas C, Tesla RJ, Prange-Kiel J, Wasser CR, Yoo KS, McDonald J, Cenik B, Ravenscroft TA, Plattner F, Rademakers R, Yu G, White CL 3rd, Herz J (2017) Lipidomic and Transcriptomic basis of Lysosomal dysfunction in Progranulin deficiency. Cell Rep 20:2565–2574

Guo X, Tang P, Chen L, Liu P, Hou C, Zhang X, Liu Y, Chong L, Li X, Li R (2017) Amyloid beta-induced redistribution of transcriptional factor EB and Lysosomal dysfunction in primary microglial cells. Front Aging Neurosci 9:228

Hamano T, Gendron TF, Causevic E, Yen SH, Lin WL, Isidoro C, Deture M, Ko LW (2008) Autophagic-lysosomal perturbation enhances tau aggregation in transfectants with induced wild-type tau expression. Eur J Neurosci 27:1119–1130

Han H, Wei W, Duan W, Guo Y, Li Y, Wang J, Bi Y, Li C (2015) Autophagy-linked FYVE protein (Alfy) promotes autophagic removal of misfolded proteins involved in amyotrophic lateral sclerosis (ALS). In Vitro Cell Dev Biol Anim 51:249–263

He K, Aizenman E (2010) ERK signaling leads to mitochondrial dysfunction in extracellular zinc-induced neurotoxicity. J Neurochem 114:452–461

Heming TA, Bidani A (1995) Effects of myristate phorbol ester on V-ATPase activity and Na(+)-H+ exchange in alveolar macrophages. J Leukoc Biol 57:600–608

Huse JT, Pijak DS, Leslie GJ, Lee VM, Doms RW (2000) Maturation and endosomal targeting of beta-site amyloid precursor protein-cleaving enzyme. Alzheimer Dis Beta-secretase J Biol Chem 275:33729–33737

Jaeger PA, Pickford F, Sun CH, Lucin KM, Masliah E, Wyss-Coray T (2010) Regulation of amyloid precursor protein processing by the Beclin 1 complex. PLoS One 5:e11102

Jian JL, Hettinghouse A, Liu CJ (2017) Progranulin acts as a shared chaperone and regulates multiple lysosomal enzymes. Genes & Diseases 4:125–126

Jin S (2006) Autophagy, mitochondrial quality control, and oncogenesis. Autophagy 2:80–84

Kambe T, Tsuji T, Hashimoto A, Itsumura N (2015) The physiological, biochemical, and molecular roles of zinc transporters in zinc homeostasis and metabolism. Physiol Rev 95:749–784

Kiedrowski L (2012) Cytosolic acidification and intracellular zinc release in hippocampal neurons. J Neurochem 121:438–450

Komatsu M, Waguri S, Chiba T, Murata S, Iwata J, Tanida I, Ueno T, Koike M, Uchiyama Y, Kominami E, Tanaka K (2006) Loss of autophagy in the central nervous system causes neurodegeneration in mice. Nature 441:880–884

Konagaya M (2015) SMON: toxicity of clioquinol and the status quo. Brain Nerve 67:49–62

Kumar DK, Choi SH, Washicosky KJ, Eimer WA, Tucker S, Ghofrani J, Lefkowitz A, McColl G, Goldstein LE, Tanzi RE, Moir RD (2016) Amyloid-beta peptide protects against microbial infection in mouse and worm models of Alzheimer's disease. Sci Transl Med 8:340ra72

Lee JH, Yu WH, Kumar A, Lee S, Mohan PS, Peterhoff CM, Wolfe DM, Martinez-Vicente M, Massey AC, Sovak G, Uchiyama Y, Westaway D, Cuervo AM, Nixon RA (2010a) Lysosomal proteolysis and autophagy require presenilin 1 and are disrupted by Alzheimer-related PS1 mutations. Cell 141:1146–1158

Lee SJ, Park MH, Kim HJ, Koh JY (2010b) Metallothionein-3 regulates lysosomal function in cultured astrocytes under both normal and oxidative conditions. Glia 58:1186–1196

Lee S, Sato Y, Nixon RA (2011) Lysosomal proteolysis inhibition selectively disrupts axonal transport of degradative organelles and causes an Alzheimer's-like axonal dystrophy. J Neurosci 31:7817–7830

Lee SJ, Seo BR, Koh JY (2015) Metallothionein-3 modulates the amyloid beta endocytosis of astrocytes through its effects on actin polymerization. Mol Brain 8:84

Lee S, Rivera OC, Kelleher SL (2017) Zinc transporter 2 interacts with vacuolar ATPase and is required for polarization, vesicle acidification, and secretion in mammary epithelial cells. J Biol Chem 292:21598–21613

Leibiger C, Deisel J, Aufschnaiter A, Ambros S, Tereshchenko M, Verheijen BM, Büttner S, Braun RJ (2018) Endolysosomal pathway activity protects cells from neurotoxic TDP-43. Microb Cell 5:212–214

Lin CL, Tseng HC, Chen RF, Chen WP, Su MJ, Fang KM, Wu ML (2011) Intracellular zinc release-activated ERK-dependent GSK-3beta-p53 and Noxa-Mcl-1 signaling are both involved in cardiac ischemic-reperfusion injury. Cell Death Differ 18:1651–1663

Liuzzi JP, Guo L, Yoo C, Stewart TS (2014) Zinc and autophagy. Biometals 27:1087–1096

Manucat-Tan NB, Saadipour K, Wang YJ, Bobrovskaya L, Zhou XF (2019) Cellular trafficking of amyloid precursor protein in Amyloidogenesis physiological and pathological significance. Mol Neurobiol 56(2):812–830

Moors TE, Hoozemans JJ, Ingrassia A, Beccari T, Parnetti L, Chartier-Harlin MC, van de Berg WD (2017) Therapeutic potential of autophagy-enhancing agents in Parkinson's disease. Mol Neurodegener 12:11

Muller UC, Deller T, Korte M (2017) Not just amyloid: physiological functions of the amyloid precursor protein family. Nat Rev Neurosci 18:281–298
Nascimbeni AC, Codogno P, Morel E (2017) Phosphatidylinositol-3-phosphate in the regulation of autophagy membrane dynamics. FEBS J 284:1267–1278
Nixon RA (2013) The role of autophagy in neurodegenerative disease. Nat Med 19:983–997
Noh KM, Kim YH, Koh JY (1999) Mediation by membrane protein kinase C of zinc-induced oxidative neuronal injury in mouse cortical cultures. J Neurochem 72:1609–1616
Park MH, Lee SJ, Byun HR, Kim Y, Oh YJ, Koh JY, Hwang JJ (2011) Clioquinol induces autophagy in cultured astrocytes and neurons by acting as a zinc ionophore. Neurobiol Dis 42:242–251
Perrett RM, Alexopoulou Z, Tofaris GK (2015) The endosomal pathway in Parkinson's disease. Mol Cell Neurosci 66:21–28
Pickford F, Masliah E, Britschgi M, Lucin K, Narasimhan R, Jaeger PA, Small S, Spencer B, Rockenstein E, Levine B, Wyss-Coray T (2008) The autophagy-related protein beclin 1 shows reduced expression in early Alzheimer disease and regulates amyloid beta accumulation in mice. J Clin Invest 118:2190–2199
Polito VA, Li H, Martini-Stoica H, Wang B, Yang L, Xu Y, Swartzlander DB, Palmieri M, di Ronza A, Lee VM, Sardiello M, Ballabio A, Zheng H (2014) Selective clearance of aberrant tau proteins and rescue of neurotoxicity by transcription factor EB. EMBO Mol Med 6:1142–1160
Rahman MA, Rhim H (2017) Therapeutic implication of autophagy in neurodegenerative diseases. BMB Rep 50:345–354
Roberson ED, Scearce-Levie K, Palop JJ, Yan F, Cheng IH, Wu T, Gerstein H, Yu GQ, Mucke L (2007) Reducing endogenous tau ameliorates amyloid beta-induced deficits in an Alzheimer's disease mouse model. Science 316:750–754
Rusmini P, Cortese K, Crippa V, Cristofani R, Cicardi ME, Ferrari V, Vezzoli G, Tedesco B, Meroni M, Messi E, Piccolella M, Galbiati M, Garrè M, Morelli E, Vaccari T, Poletti A (2019) Trehalose induces autophagy via lysosomal-mediated TFEB activation in models of motoneuron degeneration. Autophagy 15(4):631–651
Samura E, Shoji M, Kawarabayashi T, Sasaki A, Matsubara E, Murakami T, Wuhua X, Tamura S, Ikeda M, Ishiguro K, Saido TC, Westaway D, St George Hyslop P, Harigaya Y, Abe K (2006) Enhanced accumulation of tau in doubly transgenic mice expressing mutant betaAPP and presenilin-1. Brain Res 1094:192–199
Sanchez-Varo R, Trujillo-Estrada L, Sanchez-Mejias E, Torres M, Baglietto-Vargas D, Moreno-Gonzalez I, De Castro V, Jimenez S, Ruano D, Vizuete M, Davila JC, Garcia-Verdugo JM, Jimenez AJ, Vitorica J, Gutierrez A (2012) Abnormal accumulation of autophagic vesicles correlates with axonal and synaptic pathology in young Alzheimer's mice hippocampus. Acta Neuropathol 123:53–70
Sardiello M, Palmieri M, di Ronza A, Medina DL, Valenza M, Gennarino VA, Di Malta C, Donaudy F, Embrione V, Polishchuk RS, Banfi S, Parenti G, Cattaneo E, Ballabio A (2009) A gene network regulating lysosomal biogenesis and function. Science 325:473–477
Seo BR, Lee SJ, Cho KS, Yoon YH, Koh JY (2015) The zinc ionophore clioquinol reverses autophagy arrest in chloroquine-treated ARPE-19 cells and in APP/mutant presenilin-1-transfected Chinese hamster ovary cells. Neurobiol Aging 36:3228–3238
Settembre C, Fraldi A, Medina DL, Ballabio A (2013) Signals from the lysosome: a control centre for cellular clearance and energy metabolism. Nat Rev Mol Cell Biol 14:283–296
Shen D, Wang X, Xu H (2011) Pairing phosphoinositides with calcium ions in endolysosomal dynamics: phosphoinositides control the direction and specificity of membrane trafficking by regulating the activity of calcium channels in the endolysosomes. BioEssays 33:448–457
Stoka V, Turk V, Turk B (2016) Lysosomal cathepsins and their regulation in aging and neurodegeneration. Ageing Res Rev 32:22–37
Tanaka Y, Suzuki G, Matsuwaki T, Hosokawa M, Serrano G, Beach TG, Yamanouchi K, Hasegawa M, Nishihara M (2017) Progranulin regulates lysosomal function and biogenesis through acidification of lysosomes. Hum Mol Genet 26:969–988

Tarasoff-Conway JM, Carare RO, Osorio RS, Glodzik L, Butler T, Fieremans E, Axel L, Rusinek H, Nicholson C, Zlokovic BV, Frangione B, Blennow K, Ménard J, Zetterberg H, Wisniewski T, de Leon MJ (2015) Clearance systems in the brain-implications for Alzheimer disease. Nat Rev Neurol 11:457–470

Tesco G, Koh YH, Kang EL, Cameron AN, Das S, Sena-Esteves M, Hiltunen M, Yang SH, Zhong Z, Shen Y, Simpkin JW, Tanzi RE (2007) Depletion of GGA3 stabilizes BACE and enhances beta-secretase activity. Neuron 54:721–737

Trudeau KM, Colby AH, Zeng J, Las G, Feng JH, Grinstaff MW, Shirihai OS (2016) Lysosome acidification by photoactivated nanoparticles restores autophagy under lipotoxicity. J Cell Biol 214:25–34

Tsunemi T, Krainc D (2014) Zn(2)(+) dyshomeostasis caused by loss of ATP13A2/PARK9 leads to lysosomal dysfunction and alpha-synuclein accumulation. Hum Mol Genet 23:2791–2801

Venkatachalam K, Wong CO, Zhu MX (2015) The role of TRPMLs in endolysosomal trafficking and function. Cell Calcium 58:48–56

Wang Y, Liu FT, Wang YX, Guan RY, Chen C, Li DK, Bu LL, Song J, Yang YJ, Dong Y, Chen Y, Wang J (2018) Autophagic modulation by Trehalose reduces accumulation of TDP-43 in a cell model of amyotrophic lateral sclerosis via TFEB activation. Neurotox Res 34:109–120

Xia Q, Wang H, Hao Z, Fu C, Hu Q, Gao F, Ren H, Chen D, Han J, Ying Z, Wang G (2016) TDP-43 loss of function increases TFEB activity and blocks autophagosome-lysosome fusion. EMBO J 35:121–142

Xiao Q, Yan P, Ma X, Liu H, Perez R, Zhu A, Gonzales E, Tripoli DL, Czerniewski L, Ballabio A, Cirrito JR, Diwan A, Lee JM (2015) Neuronal-targeted TFEB accelerates Lysosomal degradation of APP, reducing Aβ generation and amyloid plaque pathogenesis. J Neurosci 35:12137–12151

Yamada M, Tsuji S, Takahashi H (2002) Involvement of lysosomes in the pathogenesis of CAG repeat diseases. Ann Neurol 52:498–503

Chapter 11
Zinc in Neurodegeneration

Stuart D. Portbury, Ashley I. Bush, and Paul A. Adlard

Abstract Zinc is a critical regulator in the central nervous system and has many different roles in various cellular pathways under normal conditions. As a consequence, when there is a disturbance in zinc homeostasis in the brain, as can occur with ageing, disease or injury, then this can have significant deleterious consequences for normal brain function. In this chapter, we will review the role that zinc has in the mechanisms underlying neurodegeneration, which collectively encompasses cellular processes such as autophagy, oxidative stress, protein aggregation, excitotoxicity, mitochondrial failure and neuroinflammation. These pathways are broadly involved in the initiation and potentiation of the neuronal cell death and dysfunction that occurs in a number of common disorders that occur across lifespan, such as Alzheimer's disease, Parkinson's disease, amyotrophic lateral sclerosis, traumatic brain injury and even normal ageing. Understanding the role of zinc in neurodegeneration will ultimately facilitate the development of targeted therapeutics that may promote successful brain ageing.

Keywords Neurodegeneration · Autophagy · Oxidative stress · Aggregation · Inflammation · Excitotoxicity · Mitochondria · Alzheimer's disease · Parkinson's disease · Ageing · Traumatic brain injury · Amyotrophic lateral sclerosis

11.1 Introduction

Neurodegeneration is a broad-strokes term that refers to a progressive and irreversible loss of neurons and their axonal and dendritic projections and synaptic connections. Together, this results in an impairment in brain function, and depending on the location and magnitude of the neurodegeneration, this may manifest in gross

---

S. D. Portbury · A. I. Bush · P. A. Adlard (✉)
The Florey Institute of Neuroscience and Mental Health, Kenneth Myer Building,
The University of Melbourne, Parkville, VIC, Australia
e-mail: paul.adlard@florey.edu.au

T. Fukada, T. Kambe (eds.), *Zinc Signaling*,
https://doi.org/10.1007/978-981-15-0557-7_11

deficits in higher order functions such as learning and memory, movement and speech. Neurodegeneration is often associated with age-related neurodegenerative disorders, such as Alzheimer's disease (AD), Parkinson's disease (PD) and Motor Neuron disease (MND), but is also a feature of a multitude of other devastating conditions, such as Huntington's disease (HD), Frontotemporal Dementia (FTD), multiple sclerosis and prion disease. In addition, head trauma/traumatic brain injury (TBI) can similarly result in neurodegeneration and an associated clinical phenotype.

Whilst the cumulative global burden of "neurodegeneration" in its broadest context is not typically quantified, a recent white paper (Mindgardens_Neuroscience_Network 2019) reviewed the burden of disease in Australia for two of the most common neurodegenerative disorders: AD (including "other" dementias) (prevalence: 242,870; mortality: 17,119; disability-adjusted life years: 171,319; cost: $8.6B) and PD (prevalence: 47,209; mortality: 2328; disability-adjusted life years: 32,848; cost: $1.6B). As the population continues to live longer, the burden of disease is predicted to grow significantly. Whilst there are varying reports about current and future global incidence rates and costs of these conditions, as well as less common neurodegenerative diseases such as MND, FTD and HD, it is clear that these diseases represent a profound impost on all levels of the community. Further to this, TBI actually has a substantially higher incidence rate, with estimates of around 50–60 million new cases occurring annually worldwide (Maas et al. 2017). Significantly, TBI is the leading cause of mortality in young adults and a major cause of death and disability across all ages in all countries.

The situation, therefore, is one where there is a global epidemic of neurodegeneration in the community, effectively encompassing millions of individuals at all stages of life. The consequence to the individual is an untreatable burden of disability and/or death, with an economic cost to society well in excess of a trillion dollars. Whilst considerable effort has gone into research to identify compounds that modulate single disease-specific targets, these have all largely failed in clinical trial. There remains, therefore, a significant translational gap (Cummings 2017). Given the multi-factorial nature of neurodegenerative disorders and the complexity and overlapping nature of many of the pathways involved, it has been suggested that whilst the mechanism that initiates the injury/disease processes may vary from condition to condition, there may be common pathways (Gan et al. 2018) that ultimately drive the neurodegenerative cascade across multiple disorders, such as AD, PD and TBI.

We recently reviewed the literature around TBI, chronic traumatic encephalopathy (CTE; a tau-linked neurodegenerative disorder that arises as a consequence of repeated mild injury to the brain) and AD, with the perspective that they may all share common pathologies that are potentiated by altered zinc homeostasis (Portbury and Adlard 2015). In this chapter, we will revisit the notion that zinc may be a potentially central feature of a variety of age-related neurodegenerative diseases and other brain disorders that are characterized by a neurodegenerative phenotype. Zinc is clearly involved in a diversity of cellular pathways across the body, as evidenced by the other chapters in this book. In this chapter, however, we will specifically review the role of zinc in a number of key pathways that are reported to be involved in the process of neurodegeneration, including protein misfolding and aggregation, mitochondrial dysfunction, inflammation and oxidative stress. Furthermore, we will also examine a number of small vignettes of different conditions that are characterized by a neurodegenerative phenotype and outline the role that zinc is likely to play in these conditions – such as AD, PD and TBI.

11.2 How Does Zinc Contribute to Neurodegeneration?

11.2.1 Zinc and Autophagy

The evolutionarily conserved catabolic process utilized by eukaryotic cells for the regulated degradation of damaged or redundant proteins and organelles is called autophagy (Chen and Klionsky 2011). There are three types of autophagy; chaperone-mediated autophagy, microautophagy and macroautophagy (Chen and Klionsky 2011), the latter of which is the best characterized and will be discussed herein.

Autophagy can be either selective or non-selective, the former targeting damaged or superfluous mitochondria (Narendra et al. 2008), peroxisomes (Iwata et al. 2006), lipid droplets (Singh et al. 2009) and microbes (Baxt et al. 2013), with the latter being involved with the turnover of bulk cytoplasm under starvation conditions (Kirkin et al. 2009). In each instance, however, autophagy is activated as a response to adverse cellular conditions such as the deprivation of nutrients, pathogen infection, hypoxia, radiation and oxidative stress, and as such plays a role in cellular homeostasis, development and longevity (Kroemer et al. 2010).

Whilst autophagy is generally considered an innate pro-survival mechanism, irregular or unbridled autophagy can be associated with cell death (Galluzzi et al. 2008), and has also been shown to play a negative role in neurodegenerative diseases (Nilsson and Saido 2014; Orr and Oddo 2013), ageing (Jong-ok Pyo 2013), cancer (Lozy and Karantza 2012), fatty liver (Czaja et al. 2013) and diabetes (Quan et al. 2013).

The induction of autophagy in mammalian cells is regulated by more than 30 autophagy-related genes in response to a diverse array of stimuli. The process is initiated by the formation of an isolation membrane known as the phagophore which subsequently expands and sequesters cellular constituents such as lipid droplets, organelles, cytoplasm and damaged mitochondria in a double-membrane vesicle termed an autophagosome (Mizushima 2007). The maturation of the autophagosome is completed by fusing to late endosomes and lysosomes to form an autolysosome, in which the cargo is subjected to lysosomal enzymes for degradation (Mizushima 2007).

The mechanisms by which zinc regulates the process of autophagy are yet to be clearly defined; however, there is strong evidence to suggest that zinc is a positive regulator of autophagy. *In vitro* zinc depletion studies using either the cell-permeable zinc chelator TPEN (N, N, N′N′-tetrakis(−)[2-pyridylmethyl]-ethylenediamine) or Chelex-100 have demonstrated a suppressive action of basal and induced autophagy in MCF-7 cells (Hwang et al. 2010), human hepatoma cells (Liuzzi and Yoo 2013) and astrocytes (Lee and Koh 2010). Moreover, studies in which excess zinc is provided in cell culture medium (20–200 μm) have been shown to enhance autophagy induced by tamoxifen in MCF-7 breast cancer cells (Lee and Koh 2010), in astrocytes and dopamine in PC12 cells and cultured neurons (Hung et al. 2013), and the zinc ionophore PCI-5002 was shown to radio-sensitize non-small-cell lung cancer by potentiating autophagy (Kim et al. 2011a).

These studies provide strong evidence for a critical role of zinc in autophagy flux; however, the direct mechanism of action is not clearly defined. Nonetheless, observations from studies with extracellular-signal-regulated kinase (ERK1/2) inhibitors indicate that phosphorylation of ERK1/2 is a requirement for zinc-regulated basal and induced autophagy (Liuzzi and Yoo 2013; Lee and Koh 2010). ERK1/2 is a known regulator of autophagy whereby it can activate the Beclin1-PI3K complex by phosphorylating the negative regulator of Beclin-1, Bcl-2 (Botti et al. 2006), or by promoting the disassembly of mTORC1 complex (Wang et al. 2009). Furthermore, a recent study implicated the zinc-binding low molecular weight protein metallothionein as the intracellular source for autophagy activation during oxidative stress, evidenced by the observation that astrocytes from metallothionein 3 (MT3) knockout mice exhibited reduced autophagy activation by H_2O_2 (Lee and Koh 2010).

It is conceivable that zinc may also regulate autophagy via the modulation of autophagy regulating genes. For example, zinc has been shown to induce gene expression through activation of the metal-responsive transcription factor (MTF1) (Lichtlen and Schaffner 2001); therefore, autophagy-regulating genes containing binding sites for MTF1 may be regulated by zinc. Similarly, zinc regulation of microRNA expression (Ryu et al. 2011), and zinc-mediated methylation (Sharif et al. 2012) of autophagy-related genes may also allow zinc to act as an autophagy modulator.

Evidence indicates that in addition to a gene regulatory role for zinc in autophagy, it also likely plays a role in late-stage autophagy whereby the degradation of lysosomal cargo is dependent upon zinc. For example, it has been shown that astrocytes from mice lacking MT3 have reduced expression of the known autophagy activators cathepsin D and L concurrent with accumulation of cholesterol and lipofuscin (Lee and Koh 2010), and zinc deficiency in rats resulted in elevated lipofuscin deposition in retinal epithelium (Julien et al. 2011), an indicator of incomplete lysosomal digestion, emphasizing zinc as a requirement for proper lysosomal function. Furthermore, low zinc has also been demonstrated to affect lysosome function and integrity via the blocking of cargo degradation in lysosomes. In both macrophage (Summersgill et al. 2014) and human hepatoma cells (Liuzzi and Yoo 2013), treatment with the zinc chelator TPEN resulted in compromised lysosomal function and impaired degradation of LC3II and p62/SQSTM proteins which are known to be essential autophagy markers.

Conversely, recent studies have shown that autophagy dysregulation can cause significant changes in cellular zinc homeostasis. Experiments investigating the autophagy inhibitor 3-methyladenine (3-MA) on the regulation of the zinc transporters ZnT1, ZnT2, ZnT3 and ZnT10 (Liuzzi and Yoo 2013) indicated a downregulation of these transporters. Interestingly, the expression of the ZnT10 gene has been shown to be downregulated in the brains of patients with Alzheimer's disease (Bosomworth et al. 2013), a condition associated with compromised autophagy (Nilsson and Saido 2014; Orr and Oddo 2013).

11.2.2 Zinc and Protein Aggregation

The aggregation of specific proteins into either extra- or intra-cellular "lesions" is a common phenomenon across many different neurodegenerative diseases, such as AD (extracellular Aß plaques and intracellular tau-containing neurofibrillary tangles), PD (intracellular Lewy bodies comprised of alpha-synuclein), MND (intracellular inclusions of SOD1), HD (huntingtin and ubiquitin) and many other conditions. In many cases, these proteinaceous inclusions are specific to a given disease, but may also present as comorbidities across multiple disorders (often complicating diagnoses). These aggregates are often believed to be critical components of the pathogenesis of disease, and are often targets for therapeutic modulation. Whilst we will not debate the relative toxicity of these pathological structures, which also typically occur in a defined spatiotemporal fashion across the different conditions, the process by which they can arise is relevant to this chapter.

In approximately one-third of the human proteome, the protein structures consist of key residues or combinations of residues that create specific coordination motifs that can bind different metals (such as iron and copper) (Leal et al. 2012), with approximately 3000 proteins (~10% of all those encoded) containing zinc (Kochanczyk et al. 2015). The coordination of zinc typically involves ligands from specific amino acids, including histidine (nitrogen), glutamate/aspartate (oxygen) and/or cysteine (sulphur) – and these ligands may arise from up to four proteins, with zinc typically coordinating in a tetrahedral geometry within the protein complex (Kochanczyk et al. 2015; Krezel and Maret 2016). In addition to coordination chemistry, the metal–protein interaction can also be influenced by a variety of other factors such as electrostatic forces, the thermodynamics of zinc binding to different combinations of ligands, cellular zinc availability, the stability of the zinc binding site and so on. These and other issues have previously been thoroughly reviewed (Kochanczyk et al. 2015; Krezel and Maret 2016).

This metal ion binding facilitates a diverse range of functions of proteins which have historically been classified as being either catalytic or structural in nature. More recent work (Kochanczyk et al. 2015) has suggested that further delineation into the following classes is required: catalytic zinc binding (typically high affinity within the cell), where the zinc is coordinated using three amino acid donors from one polypeptide side chain (typically histidine and aspartate/glutamate), with such zinc binding sites thought to be critical for catalytic activity and found across a range of enzymes; structural zinc binding (thermodynamically stable zinc binding and kinetically stable complex), characterized by four protein donors (typically via cysteine and histidine donors), where the coordination facilitates protein folding and the subsequent formation of defined structures which can be sites of interaction with other domains within the protein or with other cellular components; multinuclear zinc binding (thermodynamically stable zinc binding but kinetically labile complex), where there is more than one zinc molecule per binding site with associated bridging ligands typically coming from cysteine, which facilitates functions associated with the redistribution of zinc; a transport architecture in which the properties of the metal binding site are quite variable and likely dependent on the overall protein structure, but in which there is a low-to-moderate affinity for the zinc, which facilitates a mobility in the metal that is critical for proteins such as zinc transporters, which function to translocate zinc across membranes; inter-protein coordination, in which the ligands for binding are derived from multiple polypeptides, which creates an environment of protein–protein interactions that may be important for the formation or stability of specific protein complexes.

It is clear, therefore, that there are a variety of ways in which zinc can bind to proteins, with implications for both the structure and function of the protein and the zinc itself. In the context of neurodegeneration, this becomes highly relevant when considering that many of the disease-associated proteins, such as ß-amyloid, tau, alpha-synuclein and others, are metalloproteins that can bind zinc with all the consequent downstream effects on protein folding and misfolding (and hence, three-dimensional conformation of the protein), protein stability and protein deposition (Leal et al. 2012). One of the classic examples of this is found in Alzheimer's disease, where the two primary pathological structures found in the brain (the extracellular ß-amyloid plaque and, to a lesser extent, the intracellular neurofibrillary tangle (NFT)) have been reported to be modulated by zinc (Wang and Wang 2017).

The microtubule-associated protein tau is the principal component of the paired helical filaments that comprise the NFTs that characterize neurodegenerative tauopathies such as AD. Data suggest that zinc can bind to tau and promote its fibrillization and aggregation, and that zinc is critical for toxicity (Hu et al. 1863; Huang et al. 2014, 2016; Mo et al. 2009). Furthermore, the regulatory pathways (kinases and phosphatases) that control the phosphorylation of tau (which is critical for its function) are also subject to modulation by zinc (An et al. 2005; Bjorkdahl et al. 2005; Harris et al. 2004; Sun et al. 2012).

A greater body of work has focused on the amyloid precursor protein (APP; the parent protein to the ß-amyloid protein that comprises the extracellular plaques in the disease), which will bind zinc between position 170 and 188, where there are key cysteines and other ligands. This binding can result in the dimerization and aggregation of APP, with potential functional implications for the protein (Brown et al. 1997; Ciuculescu et al. 2005; Scheuermann et al. 2001). The cleavage of APP to generate ß-amyloid also involves the coordinated activities of the secretase family of proteins (alpha, beta and gamma secretase) which themselves are modulated by zinc (Cross et al. 2002; Greenough et al. 2011; Hoke et al. 2005; Lammich et al.

1999; Park et al. 2001); and the zinc-binding domain in the ß-amyloid region of APP overlaps the cleavage site of alpha-secretase, therefore potentially modulating the cleavage of the protein (Bush et al. 1994a). Finally, once ß-amyloid is formed, the subsequent aggregation and function of the protein are critically dependent on the coordination of zinc (Kepp 2017; Liu et al. 1999; Matheou et al. 2016). On the flip-side to the aggregation of ß-amyloid, considered to be a primary pathogenic event in AD, is the proteolytic degradation of the protein by enzymes such as neprilysin and insulin-degrading enzyme. These and other proteins are metallopeptidases that bind catalytic zinc via the HEXXH consensus sequence (Turner et al. 2001).

It is clear, therefore, that metals such as zinc are involved in multiple different aspects of the various pathways (at both a transcriptional, translational and post-translational level) that potentiate pathological protein aggregation to then contribute to specific anatomical, biochemical and psychiatric manifestations that collectively characterize specific neurodegenerative disorders.

11.2.3 Zinc and Oxidative Stress

Oxidative stress is characterized as a metabolic disturbance that results in an imbalance between the production of reactive oxygen species (ROS) and antioxidant defences. ROS include such chemical moieties as superoxide ion (O_2^-), hydrogen peroxide (H_2O_2), hydroxyl radical (HO^-) and peroxynitrite ($ONOO^-$). Whilst ROS is a natural by-product of cellular metabolism, under times of cellular stress, the ROS-buffering capability of the cell can become compromised, resulting in an imbalance of ROS to natural cellular defence mechanisms. The imbalance potentiates the oxidation of biomolecules, leading to oxidative damage to cells and tissues, subsequently contributing to the development of chronic disease. Zinc has been shown to play a myriad of roles in the cellular antioxidant defence system, with numerous cellular (Aimo et al. 2010; Ho and Ames 2002; Kojima-Yuasa et al. 2005; Oteiza et al. 2000; Yan et al. 2008; Zago et al. 2005) and animal (Bray et al. 1986; Bruno et al. 2007; Canali et al. 2000; Hammermueller et al. 1987; Oteiza et al. 1995; Shaheen and Elfattah 1995; Song et al. 2009a, 2010; Sullivan et al. 1980) models, indicating a requirement of homeostatic zinc maintenance to maintain optimal antioxidant defences. Of importance to antioxidant defence mechanisms, however, is the role of zinc in glutathione peroxidase regulation, metallothionein protein production, and providing a critical structural component of the antioxidant enzyme superoxide dismutase.

Glutathione (GSH) is an endogenous cellular antioxidant capable of preventing damage to important cellular components caused by ROS (Pompella et al. 2003). Zinc has been shown to regulate the cellular GSH concentration via the modulation of glutamate-cysteine ligase (GCL), the rate-limiting enzyme of GSH de novo synthesis. One such study investigated the effects of zinc against oxidative stress in cultured retinal pigment epithelial cells and demonstrated that zinc (150 μm) significantly increased GSH (70%) and the mRNA levels of GCL (Ha et al. 2006).

Supporting this observation was a recent study by Omata et al., in which GSH levels were shown to be significantly reduced in the brains of gestation day 19 foetuses from dams fed marginal zinc diets throughout gestation, and in zinc-deficient IMR-32 human neuroblast cell cultures (Omata et al. 2013). Indeed, these observations have also been extended to human studies where sub-optimal zinc intake is associated with increased oxidative stress (Prasad et al. 2007; Song et al. 2009b).

Metallothioneins (MT) are a family of cysteine-rich low molecular weight zinc-binding proteins that mediate the cellular control of oxidative stress (Ruttkay-Nedecky et al. 2013). Whilst MTs can bind a wide range of metals, its role in the uptake, transport and regulation of zinc in biological systems and its subsequent role in the control of oxidative stress have received the most attention. Under normal physiological conditions, MT binds zinc; however, under conditions of oxidative stress, zinc is liberated from its MT complex, resulting in the initiation of two major antioxidants mediating events within the cell. Firstly, the liberated zinc activates the synthesis of more MT (Maret and Krezel 2007; Ozcelik et al. 2012). Secondly, the liberation of zinc from MT cysteine residues renders the residues available for the binding of destructive oxidant radicals like superoxide and hydroxyl radicals (Kumari et al. 1998).

Additionally, MT has the capacity to bind other heavy metals such as mercury, cadmium, chromium and copper (Coyle et al. 2002; Gunther et al. 2012a, b), known to be toxic in excess amounts and can themselves initiate ROS production. MT binding of these heavy metals acts as a system of detoxification and reduces ROS (Andrews 2000).

Superoxide dismutase is an enzyme present in mammals that catalyses the dismutation, or partitioning, of the superoxide radical (O_2^-) into molecular oxygen (O_2) or hydrogen peroxide (H_2O_2) (Wang et al. 2018). There are three forms of SOD that vary based upon location and structure. SOD1 is in cytoplasm, SOD2 in mitochondria and SOD3 is extracellular. Both SOD1 (Cao et al. 2008) and SOD3 (Antonyuk et al. 2009) contain copper and zinc, whereas SOD2 (Borgstahl et al. 1996) has manganese in its reactive centre. Cellular zinc deficiency can induce mutant-like conformation of superoxide dismutase that can induce chronic endoplasmic reticulum stress (Homma et al. 2013). Additionally, misfolding and aggregation of SOD1 are implicated in neuronal death in amyotrophic lateral sclerosis (ALS), and there is evidence that metal ions are critical for correct SOD1 confirmation (Ding and Dokholyan 2008).

11.2.4 Zinc and Synaptic Function/ Learning and Memory

Whilst the majority of brain zinc is associated with macromolecules, a small number of zinc ions exist as free or chelatable zinc that is concentrated within synaptic vesicles of glutamatergic terminals via the activity of the zinc-specific transporter ZnT-3 (Paoletti et al. 2009). Neuronal ionotropic glutamate receptors mediate the majority of excitatory synaptic transmission throughout the central nervous system, are key players in synaptic plasticity and have been shown to be sensitive to extracellular zinc concentrations. Indeed, zinc is capable of directly modulating postsynaptic receptors to exert effects in both neuronal and glial cells.

The release of zinc into the synaptic cleft (from synaptic vesicles) upon cellular activation has been shown to contribute to the regulation of neuronal processes such as long-term potentiation (LTP) and long-term depression (LTD) via the activation of *N*-meythyl-D-aspartate (NMDA) receptors (Takeda and Tamano 2012). LTP is a component of synaptic plasticity that is widely accepted as the key modulator of learning and memory (Nicoll 2017). Zinc released into the synaptic cleft has also been shown to modulate synaptic transmission and plasticity via the regulation of α-amino-3-hydroxy-5-methyl-4-isoxazolepropionic acid (AMPA) receptors (Kalappa et al. 2015). Additionally, zinc has been shown to interact with targets such as the tyrosine kinase receptor TrkB (Huang et al. 2008), the zinc-sensing receptor ZnR/GPR39 (Besser et al. 2009) and p75 (Lee et al. 2008), all of which have influence on synaptic transmission and plasticity (Frederickson et al. 2005). Accordingly, given the breadth of post-synaptic regulating properties, zinc has been described as an 'atypical neurotransmitter' (Haase and Rink 2011).

Further supporting the critical modulatory role of zinc in synaptic plasticity and memory formation was a recent study utilizing ZnT-3 knockout mice, in which the ablation of synaptic zinc in glutamatergic synapses resulted in an age-dependent impairment in spatial learning and memory (Adlard et al. 2010). Additionally, a downregulation of ZnT- 3 protein levels has been shown in the cerebral cortex from mice and humans as an age-dependent occurrence (Adlard et al. 2010; Olesen et al. 2016), and a subsequent further decline in ZnT-3 is observed in people with AD (Beyer et al. 2009).

11.2.5 Zinc, Excitotoxicity and Mitochondria

A number of neurodegenerative diseases (e.g. AD, ALS, PD and HD) and other conditions (e.g. TBI, spinal cord injury and stroke) are characterized by excitotoxicity, which refers to a series of cellular events that can ultimately lead to neuronal death. It classically results from the sustained activation of ionotropic glutamate receptors (such as NMDAR and AMPAR) by exposure to abnormally high levels of glutamate (and also by other mechanisms such as exposure to kainic acid or NMDA), which causes an influx of various ions (especially calcium) into the neuron through the receptor. The downstream consequences of this are varied, and include the production of reactive oxygen and nitrogen species; increased potassium channel activity; mitochondrial depolarization and failure; the activation of calcium-dependent proteases (such as calpain), phospholipases and endonucelases; and the initiation of cell death cascades (via necrotic, pro-apoptotic and autophagy-related pathways). Whilst calcium was historically considered to be the primary mediator of this cellular cascade, it is now apparent that zinc is involved in this process, and may indeed be an important regulator of this pathway – with its effects (positive or negative) being very much concentration-dependent (Choi et al. 1989; Takeda 2011).

Zinc is co-released from synaptic vesicles, along with glutamate, during periods of synaptic excitation (and can also be liberated from intracellular stores under specific conditions). Once released, it can enter the post-synaptic neuron via a number of routes, including via voltage-sensitive calcium channels and a subset of calcium-permeable AMPA/kainate channels, to modulate a variety of proteins and cellular processes. In different regions of the hippocampus, zinc is reported to prevent calcium influx and to also prevent glutamate release via the activation of presynaptic potassium channels. Thus, zinc may regulate glutamate-induced excitotoxicity by providing negative feedback mechanisms, not only on glutamate but also on calcium (Bancila et al. 2004; Cohen-Kfir et al. 2005; Quinta-Ferreira and Matias 2004, 2005; Takeda et al. 2007a, b). Under pathological conditions (be it due to age, injury or disease), however, where there is often a disruption in normal zinc homeostasis, then zinc can potentiate excitotoxicity and disease (Sensi et al. 2009).

The abnormal flux of extracellular zinc into the neuron and/or the mobilization of zinc from intracellular stores (from various sources such as cytosolic zinc binding proteins like metallothionein and mitochondria for example) can have various consequences such as mitochondrial dysfunction, oxidative stress and cell death. In isolated mitochondria, it has been shown that zinc exerts differential effects, depending on the concentration of the metal, with sub micromolar levels of zinc inducing mitochondrial membrane depolarization, an increase in currents across the inner membrane, increased oxygen consumption and decreased ROS production (with higher levels causing decreased oxygen consumption and increased ROS) (Sensi et al. 2003). In the case of excitotoxicity, zinc is released in a calcium- and oxidative stress-dependent manner from intracellular stores and can, in concert with calcium, impact the mitochondrial membrane potential and cause opening of the mitochondrial permeability transition pore, cause overproduction of ROS and the release of pro-apoptotic factors (such as cytochrome c) (Sensi et al. 2000, 2003; Bossy-Wetzel et al. 2004; Jiang et al. 2001) (calcium and zinc signalling have also been shown to converge on pathways critical for neuronal cell death cascades (McCord and Aizenman 2013)). Indeed, it has been shown that the mobilization of zinc provides a critical link between ROS production, mitochondrial failure, calcium dyshomeostasis and neurotoxicity (Granzotto and Sensi 2015), although there have been views to the contrary (Pivovarova et al. 2014). Furthermore, mitochondria are a critical target of zinc (Ji et al. 2019; Ji and Weiss 2018). Taken together, it is clear that there are independent and synergistic effects of calcium and zinc that can initiate and potentiate a number of signalling events (such as occurs with excitotoxicity) that precipitate neuronal deficits and death, with downstream consequences for functional outcomes across ageing and in disease.

11.2.6 Zinc and Inflammation

Inflammation is a natural process of the body initiated to protect the host from tissue damage and infection in order to restore homeostasis to cells and tissue. The proper modulation of the inflammatory response to combat noxious stimuli such as stress, free radicals, cytokines or bacterial and viral antigens is profoundly influenced by zinc, and as such plays a critical role in the development of chronic inflammation resulting in the loss of tissue function.

One of the major inflammatory pathways is the nuclear factor kappa-light-chain-enhancer of activated B cells (NF-κB) signalling pathway. It acts as the master regulator of the genes controlling innate and adaptive immune responses, inflammatory processes, apoptosis, cell adhesion, proliferation, tissue remodelling and cellular stress. Thereupon, it influences the expression of the pro-inflammatory cytokines such as TNF-α, IL-1β, IL6, IL8 and MCP (Jarosz et al. 2017).

The role of zinc in activation of NF-κB is, however, complex and, at times, contradictory. For example, zinc has been shown to be a necessary component for the activation of lipopolysaccharide (LPS)-induced NF-κB signalling pathway, whereby chelation of zinc with the membrane-permeable zinc-specific chelator TPEN completely blocked the pathway (Haase et al. 2008). However, a study by Prasad et al. indicated that zinc suppresses the generation of NF-κB-regulated inflammatory cytokines by induction of A20, a zinc finger protein that is recognized as an anti-inflammatory protein known to regulate tumour necrosis factor receptor (TNFR)- and toll-like receptor (TLR)-initiated NF-κB pathways (Prasad et al. 2011). In addition, zinc supplementation was shown to be associated with downregulation of inflammatory cytokines via the reduced gene expression of IL-1β and TNF-α through the upregulation of mRNA and DNA-specific binding of A20 to inhibit NF-κB activation (Prasad et al. 2004).

Zinc inhibition of NF-κB has also been demonstrated via increasing the expression of peroxisome proliferator-activated receptor α (PPAR-α), which is known to be a mediator of lipoprotein metabolism, inflammation and glucose homeostasis. A zinc-induced increase in PPAR-α initiates a downregulation of inflammatory cytokines and adhesion molecules (Bao et al. 2010). Additionally, cyclic nucleotide phosphodiesterase (PDE) can be inhibited by zinc, whereby it elevates cyclic guanosine monophosphate (cGMP) activating protein kinase A (PKA) with a subsequent inhibition of NF-κB (von Bulow et al. 2007).

Zinc can also play a modulatory role in the inflammation process through toll-like receptor (TLR) signalling. Toll-like receptors are a class of proteins usually expressed on sentinel cells such as macrophages and dendritic cells that play a key role in the innate immune system. A study by Kitamura et al. demonstrated that stimulation with the toll-like receptor 4 agonist LPS altered the expression of zinc transporters in dendritic cells, thereby decreasing intracellular free zinc and inhibiting the upregulation of major histocompatibility complex class II and co-stimulatory molecules (Kitamura et al. 2006).

Zinc can also modulate the apoptotic and inflammatory processes of caspases. Caspases are a family of enzymes that play essential roles in programmed cell death or apoptosis. Zinc has been shown to inhibit all the major caspases involved in apoptosis, with each using unique zinc-binding stoichiometries and inhibition constants. Importantly, however, inhibition is observed at low, though varying, nanomolar concentrations as well as low nanomolar binding within the estimated range of free intracellular zinc (Eron et al. 2018). These observations suggest that zinc can regulate apoptosis through inactivation of caspases in a manner that is unique for each caspase.

Chronic inflammation is characterized by increased levels of inflammatory cytokines, and zinc can influence the production and signalling of numerous cytokines in many cell types. During the acute phase response to infection or trauma, there is a transient and rapid decline of the plasma zinc concentration due to redistribution of zinc into intracellular compartments that appears to be partially mediated and preceded by cytokine activation (Gaetke et al. 1997). Numerous studies indicate that acute illness presents with hypozincaemia and elevated cytokine production (Besecker et al. 2011; Young et al. 1996).

11.2.7 Zinc and Alzheimer's Disease

Alzheimer's disease is a progressive neurodegenerative disorder characterized by an abnormal deposition of protein aggregates in the form of extracellular plaques composed of fibrillar amyloid beta protein (Aβ), and intracellular neurofibrillary tangles composed of hyperphosphorylated tau (Lashley et al. 2018). Numerous biochemical studies demonstrate zinc to be intricately involved in the amyloid dysmetabolism of AD (Bush et al. 1993; Cornett et al. 1998; Cuajungco and Lees 1997; Danscher et al. 1997; Lovell et al. 1998; Multhaup et al. 1994), where elevated concentrations of zinc are observed in neuritic plaques and cerebrovascular amyloid deposits in both AD patients and AD-like transgenic mice (Lovell et al. 1998; Corrigan et al. 1993; Friedlich et al. 2004; Lee et al. 1999; Suh et al. 2000a; Zhang et al. 2008). Moreover, chelation of zinc favours the disaggregation of plaques (Cherny et al. 1999).

Under normal conditions, zinc acts as a co-transmitter on post-synaptic glutamate receptors (Pochwat et al. 2015), N-Methyl-D-aspartic acid receptors (NMDAR) (Amico-Ruvio et al. 2011), tyrosine kinase receptor type 2 (Hwang et al. 2005) and the zinc-sensing receptor ZnR (Hershfinkel et al. 2007). However, increased local zinc concentrations can contribute to the pathology of AD because Aβ is a metal-binding protein (Atwood et al. 2000; Bush et al. 1994b), and zinc has been shown to promote Aβ oligomerization (Curtain et al. 2001). Moreover, zinc is concentrated in synaptic vesicles throughout the cortex and hippocampus where, during synaptic transmission, zinc concentrations can rise to 300 μm (Frederickson and Bush 2001), a concentration at which zinc can precipitate Aβ oligomers at the synaptic cleft (Deshpande et al. 2009).

The observation of the presence of zinc in the synaptic cleft as a contributor to Aβ oligomerization is supported by two studies utilizing ZnT-3 knockout mice. The first study utilized the Tg2576 AD-like transgenic mouse (a mouse model that overexpresses mutant human amyloid precursor protein) cross-bred with a ZnT-3 knockout mouse, resulting in a significantly decreased plaque load (Lee et al. 2002). The second model assessed Aβ oligomer formation at the synaptic cleft of hippocampal slices in ZnT-3 knockout and wild type mice after stimulation with KCL or glutamate. There was a marked elevation in Aβ oligomer formation in the wild type mice that was dramatically reduced in the ZnT-3 knockout mice (Deshpande et al. 2009).

However, the role of zinc in the pathogenesis is not clearly defined, and indeed, multiple investigations in both human and animal models have produced disparate data. For example, post-mortem analysis of AD brain zinc concentrations has been largely contradictory with studies showing an increase in zinc levels in the AD Brain (Danscher et al. 1997; Lovell et al. 1998), decreased levels in the AD brain (Corrigan et al. 1993; Andrasi et al. 2000) or unchanged levels (Rulon et al. 2000). Similarly, in transgenic animal models of AD, dietary zinc supplementation has led to an increase in the number and size of Aβ plaques concomitant with cognitive deficits (Linkous et al. 2009; Railey et al. 2011). However, the same results have also been observed in a study of dietary zinc deficiency in the APP/PS1 transgenic mouse model of AD (Stoltenberg et al. 2007).

An alternate pathway through which zinc may contribute to AD onset and progression is through its role in the generation of hyperphosphorylated tau protein causing the polymerization and generation of neurofibrillary tangles (NFTs). Studies indicate that low micromolar concentrations of zinc can cause the aggregation of human tau fragments (Mo et al. 2009; An et al. 2005; Pei et al. 2006), and additionally, zinc is able to regulate phosphorylation of tau protein through the extracellular signal-related kinase pathway (MAP/ERK) (Kim et al. 2011b). Moreover, a recent study by Xiong et al. indicated that zinc can inactivate the major tau phosphatase, protein phosphatase 2A (PP2A), resulting in hyperphosphorylation of tau (Xiong et al. 2013). Supporting this observation is the fact that the use of zinc chelators attenuates zinc mediated tau hyperphosphorylation (Sun et al. 2012).

Collectively, these studies implicate a mechanistic role, although yet to be clearly defined, for zinc in the pathways leading to the two major hallmark pathologies of AD.

11.2.8 Zinc and Parkinson's Disease

The second most common brain disorder after AD is Parkinson's disease (PD), a long-term degenerative disease pathologically defined by the loss of dopaminergic neurons of the substantia nigra pars compacta concomitant with Lewy body deposition in neurons (Schneider and Obeso 2015). The clinical manifestations of PD are classically motor-related, with the most obvious signs being shaking, muscle rigidity, slowing of movement and difficulty with ambulation. Additionally, behaviour regulation and dementia are now recognized as a late-stage clinical manifestation of the disease (Pagonabarraga and Kulisevsky 2012).

Two recent meta-analyses indicated that PD patients have significantly decreased circulating zinc levels in serum, plasma and CSF when compared to healthy controls (Du et al. 2017; Sun et al. 2017). Whilst evidence of zinc supplementation in PD patients is scarce, there are animal models demonstrating zinc supplementation efficacy. A recent publication investigating biochemical and histological changes with rotenone-induced Parkinsonism in rats indicated that oral zinc supplementation attenuated the cell death and neuronal size reduction induced by rotenone concomitant with a reduction in lipid peroxidation and antioxidant status (Mbiydzenyuy et al. 2018). Another study utilizing a Drosophila melanogaster PD model in which the orthologue of the human Parkin gene was disrupted to induce motor abnormalities mimicking human PD symptoms showed attenuation of symptoms with dietary zinc supplementation (Saini and Schaffner 2010).

Early onset of PD in humans is associated with a mutation of PARK9, a lysosomal type 5 P-type ATPase, which has been demonstrated to reduce lysosomal zinc storage, resulting in elevated cytosolic zinc with α-synuclein accumulation, a pathological hallmark of PD (Kong et al. 2014; Tsunemi and Krainc 2014). Furthermore, in the substantia nigra, caudate nucleus and lateral putamen areas associated with PD pathology, there is an observed accumulation of zinc in PD patients (Dexter et al. 1991).

11.2.9 Zinc and Amyotrophic Lateral Sclerosis

Amyotrophic lateral sclerosis is a fatal neurodegenerative disease affecting the human motor system. It is both familial and sporadic in nature with 5–10% being familial and 90% sporadic. The most common cause of familial ALS is a mutation in the ubiquitously expressed free radical scavenging enzyme, copper/zinc superoxide dismutase (SOD1) gene (Rosen et al. 1993). SOD1, a 32-kDa homodimeric enzyme, is expressed in the cytosol and acts to reduce the intracellular concentration of superoxide radicals by catalysing their dismutation to O_2 and H_2O_2. SOD1 naturally binds zinc at the Derlin-1-binding region of SOD1; however, mutant forms of the enzyme associated with ALS have a significantly attenuated affinity for zinc (Smith and Lee 2007; Vonk and Klomp 2008), the result of which is a toxic gain of function in motor neurons when zinc is missing from its active site (Roberts et al. 2007). Whilst the mechanism of action causing neuronal injury upon the SOD1 toxic gain of function is unclear, aberrant free radical management, abnormal protein regulation and increased susceptibility to excitotoxicity are all hypothesized (Barber et al. 1762; King et al. 2016; Webster et al. 2017).

The zinc-binding metallothioneins are also postulated to play a yet to be clearly defined role in ALS progression. Metallothioneins are ubiquitous low molecular weight metal-binding proteins known to be present in humans, plants, fungi, cyanobacteria and prokaryotes. Human MTs are encoded by a multi-gene family located on chromosome 16, with four distinct mammalian MT isoforms currently identified, designated MT-1 through MT-4. In the central nervous system (CNS), MT-1 and MT-2 are primarily expressed in astrocytes and spinal glia, with MT-3 being exclusively expressed in neurons (Thirumoorthy et al. 2011). The CNS MTs play a significant role in neuronal zinc homeostasis, and as such, has been implicated in numerous neurodegenerative diseases (Juarez-Rebollar et al. 2017). It has been observed that ALS patients exhibit upregulation of metallothionein in the spinal cord (Sillevis Smitt et al. 1994), and deletion of MT-1, MT-2 and MT-3 in the Gly93 Ala mutant SOD1 transgenic mouse results in accelerated symptom onset and a reduction in survival time (Nagano et al. 2001; Puttaparthi et al. 2002). Both of these observations are consistent with the notion of metallothionein mediating a neuroprotective function via zinc binding to prevent the elevation of toxic-free zinc concentrations. Zinc therefore appears to play a multifactorial role in its contribution ALS progression, via both impaired SOD1 function and metallothionein dysregulation.

11.2.10 Zinc and Ageing

Zinc deficiency in developed countries is a common problem affecting the elderly, and is often a by-product of nutritional disorders caused by the ageing process itself and senescence of homeostatic mechanisms (Cabrera 2015). The daily intake recommended for people over the age of 70 years is 11 mg for males and 8 mg for females. However, it has been estimated that people older than 65 years have a zinc intake 50% below this recommended level (Mocchegiani et al. 2013). Indeed, a recent study of elderly Europeans revealed that 44% of them were zinc-deficient, and 20% were highly zinc-deficient (Madej et al. 2013). Inadequate dietary zinc intake in the ageing population results in low intracellular zinc ion availability, contributing to age-related conditions via the alteration of intracellular zinc homeostasis. Indeed, several dietary zinc supplementation studies in the elderly support the notion that adequate zinc intake provides beneficial health outcomes, primarily manifesting as improved resilience to infections via enhanced immune function (Prasad et al. 2007; Bao et al. 2010; Guo and Wang 2013; Meydani et al. 2007).

The combination of senescence of homeostatic mechanisms and low dietary zinc in the elderly results in low circulating zinc (in plasma), and has been associated with age-related diseases such as AD (Brewer et al. 2010) and diabetes mellitus (Kazi et al. 2008). In AD, increased local zinc concentrations are implicated in the pivotal pathological mechanisms as described above; however, decreased zinc at the systemic and cellular level likely contributes as well. For example, low ZnT-3 expression known to occur in AD (Beyer et al. 2009), coupled with zinc sequestration in senile plaques (Miller et al. 2006), could diminish the pool of readily releasable synaptic zinc contributing to cognitive decline as evidenced in the previously mentioned study utilizing aged ZnT-3 knockout mice (Adlard et al. 2010).

Of further significance is the age-related alterations in the gene expression of zinc transporter proteins that maintain cellular zinc homeostasis in the human brain. There are two classes of zinc transporter proteins, SCLA30A (ZnT) and SCLA39 (ZIP), and both have contrasting roles. The zinc -and iron-like regulatory proteins (ZIP) are primarily responsible for cellular zinc uptake into cells or export of zinc from vesicles or organelles (Cousins et al. 2006). In contrast, the zinc transporter (ZnT) proteins function to regulate both influx an efflux of zinc ions in cells and cellular compartments such as vesicles, mitochondria, and Golgi apparatus (Kambe 2012).

A recent study by Olesen et al. using microarray data from human frontal cortex demonstrated an age-related decrease in ZnT-3 and ZnT-4 expression, whereas the expression of ZIP-1, ZIP-9 and ZIP-13 increased (Olesen et al. 2016). Whilst it was known that ZnT-3 is downregulated with ageing, the finding that ZnT-4 is also downregulated with ageing was a novel one. Like ZnT-3, ZnT-4 is involved in vesicular secretory functions (Kambe 2012). ZIP-1 is ubiquitously expressed throughout the body; it is highly expressed in hippocampus, where its ablation in knockout mice has been shown to attenuate seizure-induced neuronal death (Qian et al. 2011). Little is known about the function of ZIP-9 and ZIP-13 in the brain. Nevertheless, the observation that ageing can alter the expression of genes that regulate intracellular zinc homeostasis should warrant continued scrutiny as they pertain to age-related cognitive decline and the development of neurodegenerative disorders.

11.2.11 Zinc and Traumatic Brain Injury

Traumatic brain injury arises from a multitude of situations that cause different types and severities of injuries to the brain, with parallel time-dependent clinical phenotypes and anatomical/biochemical alterations in the brain. TBI is the leading cause of mortality in young adults and a major cause of death and disability across all ages in all countries. In regard to this chapter, as there are a number of parallels that can be drawn between TBI and AD in terms of shared pathologies and functional deficits (and zinc has a number of known roles in AD), there are a number of defined pathways through which zinc and TBI can already be linked as we recently reviewed (Portbury and Adlard 2015). Indeed, whilst there is a poor understanding of the mechanistic pathways that contribute to the initiation and potentiation of acute and chronic TBI deficits, the literature suggests that zinc homeostasis may be an important factor in the pathobiology of TBI (Portbury and Adlard 2015).

Specifically, zinc has been shown to accumulate at the site of injury (Yeiser et al. 1999), with a loss of zinc from presynaptic boutons and a concomitant appearance of zinc in injured neurons (Suh et al. 2000b) (which may occur as a function of zinc translocation from the pre-synaptic compartment, but also from zinc mobilization from intracellular stores in the post-synaptic side, as discussed earlier). These alterations in zinc may induce toxicity via a number of different pathways, such as excitotoxicity, oxidative stress (which may be driven by, or driving, the cytotoxic zinc fluctuations (Li et al. 2010)) or mitochondrial dysfunction, and are suggested to contribute to the neurodegeneration that results from brain injury (Yeiser et al. 1999; Suh et al. 2000b, 2006; Choi and Koh 1998; Lee et al. 2003). As a consequence, zinc has become a target for therapeutic modulation in TBI, with a number of studies examining the utility of zinc chelators.

Indeed, Hellmich and colleagues (Hellmich et al. 2007) demonstrated that following TBI the degenerating neurons corresponded to those with high zinc levels, and a single injection of Ca-EDTA prior to TBI significantly increased the expression of neuroprotective genes and proteins post-TBI, which also then correlated with a reduced number of TUNEL-positive cells (Hellmich et al. 2004). Recent studies have also demonstrated that the alterations that occur in the ubiquitin-proteasome system following injury can also be prevented by scavenging endogenous zinc (Sun et al. 2013). Whilst chelator treatment is apparently beneficial in the acute phase post-injury, evidenced by a reduction in the number of injured hippocampal neurons at 24 h post-TBI, this does not appear to translate to longer term benefits on functional deficits, such as learning and memory (Hellmich et al. 2008). Whilst such studies have not been interrogated exhaustively, it does prompt questions around the use of chelation as a therapeutic strategy. Indeed, Yeiser and colleagues (Yeiser et al. 2002) have shown that dietary zinc levels are critical in limiting neuronal death in the more chronic phases following TBI; and the use of dietary zinc supplements has been shown to improve cognitive performance following injury and to reduce other psychiatric manifestations such as depression-like behaviour, which is a significant comorbidity in ~40% of TBI patients (Cope et al. 2011, 2012). Taken together with the wealth of knowledge around the role of zinc in normal neuronal function, the complexity of its pathways and the reported strong links between zinc homeostasis and behavioural/clinical end points such as depression, anxiety and cognitive function, it is perhaps not surprising that a unidirectional modulation in zinc would not be universally beneficial. As the change in zinc that occurs post-TBI is not likely to be linear, different strategies (chelation vs supplementation) will be required at different stages of "disease" (which is also true for neurodegenerative disorders and even ageing, where the alteration in zinc, the requirements for that metal and the need for its modulation will all vary across time). The use of a monotherapy that could maintain zinc homeostasis would, therefore, also be valuable. In this regard, our own data utilizing zinc chaperones (which do not exclude zinc from biological interactions, but rather facilitate the redistribution and maintenance of zinc homeostasis) suggest that this approach may have benefit in TBI and other brain disorders (Portbury et al. 2019).

11.3 Conclusion

In conclusion, there is a complex interplay between zinc and the brain, both in the context of normal cellular functions and in disorders of the CNS, such as neurodegenerative diseases and other conditions that result in a neurodegenerative phenotype.

Understanding the mechanisms and pathways through which zinc is involved will present opportunities for the development of novel therapeutic approaches that target zinc to ultimately result in a diminution of the functional deficits that arise as a consequence of neurodegeneration across lifespan.

References

Adlard PA, Parncutt JM, Finkelstein DI, Bush AI (2010) Cognitive loss in zinc transporter-3 knock-out mice: a phenocopy for the synaptic and memory deficits of Alzheimer's disease? J Neurosci 30(5):1631–1636

Aimo L, Cherr GN, Oteiza PI (2010) Low extracellular zinc increases neuronal oxidant production through nadph oxidase and nitric oxide synthase activation. Free Radic Biol Med 48(12):1577–1587

Amico-Ruvio SA, Murthy SE, Smith TP, Popescu GK (2011) Zinc effects on NMDA receptor gating kinetics. Biophys J 100(8):1910–1918

An WL, Bjorkdahl C, Liu R, Cowburn RF, Winblad B, Pei JJ (2005) Mechanism of zinc-induced phosphorylation of p70 S6 kinase and glycogen synthase kinase 3 beta in SH-SY5Y neuroblastoma cells. J Neurochem 92(5):1104–1115

Andrasi E, Farkas E, Gawlik D, Rosick U, Bratter P (2000) Brain iron and zinc contents of German patients with Alzheimer disease. J Alzheimers Dis 2(1):17–26

Andrews GK (2000) Regulation of metallothionein gene expression by oxidative stress and metal ions. Biochem Pharmacol 59(1):95–104

Antonyuk SV, Strange RW, Marklund SL, Hasnain SS (2009) The structure of human extracellular copper-zinc superoxide dismutase at 1.7 A resolution: insights into heparin and collagen binding. J Mol Biol 388(2):310–326

Atwood CS, Scarpa RC, Huang X, Moir RD, Jones WD, Fairlie DP, Tanzi RE, Bush AI (2000) Characterization of copper interactions with alzheimer amyloid beta peptides: identification of an attomolar-affinity copper binding site on amyloid beta1-42. J Neurochem 75(3):1219–1233

Bancila V, Nikonenko I, Dunant Y, Bloc A (2004) Zinc inhibits glutamate release via activation of pre-synaptic K-ATP channels and reduces ischaemic damage in rat hippocampus. J Neurochem 90(5):1243–1250

Bao B, Prasad AS, Beck FW, Fitzgerald JT, Snell D, Bao GW, Singh T, Cardozo LJ (2010) Zinc decreases C-reactive protein, lipid peroxidation, and inflammatory cytokines in elderly subjects: a potential implication of zinc as an atheroprotective agent. Am J Clin Nutr 91(6):1634–1641

Barber SC, Mead RJ, Shaw PJ (2006) Oxidative stress in ALS: a mechanism of neurodegeneration and a therapeutic target. Biochim Biophys Acta-Mol Basis Dis 1762(11–12):1051–1067

Baxt LA, Garza-Mayers AC, Goldberg MB (2013) Bacterial subversion of host innate immune pathways. Science 340(6133):697–701

Besecker BY, Exline MC, Hollyfield J, Phillips G, DiSilvestro RA, Wewers MD, Knoell DL (2011) A comparison of zinc metabolism, inflammation, and disease severity in critically ill infected and noninfected adults early after intensive care unit admission. Am J Clin Nutr 93(6):1356–1364

Besser L, Chorin E, Sekler I, Silverman WF, Atkin S, Russell JT, Hershfinkel M (2009) Synaptically released zinc triggers metabotropic signaling via a zinc-sensing receptor in the hippocampus. J Neurosci 29(9):2890–2901

Beyer N, Coulson DT, Heggarty S, Ravid R, Irvine GB, Hellemans J, Johnston JA (2009) ZnT3 mRNA levels are reduced in Alzheimer's disease postmortem brain. Mol Neurodegener 4:53

Bjorkdahl C, Sjogren MJ, Winblad B, Pei JJ (2005) Zinc induces neurofilament phosphorylation independent of p70 S6 kinase in N2a cells. Neuroreport 16(6):591–595

Borgstahl GE, Parge HE, Hickey MJ, Johnson MJ, Boissinot M, Hallewell RA, Lepock JR, Cabelli DE, Tainer JA (1996) Human mitochondrial manganese superoxide dismutase polymorphic variant Ile58Thr reduces activity by destabilizing the tetrameric interface. Biochemistry 35(14):4287–4297

Bosomworth HJ, Adlard PA, Ford D, Valentine RA (2013) Altered Expression of ZnT10 in Alzheimer's disease brain. PLoS One 8(5)

Bossy-Wetzel E, Talantova MV, Lee WD, Scholzke MN, Harrop A, Mathews E, Gotz T, Han JH, Ellisman MH, Perkins GA, Lipton SA (2004) Crosstalk between nitric oxide and zinc pathways to neuronal cell death involving mitochondrial dysfunction and p38-activated K+ channels. Neuron 41(3):351–365

Botti J, Djavaheri-Mergny M, Pilatte Y, Codogno P (2006) Autophagy signaling and the cogwheels of cancer. Autophagy 2(2):67–73

Bray TM, Kubow S, Bettger WJ (1986) Effect of dietary zinc on endogenous free-radical production in rat lung microsomes. J Nutr 116(6):1054–1060

Brewer GJ, Kanzer SH, Zimmerman EA, Molho ES, Celmins DF, Heckman SM, Dick R (2010) Subclinical zinc deficiency in Alzheimer's disease and Parkinson's disease. Am J Alzheimers Dis Other Dement 25(7):572–575

Brown AM, Tummolo DM, Rhodes KJ, Hofmann JR, Jacobsen JS, SonnenbergReines J (1997) Selective aggregation of endogenous beta-amyloid peptide and soluble amyloid precursor protein in cerebrospinal fluid by zinc. J Neurochem 69(3):1204–1212

Bruno RS, Song Y, Leonard SW, Mustacich DJ, Taylor AW, Traber MG, Ho E (2007) Dietary zinc restriction in rats alters antioxidant status and increases plasma F-2 isoprostanes. J Nutr Biochem 18(8):509–518

Bush AI, Multhaup G, Moir RD, Williamson TG, Small DH, Rumble B, Pollwein P, Beyreuther K, Masters CL (1993) A Novel Zinc(Ii) Binding-Site Modulates the Function of the Beta-A4 Amyloid Protein-Precursor of Alzheimers-Disease. J Biol Chem 268(22):16109–16112

Bush AI, Pettingell WH Jr, Paradis MD, Tanzi RE (1994a) Modulation of A beta adhesiveness and secretase site cleavage by zinc. J Biol Chem 269(16):12152–12158

Bush AI, Pettingell WH, Multhaup G, d Paradis M, Vonsattel JP, Gusella JF, Beyreuther K, Masters CL, Tanzi RE (1994b) Rapid induction of Alzheimer A beta amyloid formation by zinc. Science 265(5177):1464–1467

Cabrera AJ (2015) Zinc, aging, and immunosenescence: an overview. Pathobiol Aging Age Relat Dis 5:25592

Canali R, Vignolini F, Nobili F, Mengheri E (2000) Reduction of oxidative stress and cytokine-induced neutrophil chemoattractant (CINC) expression by red wine polyphenols in zinc deficiency induced intestinal damage of rat. Free Radic Biol Med 28(11):1661–1670

Cao X, Antonyuk SV, Seetharaman SV, Whitson LJ, Taylor AB, Holloway SP, Strange RW, Doucette PA, Valentine JS, Tiwari A, Hayward LJ, Padua S, Cohlberg JA, Hasnain SS, Hart PJ (2008) Structures of the G85R variant of SOD1 in familial amyotrophic lateral sclerosis. J Biol Chem 283(23):16169–16177

Chen Y, Klionsky DJ (2011) The regulation of autophagy – unanswered questions. J Cell Sci 124(Pt 2):161–170

Cherny RA, Legg JT, McLean CA, Fairlie DP, Huang X, Atwood CS, Beyreuther K, Tanzi RE, Masters CL, Bush AI (1999) Aqueous dissolution of Alzheimer's disease Abeta amyloid deposits by biometal depletion. J Biol Chem 274(33):23223–23228

Choi DW, Koh JY (1998) Zinc and brain injury. Annu Rev Neurosci 21:347–375

Choi DW, Weiss JH, Koh JY, Christine CW, Kurth MC (1989) Glutamate neurotoxicity, calcium, and zinc. Ann N Y Acad Sci 568:219–224

Ciuculescu ED, Mekmouche Y, Faller P (2005) Metal-binding properties of the peptide APP(170-188): A model of the Zn-II-binding site of amyloid precursor protein (APP). Chem Eur J 11(3):903–909

Cohen-Kfir E, Lee W, Eskandari S, Nelson N (2005) Zinc inhibition of gamma-aminobutyric acid transporter 4 (GAT4) reveals a link between excitatory and inhibitory neurotransmission. Proc Natl Acad Sci U S A 102(17):6154–6159

Cope EC, Morris DR, Scrimgeour AG, VanLandingham JW, Levenson CW (2011) Zinc supplementation provides behavioral resiliency in a rat model of traumatic brain injury. Physiol Behav 104(5):942–947

Cope EC, Morris DR, Scrimgeour AG, Levenson CW (2012) Use of zinc as a treatment for traumatic brain injury in the rat: effects on cognitive and behavioral outcomes. Neurorehabil Neural Repair 26(7):907–913

Cornett CR, Markesbery WR, Ehmann WD (1998) Imbalances of trace elements related to oxidative damage in Alzheimer's disease brain. Neurotoxicology 19(3):339–345

Corrigan FM, Reynolds GP, Ward NI (1993) Hippocampal tin, aluminum and zinc in Alzheimer's disease. Biometals 6(3):149–154

Cousins RJ, Liuzzi JP, Lichten LA (2006) Mammalian zinc transport, trafficking, and signals. J Biol Chem 281(34):24085–24089

Coyle P, Philcox JC, Carey LC, Rofe AM (2002) Metallothionein: the multipurpose protein. Cell Mol Life Sci 59(4):627–647

Cross JB, Duca JS, Kaminski JJ, Madison VS (2002) The active site of a zinc-dependent metalloproteinase influences the computed pK(a) of ligands coordinated to the catalytic zinc ion. J Am Chem Soc 124(37):11004–11007

Cuajungco MP, Lees GJ (1997) Zinc and Alzheimer's disease: is there a direct link? Brain Res Rev 23(3):219–236

Cummings J (2017) Disease modification and Neuroprotection in neurodegenerative disorders. Transl Neurodegener 6:25

Curtain CC, Ali F, Volitakis I, Cherny RA, Norton RS, Beyreuther K, Barrow CJ, Masters CL, Bush AI, Barnham KJ (2001) Alzheimer's disease amyloid-beta binds copper and zinc to generate an allosterically ordered membrane-penetrating structure containing superoxide dismutase-like subunits. J Biol Chem 276(23):20466–20473

Czaja MJ, Ding WX, Donohue TM, Friedman SL, Kim JS, Komatsu M, Lemasters JJ, Lemoine A, Lin JD, Ou JHJ, Perlmutter DH, Randall G, Ray RB, Tsung A, Yin XM (2013) Functions of autophagy in normal and diseased liver. Autophagy 9(8):1131–1158

Danscher G, Jensen KB, Frederickson CJ, Kemp K, Andreasen A, Juhl S, Stoltenberg M, Ravid R (1997) Increased amount of zinc in the hippocampus and amygdala of Alzheimer's diseased brains: a proton-induced X-ray emission spectroscopic analysis of cryostat sections from autopsy material. J Neurosci Methods 76(1):53–59

Deshpande A, Kawai H, Metherate R, Glabe CG, Busciglio J (2009) A role for synaptic zinc in activity-dependent Abeta oligomer formation and accumulation at excitatory synapses. J Neurosci 29(13):4004–4015

Dexter DT, Carayon A, Javoy-Agid F, Agid Y, Wells FR, Daniel SE, Lees AJ, Jenner P, Marsden CD (1991) Alterations in the levels of iron, ferritin and other trace metals in Parkinson's disease and other neurodegenerative diseases affecting the basal ganglia. Brain 114(Pt 4):1953–1975

Ding F, Dokholyan NV (2008) Dynamical roles of metal ions and the disulfide bond in Cu, Zn superoxide dismutase folding and aggregation. Proc Natl Acad Sci U S A 105(50):19696–19701

Du K, Liu MY, Zhong X, Wei MJ (2017) Decreased circulating Zinc levels in Parkinson's disease: a meta-analysis study. Sci Rep 7

Eron SJ, MacPherson DJ, Dagbay KB, Hardy JA (2018) Multiple mechanisms of zinc-mediated inhibition for the apoptotic caspases-3, -6, -7, and -8. ACS Chem Biol 13(5):1279–1290

Frederickson CJ, Bush AI (2001) Synaptically released zinc: physiological functions and pathological effects. Biometals 14(3–4):353–366

Frederickson CJ, Koh JY, Bush AI (2005) The neurobiology of zinc in health and disease. Nat Rev Neurosci 6(6):449–462

Friedlich AL, Lee JY, van Groen T, Cherny RA, Volitakis I, Cole TB, Palmiter RD, Koh JY, Bush AI (2004) Neuronal zinc exchange with the blood vessel wall promotes cerebral amyloid angiopathy in an animal model of Alzheimer's disease. J Neurosci 24(13):3453–3459

Gaetke LM, McClain CJ, Talwalkar RT, Shedlofsky SI (1997) Effects of endotoxin on zinc metabolism in human volunteers. Am J Phys 272(6 Pt 1):E952–E956

Galluzzi L, Morselli E, Vicencio JM, Kepp O, Joza N, Tajeddine N, Kroemer G (2008) Life, death and burial: multifaceted impact of autophagy. Biochem Soc Trans 36:786–790

Gan L, Cookson MR, Petrucelli L, La Spada AR (2018) Converging pathways in neurodegeneration, from genetics to mechanisms. Nat Neurosci 21(10):1300–1309

Granzotto A, Sensi SL (2015) Intracellular zinc is a critical intermediate in the excitotoxic cascade. Neurobiol Dis 81:25–37

Greenough MA, Volitakis I, Li QX, Laughton K, Evin G, Ho M, Dalziel AH, Camakaris J, Bush AI (2011) Presenilins promote the cellular uptake of copper and zinc and maintain copper chaperone of SOD1-dependent copper/zinc superoxide dismutase activity. J Biol Chem 286(11):9776–9786

Gunther V, Davis AM, Georgiev O, Schaffner W (2012a) A conserved cysteine cluster, essential for transcriptional activity, mediates homodimerization of human metal-responsive transcription factor-1 (MTF-1). Biochim Biophys Acta 1823(2):476–483

Gunther V, Lindert U, Schaffner W (2012b) The taste of heavy metals: gene regulation by MTF-1. Biochim Biophys Acta 1823(9):1416–1425

Guo CH, Wang CL (2013) Effects of zinc supplementation on plasma copper/zinc ratios, oxidative stress, and immunological status in hemodialysis patients. Int J Med Sci 10(1):79–89

Ha KN, Chen Y, Cai JY, Sternberg P (2006) Increased glutathione synthesis through an ARE-Nrf2-dependent pathway by zinc in the RPE: implication for protection against oxidative stress. Invest Ophthalmol Vis Sci 47(6):2709–2715

Haase H, Rink L (2011) Zinc signaling. Zinc Hum Health 76:94–117

Haase H, Ober-Blobaum JL, Engelhardt G, Hebel S, Heit A, Heine H, Rink L (2008) Zinc signals are essential for lipopolysaccharide-induced signal transduction in monocytes. J Immunol 181(9):6491–6502

Hammermueller JD, Bray TM, Bettger WJ (1987) Effect of zinc and copper deficiency on microsomal nadph-dependent active oxygen generation in rat lung and liver. J Nutr 117(5):894–901

Harris FM, Brecht WJ, Xu Q, Mahley RW, Huang YD (2004) Increased tau phosphorylation in apolipoprotein E4 transgenic mice is associated with activation of extracellular signal-regulated kinase – Modulation by zinc. J Biol Chem 279(43):44795–44801

Hellmich HL, Frederickson CJ, DeWitt DS, Saban R, Parsley MO, Stephenson R, Velasco M, Uchida T, Shimamura M, Prough DS (2004) Protective effects of zinc chelation in traumatic brain injury correlate with upregulation of neuroprotective genes in rat brain. Neurosci Lett 355(3):221–225

Hellmich HL, Eidson KA, Capra BA, Garcia JM, Boone DR, Hawkins BE, Uchida T, DeWitt DS, Prough DS (2007) Injured Fluoro-Jadepositive hippocampal neurons contain high levels of zinc after traumatic brain injury. Brain Res 1127(1):119–126

Hellmich HL, Eidson K, Cowart J, Crookshanks J, Boone DK, Shah S, Uchida T, DeWitt DS, Prough DS (2008) Chelation of neurotoxic zinc levels does not improve neurobehavioral outcorne after traumatic brain injury. Neurosci Lett 440(2):155–159

Hershfinkel M, Silverman WF, Sekler I (2007) The zinc sensing receptor, a link between zinc and cell signaling. Mol Med 13(7–8):331–336

Ho E, Ames BN (2002) Low intracellular zinc induces oxidative DNA damage, disrupts p53, NF kappa B, and AP1 DNA binding, and affects DNA repair in a rat glioma cell line. Proc Natl Acad Sci U S A 99(26):16770–16775

Hoke DE, Tan JL, Ilaya NT, Culvenor JG, Smith SJ, White AR, Masters CL, Evin GM (2005) In vitro gamma-secretase cleavage of the Alzheimer's amyloid precursor protein correlates to a subset of presenilin complexes and is inhibited by zinc. Febs J 272(21):5544–5557

Homma K, Fujisawa T, Tsuburaya N, Yamaguchi N, Kadowaki H, Takeda K, Nishitoh H, Matsuzawa A, Naguro I, Ichijo H (2013) SOD1 as a molecular switch for initiating the homeostatic ER stress response under zinc deficiency. Mol Cell 52(1):75–86

Hu JY, Zhang DL, Liu XL, Li XS, Cheng XQ, Chen J, Du HN, Liang Y (2017) Pathological concentration of zinc dramatically accelerates abnormal aggregation of full-length human Tau and

thereby significantly increases Tau toxicity in neuronal cells. Biochim Biophys Acta Mol basis Dis 1863(2):414–427
Huang YZ, Pan E, Xiong ZQ, McNamara JO (2008) Zinc-mediated transactivation of TrkB potentiates the hippocampal mossy fiber-CA3 pyramid synapse. Neuron 57(4):546–558
Huang Y, Wu Z, Cao Y, Lang M, Lu B, Zhou B (2014) Zinc binding directly regulates tau toxicity independent of tau hyperphosphorylation. Cell Rep 8(3):831–842
Huang YP, Wu ZZ, Zhou B (2016) Behind the curtain of tauopathy: a show of multiple players orchestrating tau toxicity. Cell Mol Life Sci 73(1):1–21
Hung HH, Huang WP, Pan CY (2013) Dopamine- and zinc-induced autophagosome formation facilitates PC12 cell survival. Cell Biol Toxicol 29(6):415–429
Hwang JJ, Park MH, Choi SY, Koh JY (2005) Activation of the Trk signaling pathway by extracellular zinc. Role of metalloproteinases. J Biol Chem 280(12):11995–12001
Hwang JJ, Kim HN, Kim J, Cho DH, Kim MJ, Kim YS, Kim Y, Park SJ, Koh JY (2010) Zinc(II) ion mediates tamoxifen-induced autophagy and cell death in MCF-7 breast cancer cell line. Biometals 23(6):997–1013
Iwata J, Ezaki J, Komatsu M, Yokota S, Ueno T, Tanida I, Chiba T, Tanaka K, Kominami E (2006) Excess peroxisomes are degraded by autophagic machinery in mammals. J Biol Chem 281(7):4035–4041
Jarosz M, Olbert M, Wyszogrodzka G, Mlyniec K, Librowski T (2017) Antioxidant and anti-inflammatory effects of zinc. Zinc-dependent NF-kappaB signaling. Inflammopharmacology 25(1):11–24
Ji SG, Weiss JH (2018) Zn2+-induced disruption of neuronal mitochondrial function: synergism with Ca2+, critical dependence upon cytosolic Zn2+ buffering, and contributions to neuronal injury. Exp Neurol 302:181–195
Ji SG, Medvedeva YV, Wang HL, Yin HZ, Weiss JH (2019) Mitochondrial Zn2+ accumulation: a potential trigger of hippocampal ischemic injury. Neuroscientist 25(2):126–138
Jiang DM, Sullivan PG, Sensi SL, Steward O, Weiss JH (2001) Zn2+ induces permeability transition pore opening and release of pro-apoptotic peptides from neuronal mitochondria. J Biol Chem 276(50):47524–47529
Jong-ok Pyo S-MY, Jung Y-K (2013) the interplay between autophagy and aging. Diabetes Metab J 37:333–339
Juarez-Rebollar D, Rios C, Nava-Ruiz C, Mendez-Armenta M (2017) Metallothionein in brain disorders. Oxidative Med Cell Longev 2017:5828056
Julien S, Biesemeier A, Kokkinou D, Eibl O, Schraermeyer U (2011) Zinc deficiency leads to lipofuscin accumulation in the retinal pigment epithelium of pigmented rats. PLoS One 6(12)
Kalappa BI, Anderson CT, Goldberg JM, Lippard SJ, Tzounopoulos T (2015) AMPA receptor inhibition by synaptically released zinc. Proc Natl Acad Sci U S A 112(51):15749–15754
Kambe T (2012) Molecular architecture and function of ZnT transporters. Curr Top Membr 69:199–220
Kazi TG, Afridi HI, Kazi N, Jamali MK, Arain MB, Jalbani N, Kandhro GA (2008) Copper, chromium, manganese, iron, nickel, and zinc levels in biological samples of diabetes mellitus patients. Biol Trace Elem Res 122(1):1–18
Kepp KP (2017) Alzheimer's disease: how metal ions define beta-amyloid function. Coord Chem Rev 351:127–159
Kim I, Park EJ, Seo J, Ko SJ, Lee J, Kim CH (2011a) Zinc stimulates tau S214 phosphorylation by the activation of Raf/mitogen-activated protein kinase-kinase/extracellular signal-regulated kinase pathway. Neuroreport 22(16):839–844
Kim KW, Speirs CK, Jung DK, Lu B (2011b) The zinc ionophore PCI-5002 radiosensitizes non-small cell lung cancer cells by enhancing autophagic Cell death. J Thorac Oncol 6(9):1542–1552
King AE, Woodhouse A, Kirkcaldie MT, Vickers JC (2016) Excitotoxicity in ALS: overstimulation, or overreaction? Exp Neurol 275(Pt 1):162–171
Kirkin V, McEwan DG, Novak I, Dikic I (2009) A role for ubiquitin in selective autophagy. Mol Cell 34(3):259–269

Kitamura H, Morikawa H, Kamon H, Iguchi M, Hojyo S, Fukada T, Yamashita S, Kaisho T, Akira S, Murakami M, Hirano T (2006) Toll-like receptormediated regulation of zinc homeostasis influences dendritic cell function. Nat Immunol 7(9):971–977

Kochanczyk T, Drozd A, Krezel A (2015) Relationship between the architecture of zinc coordination and zinc binding affinity in proteins – insights into zinc regulation. Metallomics 7(2):244–257

Kojima-Yuasa A, Umeda K, Olikita T, Kennedy DO, Nishiguchi S, Matsui-Yuasa I (2005) Role of reactive oxygen species in zinc deficiency-induced hepatic stellate cell activation. Free Radic Biol Med 39(5):631–640

Kong SM, Chan BK, Park JS, Hill KJ, Aitken JB, Cottle L, Farghaian H, Cole AR, Lay PA, Sue CM, Cooper AA (2014) Parkinson's disease-linked human PARK9/ATP13A2 maintains zinc homeostasis and promotes alpha-Synuclein externalization via exosomes. Hum Mol Genet 23(11):2816–2833

Krezel A, Maret W (2016) The biological inorganic chemistry of zinc ions. Arch Biochem Biophys 611:3–19

Kroemer G, Marino G, Levine B (2010) Autophagy and the integrated stress response. Mol Cell 40(2):280–293

Kumari MVR, Hiramatsu M, Ebadi M (1998) Free radical scavenging actions of metallothionein isoforms I and II. Free Radic Res 29(2):93–101

Lammich S, Kojro E, Postina R, Gilbert S, Pfeiffer R, Jasionowski M, Haass C, Fahrenholz F (1999) Constitutive and regulated alpha-secretase cleavage of Alzheimer's amyloid precursor protein by a disintegrin metalloprotease. Proc Natl Acad Sci U S A 96(7):3922–3927

Lashley T, Schott JM, Weston P, Murray CE, Wellington H, Keshavan A, Foti SC, Foiani M, Toombs J, Rohrer JD, Heslegrave A, Zetterberg H (2018) Molecular biomarkers of Alzheimer's disease: progress and prospects. Dis Model Mech 11(5)

Leal SS, Botelho HM, Gomes CM (2012) Metal ions as modulators of protein conformation and misfolding in neurodegeneration. Coord Chem Rev 256(19–20):2253–2270

Lee SJ, Koh JY (2010) Roles of zinc and metallothionein-3 in oxidative stress-induced lysosomal dysfunction, cell death, and autophagy in neurons and astrocytes. Mol Brain 3(1):30

Lee JY, Mook-Jung I, Koh JY (1999) Histochemically reactive zinc in plaques of the Swedish mutant beta-amyloid precursor protein transgenic mice. J Neurosci 19(11):RC10

Lee JY, Cole TB, Palmiter RD, Suh SW, Koh JY (2002) Contribution by synaptic zinc to the gender-disparate plaque formation in human Swedish mutant APP transgenic mice. Proc Natl Acad Sci U S A 99(11):7705–7710

Lee JY, Kim JH, Palmiter RD, Koh JY (2003) Zinc released from metallothionein-III may contribute to hippocampal CA1 and thalamic neuronal death following acute brain injury. Exp Neurol 184(1):337–347

Lee JY, Kim YJ, Kim TY, Koh JY, Kim YH (2008) Essential role for zinc-triggered p75NTR activation in preconditioning neuroprotection. J Neurosci 28(43):10919–10927

Li Y, Hawkins BE, DeWitt DS, Prough DS, Maret W (2010) The relationship between transient zinc ion fluctuations and redox signaling in the pathways of secondary cellular injury: relevance to traumatic brain injury. Brain Res 1330:131–141

Lichtlen P, Schaffner W (2001) The "metal transcription factor" MTF-1: biological facts and medical implications. Swiss Med Wkly 131(45–46):647–652

Linkous DH, Adlard PA, Wanschura PB, Conko KM, Flinn JM (2009) The effects of enhanced zinc on spatial memory and plaque formation in transgenic mice. J Alzheimers Dis 18(3):565–579

Liu ST, Howlett G, Barrow CJ (1999) Histidine-13 is a crucial residue in the zinc ion-induced aggregation of the A beta peptide of Alzheimer's disease. Biochemistry 38(29):9373–9378

Liuzzi JP, Yoo CW (2013) Role of zinc in the regulation of autophagy during ethanol exposure in human hepatoma cells. Biol Trace Elem Res 156(1–3):350–356

Lovell MA, Robertson JD, Teesdale WJ, Campbell JL, Markesbery WR (1998) Copper, iron and zinc in Alzheimer's disease senile plaques. J Neurol Sci 158(1):47–52

Lozy F, Karantza V (2012) Autophagy and cancer cell metabolism. Semin Cell Dev Biol 23(4):395–401

Maas AIR, Menon DK, Adelson PD, Andelic N, Bell MJ, Belli A, Bragge P, Brazinova A, Buki A, Chesnut RM, Citerio G, Coburn M, Cooper DJ, Crowder AT, Czeiter E, Czosnyka M, Diaz-Arrastia R, Dreier JP, Duhaime AC, Ercole A, van Essen TA, Feigin VL, Gao G, Giacino J, Gonzalez-Lara LE, Gruen RL, Gupta D, Hartings JA, Hill S, Jiang JY, Ketharanathan N, Kompanje EJO, Lanyon L, Laureys S, Lecky F, Levin H, Lingsma HF, Maegele M, Majdan M, Manley G, Marsteller J, Mascia L, McFadyen C, Mondello S, Newcombe V, Palotie A, Parizel PM, Peul W, Piercy J, Polinder S, Puybasset L, Rasmussen TE, Rossaint R, Smielewski P, Soderberg J, Stanworth SJ, Stein MB, von Steinbuchel N, Stewart W, Steyerberg EW, Stocchetti N, Synnot A, Te Ao B, Tenovuo O, Theadom A, Tibboel D, Videtta W, Wang KKW, Williams WH, Wilson L, Yaffe K, T. P. In and Investigators (2017) Traumatic brain injury: integrated approaches to improve prevention, clinical care, and research. Lancet Neurol 16(12):987–1048

Madej D, Borowska K, Bylinowska J, Szybalska A, Pietruszka B (2013) Dietary intakes of iron and zinc assessed in a selected group of the elderly: are they adequate? Rocz Panstw Zakl Hig 64(2):97–104

Maret W, Krezel A (2007) Cellular zinc and redox buffering capacity of metallthionein/thionein in health and disease. Mol Med 13(7–8):371–375

Matheou CJ, Younan ND, Viles JH (2016) The rapid exchange of zinc (2+) enables trace levels to profoundly influence amyloid-beta misfolding and dominates assembly outcomes in Cu (2+)/ Zn (2+) Mixtures. J Mol Biol 428(14):2832–2846

Mbiydzenyuy NE, Ninsiima HI, Valladares MB, Pieme CA (2018) Zinc and linoleic acid pre-treatment attenuates biochemical and histological changes in the midbrain of rats with rotenone-induced Parkinsonism. BMC Neurosci 19(1):29

McCord MC, Aizenman E (2013) Convergent Ca2+ and Zn2+ signaling regulates apoptotic Kv2.1 K+ currents. Proc Natl Acad Sci U S A 110(34):13988–13993

Meydani SN, Barnett JB, Dallal GE, Fine BC, Jacques PF, Leka LS, Hamer DH (2007) Serum zinc and pneumonia in nursing home elderly. Am J Clin Nutr 86(4):1167–1173

Miller LM, Wang Q, Telivala TP, Smith RJ, Lanzirotti A, Miklossy J (2006) Synchrotron-based infrared and X-ray imaging shows focalized accumulation of Cu and Zn co-localized with beta-amyloid deposits in Alzheimer's disease. J Struct Biol 155(1):30–37

Mindgardens_Neuroscience_Network (2019) Review of the burden of disease for neurological, mental health and substance use disorders in Australia. Mindgardens Neuroscience Network White Paper 2019

Mizushima N (2007) Autophagy: process and function. Genes Dev 21(22):2861–2873

Mo ZY, Zhu YZ, Zhu HL, Fan JB, Chen J, Liang Y (2009) Low micromolar zinc accelerates the fibrillization of human tau via bridging of Cys-291 and Cys-322. J Biol Chem 284(50):34648–34657

Mocchegiani E, Romeo J, Malavolta M, Costarelli L, Giacconi R, Diaz LE, Marcos A (2013) Zinc: dietary intake and impact of supplementation on immune function in elderly. Age (Dordr) 35(3):839–860

Multhaup G, Bush AI, Pollwein P, Masters CL (1994) Interaction between the Zinc(Ii) and the heparin-binding site of the Alzheimers-disease Beta-A4 amyloid precursor protein (App). FEBS Lett 355(2):151–154

Nagano S, Satoh M, Sumi H, Fujimura H, Tohyama C, Yanagihara T, Sakoda S (2001) Reduction of metallothioneins promotes the disease expression of familial amyotrophic lateral sclerosis mice in a dose-dependent manner. Eur J Neurosci 13(7):1363–1370

Narendra D, Tanaka A, Suen DF, Youle RJ (2008) Parkin is recruited selectively to impaired mitochondria and promotes their autophagy. J Cell Biol 183(5):795–803

Nicoll RA (2017) A brief history of long-term potentiation. Neuron 93(2):281–290

Nilsson P, Saido TC (2014) Dual roles for autophagy: degradation and secretion of Alzheimer's disease A beta peptide. BioEssays 36(6):570–578

Olesen RH, Hyde TM, Kleinman JE, Smidt K, Rungby J, Larsen A (2016) Obesity and age-related alterations in the gene expression of zinc-transporter proteins in the human brain. Transl Psychiatry 6(6):e838
Omata Y, Salvador GA, Supasai S, Keenan AH, Oteiza PI (2013) Decreased zinc availability affects glutathione metabolism in neuronal cells and in the developing brain. Toxicol Sci 133(1):90–100
Orr ME, Oddo S (2013) Autophagic/lysosomal dysfunction in Alzheimer's disease. Alzheimers Res Ther 5(5):53
Oteiza PI, Olin KL, Fraga CG, Keen CL (1995) Zinc-deficiency causes oxidative damage to proteins, lipids and DNA in rat testes. J Nutr 125(4):823–829
Oteiza PI, Clegg MS, Zago MP, Keen CL (2000) Zinc deficiency induces oxidative stress and AP-1 activation in 3T3 cells. Free Radic Biol Med 28(7):1091–1099
Ozcelik D, Naziroglu M, Tuncdemir M, Celik O, Ozturk M, Flores-Arce MF (2012) Zinc supplementation attenuates metallothionein and oxidative stress changes in kidney of streptozotocin-induced diabetic rats. Biol Trace Elem Res 150(1–3):342–349
Pagonabarraga J, Kulisevsky J (2012) Cognitive impairment and dementia in Parkinson's disease. Neurobiol Dis 46(3):590–596
Paoletti P, Vergnano AM, Barbour B, Casado M (2009) Zinc at glutamatergic synapses. Neuroscience 158(1):126–136
Park IH, Jung MW, Mori H, Mook-Jung I (2001) Zinc enhances synthesis of presenilin 1 in mouse primary cortical culture. Biochem Biophys Res Commun 285(3):680–688
Pei JJ, An WL, Zhou XW, Nishimura T, Norberg J, Benedikz E, Gotz J, Winblad B (2006) P70S6 kinase mediates tau phosphorylation and synthesis. FEBS Lett 580(1):107–114
Pivovarova NB, Stanika RI, Kazanina G, Villanueva I, Andrews SB (2014) The interactive role of zinc and calcium in mitochondrial dysfunction and neurodegeneration. J Neurochem 128(4):592–602
Pochwat B, Nowak G, Szewczyk B (2015) Relationship between Zinc (Zn (2+)) and glutamate receptors in the processes underlying neurodegeneration. Neural Plast 2015:591563
Pompella A, Visvikis A, Paolicchi A, De Tata V, Casini AF (2003) The changing faces of glutathione, a cellular protagonist. Biochem Pharmacol 66(8):1499–1503
Portbury SD, Adlard PA (2015) Traumatic brain injury, chronic traumatic encephalopathy, and Alzheimer's disease: common pathologies potentiated by altered zinc homeostasis. J Alzheimers Dis 46(2):297–311
Portbury SD, Sedjahtera A, Perrones K, Sgambelloni C, Zhang M, Crack PJ, Finkelstein DI, Adlard PA (2019) Metal chaperones: a novel therapeutic strategy for brain injury? Brain Inj 33(3):305–312
Prasad AS, Bao B, Beck FW, Kucuk O, Sarkar FH (2004) Antioxidant effect of zinc in humans. Free Radic Biol Med 37(8):1182–1190
Prasad AS, Beck FWJ, Bao B, Fitzgerald JT, Snell DC, Steinberg JD, Cardozo LJ (2007) Zinc supplementation decreases incidence of infections in the elderly: effect of zinc on generation of cytokines and oxidative stress. Am J Clin Nutr 85(3):837–844
Prasad AS, Bao B, Beck FW, Sarkar FH (2011) Zinc-suppressed inflammatory cytokines by induction of A20-mediated inhibition of nuclear factorkappaB. Nutrition 27(7–8):816–823
Puttaparthi K, Gitomer WL, Krishnan U, Son M, Rajendran B, Elliott JL (2002) Disease progression in a transgenic model of familial amyotrophic lateral sclerosis is dependent on both neuronal and non-neuronal zinc binding proteins. J Neurosci 22(20):8790–8796
Qian J, Xu K, Yoo J, Chen TT, Andrews G, Noebels JL (2011) Knockout of Zn transporters Zip-1 and Zip-3 attenuates seizure-induced CA1 neurodegeneration. J Neurosci 31(1):97–104
Quan W, Jung HS, Lee MS (2013) Role of autophagy in the progression from obesity to diabetes and in the control of energy balance. Arch Pharm Res 36(2):223–229
Quinta-Ferreira ME, Matias CM (2004) Hippocampal mossy fiber calcium transients are maintained during long-term potentiation and are inhibited by endogenous zinc. Brain Res 1004(1–2):52–60

Quinta-Ferreira ME, Matias CM (2005) Tetanically released zinc inhibits hippocampal mossy fiber calcium, zinc and synaptic responses (vol 1047, pg 1, 2005). Brain Res 1063(1):102–104
Railey AM, Groeber CM, Flinn JM (2011) The effect of metals on spatial memory in a transgenic mouse model of Alzheimer's disease. J Alzheimers Dis 24(2):375–381
Roberts BR, Tainer JA, Getzoff ED, Malencik DA, Anderson SR, Bomben VC, Meyers KR, Karplus PA, Beckman JS (2007) Structural characterization of zinc-deficient human superoxide dismutase and implications for ALS. J Mol Biol 373(4):877–890
Rosen DR, Siddique T, Patterson D, Figlewicz DA, Sapp P, Hentati A, Donaldson D, Goto J, Oregan JP, Deng HX, Rahmani Z, Krizus A, Mckennayasek D, Cayabyab A, Gaston SM, Berger R, Tanzi RE, Halperin JJ, Herzfeldt B, Vandenbergh R, Hung WY, Bird T, Deng G, Mulder DW, Smyth C, Laing NG, Soriano E, Pericakvance MA, Haines J, Rouleau GA, Gusella JS, Horvitz HR, Brown RH (1993) Mutations in Cu/Zn superoxide-dismutase gene are associated with familial amyotrophic-lateral-sclerosis. Nature 362(6415):59–62
Rulon LL, Robertson JD, Lovell MA, Deibel MA, Ehmann WD, Markesber WR (2000) Serum zinc levels and Alzheimer's disease. Biol Trace Elem Res 75(1–3):79–85
Ruttkay-Nedecky B, Nejdl L, Gumulec J, Zitka O, Masarik M, Eckschlager T, Stiborova M, Adam V, Kizek R (2013) The role of metallothionein in oxidative stress. Int J Mol Sci 14(3):6044–6066
Ryu MS, Langkamp-Henken B, Chang SM, Shankar MN, Cousins RJ (2011) Genomic analysis, cytokine expression, and microRNA profiling reveal biomarkers of human dietary zinc depletion and homeostasis. Proc Natl Acad Sci U S A 108(52):20970–20975
Saini N, Schaffner W (2010) Zinc supplement greatly improves the condition of parkin mutant Drosophila. Biol Chem 391(5):513–518
Scheuermann S, Hambsch B, Hesse L, Stumm J, Schmidt C, Beher D, Bayer TA, Beyreuther K, Multhaup G (2001) Homodimerization of amyloid precursor protein and its implication in the amyloidogenic pathway of Alzheimer's disease. J Biol Chem 276(36):33923–33929
Schneider SA, Obeso JA (2015) Clinical and pathological features of Parkinson's disease. Curr Top Behav Neurosci 22:205–220
Sensi SL, Yin HZ, Weiss JH (2000) AMPA/kainate receptor-triggered Zn2+ entry into cortical neurons induces mitochondrial Zn2+ uptake and persistent mitochondrial dysfunction. Eur J Neurosci 12(10):3813–3818
Sensi SL, Ton-That D, Sullivan PG, Jonas EA, Gee KR, Kaczmarek LK, Weiss JH (2003) Modulation of mitochondrial function by endogenous Zn2+ pools. Proc Natl Acad Sci U S A 100(10):6157–6162
Sensi SL, Paoletti P, Bush AI, Sekler I (2009) Zinc in the physiology and pathology of the CNS. Nat Rev Neurosci 10(11):780–U738
Shaheen AA, Elfattah AAA (1995) Effect of dietary zinc on lipid-peroxidation, glutathione, protein thiols levels and superoxide-dismutase activity in rat-tissues. Int J Biochem Cell Biol 27(1):89–95
Sharif R, Thomas P, Zalewski P, Fenech M (2012) The role of zinc in genomic stability. Mutat Res Fundam Mol Mech Mutagen 733(1–2):111–121
Sillevis Smitt PA, Mulder TP, Verspaget HW, Blaauwgeers HG, Troost D, de Jong JM (1994) Metallothionein in amyotrophic lateral sclerosis. Biol Signals 3(4):193–197
Singh R, Xiang YQ, Wang YJ, Baikati K, Cuervo AM, Luu YK, Tang Y, Pessin JE, Schwartz GJ, Czaja MJ (2009) Autophagy regulates adipose mass and differentiation in mice. J Clin Investig 119(11):3329–3339
Smith AP, Lee NM (2007) Role of zinc in ALS. Amyotroph Lateral Scler 8(3):131–143
Song Y, Chung CS, Bruno RS, Traber MG, Brown KH, King JC, Ho E (2009a) Dietary zinc restriction and repletion affects DNA integrity in healthy men. Am J Clin Nutr 90(2):321–328
Song Y, Leonard SW, Traber MG, Ho E (2009b) Zinc deficiency affects DNA damage, oxidative stress, antioxidant defenses, and DNA repair in rats. J Nutr 139(9):1626–1631
Song Y, Elias V, Loban A, Scrimgeour AG, Ho E (2010) Marginal zinc deficiency increases oxidative DNA damage in the prostate after chronic exercise. Free Radic Biol Med 48(1):82–88

Stoltenberg M, Bush AI, Bach G, Smidt K, Larsen A, Rungby J, Lund S, Doering P, Danscher G (2007) Amyloid plaques arise from zinc-enriched cortical layers in APP/PS1 transgenic mice and are paradoxically enlarged with dietary zinc deficiency. Neuroscience 150(2):357–369

Suh SW, Chen JW, Motamedi M, Bell B, Listiak K, Pons NF, Danscher G, Frederickson CJ (2000a) Evidence that synaptically-released zinc contributes to neuronal injury after traumatic brain injury. Brain Res 852(2):268–273

Suh SW, Jensen KB, Jensen MS, Silva DS, Kesslak PJ, Danscher G, Frederickson CJ (2000b) Histochemically-reactive zinc in amyloid plaques, angiopathy, and degenerating neurons of Alzheimer's diseased brains. Brain Res 852(2):274–278

Suh SW, Frederickson CJ, Danscher G (2006) Neurotoxic zinc translocation into hippocampal neurons is inhibited by hypothermia and is aggravated by hyperthermia after traumatic brain injury in rats. J Cereb Blood Flow Metab 26(2):161–169

Sullivan JF, Jetton MM, Hahn HKJ, Burch RE (1980) Enhanced lipid-peroxidation in liver-microsomes of zinc-deficient rats. Am J Clin Nutr 33(1):51–56

Summersgill H, England H, Lopez-Castejon G, Lawrence CB, Luheshi NM, Pahle J, Mendes P, Brough D (2014) Zinc depletion regulates the processing and secretion of IL-1 beta. Cell Death Dis 5

Sun XY, Wei YP, Xiong Y, Wang XC, Xie AJ, Wang XL, Yang Y, Wang Q, Lu YM, Liu R, Wang JZ (2012) Synaptic released zinc promotes tau hyperphosphorylation by inhibition of protein phosphatase 2A (PP2A). J Biol Chem 287(14):11174–11182

Sun KJ, Zhu L, Wang HD, Ji XJ, Pan H, Chen M, Lu TJ, Fan YW, Cheng HL, Hang CH, Shi JX (2013) Zinc as mediator of ubiquitin conjugation following traumatic brain injury. Brain Res 1506:132–141

Sun HL, Liu XX, Ge HN, Wang T, Wang Y, Li WJ (2017) Association between serum zinc levels and the risk of Parkinson's disease: a meta-analysis. Biol Trace Elem Res 179(1):45–51

Takeda A (2011) Insight into glutamate excitotoxicity from synaptic zinc homeostasis. Int J Alzheimer's Dis Article ID 491597

Takeda A, Tamano H (2012) Proposed glucocorticoid-mediated zinc signaling in the hippocampus. Metallomics 4(7):614–618

Takeda A, Fuke S, Minami A, Oku N (2007a) Role of zinc influx via AMPA/kainate receptor activation in metabotropic glutamate receptor-mediated calcium release. J Neurosci Res 85(6):1310–1317

Takeda A, Minami A, Sakurada N, Nakajima S, Oku N (2007b) Response of hippocampal mossy fiber zinc to excessive glutamate release. Neurochem Int 50(2):322–327

Thirumoorthy N, Sunder AS, Kumar KTM, Kumar MS, Ganesh GNK, Chatterjee M (2011) A review of metallothionein isoforms and their role in pathophysiology. World J Surg Oncol 9

Tsunemi T, Krainc D (2014) Zn(2)(+) dyshomeostasis caused by loss of ATP13A2/PARK9 leads to lysosomal dysfunction and alpha-synuclein accumulation. Hum Mol Genet 23(11):2791–2801

Turner AJ, Isaac RE, Coates D (2001) The neprilysin (NEP) family of zinc metalloendopeptidases: genomics and function. BioEssays 23(3):261–269

von Bulow V, Dubben S, Engelhardt G, Hebel S, Plumakers B, Heine H, Rink L, Haase H (2007) Zinc-dependent suppression of TNF-alpha production is mediated by protein kinase A-induced inhibition of Raf-1, I kappa B kinase beta, and NF-kappa B. J Immunol 179(6):4180–4186

Vonk WIM, Klomp LWJ (2008) Role of transition metals in the pathogenesis of amyotrophic lateral sclerosis. Biochem Soc Trans 36:1322–1328

Wang P, Wang ZY (2017) Metal ions influx is a double edged sword for the pathogenesis of Alzheimer's disease. Ageing Res Rev 35:265–290

Wang JR, Whiteman MW, Lian HQ, Wang GX, Singh A, Huang DY, Denmark T (2009) A non-canonical MEK/ERK signaling pathway regulates autophagy via regulating Beclin 1. J Biol Chem 284(32):21412–21424

Wang Y, Branicky R, Noe A, Hekimi S (2018) Superoxide dismutases: dual roles in controlling ROS damage and regulating ROS signaling. J Cell Biol 217(6):1915–1928

Webster CP, Smith EF, Shaw PJ, De Vos KJ (2017) Protein homeostasis in amyotrophic lateral sclerosis: therapeutic opportunities? Front Mol Neurosci 10:123

Xiong Y, Jing XP, Zhou XW, Wang XL, Yang Y, Sun XY, Qiu M, Cao FY, Lu YM, Liu R, Wang JZ (2013) Zinc induces protein phosphatase 2A inactivation and tau hyperphosphorylation through Src dependent PP2A (tyrosine 307) phosphorylation. Neurobiol Aging 34(3):745–756
Yan M, Song Y, Wong CP, Hardin K, Ho E (2008) Zinc deficiency alters DNA damage response genes in normal human prostate epithelial cells. J Nutr 138(4):667–673
Yeiser EC, Lerant AA, Casto RM, Levenson CW (1999) Free zinc increases at the site of injury after cortical stab wounds in mature but not immature rat brain. Neurosci Lett 277(2):75–78
Yeiser EC, Vanlandingham JW, Levenson CW (2002) Moderate zinc deficiency increases cell death after brain injury in the rat. Nutr Neurosci 5(5):345–352
Young B, Ott L, Kasarskis E, Rapp R, Moles K, Dempsey RJ, Tibbs PA, Kryscio R, McClain C (1996) Zinc supplementation is associated with improved neurologic recovery rate and visceral protein levels of patients with severe closed head injury. J Neurotrauma 13(1):25–34
Zago MP, Mackenzie GG, Adamo AM, Keen CL, Oteiza PI (2005) Differential modulation of MAP kinases by zinc deficiency in IMR-32 cells: role of H_2O_2. Antioxid Redox Signal 7(11–12):1773–1782
Zhang LH, Wang X, Stoltenberg M, Danscher G, Huang LP, Wang ZY (2008) Abundant expression of zinc transporters in the amyloid plaques of Alzheimer's disease brain. Brain Res Bull 77(1):55–60

Chapter 12
Role of Zinc Transporters in Type 2 Diabetes and Obesity

Ayako Fukunaka and Yoshio Fujitani

Abstract Whereas zinc ions play an important role in insulin biosynthesis and storage, many previous studies have reported the possible involvement of zinc deficiency in the pathogenesis of diabetes. Zinc can counteract the deleterious effects of oxidative stress, which contributes to increased insulin sensitivity, and can also protect pancreatic β cells from glucolipotoxicity. The finding that *SLC30A8*/ZnT8 function modulates the susceptibility to type 2 diabetes provides new insights into the roles of zinc in diabetes and obesity. Here, we discuss the roles of zinc in glucose homeostasis and obesity, based on our knowledge of the functions of the zinc transporters ZnT8 and ZIP13 on metabolic abnormalities in mice and humans.

Keywords Zinc · Insulin · Zinc transporter · ZnT8 · *SLC30A8* · ZIP13 · Diabetes · Obesity · Beige adipocyte · Browning

12.1 Zinc and Pancreatic β Cells

Zinc was reported as an essential factor for insulin crystal formation in the 1930s (Scott 1934), and the zinc content in the pancreas of diabetic patients was shown to be reduced by 50% (Scott and Fisher 1938). These findings suggested a close association between zinc and diabetes. Follow-up studies have shown that insulin exists as a hexamer consisting of six insulin and two zinc molecules (Dodson et al. 1979; Dunn 2005) (Fig. 12.1). Crystallized insulin in insulin secretory granules is observed as "dense core granules" upon electron-microscopic analysis (Hou et al. 2009; Tamaki et al. 2013).

After folding, the proinsulin protein forms dimers via electrostatic interactions. The residues B24–26 (amino acids 24–26 of the B chain) are thought to be impor-

A. Fukunaka · Y. Fujitani (✉)
Laboratory of Developmental Biology and Metabolism, Institute for Molecular and Cellular Regulation, Gunma University, Maebashi, Japan
e-mail: fujitani@gunma-u.ac.jp

T. Fukada, T. Kambe (eds.), *Zinc Signaling*,
https://doi.org/10.1007/978-981-15-0557-7_12

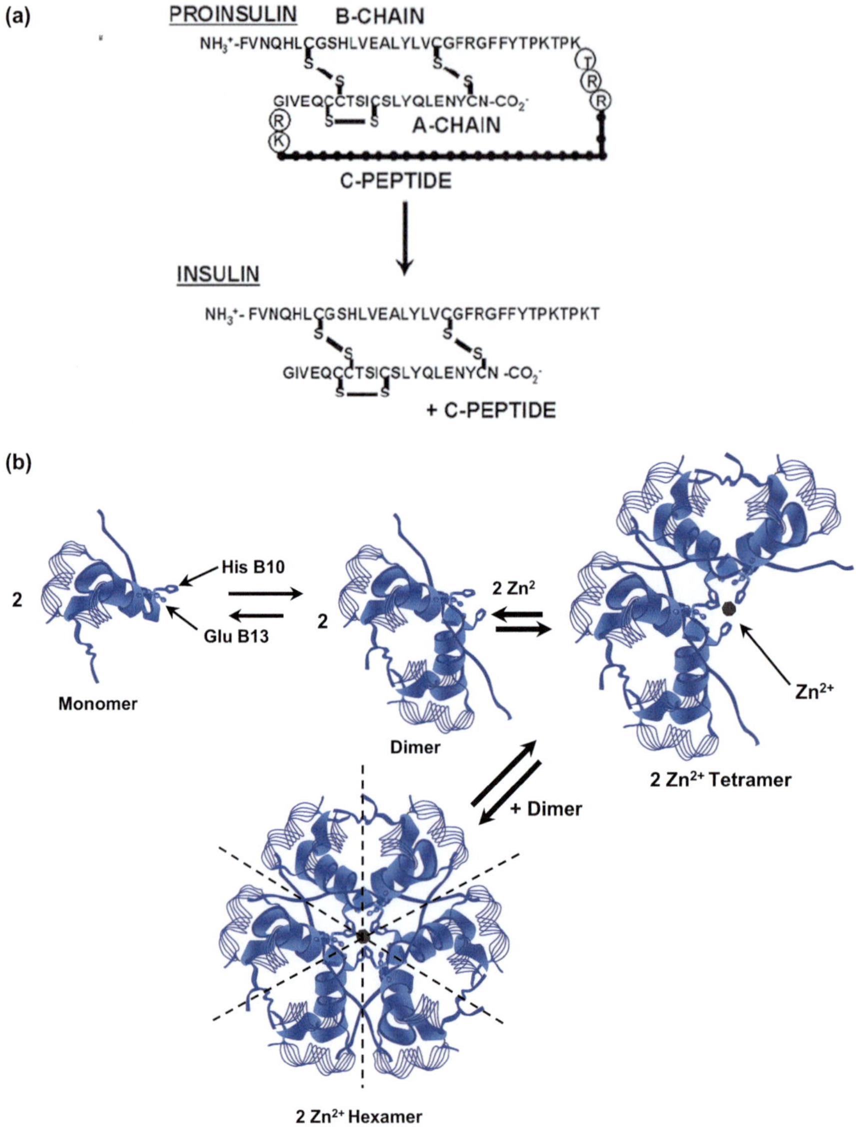

Fig. 12.1 Biosynthesis of insulin (**a**) The insulin precursor preproinsulin contains a signal sequence that is proteolytically cleaved to yield proinsulin, in which the C-peptide links the A and B chains of mature insulin. Cleavage of the C-peptide converts proinsulin into insulin. (**b**) Association of insulin monomers in the presence of zinc. Insulin monomer readily associates into dimers and then aggregates into hexameric forms in the presence of zinc. His (B10) is implicated in zinc-mediated insulin hexamer formation

tant for this dimer formation (Zoete et al. 2005). In fact, amino acid substitutions in these residues yielded a series of ultra-rapid-acting insulin analogues, which are now widely used in clinical practice (Brange et al. 1990; Kang et al. 1991). The electrostatically coupled proinsulin dimers and zinc bind with B10 histidine, resulting in proinsulin hexamer formation (Havu et al. 1977) (Fig. 12.1). This insulin hexamer formation is accomplished by prohormone convertase-mediated C-peptide dissociation (Steiner et al. 1996). Because the C-peptide portion of proinsulin is located outside of the proinsulin hexamer, the formation of the hexamer has been considered to contribute to C-peptide dissociation (Dunn 2005). Insulin crystals are formed under specific conditions within insulin granules in which both insulin and zinc exist in high concentrations and the acidic pH is maintained (Emdin et al. 1980; Hutton 1989). Given that the addition of zinc into commercial insulin products improves their stability, hexamer formation and crystallization are thought to similarly contribute to insulin stabilization in vivo. Once insulin crystals are secreted from pancreatic β cells, they are believed to rapidly dissociate into monomers and flow into the bloodstream (Fig. 12.2).

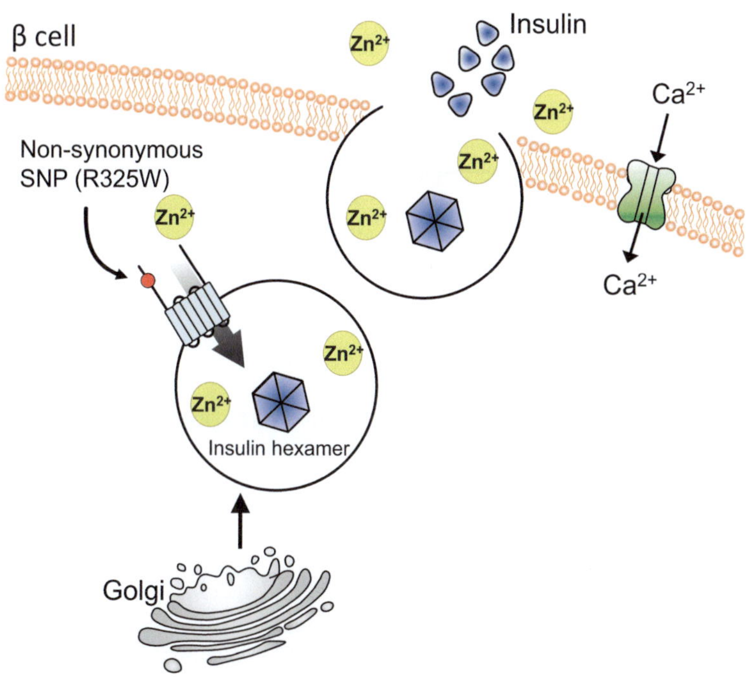

Fig. 12.2 ZnT8 is involved in zinc transport into insulin secretory granules. Converting the 325th tryptophan (W) to arginine (R) changes the zinc transportability and increases the risk of developing type 2 diabetes

12.2 Zinc Deficiency and Diabetes

The association between zinc deficiency and diabetes is widely studied in mouse models. Zinc-deficient mice showed reduced numbers of insulin granules in their pancreatic β cells (Boquist and Lernmark 1969) and had impaired glucose-stimulated insulin secretion (Huber and Gershoff 1973). Because pancreatic β cells actively synthesize large amounts of ATP, they are susceptible to oxidative stress exposure and are thereby damaged (Donath et al. 2005). As zinc is an essential component of antioxidant enzymes, including Cu, Zn superoxide dismutase (Mysore et al. 2005) and catalase (Marklund et al. 1982), zinc deficiency could worsen the condition of pancreatic β cells under oxidative stress. These findings support the notion that zinc deficiency may lead to impaired pancreatic β-cell function and survival.

Zinc supplementation also improved insulin sensitivity in obese diabetic *db/db* and *ob/ob* mice (Begin-Heick et al. 1985; Simon and Taylor 2001). A recent study by Adachi et al. demonstrated that 4 weeks of oral administration of zinc (II) complexes to obese diabetic KKAy mice significantly improved their glucose tolerance and insulin sensitivity, which has been reported to be associated with increased plasma adiponectin levels (Adachi et al. 2006). The insulin-sensitizing effect of zinc has been explained by inhibition of the tyrosine phosphatase activity of protein tyrosine phosphatase 1B (PTP1B) (Haase and Maret 2005). The enzymatic activity of PTP1B is inhibited by zinc ions. The concentration of available cellular zinc is in the nanomolar or picomolar range, and hence inhibition of PTPs by zinc occurs within this low range of available zinc (Haase and Maret 2003). Experiments in cell culture provided evidence for a physiological role of zinc in modulating PTP activity. The incubation of cells with zinc increases tyrosine phosphorylation of the insulin receptor, and its effects are mediated by the inhibition of dephosphorylation rather than by enhanced phosphorylation. Thus, zinc appears to show antidiabetic activity not only through potentiating β-cell function but also through insulin-sensitizing effects on the liver and other insulin-sensitive tissues.

12.3 Physiological Functions of Zinc Transporters

Zinc homeostasis is controlled by zinc transporters, which are classified into two categories according to their direction of zinc transport. The ZnT family (Zn transporter, SLC30A), which transports zinc from the cytoplasm to the extracellular space or to intracellular organelles, and the ZIP family (Zrt-, Irt-like protein), which transports zinc from the extracellular space or intracellular organelles to the cytoplasm. There are 9 types of ZnT and 14 types of ZIP in mammals, and each has been found to work in cooperation with each other or in a tissue-specific manner. The phenotypes of these zinc transporter knockout (KO) mice and variants in humans have been characterized (Fukunaka et al. 2017; Hara et al. 2017; Takagishi et al.

Table 12.1 Zinc transporters associated with obesity and diabetes

Protein name	Localization	Phenotype of knockout mice or human patient	Zinc signal	References
ZIP13	Golgi	Impaired bone formation and growth retardation (KO mice)	TGF-β/BMP signal	Fukada (2008)
		Ehlers-Danlos syndrome(SCD-EDS), increased beige fat		Fukunaka (2017)
ZIP14	Cellular membrane	Adipocyte hypertrophy, obesity-induced insulin resistance	cAMP-CREB signal	Troche 2016
		Liver iron overload in hemochromatosis		Jenkitkasemwong (2015)
ZnT7	Early secretory pathway	Low body weight, impaired glucose tolerance, insulin resistant (KO mice)	PI3K/Akt signal, MAPK/ERK signal	Huang (2012)
ZnT8	Insulin granule	Autoantigen in type 1 diabetes	Insulin clearance	Wenzlau et al. (2007)
		Impaired glucose tolerance (ZnT8 KO mice, *rs13266634*)		Sladek (2007) and Tamaki et al. (2013)

2017). Surprisingly, although there are 23 types of zinc transporters, their phenotypes in each zinc transporter-deficient mouse as well as human are not redundant, suggesting that the zinc signal transported via each zinc transporter is specific and may exert a specific physiological function by affecting a specific target molecule. Table 12.1 shows the phenotypes of zinc transporter KO mice which have been reported to date in association with obesity and diabetes (Fukunaka et al. 2017; Huang et al. 2012; Jenkitkasemwong et al. 2015; Sladek et al. 2007; Tamaki et al. 2013; Troche et al. 2016; Wenzlau et al. 2007).

12.4 Structure and Function of ZnT8

ZnT8 is a member of the ZnT family, which has a predicted membrane topology of six membrane-spanning domains with both N- and C-terminal ends, four conserved amino acid residues within transmembrane domains (TMDs) II and V, and a cytoplasmic His-rich loop between TMDs IV and V (Kambe 2012). The expression of ZnT8 is 2–3 orders of magnitude higher in the islets than in all other tissues analyzed (Nicolson et al. 2009). Thus, ZnT8 is responsible for transporting zinc into insulin-containing granules in islet β cells, where insulin is stored as a hexamer bound with two zinc ions before secretion (Figs. 12.1b and 12.2). ZnT8 is also expressed in other tissues, including islet α cells, pancreatic polypeptide cells, and a subset of endocrine cells in the thyroid and adrenal glands (Tamaki et al. 2009).

The islet-specific expression of ZnT8 was shown to be regulated by the β-cell enriched transcriptional factor Pdx-1 through an intronic enhancer (Pound et al. 2011).

The expression of ZnT transporters is transcriptionally regulated by various stimuli. In β cells, it has been shown that glucose and cytokines play roles in the regulation of zinc transporter expression. The cytokines IL-1β and TNF-α decrease the expression of zinc transporters, particularly ZnT8 (El Muayed et al. 2010), suggesting that the expression of ZnT8 may be decreased during diabetes that accompanies pancreatitis. Moreover, ZnT8 expression was found to be remarkably downregulated in the early stages of diabetes in diabetic model mice, such as *db/db* and *Akita* (Tamaki et al. 2009). These findings suggested that the expression of ZnT8 is reduced during the course of hyperglycemia and inflammation in type 2 diabetes.

Several studies have shown that ZnT8 is a bona fide zinc transporter, using ZnT8-overexpressed cells (Lefebvre et al. 2012), ZnT8-knockdown cells (Fu et al. 2009), and ZnT8-KO mice (Nicolson et al. 2009). The mechanism of zinc transport is still not fully understood. However, experiments using fluorescence-based functional measurements have shown that ZnT transporters function as Zn^{2+}/H^{+} exchangers (Ohana et al. 2009). ZnT8 may also use the same mechanism because ZnT8 is located in insulin granules in which is having a lower pH than cytoplasm.

Homology modeling studies based on the bacterial zinc transporter and ZnT homolog YiiP have shown that ZnT8 forms homodimers which are held together by

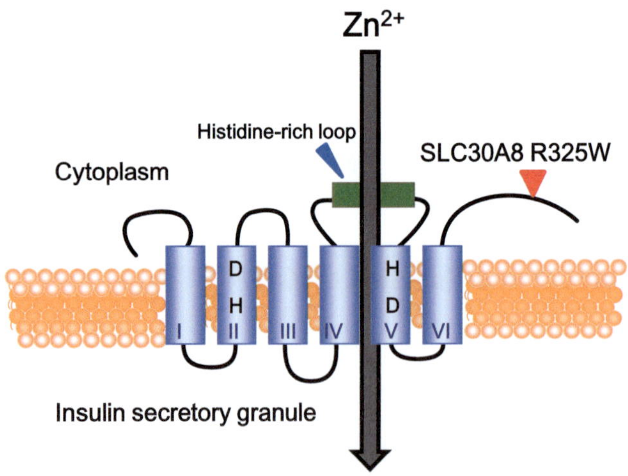

Fig. 12.3 Structure of Zinc transporter 8. ZnT8 encoded by *SLC30A8* is localized on the membranes of insulin secretory granules. Like other ZnT family members, it contains six transmembrane domains and a histidine-rich loop between TMDs IV and V, possibly for zinc binding. Insulin is stored inside secretory granules, bound to zinc ions. Zinc is released, together with insulin, upon glucose-induced exocytosis. A nonsynonymous single nucleotide polymorphism affecting amino acid 325 on ZnT8 has been implicated in type 2 diabetes from genome-wide association studies

four zinc ions located at the interface of the cytoplasmic domains, with two transmembrane domains swinging out to yield a Y-shaped structure (Nicolson et al. 2009). Genome-wide association studies demonstrated that the nonsynonymous single nucleotide polymorphism (SNP) *rs1326634* in the *SLC30A8* gene, resulting in the replacement of tryptophan-325 with arginine, increases the risk of type 2 diabetes (Figs. 12.2 and 12.3) (Diabetes Genetics Initiative of Broad Institute of et al. 2007; Scott et al. 2007; Sladek et al. 2007). To analyze the potential physiological differences between the low-risk and increased-risk variants of ZnT8, Guy Rutter's group showed that the increased-risk R-form of ZnT8 is less active as a zinc transporter, using fluorescent dyes to monitor the vacuolar accumulation of zinc (Nicolson et al. 2009). More recently, as a result of the reexamination of zinc transport activity by liposome analysis using purified human ZnT8 proteins and analysis using human islets, it was concluded that the risk allele type has higher zinc transport activity (Merriman et al. 2016).

12.5 The Role of ZnT8 in Glucose Homeostasis

As mentioned above, genome-wide analysis reported that ZnT8 is associated with type 2 diabetes. To clarify the physiological roles of ZnT8, whole-body ZnT8-deficient mice have been created by various research groups (Lemaire et al. 2009; Mitchell et al. 2016; Nicolson et al. 2009; Pound et al. 2009, 2012; Tamaki et al. 2013; Wijesekara et al. 2010). However, there has been a lack of consensus on the effects of *SlC30A8* deletion on glucose tolerance, insulin secretion, and circulation in mice, as summarized in Table 12.2 (O'Halloran et al. 2013). Whereas some studies reported that deletion of *SLC30A8*/ZnT8 results in impaired glucose tolerance (Nicolson et al. 2009; Wijesekara et al. 2010), others reported that loss of ZnT8 results in essentially no change in glucose tolerance (Lemaire et al. 2009; Pound et al. 2009). These different results may have been caused by differences in the study designs or the genetic backgrounds of the mice generated (Pound et al. 2012; Tamaki et al. 2013).

We also decided to create and analyze mice that specifically lack ZnT8 in pancreatic β cells (βZnT8-KO mice) (Tamaki et al. 2013). In line with previous reports, the formation of insulin crystal structures was incomplete in pancreatic β cells of βZnT8-KO mice. Thus, ZnT8-mediated zinc transport into insulin granules was shown to be important for insulin crystallization (Table 12.2). In the glucose tolerance test, βZnT8-KO mice were found to have mild glucose intolerance; hence it was initially predicted that insulin secretion was impaired in addition to impaired insulin crystal formation. In fact, insulin concentrations in the peripheral blood after glucose loading were reduced in βZnT8-KO mice (Table 12.2). However, when analyzed using isolated mouse islets, insulin secretion was unexpectedly enhanced in βZnT8-KO mice. To explain this discrepancy, the pancreas perfusion experiment was performed. Enhanced insulin secretion was still noted in ZnT8-KO islets upon pancreas perfusion, further supporting that insulin secretion is actually increased

Table 12.2 Summary of ZnT8-KO mouse studies

Study	Deletion location	Deletion strategy	Genetic background	Crystallization of insulin granules	Proinsulin concentration	Glucose tolerance	Glucose-stimulated insulin secretion	Insulin level in vivo
Nicolson et al. (2009)	Whole body	Exon 1	Mixed	Loss of dense core granule/rod-shaped cores	Unchanged	Impaired (only males)	Increased	Impaired
Pound et al. (2009)	Whole body	Exon 3	Mixed	Not examined	Not examined	Normal	Decreased	Impaired
Lemaire et al. (2009)	Whole body	Exon 1	Mixed	Loss of dense core granule	Unchanged	Normal	Unchanged	Unchanged
Wijesekara et al. (2010)	Pancreatic β-cell specific (Ins2-Cre)	Exon 1	Mixed	Loss of dense core granule/rod-shaped cores	Increased	Impaired	Decreased	Unchanged
Pound et al. (2012)	Whole body	Exon 3	C57BL/6J	Unchanged	Decreased	Impaired	Decreased	Unchanged
Tamaki et al. (2013)	Pancreatic β-cell specific (Ins2-Cre)	Exon 5	C57BL/6J	Loss of dense core granules	Increased	Impaired	Increased	Impaired
Mitchell et al. (2016)	Pancreatic β-cell specific (Ins1-Cre)	Exon 1	C57BL/6J	Loss of dense core granule/rod-shaped cores	Not examined	Impaired	Unchanged	Impaired

in vivo in βZnT8-KO islets. Finally, a comparison of the results obtained by pancreas perfusion versus pancreas-liver perfusion demonstrated that a large proportion of the secreted insulin was degraded during the first liver passage. In addition, in βZnT8-KO mice and humans carrying *rs13266634*, which is a major risk allele of *SLC30A8*, the insulin degradation rate calculated by the C-peptide/insulin ratio was enhanced (Tamaki et al. 2013). Our results hence suggest that *SLC30A8*, which is expressed in pancreatic β cells, regulates insulin degradation in the liver, and that dysregulation of this system may be involved in the increased risk of developing type 2 diabetes.

12.6 Reconsideration of *SLC30A8* Activity and the Risk of Diabetes

Recently, the analysis of several rare truncating variants of *SLC30A8* has demonstrated that haploinsufficiency of ZnT8 reduces the risk of developing diabetes by 65% (Flannick et al. 2014). Functional analysis of each of the variants and the corresponding ZnT8 proteins showed a functional decline, such as decreased expression levels and decreased protein stability, and these variants reduced the risk of developing diabetes in the case-control study. Given the recently reported finding that the protein encoded by the risk allele of ZnT8 has higher zinc transport activity than that encoded by the normal allele (Merriman et al. 2016), it is reasonable to conclude that at least in humans, subjects expressing ZnT8 with reduced function have a lower risk of diabetes. In accordance with this, most recently, knock-in mice with the human ZnT8 risk allele were analyzed and were reported to have an increased ability to secrete insulin by enhancing pancreatic β-cell function. However, the mechanism at present as to why a reduction in the function of human ZnT8 lowers the risk of developing diabetes remains unknown. In studies using mice, as mentioned above, there have been multiple reports of impaired glucose tolerance in the loss-of-function models of ZnT8 published so far, but on the contrary, none of the phenotypes associated with diabetes have been reported to be improved (Table 12.2). Further analyses are needed to clarify the reasons for these discrepancies.

12.7 Zinc Transporter ZIP13 Inhibits Adipocyte Browning

In association with obesity, attention has been focused on the phenomenon of the browning of fat cells in recent years (Kajimura et al. 2015). The browning of fat cells/adipocytes means transforming the energy storing white fat cells into beige fat cells which are inducible brown fat cells that consume energy. Beige adipocytes, which are rich in mitochondria compared with white adipocytes, are brown and have a high iron content. It has been shown that beige adipocytes induced in response to cold stimulation are also present in human adults, whereas human beige adipocytes have been reported to decrease with aging and obesity (Leitner et al. 2017). Based on these findings, it may be possible to achieve a healthier lifespan by elucidating the mechanism of increase and activation of beige adipocytes and developing a therapy based on its control (Kajimura et al. 2015).

Following the above analysis of ZnT8, we analyzed the possibility that zinc transporters other than ZnT8 may be involved in the onset and pathogenesis of lifestyle-associated diseases. It has been reported that lipoatrophy is observed in patients with Ehlers-Danlos syndrome who have ZIP13 loss-of-function mutations (Fukada et al. 2008). Thus, we decided to analyze the role of ZIP13 in murine adipose tissues. Surprisingly, *Zip13*-KO mice had increased beige fat cells in subcutaneous adipose tissue (adipocyte browning), whereby an increase in energy consumption was observed. Furthermore, *Zip13*-KO mice did not gain weight even when fed a high-fat diet, and thereby their glucose tolerance and insulin sensitivity were improved compared with control mice. The expression levels of brown adipocyte marker genes were found to be significantly increased in adipose cells derived from *Zip13*-KO mice in a cell-autonomous manner (Fukunaka et al. 2017). These results indicate that ZIP13 may be a molecule responsible for regulating whole body metabolism by controlling the fate of white and beige fat cells. Further analysis demonstrated that the overexpression of ZIP13 in *Zip13*-KO cells can suppress adipocyte browning, but a mutant of ZIP13 lacking the ability to transport zinc cannot suppress adipocyte browning. In addition, it was found that when an excess amount of zinc is added to the medium, the browning cannot be suppressed despite the increase in the amount of zinc in the cytoplasm. Furthermore, the browning could not be suppressed even when other zinc transporters were expressed in *Zip13*-KO cells. From these results, it was concluded that the zinc signals mediated by ZIP13 regulate the quality (specificity) of ZIP13 as well as its quantity very precisely and regulate the suppression of adipocyte browning. We are currently working toward elucidating the mechanism of determination of zinc signal specificity by ZIP13, using the beige fat cell differentiation system as a model.

To date, the involvement of zinc in the functions of adipocytes has been assumed, but the role of zinc in determining the whitening versus browning of adipocytes has not been reported. As ZIP13 is involved in the transport of zinc from the Golgi to the cytoplasm, our results show that the transport of zinc from organelles to the cytoplasm via ZIP13 controls adipocyte browning by a completely novel mechanism, which might be conserved not only in mice but also in humans. Elucidation of the mechanism of suppression of adipocyte browning by ZIP13 has the potential to lead to novel therapeutic strategies against obesity.

12.8 Therapeutic Application of Zinc Transporters Toward Obesity and Diabetes

The number of patients with type 2 diabetes continues to increase in both developed and developing countries, and elucidation of the pathogenesis of diabetes as well as the development of new therapeutic strategies are urgently needed. Obesity and type 2 diabetes are associated with both hereditary factors and lifestyle. Our studies indicated that the regulation of zinc transporters may enable manipulation of hereditary factors (Fukunaka et al. 2017) (Fig. 12.4). In particular, as βZnT8-KO mice and *Zip13*-KO mice cannot restore their respective phenotypes by simple zinc supplementation, we hope that new treatments will be developed in the future through clarification of the method of control of each zinc transporter. We are currently working on elucidation of the mechanism of determination of the specificity of zinc signals facilitated by ZIP13, and this elucidation will clarify the ZIP13-zinc signaling pathway that enhances only adipocyte browning without leading to Ehlers-Danlos syndrome (Fig. 12.4).

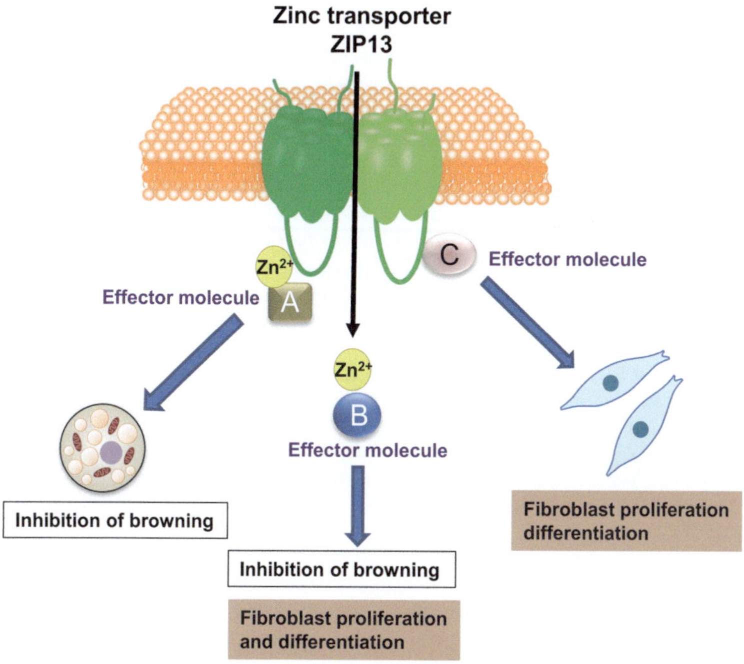

Fig. 12.4 ZIP13-mediated zinc signaling specificity. ZIP13-mediated zinc signaling regulates the inhibition of adipocyte browning and fibroblast proliferation and differentiation. We are now investigating the mechanism as to how the specificity of zinc signaling is determined by ZIP13, via identifying putative effector molecules (*A*, *B*, and *C* in the figure) linking zinc signals to each biological effect

Acknowledgments Our works on ZnT8 and ZIP13 were supported by grants from the Ministry of Education, Culture, Sports, Science and Technology in Japan (to A.F. and Y.F.), the SECOM Science and Technology Foundation, and the Naito Foundation (to Y.F.).

References

Adachi Y, Yoshida J, Kodera Y, Kiss T, Jakusch T, Enyedy EA, Yoshikawa Y, Sakurai H (2006) Oral administration of a zinc complex improves type 2 diabetes and metabolic syndromes. Biochem Biophys Res Commun 351:165–170

Begin-Heick N, Dalpe-Scott M, Rowe J, Heick HM (1985) Zinc supplementation attenuates insulin secretory activity in pancreatic islets of the ob/ob mouse. Diabetes 34:179–184

Boquist L, Lernmark A (1969) Effects on the endocrine pancreas in Chinese hamsters fed zinc deficient diets. Acta Pathol Microbiol Scand 76:215–228

Brange J, Owens DR, Kang S, Volund A (1990) Monomeric insulins and their experimental and clinical implications. Diabetes Care 13:923–954

Diabetes Genetics Initiative of Broad Institute of Harvard, MIT, Lund University, Novartis Institutes of BioMedical Research, Saxena R, Voight BF, Lyssenko V, Burtt NP, de Bakker PI, Chen H, Roix JJ et al (2007) Genome-wide association analysis identifies loci for type 2 diabetes and triglyceride levels. Science 316:1331–1336

Dodson EJ, Dodson GG, Hodgkin DC, Reynolds CD (1979) Structural relationships in the two-zinc insulin hexamer. Can J Biochem 57:469–479

Donath MY, Ehses JA, Maedler K, Schumann DM, Ellingsgaard H, Eppler E, Reinecke M (2005) Mechanisms of beta-cell death in type 2 diabetes. Diabetes 54(Suppl 2):S108–S113

Dunn MF (2005) Zinc-ligand interactions modulate assembly and stability of the insulin hexamer – a review. Biometals 18:295–303

El Muayed M, Billings LK, Raja MR, Zhang X, Park PJ, Newman MV, Kaufman DB, O'Halloran TV, Lowe WL Jr (2010) Acute cytokine-mediated downregulation of the zinc transporter ZnT8 alters pancreatic beta-cell function. J Endocrinol 206:159–169

Emdin SO, Dodson GG, Cutfield JM, Cutfield SM (1980) Role of zinc in insulin biosynthesis. Some possible zinc-insulin interactions in the pancreatic B-cell. Diabetologia 19:174–182

Flannick J, Thorleifsson G, Beer NL, Jacobs SB, Grarup N, Burtt NP, Mahajan A, Fuchsberger C, Atzmon G, Benediktsson R et al (2014) Loss-of-function mutations in SLC30A8 protect against type 2 diabetes. Nat Genet 46:357–363

Fu Y, Tian W, Pratt EB, Dirling LB, Shyng SL, Meshul CK, Cohen DM (2009) Down-regulation of ZnT8 expression in INS-1 rat pancreatic beta cells reduces insulin content and glucose-inducible insulin secretion. PLoS One 4:e5679

Fukada T, Civic N, Furuichi T, Shimoda S, Mishima K, Higashiyama H, Idaira Y, Asada Y, Kitamura H, Yamasaki S et al (2008) The zinc transporter SLC39A13/ZIP13 is required for connective tissue development; its involvement in BMP/TGF-beta signaling pathways. PLoS One 3:e3642

Fukunaka A, Fukada T, Bhin J, Suzuki L, Tsuzuki T, Takamine Y, Bin BH, Yoshihara T, Ichinoseki-Sekine N, Naito H et al (2017) Zinc transporter ZIP13 suppresses beige adipocyte biogenesis and energy expenditure by regulating C/EBP-beta expression. PLoS Genet 13:e1006950

Haase H, Maret W (2003) Intracellular zinc fluctuations modulate protein tyrosine phosphatase activity in insulin/insulin-like growth factor-1 signaling. Exp Cell Res 291:289–298

Haase H, Maret W (2005) Protein tyrosine phosphatases as targets of the combined insulinomimetic effects of zinc and oxidants. Biometals 18:333–338

Hara T, Takeda TA, Takagishi T, Fukue K, Kambe T, Fukada T (2017) Physiological roles of zinc transporters: molecular and genetic importance in zinc homeostasis. J Physiol Sci 67:283–301

Havu N, Lundgren G, Falkmer S (1977) Zinc and manganese contents of micro-dissected pancreatic islets of some rodents. A microchemical study in adult and newborn guinea pigs, rats, Chinese hamsters and spiny mice. Acta Endocrinol 86:570–577

Hou JC, Min L, Pessin JE (2009) Insulin granule biogenesis, trafficking and exocytosis. Vitam Horm 80:473–506

Huang L, Kirschke CP, Lay YA, Levy LB, Lamirande DE, Zhang PH (2012) Znt7-null mice are more susceptible to diet-induced glucose intolerance and insulin resistance. J Biol Chem 287:33883–33896

Huber AM, Gershoff SN (1973) Effect of zinc deficiency in rats on insulin release from the pancreas. J Nutr 103:1739–1744

Hutton JC (1989) The insulin secretory granule. Diabetologia 32:271–281

Jenkitkasemwong S, Wang CY, Coffey R, Zhang W, Chan A, Biel T, Kim JS, Hojyo S, Fukada T, Knutson MD (2015) SLC39A14 is required for the development of hepatocellular Iron overload in murine models of hereditary hemochromatosis. Cell Metab 22:138–150

Kajimura S, Spiegelman BM, Seale P (2015) Brown and Beige fat: physiological roles beyond heat generation. Cell Metab 22:546–559

Kambe T (2012) Molecular architecture and function of ZnT transporters. Curr Top Membr 69:199–220

Kang S, Brange J, Burch A, Volund A, Owens DR (1991) Subcutaneous insulin absorption explained by insulin's physicochemical properties. Evidence from absorption studies of soluble human insulin and insulin analogues in humans. Diabetes Care 14:942–948

Lefebvre B, Vandewalle B, Balavoine AS, Queniat G, Moerman E, Vantyghem MC, Le Bacquer O, Gmyr V, Pawlowski V, Kerr-Conte J et al (2012) Regulation and functional effects of ZNT8 in human pancreatic islets. J Endocrinol 214:225–232

Leitner BP, Huang S, Brychta RJ, Duckworth CJ, Baskin AS, McGehee S, Tal I, Dieckmann W, Gupta G, Kolodny GM et al (2017) Mapping of human brown adipose tissue in lean and obese young men. Proc Natl Acad Sci USA 114:8649–8654

Lemaire K, Ravier MA, Schraenen A, Creemers JW, Van de Plas R, Granvik M, Van Lommel L, Waelkens E, Chimienti F, Rutter GA et al (2009) Insulin crystallization depends on zinc transporter ZnT8 expression, but is not required for normal glucose homeostasis in mice. Proc Natl Acad Sci USA 106:14872–14877

Marklund SL, Westman NG, Lundgren E, Roos G (1982) Copper- and zinc-containing superoxide dismutase, manganese-containing superoxide dismutase, catalase, and glutathione peroxidase in normal and neoplastic human cell lines and normal human tissues. Cancer Res 42:1955–1961

Merriman C, Huang Q, Rutter GA, Fu D (2016) Lipid-tuned zinc transport activity of human ZnT8 protein correlates with risk for type-2 diabetes. J Biol Chem 291:26950–26957

Mitchell RK, Hu M, Chabosseau PL, Cane MC, Meur G, Bellomo EA, Carzaniga R, Collinson LM, Li WH, Hodson DJ et al (2016) Molecular genetic regulation of Slc30a8/ZnT8 reveals a positive association with glucose tolerance. Mol Endocrinol 30:77–91

Mysore TB, Shinkel TA, Collins J, Salvaris EJ, Fisicaro N, Murray-Segal LJ, Johnson LE, Lepore DA, Walters SN, Stokes R et al (2005) Overexpression of glutathione peroxidase with two isoforms of superoxide dismutase protects mouse islets from oxidative injury and improves islet graft function. Diabetes 54:2109–2116

Nicolson TJ, Bellomo EA, Wijesekara N, Loder MK, Baldwin JM, Gyulkhandanyan AV, Koshkin V, Tarasov AI, Carzaniga R, Kronenberger K et al (2009) Insulin storage and glucose homeostasis in mice null for the granule zinc transporter ZnT8 and studies of the type 2 diabetes-associated variants. Diabetes 58:2070–2083

O'Halloran TV, Kebede M, Philips SJ, Attie AD (2013) Zinc, insulin, and the liver: a menage a trois. J Clin Invest 123:4136–4139

Ohana E, Hoch E, Keasar C, Kambe T, Yifrach O, Hershfinkel M, Sekler I (2009) Identification of the Zn2+ binding site and mode of operation of a mammalian Zn2+ transporter. J Biol Chem 284:17677–17686

Pound LD, Sarkar SA, Benninger RK, Wang Y, Suwanichkul A, Shadoan MK, Printz RL, Oeser JK, Lee CE, Piston DW et al (2009) Deletion of the mouse Slc30a8 gene encoding zinc transporter-8 results in impaired insulin secretion. Biochem J 421:371–376

Pound LD, Sarkar SA, Cauchi S, Wang Y, Oeser JK, Lee CE, Froguel P, Hutton JC, O'Brien RM (2011) Characterization of the human SLC30A8 promoter and intronic enhancer. J Mol Endocrinol 47:251–259

Pound LD, Sarkar SA, Ustione A, Dadi PK, Shadoan MK, Lee CE, Walters JA, Shiota M, McGuinness OP, Jacobson DA et al (2012) The physiological effects of deleting the mouse SLC30A8 gene encoding zinc transporter-8 are influenced by gender and genetic background. PLoS One 7:e40972

Scott DA (1934) Crystalline insulin. Biochem J 28:1592–1602 1591

Scott DA, Fisher AM (1938) The insulin and the zinc content of normal and diabetic pancreas. J Clin Invest 17:725–728

Scott LJ, Mohlke KL, Bonnycastle LL, Willer CJ, Li Y, Duren WL, Erdos MR, Stringham HM, Chines PS, Jackson AU et al (2007) A genome-wide association study of type 2 diabetes in Finns detects multiple susceptibility variants. Science 316:1341–1345

Simon SF, Taylor CG (2001) Dietary zinc supplementation attenuates hyperglycemia in db/db mice. Exp Biol Med (Maywood) 226:43–51

Sladek R, Rocheleau G, Rung J, Dina C, Shen L, Serre D, Boutin P, Vincent D, Belisle A, Hadjadj S et al (2007) A genome-wide association study identifies novel risk loci for type 2 diabetes. Nature 445:881–885

Steiner DF, Rouille Y, Gong Q, Martin S, Carroll R, Chan SJ (1996) The role of prohormone convertases in insulin biosynthesis: evidence for inherited defects in their action in man and experimental animals. Diabetes Metab 22:94–104

Takagishi T, Hara T, Fukada T (2017) Recent advances in the role of SLC39A/ZIP zinc transporters in vivo. Int J Mol Sci:18

Tamaki M, Fujitani Y, Uchida T, Hirose T, Kawamori R, Watada H (2009) Downregulation of ZnT8 expression in pancreatic beta-cells of diabetic mice. Islets 1:124–128

Tamaki M, Fujitani Y, Hara A, Uchida T, Tamura Y, Takeno K, Kawaguchi M, Watanabe T, Ogihara T, Fukunaka A et al (2013) The diabetes-susceptible gene SLC30A8/ZnT8 regulates hepatic insulin clearance. J Clin Invest 123:4513–4524

Troche C, Aydemir TB, Cousins RJ (2016) Zinc transporter Slc39a14 regulates inflammatory signaling associated with hypertrophic adiposity. Am J Physiol Endocrinol Metab 310:E258–E268

Wenzlau JM, Juhl K, Yu L, Moua O, Sarkar SA, Gottlieb P, Rewers M, Eisenbarth GS, Jensen J, Davidson HW et al (2007) The cation efflux transporter ZnT8 (Slc30A8) is a major autoantigen in human type 1 diabetes. Proc Natl Acad Sci USA 104:17040–17045

Wijesekara N, Dai FF, Hardy AB, Giglou PR, Bhattacharjee A, Koshkin V, Chimienti F, Gaisano HY, Rutter GA, Wheeler MB (2010) Beta cell-specific Znt8 deletion in mice causes marked defects in insulin processing, crystallisation and secretion. Diabetologia 53:1656–1668

Zoete V, Meuwly M, Karplus M (2005) Study of the insulin dimerization: binding free energy calculations and per-residue free energy decomposition. Proteins 61:79–93

Chapter 13
Zinc Signals in Immunology

Martina Maywald and Lothar Rink

Abstract Zinc is known to be an essential trace element that is highly important for all proliferating cells in the human body, especially for the immune system. Free zinc influences several signaling pathways, such as Toll-like receptor 4 or T cell receptor signaling, by binding reversibly to regulatory sites in signaling proteins, resulting in a change of free zinc concentrations that can affect signal transduction, and thus cellular responses can be altered. Zinc signals have been observed in cells of the innate as well as of the adaptive immune system, i.e., neutrophil granulocytes, mast cells, monocytes, dendritic cells, and T cells, mostly in changes of the cytoplasmic zinc concentration. To characterize zinc signals, one can distinguish them by the timescale they take place. First, zinc signals can occur within a few seconds to minutes and are, therefore, called fast zinc signals. Second, a slightly slower type of zinc signal is known and described as "zinc wave." Third, some zinc signals occur on a timescale significantly longer than the others. In these cases, the signals are typically involved in altered expression of proteins involved in zinc homeostasis.

Zinc signals occurring in different cell types and signaling pathways that are mentioned in this chapter are classified regarding the specific discrimination of fast zinc signal, zinc wave, and late zinc signal.

Keywords Zinc · Zinc homeostasis · Signal transduction · TLR-4 · Adaptive and innate immunity · Zinc deficiency

---

M. Maywald · L. Rink (✉)
Institute of Immunology, Aachen, Germany
e-mail: lrink@ukaachen.de

T. Fukada, T. Kambe (eds.), *Zinc Signaling*,
https://doi.org/10.1007/978-981-15-0557-7_13

13.1 Zinc Signals in Immunology

In 1934, the essentiality of zinc for animals has been recognized, but it was not until the early 1960s that zinc was known as essential for humans. Compared to other metal ions with similar chemical properties, zinc is relatively harmless (Fosmire 1990). Studies have shown that only exposure to high doses has toxic effects, making acute zinc intoxication a rare event. Due to the systemic homeostasis and efficient regulatory mechanisms on the cellular level, the uptake of cytotoxic doses of exogenous zinc is prevented. Nevertheless, intracellular zinc plays an important role in single cells and organ systems, since various functions and signaling pathways in virtually all immune cells are modulated by zinc in vitro and in vivo (Prasad 2009). Regarding this, it is quite obvious that there is strong evidence between zinc deficiency and several infectious diseases such as shigellosis, acute cutaneous leishmaniosis, malaria, human immunodeficiency virus (HIV), tuberculosis, measles, and pneumonia (Gammoh and Rink 2017).

In this chapter, we discuss the importance of zinc signals in the immune system and the interactions of zinc with major signaling pathways responsible for the regulation of cellular functions, as well as the influence of zinc deficiency and zinc supplementation on zinc signaling in various immune cells.

13.2 Zinc Homeostasis and the Immune System

Zinc is considered to be relatively nontoxic to humans, and the human body contains a total amount of 2–4 g zinc (Haase and Rink 2010). In general, there is no specialized zinc storage system in the human body, and thus a daily intake of zinc is necessary to achieve a steady state for proper immune function. The distribution of zinc in the human body differs significantly between various tissues and organs, ranging from 10 μg/g dry weight in the brain (1.5% of the total body zinc) up to 100 μg/g in bones (29.0% of the total body zinc). Other organs contain estimable concentrations of zinc including liver, prostate, kidney, skin, lung, heart, pancreas, and the gastrointestinal tract. Oral uptake of zinc results in absorption throughout the small intestine and distribution subsequently occurs via serum, where zinc concentrations merely reach 10–18 μM, comparable to around 1 μg/g (0.1% of the total body zinc) (Mills 1989). It is abundantly bound with low affinity to albumin (60%), with high affinity to α_2-macroglobulin (30%), and to transferrin (10%) (Rink and Gabriel 2000). Thus, it is a predominantly intracellular ion (Haase and Maret 2009), distributed between the cell nucleus (30–40%), cytoplasm, organelles, and vesicles (50%) (Vallee and Falchuk 1993) including zinc-specific vesicles called "zincosomes" (Beyersmann and Haase 2001). The latter segregate high amounts of zinc upon stimulation, as shown for growth factors (Haase and Maret 2003). Besides storage in specific organelles, the so-called buffering and muffling reactions are also highly important for intracellular zinc homeostasis. Efficient zinc buffering inside the cell is indispensable to avoid zinc intoxication and zinc deficiency regulated by coordinating expression of zinc binding proteins and zinc transporters (Kimura and Kambe 2016). Intracellular zinc-binding proteins include members of the metallothionein (MT) family and members of the S100 family (Brophy et al. 2012). The latter is still under debate; however, binding of intracellular zinc by calprotectin, a heterodimer of S100A8 and A9, in, e.g., myeloid cells is likely, as the decrease in intracellular zinc during monopoiesis is paralleled by an increase in S100A8 and A9 expression, but no changes were found for MT-1 levels (Dubben et al. 2010).

MTs are proteins of 6–7 kDa that play an important role in zinc homeostasis by complexing about 20% of intracellular zinc. MTs have high cysteine content and thus are able to complex metal ions decreasing their cytotoxicity, and they can scavenge reactive oxygen species (ROS). Within the four MT classes, MT-1 and MT-2 are expressed ubiquitous throughout the body, whereas MT-3 and MT-4 are expressed cell type specific (Kimura and Kambe 2016).

Although all zinc ions are bound in similar tetrathiolate coordination environments, the affinities of the sites for zinc range with a logK from 7.7 to 11.8. Thus, MT can act as a cellular zinc buffer over several orders of magnitude that range from picomolar to nanomolar concentrations due to different affinities of the metal ion binding sites (Krezel and Maret 2007). The dynamic regulation of cellular zinc homeostasis results from the synthesis of the apo-form thionein (T) in response to raised intracellular zinc levels by triggering the metal-response element binding transcription factor-1 (MTF-1). MTF-1 is a cellular zinc ion sensor and regulates transcription of genes, whose promoters contain metal-response elements, because at least two of its six zinc fingers are not constitutively bound to zinc and sense available free zinc by reversible binding, resulting in stabilization of the zinc-finger domains (Radtke et al. 1993). Furthermore, oxidation of cysteine residues alters the number of metal binding sites and thus connecting redox and zinc metabolism (Laity and Andrews 2007).

The S100 protein family includes 24 members, knowing to modulate apoptosis, transcription, and enzyme activities (Donato et al. 2013). Zinc binding S100 proteins include S100B, S100A1, S100A2, S100A3, S100A5, S100A7, S100A8/9, S100A12, and S100A16. Calprotectin (S100A8/9) is the most abundant protein of neutrophils. Interestingly, recent studies showed a correlation between increased plasma levels during severe inflammatory diseases and serum hypozincemia (Bao et al. 2010; Lienau et al. 2018).

Despite the fact that the plasma zinc pool is very small, it is highly mobile and immunologically important. Like serum zinc, the vast majority of cellular zinc is bound to proteins leaving only a minor loosely bound or unbound part of intracellular zinc, which is also referred to as labile, mobile, or "free zinc." "Free zinc" is an operative term, used to distinguish zinc involved in signal transduction from the tightly protein-bound zinc which is thermodynamically unavailable. The term is used in this chapter with the understanding that it is chemically incorrect, because zinc ions will never be completely without ligands but form complexes with amino acids, glutathione, phosphate, or other low molecular weight ligands, when it is not bound to proteins.

So far, more than 300 enzymes and even more proteins are identified to contain zinc. Based on a screening for well-known zinc binding sequences, it is estimated that up to 10% of the encoded proteins in the human body may contain zinc (Andreini et al. 2006). The zinc proteome consists of two major groups of comparable size, which are enzymes and transcription factors. Together, these form over 90% of the zinc proteome, indicating that the vast majority of zinc is required for catalysis and transcriptional regulation (Andreini et al. 2009).

Tightly protein-bound zinc is required for catalytic, co-catalytic, and structural functions of enzymes (Vallee and Falchuk 1993). In some enzymes it is important for structural integrity, in others it is the central ion for enzymatic activity, but sometimes both functions are involved. Moreover, zinc is involved in the stabilization of structural domains, such as in zinc fingers and related structures, and enables protein-protein interaction or interaction of nucleic acid and many transcription factors (Maret 2006).

But first of all the understanding of how the mechanism of zinc uptake into the cell takes place is important. It has been shown that exogenous zinc enters the cell within minutes (Wellinghausen et al. 1996). Cellular zinc underlies an efficient homeostatic control to avoid excessive accumulation. Therefore, two families of eukaryotic zinc transporter are known to date: the Zip and ZnT proteins. The Zrt-like, Irt-like protein (Zip) family comprise 14 genes designate solute carrier family 39 (SLC39) A1 to A14 that transport zinc into the cytosol, whereas the zinc transporter (ZnT) family of 10 genes (SLC30A1 to SLC30A10) transport zinc in the opposite direction (Cousins et al. 2006; Eide 2006). The same transporter families regulate the intracellular distribution of zinc into cellular organelles as the endoplasmic reticulum, mitochondria, and Golgi, respectively (see Chap. 3). In addition to that, different mechanisms for zinc uptake such as facilitated diffusion via amino acids and anionic exchange or calcium-conducting channels have been reported (Bentley 1992; Hogstrand et al. 1996). To date, additionally members of four distinct superfamilies have been shown to transport zinc through the cell membrane, as voltage-gated calcium-channels (VGCC) (Gyulkhandanyan et al. 2006), glutamatergic receptors (Jia et al. 2002), nicotinic acetylcholine receptors (Ragozzino et al. 2000), and transient receptor potential channels (Dong et al. 2008).

However, free zinc contributes to the regulation of numerous cellular processes and influences several signaling pathways by binding reversibly to regulatory sites in signaling proteins. This results in a change of the free zinc concentration that can affect signaling pathways, and thus cellular responses can be altered (Ibs and Rink 2003). Furthermore, it shows intracellular fluctuation after stimulation (Haase et al. 2006a), assuming a complex interaction between zinc homeostasis and signal transduction, where zinc may have similar functions as the second messenger calcium (Williams 1984; Colvin et al. 2010).

Free zinc is crucial for the appropriate development and function of the innate as well as the adaptive immunity, but the importance of zinc for proper immune function is most obvious in zinc-deficient individuals.

Whereas intoxication by excessive exposure is rare, zinc deficiency is widespread and has been known for a detrimental impact on growth, neuronal development, and immunity, and in severe cases its consequences are lethal. Human zinc deficiency has been reported first time in 1961, when Iranian males were diagnosed with symptoms including growth retardation, skin abnormalities, hypogonadism, and mental lethargy led back to nutritional zinc deficiency (Prasad et al. 1961). Later additional studies showed remarkably similar features and manifested zinc deficiency as a potentially widespread problem in developing countries as well as in industrial nations (Prasad et al. 1963; Sandstead 1991).

Zinc deficiency manifests itself on different levels and can be subdivided into either severe or marginal zinc deficiency. Moreover, zinc deficiency can be either inherited or acquired. On the one hand, severe zinc deficiency is caused by metabolic disorders regarding zinc uptake as well as by other sources such as parenteral nutrition without zinc.

Zinc deficiency results in complex immune defects, leading to various immune disorders (see Chap. 11). This is observed in its most severe form in the zinc malabsorption syndrome acrodermatitis enteropathica, an autosomal recessive disorder characterized by periorificial and acral dermatitis, mucocutaneous lesions, failure to thrive, alopecia, diarrhea, neuropsychological disturbances, weight loss, and frequent severe infections with fungi, viruses, and bacteria that base in most cases on a mutation of the intestinal zinc uptake protein Zip 4 (Kury et al. 2002; Wang et al. 2002).

Acquired severe zinc deficiency has been observed in patients receiving total parental nutrition without zinc supplementation, following excessive alcohol consumption, severe malabsorption, and iatrogenic causes like treatment with histidine or penicillamine (Prasad 1985). The arising symptoms are mostly similar to those in acrodermatitis enteropathica. Patients who become zinc deficient while receiving total parenteral nutrition without zinc supplementation develop abnormalities, which include lymphopenia, decreased ratios of $CD4^+$-to-$CD8^+$ T cells, decreased NK cell activity (Rolles et al. 2017a), and increased monocyte cytotoxicity, but are readily corrected by proper zinc supplementation (Allen et al. 1983).

Another group of inherited disorders of zinc metabolism is mentioned in some reports. Here the zinc plasma level lies above 3 μg/mL, more than three times the physiological level, while iron and copper levels stay normal (Failla et al. 1982; Fessatou et al. 2005; Smith et al. 1976). Even though this exceeds the amount found in serum after zinc intoxication, symptoms range from none to severe anemia, growth failure, to systemic inflammation (Fessatou et al. 2005; Smith et al. 1976; Saito et al. 2002; Sampson et al. 1997). The increased zinc level has been attributed to excessive binding to serum proteins like albumin (Failla et al. 1982; Smith et al. 1976) or to an overexpression of calprotectin, a zinc binding S100 protein (Fessatou et al. 2005; Saito et al. 2002). Therefore, the large amount of zinc in the serum of these patients is attached to proteins, and thus biologically available zinc may be depleted (Sampson et al. 1997). Moderate zinc deficiency can also accrue as a consequence of sickle cell disease (Prasad 1981), where hyperzincuria and a high protein turnover due to increased hemolysis lead to moderate zinc deficiency in these patients. This causes clinical manifestations typical for zinc deficiency, such as growth retardation, hypogonadism in males, hyperammonemia, and cell-mediated immune disorder (Prasad 2002) connected with thymic atrophy (Dardenne et al. 1984).

Affected subjects show reduced lymphocyte proliferation response to mitogens, anergy, thymic atrophy, a selective decrease in $CD4^+$ T cells, and deficient thymic hormone activity (Prasad 2000). In general, zinc deficiency leads to an increased susceptibility to infection and parasitic disease and is furthermore the fifth leading cause of mortality and morbidity in developing countries (WHO 2002).

Marginal zinc deficiency can be characterized by slight weight loss, oligospermia, and hyperammonemia (Prasad 1985). It may be caused by nutritional zinc deficiency due to high consumption levels of zinc-chelating phosphates, lignins, and phytates, which counteract zinc absorption by malabsorption syndrome or by sickle cell anemia.

Moreover, parallels exists between the declined immune function in elderly and zinc deficiency even in industrialized countries. A significant percentage show a reduced serum zinc level, and zinc supplementation studies indicate that this deficiency contributes to increased susceptibility to infectious diseases. It also occurs frequently during pregnancy and lactation, due to rapid growth, and in vegetarians or persons with renal insufficiency (Brieger and Rink 2010; Haase and Rink 2010).

The overall frequency of zinc deficiency worldwide is expected to be higher than 20% (Wuehler et al. 2005). In developing countries, it may affect more than two billion people (Prasad et al. 1961; Cavdar et al. 1983; Prasad et al. 1963). Furthermore, it has been estimated that only 42% of the elderly (≥71 years) in the United States have adequate zinc intake (Briefel et al. 2000). This widespread occurrence combined with the variety of clinical manifestations makes zinc deficiency a serious nutritional problem.

Until now, no biomarker of zinc status exists, although serum and plasma zinc concentrations, hair zinc concentration, and urinary zinc excretion can be considered as potentially useful. However, zinc status is highly impacted by the immune status itself, diet, absorption, and conserving mechanisms via gastrointestinal tract and kidneys (Lowe et al. 2013), which makes the diagnosis of suboptimal zinc status very complicated. In this regard, a recent study published a biochemically validated food frequency questionnaire to estimate zinc status in humans (Trame et al. 2018) that might find application in routine use in future times.

Detractions in the zinc homeostasis affect multiple aspects of the immune system including hematopoiesis, cell cycle progression, and immune-regulating molecules. Intracellular killing, cytokine production, and ROS synthesis are all affected by zinc deficiency. It also affects adversely the growth and function of T and B cells, which occurs through dysregulation of basic biological functions at the cellular level. Moreover, zinc is needed for DNA synthesis, RNA transcription, cell division, and cell activation. Programmed cell death (apoptosis) is also elevated in the absence of adequate zinc levels. Also secretion and function of cytokines, the basic messengers of the immune system, are adversely affected by zinc deficiency. The ability of zinc to function as a pro-antioxidant and to stabilize membranes suggests that it has a role in prevention of free radical induced injury during inflammatory processes.

All in all zinc signals contribute to the regulation of virtually all immune cells (see Chap. 11), numerous cellular processes, and influence several signaling pathways by binding reversibly to regulatory sites in signaling proteins. Thus, zinc is crucial for the appropriate development and function of the innate as well as the adaptive immunity that is further described below.

13.3 Types of Zinc Signals

Free zinc influences several signaling pathways by binding reversibly to regulatory sites in signaling proteins, resulting in a change of free zinc concentrations that can affect signaling pathways, and thus cellular responses can be altered (Haase and Rink 2009a).

It has been shown that activation of T cells causes an intracellular redistribution of zinc and an activation of protein kinase C (PKC) (Csermely et al. 1987; Csermely et al. 1988). Since this time, zinc signals have been observed in different immune cells, i.e., T cells (Kaltenberg et al. 2010; Yu et al. 2011; Lee et al. 2008; Aydemir et al. 2009), dendritic cells (DCs) (Kitamura et al. 2006b), monocytes (Haase et al. 2008a), neutrophil granulocytes (Hasan et al. 2013), mast cells (Yamasaki et al. 2007), and NK cells (Rolles et al. 2018) mostly in changes of the cytoplasmic zinc concentration (Haase and Rink 2009a).

One possibility to characterize zinc signals is by the timescale they take place. First, zinc signals can occur within a few seconds to minutes by triggering receptors like Toll-like receptor (TLR)-4. Zinc acts as second messenger, comparable to calcium, and influences several signaling pathways directly. Therefore, the zinc signal is independent from the synthesis of proteins like Zips and ZnTs and is thus called fast zinc signal.

Second, a slightly slower type of zinc signal is known and described as "zinc wave" investigated in mast cells. Here cross-linking of FcεRI induces a release of free zinc from the perinuclear area, including the endoplasmic reticulum (Yamasaki et al. 2007, 2012). Comparable to the fast zinc signal, zinc also acts as second messenger but is induced indirectly depending on calcium influx (see Chap. 5). However, zinc signals and zinc wave also have a function as second messenger; the altered cytoplasmic zinc concentration influences other signaling pathways also, comparable to the role of late zinc signals.

Third, some zinc signals occur on a timescale significantly longer than the others. In these cases, the late signals are typically involved in altered expression of proteins involved in zinc homeostasis, like maturation of monocytes and DCs or cytokine expression (Dubben et al. 2010; Kitamura et al. 2006b). Furthermore the altered zinc homeostasis results in an influence of several signaling pathways, but zinc itself does not function as a second messenger.

Examples for each classification of zinc signals occurring in different cell types and signaling pathways are mentioned in the following section below and are summarized in Table 13.1.

Table 13.1 Characterization of zinc signals concerning the timescale they take place

Zinc signal	Duration	Effect
Fast zinc signal	Sec-min	Influence of cGMP by inhibition of PDE in monocytes/macrophages
		MKP inhibition in monocytes/macrophages
		PMA-induced NET-formation in PMN
		Expression of IL-8 in epithelial cells
		Lck recruitment to TCR-activating complex
		Zinc release from lysosomes in T cells
		Triggered T cell activation by APC
		Homodimerization and activation of Lck in T cells
		Redistribution of zinc from nucleus/mitochondria to cytosol/microsomes in T cells
Zinc wave	Min	Zinc release from perinuclear area in mast cells
Altered zinc homeostasis (late zinc signal)	Hours	Adherence of monocytes to endothelial cells
		Influence of mRNA transcription, e.g., IL-2, A20 in T cells, monocytes/macrophages
		Alteration of M1/M2 differentiation by inhibition of STAT6 phosphorylation
		MAPK activation in airway epithelial cells and monocytes/macrophages
		Negative regulation of TRIF pathway in macrophages
		Change of Zip/ZnT expression in DCs and T cells
		Inhibition of adenylate cyclase transcription in T cells
		Influence of cytokine production, e.g., IL-2 in T cells
		Upregulation of Akt phosphorylation in T cells
		PTEN inhibition in T cells
		STAT3/STAT6 phosphorylation in B and T cells
		Induction/stabilization of regulatory T cells
		Inhibition pro-inflammatory Th9 cell differentiation
		Reduced cytokine production, e.g., IFN-γ in T cells
		Inhibition of PTEN in T cells
		Induction of NK cell killing and granzyme expression
		Inhibition of caspase activity
		Epigenetic modifications due to inhibition of Sirt1

Zinc signals, occurring within a few seconds to minutes, are called fast zinc signals and are independent from the synthesis of proteins like Zips and ZnTs. A slightly slower type of zinc signal is known and described as a "zinc wave" occurring after a few minutes. Zinc signals that take place on a timescale significantly longer (hours) are typically involved in altered expression of proteins involved in zinc homeostasis

13.4 Zinc Signal and the Innate Immunity

The innate immunity as the first line of defense is a highly important natural protection system against harmful substances, consisting of different cell types like monocytes/macrophages, granulocytes, DCs, mast cells, and natural killer (NK) cells, as well as countless soluble proteins like complement proteins, acute phase proteins, cytokines, and chemokines. It is not highly specific like the adaptive immunity but reacts on different antigens in the same way. Its cells are activated by conserved structures in pathogens, called pathogen-associated molecular patterns (PAMPs), that are recognized by conserved receptors triggering processes like cytokine production, killing of target cells, or antigen presentation to cells of the adaptive immune system. Compared to the number of processes regulated on the cellular level, only limited numbers of signaling pathways exist. These pathways, depending on the cell type, are induced by different receptors and lead to the expression of cell type specific genes. Immune cells as well as a myriad of other cell types share these identical pathways that are discussed below.

In case of an infection, polymorphonuclear leukocytes (PMN) are the first cells actively entering the infected tissue by following a concentration gradient of chemical messengers, a mechanism known as chemotaxis. It is reported that this and other cellular functions are disturbed by altered zinc levels. Zinc deficiency leads to reduced PMN chemotaxis (Ibs and Rink 2003; Shankar and Prasad 1998), whereas elevated zinc concentrations of about 500 μM are reported to induce chemotactic activity directly in vitro (Hujanen et al. 1995). PMN are phagocytes that are essential to mortify pathogens by phagocytosis and generation of the respiratory burst. Zinc deficiency leads to a decreased phagocytosis (Hasan et al. 2016) that can be restored by zinc supplementation (Sheikh et al. 2010). These effects depend on an altered zinc homeostasis.

PMN kill phagocytosed pathogens through the rapid production and release of ROS, including superoxide anion, hydrogen peroxide, and hypochlorous acid. Superoxide anions are produced by NADPH oxidase, which can be inhibited by zinc deficiency as well as by zinc excess (DeCoursey et al. 2003; Chasapis et al. 2012; Hasegawa et al. 2000). Another neutrophil granulocyte function is influenced due to altered zinc concentrations, namely, NETosis. NETosis is described as the release of a matrix composed of DNA, chromatin, and granule proteins to capture extracellular bacteria within so-called neutrophil extracellular traps (NETs). NETosis can be induced by PKC activator 12-myristate 13-acetate (PMA) and depends on the production of ROS by NADPH oxidase. PMA treatment leads to a fast zinc signal in neutrophil granulocytes that is an essential component of the ROS-dependent signal transduction leading to NET formation. Recently, it was shown that zinc chelation abrogates NET formation in vitro (Hasan et al. 2013). Moreover, chemotaxis, phagocytosis, oxidative burst, degranulation, and cytokine production are impaired (Hasan et al. 2016).

Like PMN, macrophages also take up pathogens or cellular debris by phagocytosis and kill pathogens by oxidative burst, which are both zinc-dependent. Monocytes produce various cytokines, e.g., pro-inflammatory cytokines such as IL-1β, IL-6, and TNF-α. Therefore, zinc signals are essential and can also be directly induced by stimulation with high extracellular zinc concentrations (Wellinghausen and Rink 1998; Wellinghausen et al. 1997a). Monocytes develop from myeloid precursor cells and can move quickly to sites of infection in the tissues and differentiate into macrophages and DCs to elicit an immune response. Zinc signals have been shown recently to alter M1/M2 differentiation by inhibition of STAT6 phosphorylation (Dierichs et al. 2017).

In general, cellular maturation is described to be dependent on long-term reduction of the intracellular zinc (Kitamura et al. 2006a). The differentiation of monocytes is calcitriol-dependent and can be enhanced by zinc sequestration. This indicates a negative regulatory role for this ion (Dubben et al. 2010), depending on an altered zinc homeostasis. Additionally, recently similar observations have been found for granulopoiesis as well (Tillmann et al. in preparation).

After their development, monocytes circulate in the blood stream, from where they start to migrate into the tissue, where the maturation into resident macrophages takes place. The migration from the blood into the tissue starts by adherence to endothelial cells, which can be augmented by zinc, and stimulation of monocytes with chemoattractant protein-1 (MCP-1) or PMA induces an intracellular zinc signal promoting adhesion to endothelial cells (Chavakis et al. 1999; Kojima et al. 2007). These effects depend on an altered zinc homeostasis.

Monocytes/macrophages and DCs take up and present antigens to T cells. They coordinate the immune response by producing cytokines regarding to a sufficient stimulation that leads to an activation of both cell types, like lipopolysaccharide (LPS), which is sensed by TLR-4. Thus, TLR-4 signaling leads to secretion of pro-inflammatory cytokines and maturation and antigen presentation in DCs and macrophages.

In TLR-4 signaling zinc has a double-edged effect on the cytokine secretion in monocytes (see Fig. 13.1). Herein fast zinc signals that are moderately increased are involved in TLR-4 signal transduction, whereas longtime higher concentrations are inhibitory. Additionally, long-term zinc deficiency augments pro-inflammatory cytokine production, which is reversible if zinc is reconstituted (Wessels et al. 2013). A fast zinc signal has been observed within less than 2 min by binding of bacterial LPS to TLR-4 that is involved in the activation of mitogen-activated protein kinase (MAPK) signaling. MAPK activation via zinc can be observed in lots of cell types. The MAPK extracellular regulated kinase (ERK) is shown to be activated by zinc treatment of fibroblasts (Hansson 1996), as well as of neurons or neuroblastoma cells (An et al. 2005; Park and Koh 1999). In addition, this effect is also described in the activation of further MAPKs like p38 and c- Jun N-terminal kinase (JNK) (An et al. 2005; LaRochelle et al. 2001). Similar observations have been made with other TLR-ligands, such as Pam3CSK4, insulin, and TNF-α.

Even though no direct effect of zinc on MAPK activity is reported, a fast zinc signal is reported to lead to an inhibition of MAP-kinase phosphatases (MKP) that dephosphorylate both tyrosine and threonine residues in activated MAPKs. Thus,

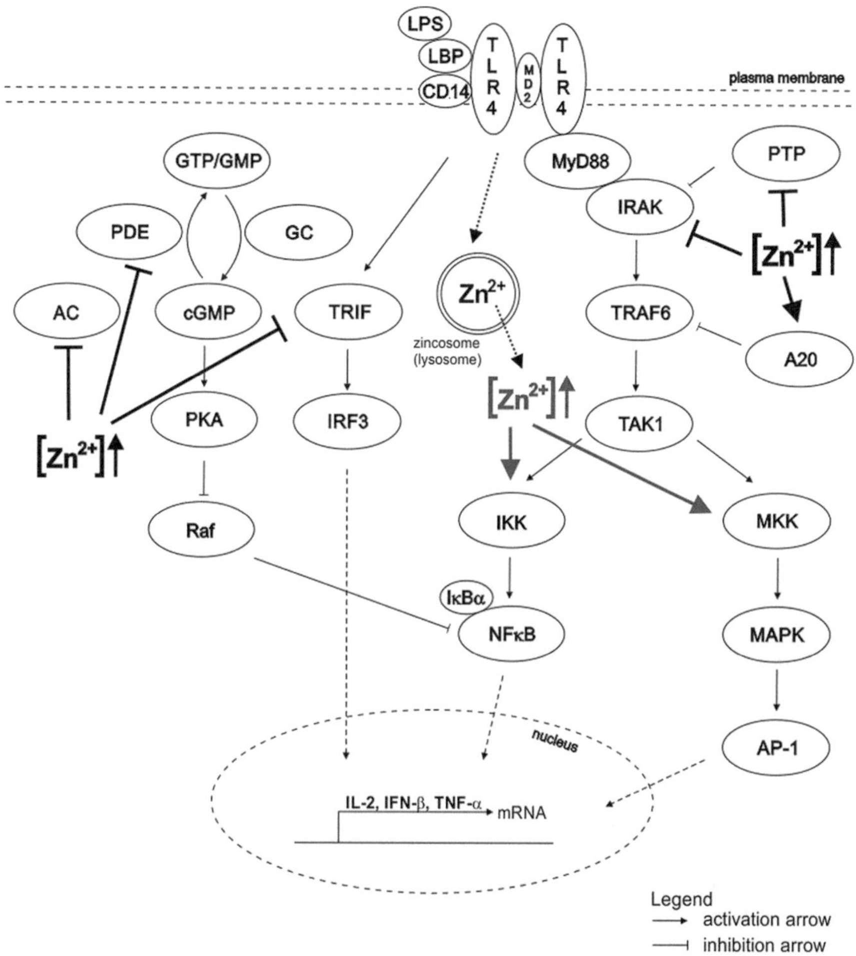

Fig. 13.1 Zinc and Toll-like receptor 4 (TLR-4) signaling. Triggering of TLR-4 induces fast zinc release (dotted arrows) from zincosomes. These physiological fast zinc signals (highlighted in green) are required for activation of nuclear factor kappa B (NF-κB) and mitogen-activated protein kinase (MAPK) signaling (activation arrow) leading to translocation (dashed arrows) into the nucleus. Longtime higher concentrations of zinc, due to the use of ionophores in experimental procedures (highlighted in red), inhibit TLR-4 signaling (inhibition arrow). Suggested mechanisms are a direct inhibition of interleukin-1 receptor-associated kinase (IRAK), an upregulation of A20 that removes the activating ubiquitination of tumor necrosis factor receptor-associated factor 6 (TRAF6), an inhibition of cyclic nucleotide phosphodiesterases (PDE) or an inhibition of TIR-domain-containing adapter-inducing interferon-β (TRIF). The inhibition of PDE leads to a rise in cyclic guanosine monophosphate (cGMP) and cross-activation of protein kinase A (PKA), resulting in Raf inhibition and thereby TLR-mediated NF-κB activation. Furthermore increased intracellular zinc concentrations inhibit adenylate cyclase (AC). *IKK* I kappa B kinase, *IRF3* Interferon regulatory factor 3, *MKK* MAPK kinase, *PKA* protein kinase A, *PTP* protein tyrosine phosphatases *AP-1* activator protein 1, *GMP* guanosine monophosphate, *GTP* guanosine trisphosphate

indirectly zinc causes the inhibition of ERK1/2 and dephosphorylation of p38 in human macrophages (Haase et al. 2008a). The zinc-dependent inhibition of MKP is involved in ERK activation during oxidative stress in neurons (Ho et al. 2008) and is one mechanism by which zinc stimulates expression of IL-8 in epithelial cells (Kitamura et al. 2006b). However, not all of the zinc effects on MAPK phosphorylation and activation can be explained simply by zinc inhibition of MKP because zinc also affects targets upstream of the MAPKs. For instance, MAPK activation in airway epithelial cells also depends on a longtime zinc effect on tyrosine phosphorylation upstream of ERK (Wu et al. 2002).

However, the effect of zinc on MAPKs is not always agonistic. Low concentrations of zinc are shown to activate ERK in rat glioma cells, whereas higher concentrations reduced the activating phosphorylation (Haase and Maret 2005). One potent molecular mechanism could be that the zinc transporter (ZnT)-1 homolog CDF (cation diffusion facilitator)-1 mediates zinc efflux, leading to activation of Ras/Raf/MEK/ERK pathway (Bruinsma et al. 2002). Here zinc may promote the inhibitory phosphorylation of kinase suppressor of Ras (KSR). KSR works as a scaffolding protein that stabilizes the interaction of Ras, Raf, and MEK (Yoder et al. 2004). Another molecular mechanism could be the CDF/ZnT-1 binding to Raf-1, promoting its biological activity which is inhibited by zinc (Jirakulaporn and Muslin 2004). Therefore, an altered zinc homeostasis influences the signaling pathways and the threshold of activation.

Also extended concentrations of zinc inhibit TLR-4 signaling. In this connection the first suggested mechanism is a direct inhibition of the interleukin-1 receptor-associated kinase (IRAK). Second, the late zinc signal leads to upregulation of mRNA of the cytoplasmic zinc finger protein A20 (Prasad et al. 2004) that removes the activating ubiquitination of tumor necrosis factor receptor-associated factor 6 (TRAF6), which could result in the disturbed TLR-4 signaling. In TLR signaling TRAF6 ubiquitination leads to proteasome-independent activation of the transforming growth factor beta-activated kinase 1 (TAK1), which in turn phosphorylates IκB kinases (IKKs) and MAPK kinases (MKKs) (Deng et al. 2000; Wang et al. 2001). A20 is induced by pro-inflammatory stimuli and is involved in the termination of TLR-induced activity of NF-κB and pro-inflammatory gene expression in macrophages (Boone et al. 2004). Moreover, it de-ubiquitinates TRAF6, inactivating this protein. Therefore, upregulation of A20 by zinc leads to diminished NF-κB and MAPK activation in response to TLR stimulation. Third, the inhibition of the cyclic nucleotide phosphodiesterases (PDE) via a fast zinc signal is named, which results in an increase in cyclic guanosine monophosphate (cGMP) synthesized by guanylate cyclases (GCs). This effect is due to cross-activation of protein kinase A (PKA) by elevated cGMP concentrations. In human monocytes, degradation of cyclic nucleotides is mediated by PDEs 1, 3, and 4 that are all inhibited by zinc (von Bulow et al. 2005). This leads to inhibitory phosphorylation of Raf-1 resulting in its inactivation. But because Raf-1 is necessary for TLR-4 mediated NF-κB activation, the zinc/cGMP/PKA/Raf pathway blocks the activation of this transcription factor and thereby transcription of genes for pro-inflammatory cytokines (von Bulow et al. 2007). In addition to PDEs, the adenylate cyclase (AC) is inhibited by the late zinc signal leading to a reduced cAMP level, whereas the synthesis of cGMP by soluble GC is unaffected. So far, two different mechanisms for zinc inhibition of AC have been presented: first, zinc may influence enzymatic activity by altering the conformation of AC (Klein et al. 2004) or, alternatively, by inhibiting the activation of AC via the heterotrimeric G-protein alpha subunit (Gao et al. 2005).

Moreover, during a longer timescale zinc affects the response of DCs to LPS. The treatment of murine DCs with LPS leads to a reduction of free intracellular zinc, based on a TIR-domain-containing adaptor protein inducing IFN-β (TRIF)-dependent change in the expression pattern of several zinc transporters. This involves a reduction of the Zrt-like, Irt-like proteins (Zip) 6 and 10 gene expression and upregulates expression of ZnT1, ZnT4, and ZnT6. The reduction in free zinc is important for maturation of DCs (Kitamura et al. 2006b). But also chelation of zinc influences TLR-4 signaling in macrophages. Stimulation with the zinc chelator TPEN augments TLR4-mediated production of IFN-β and subsequent synthesis of inducible NO synthase and production of NO. The effect is based on zinc acting as a negative regulator of the TRIF pathway via reducing IFN regulatory factor 3 activation (Brieger et al. 2013).

Zinc is involved in the regulation of LPS signaling in cells of the innate immune system, but effects depending on zinc are not limited to the response to LPS. Also a role for free zinc in regulation of PKC activity during monocyte chemoattractant protein (MCP)-1-induced adhesion is known (Kojima et al. 2007). In this context it has to be investigated which other receptors depend on zinc for the cellular signaling. In vivo, zinc deficiency will impair the function of monocytes, in particular their ability to mount an adequate cytokine response during infection, but because zinc is equally involved in negative regulation of the same pathways, a disturbance of zinc homeostasis could also affect the limitation of pro-inflammatory cytokine production, leading to an overproduction. Although this seems contradictory, it is supported by experimental evidence. For instance, in the elderly a reduced spontaneous cytokine production is observed by longtime zinc supplementation, but simultaneously an improved cytokine response to PAMPs (Kahmann et al. 2008).

NK cells also play a major role in immunity against infections and tumor development. They are a subset of lymphocytes that mediate the killing of either infected or transformed cells. Therefore, they use a variety of different receptors. They act mainly by two different mechanisms: the first one bases on the availability of major histocompatibility complex class I (MHC-I) molecules or other cell-surface proteins on target cells screened by NK cells. The absence of those molecules indicates an attempted evasion of T cell immune surveillance. The second mechanism is the antibody-dependent cell-mediated cytotoxicity (ADCC), which is based on the recognition of antibodies that are bound to the target cell surface. In both cases target cells are killed by release of the content of cytotoxic granules, such as Perforin and Granzyme B.

Here zinc is involved in the recognition of MHC-I on target cells, mainly human leukocyte antigen C, by p58 killer cell inhibitory receptors on NK cells for the inhibition of the killing activity (Rajagopalan et al. 1995; Rajagopalan and Long 1998). In general, the NK cell number as well as activity is dependent on the serum zinc level. The lytic activity of NK cells is decreased by zinc deficiency, probably due to decreased stimulation from T cells via the cytokine IL-2. Furthermore, the relative number of precursors of cytolytic cells is decreased during zinc deficiency as well as the lytic activity of NK cells itself (Prasad 2000; Rolles et al. 2018), but moderate zinc supplementation increases differentiation of $CD34^+$ progenitors toward NK cells and their cytotoxic activity (Muzzioli et al. 2007). However, merely the inhibitory signal is zinc-dependent, whereas the human leukocyte antigen C interaction and positive signals do not require zinc. Thus, zinc is needed to maintain the normal function of NK cells, and zinc deficiency may result in functional loss and evoke nonspecific killing.

To date, intracellular zinc signals have not been investigated in NK cells in detail; however, a recent study uncovered changes in the intracellular zinc level after cellular activation (Rolles et al. 2019). Lines of evidence connect major signaling pathways in NK cells to zinc. A multitude of different receptors are required for fine-tuning of NK cell activity (Yokoyama and Plougastel 2003). A common feature of many of these receptors is the presence of characteristic tyrosine phosphorylation sites that either activate (immunoreceptor tyrosine-based activation motif; ITAM) or inhibit (immunoreceptor tyrosine-based inhibitory motif; ITIM) NK cell function. In this respect, one feature of zinc is of particular importance: its ability to inhibit protein tyrosine phosphatases (PTPs) by a fast zinc signal (Brautigan et al. 1981). This inhibition seems to be a common feature of PTPs, in which zinc interacts with the highly conserved catalytic domain, possibly by binding to the active site cysteine (Haase and Maret 2005). Notably, a nucleophile attack by the catalytically active cysteine residue is a shared feature of PTPs and MKPs, another target of zinc that has been mentioned above. On the other hand, serine/threonine phosphatases have a dinuclear metal center in which water acts as the nucleophile (Barford et al. 1998), and no reports exist about a specific inhibition of these enzymes by zinc so far. Several PTPs are inhibited by low concentrations of zinc, including T cell PTP, with a half maximal inhibitory concentration (IC_{50}) of 200 nM (Maret et al. 1999), and PTP1B and SHP-1, with an IC_{50} of 17 nM and 93 nM, respectively (Haase and Maret 2003). These values are close to the concentration range of physiological levels of free zinc and presume a partial inhibition of PTPs by free zinc even at basal levels. These effects, therefore, depend on the extracellular zinc concentrations and not zinc signals.

Cell surface antibody receptors, called Fc receptors, allow different immune cells to utilize antibodies for pathogen recognition, like in mast cells. The triggering of the high affinity IgE-specific FcεR on mast cells induces degranulation. Notably, the secretory mast cell granules are rich in zinc that is released into the cellular environment together with a variety of immunological mediators (Gustafson 1967; Ho et al. 2004). The relationship between activation of FcRs and zinc status has been investigated in mast cells, but so far the results are undetermined. Measurements with fluorescent probes show a reduction of free zinc in response to treatments that induce degranulation including the triggering of FcεRs (Ho et al. 2004). On the other hand, other studies (Yamasaki et al. 2007; Yamasaki et al. 2012) report that the cross-linking of FcεRI induces a release of free zinc from the perinuclear area, including the endoplasmic reticulum in mast cells. This phenomenon is known as zinc wave (see Chap. 5), which depends on calcium influx and MAPK.

This discrepancy might be due to the use of different fluorophores in the two studies, detecting different pools of cellular zinc. Whereas the decrease of zinc is attributed to a loss of granule zinc, the increase seems to be cytosolic, originating from the perinuclear region (Yamasaki et al. 2007). Furthermore, the treatment with the membrane permeable zinc-chelator TPEN (N,N,N′,N′-tetrakis-(2-pyridylmethyl) ethylenediamine) activates NF-κB. The TNF-α-induced NF-κB activation is blocked by treatment with zinc/pyrithione (Ho et al. 2004). A later study states a blockade of NF-κB activation in response to FcεR cross-linking by TPEN administration, and this effect can be reversed by zinc treatment (Kabu et al. 2006). Thus, it remains to be analyzed under which circumstances zinc inhibits NF-κB and caspase-3 activity or is involved in the activation of NF-κB, ERK, JNK, and PKCβI and is thereby required for degranulation and cytokine production in mast cells in vitro as well as allergic reactions in vivo (Yamasaki et al. 2007).

Despite the fact that the innate immunity is the first stage in response of the immune system, observations relating to the influence of zinc signals are also associated with the function of the adaptive immunity.

13.5 Zinc Signal and the Adaptive Immunity

Influences of zinc have also been identified regarding the adaptive immunity with its two parts, humoral (B cells) and cellular immunity (T cells). B and T cells are produced as precursors and educated to recognize their specific antigen in the thymus (T cells) or bone marrow (B cells). The resulting naïve lymphocytes differentiate after antigen contact into effector cells and memory cells, in which the latter are the basis for the immunological memory and the stronger reaction to a known antigen as a secondary response. Zinc is highly important especially concerning the development of T cells and B cells (Osati-Ashtiani et al. 1998).

T cells are effector cells as well as important cells for the regulation of the specific immune system. Zinc not only influences NK cell-mediated killing, as mentioned above, it also affects the activity of cytolytic T cells (Mingari et al. 1998). Here the relative amount of $CD8^+$ $CD73^+$ T cells is decreased during zinc deficiency (Prasad 2000). These cells are predominantly precursors of cytotoxic T cells ($CD8^+$), and CD73 is needed for antigen recognition and proliferation as well as cytolytic process generation (Beck et al. 1997). Furthermore, zinc is involved in the development of T cells, because zinc deficiency results in various T cell defects manifested in thymus atrophy and lymphopenia (King et al. 2005).

Zinc acts as an essential cofactor for various enzymes, like the thymic hormone thymulin (a nonapeptide). Thymulin is a hormone that is produced by the thymus, released by thymic epithelial cells (Dardenne et al. 1984), and induces markers of differentiation in immature T cells (Saha et al. 1995). Besides these intrathymic functions on thymocytes and immature T cells, thymulin acts on mature T cells in the periphery. It also modulates the cytokine release by peripheral blood mononuclear cells (PBMC), and proliferation of $CD8^+$ T cells in combination with interleukin (IL)-2 (Coto et al. 1992). Therefore, zinc influences immature and mature T cells through the activation of thymulin. As a consequence of zinc deficiency, T cell proliferation decreases after mitogen stimulation (Dowd et al. 1986), whereas zinc supplementation is able to reverse the zinc deficiency induced changes in the thymus and on peripheral cells. This can also be observed in patients with acquired immune deficiency syndrome (Mocchegiani et al. 1995). In contrast to other lymphocyte populations, a direct effect of zinc on T cells can be observed. For instance, zinc induces blast transformation in human lymphocytes (Berger and Skinner 1974; Sood et al. 1999) and ensures the expression of the high-affinity receptor for IL-2 on mature T cells (Tanaka 1989), one effect resulting in decreased proliferation of T cells in zinc deficiency (Crea 1990). Moreover, IL-2, the soluble IL-2 receptor (sIL-2R), and IFN-γ can be induced by high concentrations of zinc in human PBMC (Driessen et al. 1994; Wellinghausen et al. 1997b).

In addition, zinc deficiency results in disturbed polarization of mature T cells. Zinc deprivation inhibits the polarization into Th1 cells and therefore changes the Th1/Th2 ratio toward Th2 cells, leading to unbalanced cell-mediated immune responses (Prasad 2000). It is observed that the production of IFN-γ and IL-2 (products of Th1 cells) is decreased during zinc deficiency, whereas production of IL-4, IL-6, and IL-10 (products of Th2 cells) is not affected. Accordingly, the risk for infections and Th2-driven allergies is increased. Furthermore, the functional impairment of T cell-mediated responses during zinc deficiency favors the development of autoimmune diseases (Honscheid et al. 2009). Consistently, both zinc deficiency and increased zinc levels influence T cell function. Increasing zinc levels during T cell activation are postulated to calibrate T cell receptor (TCR) signaling leading to T cell responses following suboptimal stimuli (Yu et al. 2011). This effect depends on a fast zinc signal. In contrast, physiological zinc-level-exceeding zinc concentrations, ranging from 50 to 100 μM, inhibit T cell activity as demonstrated by suppressed IL-1β-stimulated IFN-γ expression (Wellinghausen et al. 1997b). Furthermore, an induction and stabilization of regulatory T cells (Treg) can be observed in vitro and in vivo (Rosenkranz et al. 2015, 2016a; Maywald et al. 2017), whereas the pro-inflammatory Th17 and Th9 cells are dampened (Rosenkranz et al. 2016a; Maywald et al. 2018). This is of great importance, since the discovery of Treg offers a new paradigm for transplantation medicine since intra-graft Treg frequency seems to correlate with clinical graft acceptance, survival, and function (Graca et al. 2002; Hanidziar and Koulmanda 2010). Treg function and maturation is highly TGF-β1 dependent (Horwitz et al. 2008) by triggering the TGF-β1-dependent Smad signaling pathway. Recently, TGF-β-induced Smad signaling was shown to be intensified by zinc administration contributing to higher Treg cell induction (Maywald et al. 2017). Since Smad-binding elements are prominent in the conserved non-coding DNA sequence (CNS) 1 region of the Foxp3 promoter (Takaki et al. 2008), and zinc promoted Foxp3 stability by preventing proteasomal degradation caused by the histone deacetylase Sirt1 (Rosenkranz et al. 2016b), the increased numbers of Treg after zinc supplementation result from a synergism of both triggered mechanisms.

CK2-mediated activation of zinc transporter Zip7 is mentioned as another mechanism controlling Treg function. CK2 mediates phosphorylation and subsequent activation of Zip7 resulting in Zip7-mediated zinc release from the ER triggering activation of multiple downstream pathways enhancing cell proliferation and migration. Regarding Treg function, genetic ablation of the β-subunit of CK2 is addressed to insufficient potential to suppress allergic immune responses in the lungs mediated by Th2 cells in vivo (Ulges et al. 2015). Therefore, inappropriate CK2 function impairs cell-specific immunological tolerance which is partly due to altered zinc homeostasis.

Furthermore, studies highlighted that Treg development and survival is dependent on specific transcription factor expression, as IRF-1 and KLF-10. IRF-1 deficiency results in a prominent increase in highly differentiated and activated Treg cells in vivo (Fragale et al. 2008). IRF-1 plays a direct role in the generation and expansion of Treg cells by specifically repressing Foxp3 activity that can be dampened by zinc supplementation supporting the pro-tolerogenic immune reaction (Maywald and Rink 2016).

KLF-10 is important for appropriate Treg function, because animals carrying a disruption in KLF-10 no longer show Foxp3 activation (Cao et al. 2009). KLF-10-deficient Treg cells display impaired cell differentiation, altered cytokine profiles with enhanced Th1, Th2, and Th17 cytokine expression. Furthermore, a reduced capacity for suppression by wild-type co-cultured T effector cells as well as accelerated atherosclerosis in immunodeficient atherosclerotic mice were exhibited (Cao et al. 2009).

In contrast to T cell stimulation, T cell inhibition by zinc excess can also be observed in vivo, leading to the assumption that T cell activity is critically regulated by the zinc concentration. This might be a reason why some diseases with a T cell pathology, like rheumatoid arthritis, type 1 diabetes, cancer, neurodegenerative diseases, and depression (Alder et al. 2012; Szewczyk 2013; Stelmashook et al. 2014; Xin et al. 2015; Bredholt and Frederiksen 2016; Ressnerova et al. 2016; Maret 2017), are associated with moderate zinc deficiency. Some clinical trials show a reduction of the pain score while on zinc supplementation, leading to a presumption that zinc deficiency increases allo- or autoreactivity, whereas it is inhibited by high zinc dosages.

Important molecular zinc targets responsible for the above-mentioned zinc effects comprise receptor proteins, kinases, phosphatases, caspases, and transcription factors, which can be activated or inactivated by zinc (Haase and Rink 2009b). Furthermore, it functions as a second messenger in signal transduction contributing to its immunomodulating capacity (Haase et al. 2008b).

In T cells, fast zinc signals have been observed in response to TCR or interleukin-2 receptor (IL-2R) stimulation (Kaltenberg et al. 2010; Yu et al. 2011). Regarding the TCR, the signal seems not to be dependent on direct activation of this receptor. Initiation of the signaling from the TCR occurs by specific interaction with an antigen-loaded MHC molecule on the surface of a neighboring antigen-presenting

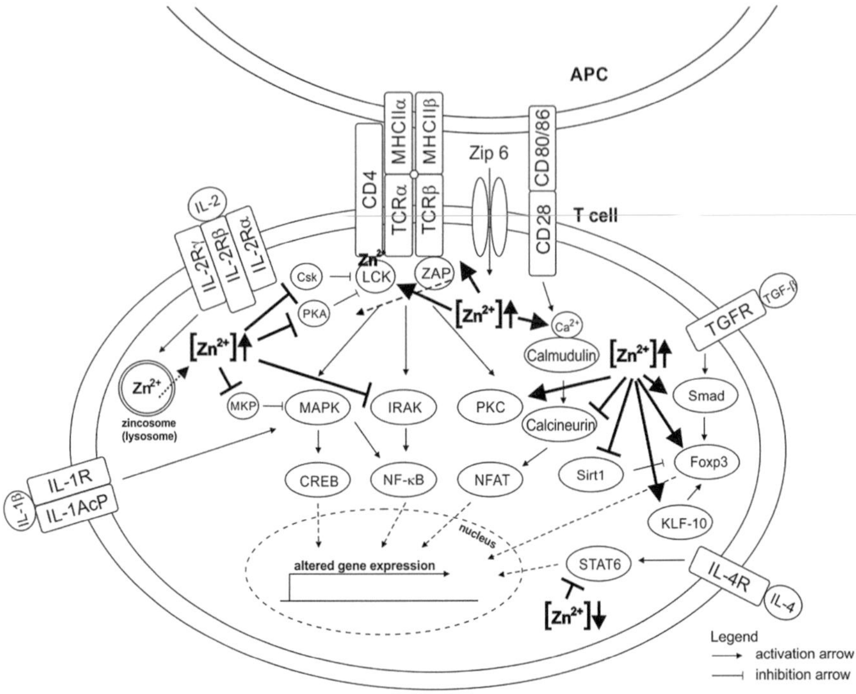

Fig. 13.2 Zinc and T cell signaling. Overview of T cell receptor (TCR) signaling, Interleukin-*1* receptor (IL-1R), IL-2R, IL-4R, and TGF-β signaling in T cells. A direct induced zinc signal via Zip 6, by APC mediated T cell activation ([Zn]) is known to activate TCR signaling, by augmented ZAP phosphorylation, sustained Ca^{2+} influx and inducing the active TCR signaling complex by binding of lymphocyte-specific protein tyrosine kinase LCK. IL-2R triggering in experiments leads to a zinc release (dotted arrows) from zincosomes, resulting in an increased intracellular zinc concentration ([Zn]). Zinc mediates the inhibition (inhibition arrow) of signaling via interleukin-1 receptor-associated kinase (IRAK), or c-Src tyrosine kinase (Csk)/protein kinase A (PKA) mediated inhibition LCK in TCR signaling, but induces (activation arrow) zeta-chain (TCR)-associated protein kinase (ZAP), protein kinase C (PKC) activity, mitogen-activated protein kinase (MAPK) signaling, nuclear factor kappa B (NF-kB) phosphorylation, KLF-10 expression, Smad Signaling, and subsequent Foxp3 expression. Increased intracellular zinc concentrations lead to an inhibition of calcineurin (CN) and avoid translocation into the nucleus (dashed arrows) of nuclear factor of activated T cells (NFAT). Besides zinc signals, zinc deficiency ([Zn]) also influences T cell signaling, i.e., IL-4R signaling. Zinc deficiency leads to a decreased phosphorylation of signal transducer and activator of transcription 6 (STAT6). *MKP* MAP-kinase phosphatase, *CREB* cyclic adenosine monophosphate response element-binding protein

cell (APC) by formation of a functional immunologic synapse (see Fig. 13.2). This results in an immediate influx of zinc from the extracellular environment through the transporter Zip 6 after T cell stimulation (Yu et al. 2011). In contrast to this, triggering of the IL-2R results in a release of zinc from lysosomes within 2 min after T cell stimulation (Kaltenberg et al. 2010). Thus, an altered intracellular zinc level is induced, leading to an influence of, e.g., dual-specificity phosphatases (DUSP) or protein phosphatase 2 (PP2A) by a fast zinc signal, as well as signal transduction and protein expression by late zinc signals, such as diminished dephosphorylation of MEK and ERK or elevated transcription of Zip 6, leading to an increased MT expression, which supports T cell proliferation and is of particular importance for T cell survival and expansion in the elderly (Kaltenberg et al. 2010; Yu et al. 2011; Lee et al. 2008).

Concerning the TCR, zinc signal influences assembly of the TCR activating complex by reduced recruitment of SHP-1 to the TCR activation complex and, thus, increased activation of the lymphocyte protein tyrosine kinase (Lck). TCR has no intrinsic kinase activity and depends on a Src-family tyrosine kinase, Lck, for signal transduction. Lck is one of the first kinases activated and is essential for phosphorylation of the 10 ITAM motifs of the T cell antigen receptor-signaling complex and augments phosphorylation of the kinase ZAP70 (Palacios and Weiss 2004). The activation of ZAP70 phosphorylates downstream targets that activate MAPK pathways and cause T cell activation.

Lck expression is upregulated in T cells of zinc-deficient mice (Lepage et al. 1999; Moore et al. 2001), and a number of additional lines of evidence also point to a direct regulation of Lck and sustains calcium influx (Yu et al. 2011). Zinc-provoked Lck activation occurs by linking two distinct protein interface sites. The N-terminal region of Lck is bound to the intracellular domains of the membrane proteins CD4 and CD8 (Huse et al. 1998; Lin et al. 1998). In a so-called "zinc clasp structure," this interaction is stabilized by a zinc ion that is bound to two cysteine residues from each protein at the interface site between Lck and CD4/CD8 (Kim et al. 2003). Due to the fact that CD4 and CD8 bind to MHC with their extracellular domains, too, they are thus recruited in close proximity of the TCR signaling complex, and this brings Lck close to its substrates.

Another zinc-dependent interface site is necessary for the homodimerization of Lck. Therefore, two zinc ions at the dimer interface of the SH3 domains stabilize the complex (Romir et al. 2007). The activation of Lck is a complex event, depending on two tyrosine phosphorylation sites. The first is tyrosine 394 situated in the so-called activation loop. It is transphosphorylated between Lck molecules upon activation, increasing kinase activity (Palacios and Weiss 2004; Brautigan et al. 1981). The Lck activation is due to the zinc-induced homodimerization by a fast zinc signal, bringing two Lck molecules into close proximity. The second phosphorylation occurs in the C terminal negative regulatory site at tyrosine 505. According to the phosphorylation, the kinase assumes a closed, inactive conformation. The transmembrane PTP CD45 selectively dephosphorylates tyrosine 505, thus keeping Lck in a primed state. TCR signaling and Lck phosphorylation are regulated by numerous PTPs (Mustelin and Tasken 2003) that are all potential targets for zinc-inhibition. Consequently, it cannot be predicted if this would result in preferential dephosphorylation of an activating or inactivating tyrosine by zinc in vivo. Zinc can activate Lck directly in the absence of phosphatases, most likely as a result of dimerization. Zinc-dependent homodimerization as well as heterodimerization are specific for Lck, because the cysteine residues that form both intermolecular zinc-binding sites are not present in other Src kinase family members but are unique for this kinase (Romir et al. 2007; Kim et al. 2003).

An inhibition of TCR signaling and T cell activation result from a phosphorylation of the inactivating tyrosine 505 of Lck via c-Src tyrosine kinase COOH-terminal Srk kinase (Csk) (Chow et al. 1993). Zinc can interfere with these events in several ways. First, the fast zinc signal leads to an inhibition of Csk by reversible binding of zinc that is mediated by substituting for Mg^{2+} at one of its binding sites (Klein et al. 2002). Second, the activation of Csk in T cells can be observed via phosphorylation by PKA (Vang et al. 2001). Although zinc has no direct influence on PKA activity, it can inhibit AC transcription by a late zinc signal, and thereby the formation of the PKA activator cAMP. Furthermore, also fast zinc signals influence cAMP by inhibition of PDE that can block degradation of cyclic nucleotides, leading to activation of PKA (von Bulow et al. 2007). An inhibition of Csk and AC promote TCR signaling, while PDE inhibition antagonizes it. But the outcome on TCR signaling resulting from the modulation of this pathway by zinc in vivo remains to be investigated.

In response to TCR stimulation, another transcription factor, nuclear factor of activated T cells (NFAT), mediates the expression of plenty of genes, like IL-2. In resting cells, NFAT proteins are constitutively phosphorylated and are found in the cytoplasm. In response to TCR/CD28-mediated calcium signaling, NFAT is dephosphorylated by calcineurin (CN), a Ca^{2+}/calmodulin-dependent serine/threonine phosphatase, and translocates into the nucleus (Macian 2005). Iron and zinc are essential cofactors for the catalytic domain of CN, containing a Fe^{2+}-Zn^{2+} binuclear center, but merely an inhibition of CN by zinc is described. Also, in vitro experiments show an inhibition at physiologically relevant zinc concentrations, ranging from 10 nM to 10 μM (Huang et al. 2008; Takahashi et al. 2003). One of the kinases that keeps CN in its phosphorylated, inactive state in resting cells is phosphatidylinositol-3-kinase (PI3K) which is a negative regulator of Glycogen synthase kinase 3 beta (GSK-3β). Therefore, augmented PI3K activity can abrogate NFAT activity (Macian 2005).

An increased GSK-3β phosphorylation via a pathway involving PI3K is known. Here zinc acts through the PI3K pathway in various cell types (Eom et al. 2001; LaRochelle et al. 2001; Tang and Shay 2001). One possible explanation for this mechanism is an increased enzyme degradation of phosphatase and tensin homolog deleted on chromosome 10 (PTEN) (Wu et al. 2003). A later study examining the IL-2-induced PI3K/Akt signaling pathway could not verify these results but showed an inhibition of PTEN (Plum et al. 2014). Here it has been indicated that late zinc signals seem to be necessary for the regulation of PI3K/Akt pathway, because under zinc deficiency the IL-2-induced Akt phosphorylation is diminished. Late zinc signals are known to upregulate phosphorylation of Akt at Ser473 and inhibit PTEN at sub-nanomolar concentrations (IC_{50} 0.59 nM). This inhibition seems to be mediated by binding of zinc ions at its cysteine thiol at position 124 (cys124) that is essential for the catalytic activity of PTEN (Plum et al. 2014). Thus, a modulation of this pathway occurs upstream of Akt, but downstream of Jak1, because STAT5 signaling, which is also downstream of Jak1, is not influenced by zinc.

PTEN in general functions as a dephosphorylase of phosphatidylinositol 3,4,5-trisphosphate, a product of PI3K, mediating the activation of PDK-1/Akt/GSK-3β. These results suggest a comparable activation of PI3K via zinc occurring in T cells. Inhibition of CN by zinc results in NFAT inactivation and reduction of TCR-mediated transcription simultaneously, whereas activation of PI3K signaling acts agonistic.

In addition to TCR signaling, zinc signals also affect signals originating from the IL-1 receptor. High zinc concentrations of about 100 μM inhibit IL-1β-stimulated IFN-γ production in primary human T cells and IL-1 dependent proliferation of murine T cells. Zinc incubation leads to a reduced activity of IL-1 receptor-associated kinase (IRAK) that is a central kinase in the signaling pathways downstream of the IL-1 receptor. Zinc supplementation to zinc-deficient T cells has also regulatory effects on cytokine production. Zinc supplementation leads to a fast rise in intracellular zinc levels, i.e., zincflux, due to an increased expression of cell membrane-located zinc transporters Zip 10 and Zip 12 because of the former zinc deficiency condition. Thus, IL-2 production and IL-2 mRNA expression are highly increased after IL-1β stimulation, concerning phosphorylation of MAPK p38 and NF-κB subunit p65 (Daaboul et al. 2012).

Moreover, the IRAK family of serine/threonine kinases is involved in other pathways, too. IRAK4 participates in TCR-mediated NF-κB activation (Suzuki et al. 2006), although this observation is debated (Kawagoe et al. 2007). Hence, zinc can inhibit two central signaling pathways (TCR and IL-1 receptor) that are essential in the activation of T cells, by its effect on IRAK, making this a major target for negative regulation of T cells by zinc supplementation. But the role of IRAK is not merely limited to T cells. The high rate of similarity between TLR-4 and IL-1 receptor signal transduction suggests IRAK inhibition for another mechanism by which zinc negatively influences TLR-4 signaling.

Zinc as a mediator of T cell signal transduction has been known since the first reports of a potential role of this ion in signaling. Its interaction with PKC is identified as the biochemical basis of these observations (Csermely and Somogyi 1989). The PKC is a family of serine/threonine kinases that comprise of several isoforms. They are differentiated into classical PKCs, which are activated by cofactors like Ca^{2+} and diacylglycerol, novel PKCs, which bind diacylglycerol but no Ca^{2+}, and atypical PKCs, which do not interact with either cofactor. Each isoform has specific roles in the regulation of cellular functions; e.g., in T cells, PKCθ is involved in the activation of several transcription factors in response to TCR/CD28 stimulation. In addition, a role for PKCα in proliferation and IL-2 production of T cells is known. Furthermore, several PKC isoforms are involved in survival of B cells, pre-B cell development, and induction of tolerance toward self-antigens (Tan and Parker 2003).

PKCs have four conserved domains (C1-C4) in common. The diacylglycerol binding C1 domain in the N-terminal regulatory part of PKCβ1 contains two homologous regions with six cysteines and two histidines, forming a total of four Cys(3)His zinc binding motifs (Hubbard et al. 1991). No information about a differential effect of zinc on the different isoforms is available, and varying forms of C1 domains are present in conventional, novel, and atypical PKCs, indicating zinc binding to all known PKC isoforms (Tan and Parker 2003).

Late zinc signals affect multiple steps during PKC activation, for example, augmented PKC kinase activity, increased affinity to phorbol esters, and enhanced binding to the cytoskeleton and plasma membrane (Forbes et al. 1990; Wellinghausen et al. 1997b). An inhibition of these events can be mediated by membrane-permeable zinc chelators, such as TPEN. PKC itself can be a source for zinc release; thus, interaction between PKC and zinc is not limited to an effect of zinc on the PKC activation. PKC activation by lipid second messengers or thiol oxidation leads to measurable zinc release from the regulatory domain (Knapp and Klann 2000; Korichneva et al. 2002). In addition to that, PKC regulates the intracellular free zinc concentration and distribution. In T cells, phorbol ester treatment leads to a fast zinc signal resulting in redistribution of zinc from the nucleus and mitochondria to the cytosol and microsomes (Csermely et al. 1987). Furthermore, treatment with phorbol esters, like PMA, leads to an increase of free zinc in the human T cells, whereas, e.g., PHA stimulation leads to calcium increase (Haase et al. 2008a), and the phorbol ester-induced differentiation of HL-60 cells into macrophages comprises a transitional increase of nuclear zinc depending on PKCβ (Glesne et al. 2006).

In general, a forecast of zinc effects on T cells is impossible because of the high number of zinc-related signaling pathways. Several studies concerning zinc status and T cell function in vivo stated an increase of the delayed type hypersensitivity reaction upon correction of zinc deficiency (Haase et al. 2006b). On the other hand, zinc supplementation diminishes the allogeneic reaction in the MLCs (Faber et al. 2004) and stabilizes regulatory T cell function (Rosenkranz et al. 2016b), indicating that zinc may have multiple, opposing functions, depending on its concentration and certainly also on the interaction with multiple other environmental factors.

B cells represent the main cells of humoral immunity and differentiate to antibody-producing plasma cells after stimulation. Nearly all of the zinc regulated signaling pathways discussed above are also important in B cells, e.g., tyrosine phosphorylation, PKC, MAPK, and activation of the transcription factors NFAT and NF-κB. In contrast to many of the cell types discussed before, mature B cell proliferation and function are not as dependent on the organisms' zinc status as it is in T cells. Therefore, the influence of zinc deficiency on B cells is not comparable to the situation of T cells (Fraker and Telford 1997). Zinc itself seems to have no direct influence on the activity of B cells (Crea 1990). However, zinc deficient patients, like elderly and hemodialysis patients, show a reduced response to vaccination (Cakman et al. 1996; Bonomini et al. 1993). Rather, lymphopoiesis and pre-B cell development are mainly affected by zinc deprivation in vivo (Fraker and King 2004). Zinc deficiency causes an exceeded loss of lymphoid tissue compared to other tissues, and thus additionally to B cell, T cell development is impaired (Keen and Gershwin 1990). B cells and their precursors (especially pre-B cells and immature B cells) are reduced in absolute number during zinc deficiency, whereas changes among mature B cells are only slight.

During development lymphocytes are sensitive for apoptosis signals due to positive and negative selection mechanisms in the primary lymphoid organs, whereas mature B and T cells are apoptosis resistant, and therefore inactivated (known as "anergy") in case of autoreactivity. A strict selection guarantees functionality and avoids autoreactive cells by eliminating the majority of newly formed cells by apoptosis. Zinc deficiency increases the rate of apoptosis within the B cell population

(Fraker et al. 2000) and leads to cell depletion. Contrarily to zinc deficiency low zinc levels have no influence on the cell cycle status of precursor B cells and only modest influence on cycling pro-B cells (King and Fraker 2000). Hence, there are fewer naïve B cells during zinc deficiency that can react on neoantigens. Taking into account that the number of T cells is reduced during zinc deficiency, too, and that the most antigens are T cell dependent, it is probable that while zinc deficiency, the body is unable to respond with antibody production in response to neoantigens. This assumption is consistent with findings showing a disturbed antibody production by B cells during zinc depletion (DePasquale-Jardieu and Fraker 1984). In accordance to this, a late zinc signal is essential for STAT6 phosphorylation via IL-4 induction, because STAT6 phosphorylation is diminished by zinc deficiency in B cells, as well as in T cells (Gruber et al. 2013). Thus, T cell reactivity is impaired, and B cell activity and antibody production are affected as well since IL-4 promotes the activation of early B cells and the immunoglobulin class switch toward IgE, and thereby the further antibody specification, leading to a higher risk of parasitic infection (Kopf et al. 1993). Furthermore, zinc signals are important for IL-6 induced STAT3 phosphorylation, increased due to zinc deficiency. IL-6 is responsible for the activation and final differentiation of B cells into plasma cells and IL-6 overproduction is associated with autoantibody production. Many diseases, e.g., RA and plasma cell neoplasias, go along with reduced serum zinc levels, indicating potential co-effects of IL-6 overproduction and enhanced susceptibility of B cells due to zinc deficiency (Gruber et al. 2013).

A further mechanism impairing adequate immune responses might be epigenetic changes of DNA. The epigenome is highly influenced either by environmental changes and daily nutrition (Wessels 2014). In line with that, histone-modifying enzymes as histone deacetylases are known to be regulated zinc-dependent (Rosenkranz et al. 2016b). Hence, the organism's zinc status is important for adequate epigenetic modification of DNA. This indicates that a strict regulation is necessary to keep the immune system balanced.

In addition, studies reveal that antibodies as a response to T cell-dependent antigens are more sensitive to zinc deficiency than antibody production in response to T cell-independent antigens (Moulder and Steward 1989). For example, zinc-deficient mice show reduced antibody recall responses to antigens for which they were immunized before. This effect has been observed in T cell-independent as well as in T cell-dependent systems. Thus, immunologic memory also seems to be influenced by zinc (Fraker et al. 1986). However, mature B cells are more resistant to zinc deficiency due to high Bcl2 level; B cell memory is less affected than the primary response (Fraker et al. 2000).

Late zinc signals influence several regulatory proteins, such as those from the Bcl/Bax family (Truong-Tran et al. 2001), and furthermore several aspects of apoptotic signal transduction. In this connection, zinc directly regulates the acidity of enzymes in the apoptotic cascade. First, the calcium-dependent endonuclease, which mediates DNA fragmentation, is inhibited by zinc. However, this target is beyond the point of no return for programmed cell death, and an inhibition could explain a suppression of DNA fragmentation during apoptosis, but not the effect on cellular survival.

Another important group of enzymes in apoptosis are cysteine-aspartic acid proteases, also known as caspases. They form a cascade to transduce initial apoptotic signals to the effector enzymes that mediate the organized cellular destruction which is characteristic for programmed cell death. In this process, inactive procaspases are activated by proteolytic cleavage. An additional regulatory mechanism is shown by inhibition of caspase-3, caspase-6, and caspase-8 by application of low micromolar zinc concentrations (Stennicke and Salvesen 1997). A half maximal inhibitory concentration for caspase-3 was found below 10 nM (Maret et al. 1999). This value is clearly in the physiological range of free intracellular zinc, leading to the suggestion that endogenous zinc can inhibit caspase-3. Furthermore, studies showed a blockade of caspase-3-dependent apoptosis occurring in a rat heterotrophic heart transplant model leading to an increased allograft survival (Kown et al. 2000; Kown et al. 2002).

It is interesting that the function of mature B cells, which utilize the same signaling pathway like other immune cells, seems to be affected by zinc to a lesser magnitude. Even the reduced antibody production during zinc deficiency is based on reduced B cell numbers whereas it is unaffected on a per-cell basis (Cook-Mills and Fraker 1993). This indicates an effect on cellular development rather than on function. One reason might be a difference in zinc homeostasis, making mature B cells less susceptible to conditions of limited zinc availability. Although B cells are highly susceptible to apoptosis during development, and zinc is one factor that influences these signals, mature B cells can tolerate comparable conditions due to changes in zinc-regulating proteins, but also by changing the expression patterns of several other factors that regulate the cellular responsiveness to apoptotic signals. One important zinc-regulating protein in B cells is the zinc importer Zip10. Zip10 either is essential for cell survival during early B-cell development and for adequate B-cell receptor (BCR) signaling (Miyai et al. 2014; Hojyo et al. 2014). If Zip10 is lost during an early B cell stage, cell survival is abrogated. Eventually, this leads to an absence of mature B cells in vivo, provoking spleno-atrophy and reduced Ig level. The absence of Zip10 also causes an impaired T cell-dependent and T cell-independent immune response, respectively. Moreover, mature B cells proliferated poorly in response to BCR cross-linking due to the dysregulated BCR signaling. On the molecular level, this might be due to a disturbed JAK-STAT signaling pathway, which modulates Zip10 expression (Miyai et al. 2014). Additionally, a disturbed regulation of the BCR signal strength is likely, which might be due to the positive regulator function of Zip10 regarding CD45R phosphatase activity. In the case of Zip10 malfunction, CD45R phosphatase activity is reduced resulting in hyperactivation of tyrosine-protein kinase Lyn that consequently provokes an altered threshold for BCR signaling (Hojyo et al. 2014). Besides Zip10, a recent study uncovered the zinc transporter Zip7 to be essential for B cell development, since a mutation results in an autosomal recessive disease characterized by absent B cells, agammaglobulinemia, and early onset infections (Anzilotti et al. 2019).

Summing up, nearly all immune cells are directly affected by zinc, especially by zinc deficiency, indicating that zinc is a main regulator of cellular function and signal transduction. But the majority of these effects are characterized by investigations at isolated signaling pathways in vitro. Thus, further efforts have to elucidate the in vivo situation, i.e., which effects are actually physiologically relevant.

References

Alder H, Taccioli C, Chen H, Jiang Y, Smalley KJ, Fadda P, Ozer HG, Huebner K, Farber JL, Croce CM, Fong LY (2012) Dysregulation of miR-31 and miR-21 induced by zinc deficiency promotes esophageal cancer. Carcinogenesis 33(9):1736–1744. https://doi.org/10.1093/carcin/bgs204

Allen JI, Perri RT, McClain CJ, Kay NE (1983) Alterations in human natural killer cell activity and monocyte cytotoxicity induced by zinc deficiency. J Lab Clin Med 102(4):577–589

An WL, Bjorkdahl C, Liu R, Cowburn RF, Winblad B, Pei JJ (2005) Mechanism of zinc-induced phosphorylation of p70 S6 kinase and glycogen synthase kinase 3beta in SH-SY5Y neuroblastoma cells. J Neurochem 92(5):1104–1115. https://doi.org/10.1111/j.1471-4159.2004.02948.x

Andreini C, Banci L, Bertini I, Rosato A (2006) Counting the zinc-proteins encoded in the human genome. J Proteome Res 5(1):196–201. https://doi.org/10.1021/pr050361j

Andreini C, Bertini I, Rosato A (2009) Metalloproteomes: a bioinformatic approach. Acc Chem Res 42(10):1471–1479. https://doi.org/10.1021/ar900015x

Anzilotti C et al (2019) An essential role for the Zn2+ transporter ZIP7 in B cell development. Nat Immunol 20(3):350–361. https://doi.org/10.1038/s41590-018-0295-8

Aydemir TB, Liuzzi JP, McClellan S, Cousins RJ (2009) Zinc transporter ZIP8 (SLC39A8) and zinc influence IFN-gamma expression in activated human T cells. J Leukoc Biol 86(2):337–348. https://doi.org/10.1189/jlb.1208759

Bao B, Prasad AS, Beck FW, Fitzgerald JT, Snell D, Bao GW, Singh T, Cardozo LJ (2010) Zinc decreases C-reactive protein, lipid peroxidation, and inflammatory cytokines in elderly subjects: a potential implication of zinc as an atheroprotective agent. Am J Clin Nutr 91(6):1634–1641

Barford D, Das AK, Egloff MP (1998) The structure and mechanism of protein phosphatases: insights into catalysis and regulation. Annu Rev Biophys Biomol Struct 27:133–164. https://doi.org/10.1146/annurev.biophys.27.1.133

Beck FW, Kaplan J, Fine N, Handschu W, Prasad AS (1997) Decreased expression of CD73 (ecto-5′-nucleotidase) in the CD8+ subset is associated with zinc deficiency in human patients. J Lab Clin Med 130(2):147–156

Bentley PJ (1992) Influx of zinc by channel catfish (*Ictalurus punctatus*): uptake from external environmental solutions. Comp Biochem Physiol C, Comp Pharmacol Toxicol 101(2):215–217

Berger NA, Skinner AM (1974) Characterization of lymphocyte transformation induced by zinc ions. J Cell Biol 61(1):45–55

Beyersmann D, Haase H (2001) Functions of zinc in signaling, proliferation and differentiation of mammalian cells. Biometals Int J Role Metal Ions Biol Biochem Med 14(3–4):331–341

Bonomini M, Di Paolo B, De Risio F, Niri L, Klinkmann H, Ivanovich P, Albertazzi A (1993) Effects of zinc supplementation in chronic haemodialysis patients. Nephrology, dialysis, transplantation : official publication of the. Eur Dial Transplant Assoc Eur Renal Assoc 8(10):1166–1168

Boone DL, Turer EE, Lee EG, Ahmad RC, Wheeler MT, Tsui C, Hurley P, Chien M, Chai S, Hitotsumatsu O, McNally E, Pickart C, Ma A (2004) The ubiquitin-modifying enzyme A20 is required for termination of toll-like receptor responses. Nat Immunol 5(10):1052–1060. https://doi.org/10.1038/ni1110

Brautigan DL, Bornstein P, Gallis B (1981) Phosphotyrosyl-protein phosphatase. Specific inhibition by Zn. J Biol Chem 256(13):6519–6522

Bredholt M, Frederiksen JL (2016) Zinc in multiple sclerosis: a systematic review and meta-analysis. ASN Neuro:8(3). https://doi.org/10.1177/1759091416651511

Briefel RR, Bialostosky K, Kennedy-Stephenson J, McDowell MA, Ervin RB, Wright JD (2000) Zinc intake of the U.S. population: findings from the third National Health and nutrition examination survey, 1988-1994. J Nutr 130(5S Suppl):1367s–1373s

Brieger A, Rink L (2010) Zink und Immunfunktion. Ernährung Med 25:156–160

Brieger A, Rink L, Haase H (2013) Differential regulation of TLR-dependent MyD88 and TRIF signaling pathways by free zinc ions. J Immunol 191(4):1808–1817. https://doi.org/10.4049/jimmunol.1301261

Brophy MB, Hayden JA, Nolan EM (2012) Calcium ion gradients modulate the zinc affinity and antibacterial activity of human calprotectin. J Am Chem Soc 134(43):18089–18100. https://doi.org/10.1021/ja307974e

Bruinsma JJ, Jirakulaporn T, Muslin AJ, Kornfeld K (2002) Zinc ions and cation diffusion facilitator proteins regulate Ras-mediated signaling. Dev Cell 2(5):567–578

Cakman I, Rohwer J, Schutz RM, Kirchner H, Rink L (1996) Dysregulation between TH1 and TH2 T cell subpopulations in the elderly. Mech Ageing Dev 87(3):197–209

Cao Z, Wara AK, Icli B, Sun X, Packard RR, Esen F, Stapleton CJ, Subramaniam M, Kretschmer K, Apostolou I, von Boehmer H, Hansson GK, Spelsberg TC, Libby P, Feinberg MW (2009) Kruppel-like factor KLF10 targets transforming growth factor-beta1 to regulate CD4(+) CD25(−) T cells and T regulatory cells. J Biol Chem 284(37):24914–24924. https://doi.org/10.1074/jbc.M109.000059

Cavdar AO, Arcasoy A, Cin S, Babacan E, Gozdasoglu S (1983) Geophagia in Turkey: iron and zinc deficiency, iron and zinc absorption studies and response to treatment with zinc in geophagia cases. Prog Clin Biol Res 129:71–97

Chasapis CT, Loutsidou AC, Spiliopoulou CA, Stefanidou ME (2012) Zinc and human health: an update. Arch Toxicol 86(4):521–534. https://doi.org/10.1007/s00204-011-0775-1

Chavakis T, May AE, Preissner KT, Kanse SM (1999) Molecular mechanisms of zinc-dependent leukocyte adhesion involving the urokinase receptor and beta2-integrins. Blood 93(9):2976–2983

Chow LM, Fournel M, Davidson D, Veillette A (1993) Negative regulation of T-cell receptor signalling by tyrosine protein kinase p50csk. Nature 365(6442):156–160. https://doi.org/10.1038/365156a0

Colvin RA, Holmes WR, Fontaine CP, Maret W (2010) Cytosolic zinc buffering and muffling: their role in intracellular zinc homeostasis. Metallomics : Integr Biometals Sci 2(5):306–317

Cook-Mills JM, Fraker PJ (1993) Functional capacity of the residual lymphocytes from zinc-deficient adult mice. Br J Nutr 69(3):835–848

Coto JA, Hadden EM, Sauro M, Zorn N, Hadden JW (1992) Interleukin 1 regulates secretion of zinc-thymulin by human thymic epithelial cells and its action on T-lymphocyte proliferation and nuclear protein kinase C. Proc Natl Acad Sci U S A 89(16):7752–7756

Cousins RJ, Liuzzi JP, Lichten LA (2006) Mammalian zinc transport, trafficking, and signals. J Biol Chem 281(34):24085–24089. https://doi.org/10.1074/jbc.R600011200

Crea A (1990) Zinc et système immunitaire. Ann Med Interne (Paris) (141):447–451

Csermely P, Somogyi J (1989) Zinc as a possible mediator of signal transduction in T lymphocytes. Acta Physiol Hung 74(2):195–199

Csermely P, Gueth S, Somogyi J (1987) The tumor promoter tetradecanoyl-phorbol-acetate (TPA) elicits the redistribution of zinc in subcellular fractions of rabbit thymocytes measured by X-ray fluorescence. Biochem Biophys Res Commun 144(2):863–868

Csermely P, Szamel M, Resch K, Somogyi J (1988) Zinc can increase the activity of protein kinase C and contributes to its binding to plasma membranes in T lymphocytes. J Biol Chem 263(14):6487–6490

Daaboul D, Rosenkranz E, Uciechowski P, Rink L (2012) Repletion of zinc in zinc-deficient cells strongly up-regulates IL-1beta-induced IL-2 production in T-cells. Metallomics Integr Biometals Sci 4(10):1088–1097. https://doi.org/10.1039/c2mt20118f

Dardenne M, Savino W, Wade S, Kaiserlian D, Lemonnier D, Bach JF (1984) In vivo and in vitro studies of thymulin in marginally zinc-deficient mice. Eur J Immunol 14(5):454–458. https://doi.org/10.1002/eji.1830140513

DeCoursey TE, Morgan D, Cherny VV (2003) The voltage dependence of NADPH oxidase reveals why phagocytes need proton channels. Nature 422(6931):531–534. https://doi.org/10.1038/nature01523

Deng L, Wang C, Spencer E, Yang L, Braun A, You J, Slaughter C, Pickart C, Chen ZJ (2000) Activation of the IkappaB kinase complex by TRAF6 requires a dimeric ubiquitin-conjugating enzyme complex and a unique polyubiquitin chain. Cell 103(2):351–361

DePasquale-Jardieu P, Fraker PJ (1984) Interference in the development of a secondary immune response in mice by zinc deprivation: persistence of effects. J Nutr 114(10):1762–1769

Dierichs L, Kloubert V, Rink L (2017) Cellular zinc homeostasis modulates polarization of THP-1-derived macrophages. Eur J Nutr:1–9

Donato R, R Cannon B, Sorci G, Riuzzi F, Hsu K, J Weber D, L Geczy C (2013) Functions of S100 proteins. Curr Mol Med 13(1):24–57

Dong XP, Cheng X, Mills E, Delling M, Wang F, Kurz T, Xu H (2008) The type IV mucolipidosis-associated protein TRPML1 is an endolysosomal iron release channel. Nature 455(7215):992–996. https://doi.org/10.1038/nature07311

Dowd PS, Kelleher J, Guillou PJ (1986) T-lymphocyte subsets and interleukin-2 production in zinc-deficient rats. Br J Nutr 55(1):59–69

Driessen C, Hirv K, Rink L, Kirchner H (1994) Induction of cytokines by zinc ions in human peripheral blood mononuclear cells and separated monocytes. Lymphokine Cytokine Res 13(1):15–20

Dubben S, Honscheid A, Winkler K, Rink L, Haase H (2010) Cellular zinc homeostasis is a regulator in monocyte differentiation of HL-60 cells by 1 alpha,25-dihydroxyvitamin D3. J Leukoc Biol 87(5):833–844. https://doi.org/10.1189/jlb.0409241

Eide DJ (2006) Zinc transporters and the cellular trafficking of zinc. Biochim Biophys Acta 1763(7):711–722. https://doi.org/10.1016/j.bbamcr.2006.03.005

Eom SJ, Kim EY, Lee JE, Kang HJ, Shim J, Kim SU, Gwag BJ, Choi EJ (2001) Zn(2+) induces stimulation of the c-Jun N-terminal kinase signaling pathway through phosphoinositide 3-kinase. Mol Pharmacol 59(5):981–986

Faber C, Gabriel P, Ibs KH, Rink L (2004) Zinc in pharmacological doses suppresses allogeneic reaction without affecting the antigenic response. Bone Marrow Transplant 33(12):1241–1246. https://doi.org/10.1038/sj.bmt.1704509

Failla ML, van de Veerdonk M, Morgan WT, Smith JC Jr (1982) Characterization of zinc-binding proteins of plasma in familial hyperzincemia. J Lab Clin Med 100(6):943–952

Fessatou S, Fagerhol MK, Roth J, Stamoulakatou A, Kitra V, Hadarean M, Paleologos G, Chandrinou H, Sampson B, Papassotiriou I (2005) Severe anemia and neutropenia associated with hyperzincemia and hypercalprotectinemia. J Pediatr Hematol Oncol 27(9):477–480

Forbes IJ, Zalewski PD, Giannakis C, Petkoff HS, Cowled PA (1990) Interaction between protein kinase C and regulatory ligand is enhanced by a chelatable pool of cellular zinc. Biochim Biophys Acta 1053(2–3):113–117

Fosmire GJ (1990) Zinc toxicity. Am J Clin Nutr 51(2):225–227

Fragale A, Gabriele L, Stellacci E, Borghi P, Perrotti E, Ilari R, Lanciotti A, Remoli AL, Venditti M, Belardelli F, Battistini A (2008) IFN regulatory factor-1 negatively regulates CD4+ CD25+ regulatory T cell differentiation by repressing Foxp3 expression. J Immunol 181(3):1673–1682

Fraker PJ, King LE (2004) Reprogramming of the immune system during zinc deficiency. Annu Rev Nutr 24:277–298. https://doi.org/10.1146/annurev.nutr.24.012003.132454

Fraker PJ, Telford WG (1997) A reappraisal of the role of zinc in life and death decisions of cells. Proc Soc Exp Biol Med Soc Exp Biol Med (New York, NY) 215(3):229–236

Fraker PJ, Gershwin ME, Good RA, Prasad A (1986) Interrelationships between zinc and immune function. Fed Proc 45(5):1474–1479

Fraker PJ, King LE, Laakko T, Vollmer TL (2000) The dynamic link between the integrity of the immune system and zinc status. J Nutr 130(5S Suppl):1399s–1406s
Gammoh NZ, Rink L (2017) Zinc in infection and inflammation. Nutrients 9(6):624. https://doi.org/10.3390/nu9060624
Gao X, Du Z, Patel TB (2005) Copper and zinc inhibit Galphas function: a nucleotide-free state of Galphas induced by Cu2+ and Zn2+. J Biol Chem 280(4):2579–2586. https://doi.org/10.1074/jbc.M409791200
Glesne D, Vogt S, Maser J, Legnini D, Huberman E (2006) Regulatory properties and cellular redistribution of zinc during macrophage differentiation of human leukemia cells. J Struct Biol 155(1):2–11. https://doi.org/10.1016/j.jsb.2005.09.012
Graca L, Thompson S, Lin CY, Adams E, Cobbold SP, Waldmann H (2002) Both CD4(+)CD25(+) and CD4(+)CD25(−) regulatory cells mediate dominant transplantation tolerance. J Immunol 168(11):5558–5565
Gruber K, Maywald M, Rosenkranz E, Haase H, Plumakers B, Rink L (2013) Zinc deficiency adversely influences interleukin-4 and interleukin-6 signaling. J Biol Regul Homeost Agents 27(3):661–671
Gustafson GT (1967) Heavy metals in rat mast cell granules. Lab Investig; J Techl Methods Pathol 17(6):588–598
Gyulkhandanyan AV, Lee SC, Bikopoulos G, Dai F, Wheeler MB (2006) The Zn2+−transporting pathways in pancreatic beta-cells: a role for the L-type voltage-gated Ca2+ channel. J Biol Chem 281(14):9361–9372. https://doi.org/10.1074/jbc.M508542200
Haase H, Maret W (2003) Intracellular zinc fluctuations modulate protein tyrosine phosphatase activity in insulin/insulin-like growth factor-1 signaling. Exp Cell Res 291(2):289–298
Haase H, Maret W (2005) Fluctuations of cellular, available zinc modulate insulin signaling via inhibition of protein tyrosine phosphatases. J Trace Elem Med Biol : Organ Soc Minerals Trace Elem (GMS) 19(1):37–42. https://doi.org/10.1016/j.jtemb.2005.02.004
Haase H, Maret W (2009) Zinc in cellular and molecular biology of metals
Haase H, Rink L (2009a) Functional significance of zinc-related signaling pathways in immune cells. Annu Rev Nutr 29:133–152. https://doi.org/10.1146/annurev-nutr-080508-141119
Haase H, Rink L (2009b) The immune system and the impact of zinc during aging. Immun Ageing 6:9. https://doi.org/10.1186/1742-4933-6-9
Haase H, Rink L (2010) Das essentielle Spurenelement Zink. Biol Unserer Zeit 40:314–321
Haase H, Hebel S, Engelhardt G, Rink L (2006a) Flow cytometric measurement of labile zinc in peripheral blood mononuclear cells. Anal Biochem 352(2):222–230. https://doi.org/10.1016/j.ab.2006.02.009
Haase H, Mocchegiani E, Rink L (2006b) Correlation between zinc status and immune function in the elderly. Biogerontology 7(5–6):421–428. https://doi.org/10.1007/s10522-006-9057-3
Haase H, Ober-Blobaum JL, Engelhardt G, Hebel S, Heit A, Heine H, Rink L (2008a) Zinc signals are essential for lipopolysaccharide-induced signal transduction in monocytes. J Immunol 181(9):6491–6502
Haase H, Ober-Blobaum JL, Engelhardt G, Hebel S, Heit A, Heine H, Rink L (2008b) Zinc signals are essential for lipopolysaccharide-induced signal transduction in monocytes. J Immunol 181(9):6491–6502
Hanidziar D, Koulmanda M (2010) Inflammation and the balance of Treg and Th17 cells in transplant rejection and tolerance. Curr Opin Organ Transplant 15(4):411–415. https://doi.org/10.1097/MOT.0b013e32833b7929
Hansson A (1996) Extracellular zinc ions induces mitogen-activated protein kinase activity and protein tyrosine phosphorylation in bombesin-sensitive Swiss 3T3 fibroblasts. Arch Biochem Biophys 328(2):233–238. https://doi.org/10.1006/abbi.1996.0168
Hasan R, Rink L, Haase H (2013) Zinc signals in neutrophil granulocytes are required for the formation of neutrophil extracellular traps. Innate Immun 19(3):253–264. https://doi.org/10.1177/1753425912458815

Hasan R, Rink L, Haase H (2016) Chelation of free Zn2+ impairs chemotaxis, phagocytosis, oxidative burst, degranulation, and cytokine production by neutrophil granulocytes. Biol Trace Elem Res 171(1):79–88

Hasegawa H, Suzuki K, Suzuki K, Nakaji S, Sugawara K (2000) Effects of zinc on the reactive oxygen species generating capacity of human neutrophils and on the serum opsonic activity in vitro. Luminescence :J Biol Chem Luminescence 15(5):321–327. https://doi.org/10.1002/1522-7243(200009/10)15:5<321::aid-bio605>3.0.co;2-o

Ho LH, Ruffin RE, Murgia C, Li L, Krilis SA, Zalewski PD (2004) Labile zinc and zinc transporter ZnT4 in mast cell granules: role in regulation of caspase activation and NF-kappaB translocation. J Immunol 172(12):7750–7760

Ho Y, Samarasinghe R, Knoch ME, Lewis M, Aizenman E, DeFranco DB (2008) Selective inhibition of mitogen-activated protein kinase phosphatases by zinc accounts for extracellular signal-regulated kinase 1/2-dependent oxidative neuronal cell death. Mol Pharmacol 74(4):1141–1151. https://doi.org/10.1124/mol.108.049064

Hogstrand C, Verbost PM, Bonga SE, Wood CM (1996) Mechanisms of zinc uptake in gills of freshwater rainbow trout: interplay with calcium transport. Am J Phys 270(5. Pt 2):R1141–R1147

Hojyo S, Miyai T, Fujishiro H, Kawamura M, Yasuda T, Hijikata A, Bin BH, Irie T, Tanaka J, Atsumi T, Murakami M, Nakayama M, Ohara O, Himeno S, Yoshida H, Koseki H, Ikawa T, Mishima K, Fukada T (2014) Zinc transporter SLC39A10/ZIP10 controls humoral immunity by modulating B-cell receptor signal strength. Proc Natl Acad Sci U S A 111(32):11786–11791. https://doi.org/10.1073/pnas.1323557111

Honscheid A, Rink L, Haase H (2009) T-lymphocytes: a target for stimulatory and inhibitory effects of zinc ions. Endocr Metab Immune Disord Drug Targets 9(2):132–144

Horwitz DA, Zheng SG, Wang J, Gray JD (2008) Critical role of IL-2 and TGF-beta in generation, function and stabilization of Foxp3+CD4+ Treg. Eur J Immunol 38(4):912–915. https://doi.org/10.1002/eji.200738109

Huang J, Zhang D, Xing W, Ma X, Yin Y, Wei Q, Li G (2008) An approach to assay calcineurin activity and the inhibitory effect of zinc ion. Anal Biochem 375(2):385–387. https://doi.org/10.1016/j.ab.2007.12.016

Hubbard SR, Bishop WR, Kirschmeier P, George SJ, Cramer SP, Hendrickson WA (1991) Identification and characterization of zinc binding sites in protein kinase C. Science 254(5039):1776–1779

Hujanen ES, Seppa ST, Virtanen K (1995) Polymorphonuclear leukocyte chemotaxis induced by zinc, copper and nickel in vitro. Biochim Biophys Acta 1245(2):145–152

Huse M, Eck MJ, Harrison SC (1998) A Zn2+ ion links the cytoplasmic tail of CD4 and the N-terminal region of Lck. J Biol Chem 273(30):18729–18733

Ibs KH, Rink L (2003) Zinc-altered immune function. J Nutr 133(5 Suppl 1):1452s–1456s

Jia Y, Jeng JM, Sensi SL, Weiss JH (2002) Zn2+ currents are mediated by calcium-permeable AMPA/kainate channels in cultured murine hippocampal neurones. J Physiol 543(Pt 1):35–48

Jirakulaporn T, Muslin AJ (2004) Cation diffusion facilitator proteins modulate Raf-1 activity. J Biol Chem 279(26):27807–27815. https://doi.org/10.1074/jbc.M401210200

Kabu K, Yamasaki S, Kamimura D, Ito Y, Hasegawa A, Sato E, Kitamura H, Nishida K, Hirano T (2006) Zinc is required for fc epsilon RI-mediated mast cell activation. J Immunol 177(2):1296–1305

Kahmann L, Uciechowski P, Warmuth S, Plumakers B, Gressner AM, Malavolta M, Mocchegiani E, Rink L (2008) Zinc supplementation in the elderly reduces spontaneous inflammatory cytokine release and restores T cell functions. Rejuvenation Res 11(1):227–237. https://doi.org/10.1089/rej.2007.0613

Kaltenberg J, Plum LM, Ober-Blobaum JL, Honscheid A, Rink L, Haase H (2010) Zinc signals promote IL-2-dependent proliferation of T cells. Eur J Immunol 40(5):1496–1503. https://doi.org/10.1002/eji.200939574

Kawagoe T, Sato S, Jung A, Yamamoto M, Matsui K, Kato H, Uematsu S, Takeuchi O, Akira S (2007) Essential role of IRAK-4 protein and its kinase activity in toll-like receptor-mediated

immune responses but not in TCR signaling. J Exp Med 204(5):1013–1024. https://doi.org/10.1084/jem.20061523
Keen CL, Gershwin ME (1990) Zinc deficiency and immune function. Annu Rev Nutr 10:415–431. https://doi.org/10.1146/annurev.nu.10.070190.002215
Kim PW, Sun ZY, Blacklow SC, Wagner G, Eck MJ (2003) A zinc clasp structure tethers Lck to T cell coreceptors CD4 and CD8. Science 301(5640):1725–1728. https://doi.org/10.1126/science.1085643
Kimura T, Kambe T (2016) The functions of metallothionein and ZIP and ZnT transporters: an overview and perspective. Int J Mol Sci 17(3):336
King LE, Fraker PJ (2000) Variations in the cell cycle status of lymphopoietic and myelopoietic cells created by zinc deficiency. J Infect Dis 182(Suppl 1):S16–S22. https://doi.org/10.1086/315923
King LE, Frentzel JW, Mann JJ, Fraker PJ (2005) Chronic zinc deficiency in mice disrupted T cell lymphopoiesis and erythropoiesis while B cell lymphopoiesis and myelopoiesis were maintained. J Am Coll Nutr 24(6):494–502
Kitamura H, Morikawa H, Kamon H, Iguchi M, Hojyo S, Fukada T, Yamashita S, Kaisho T, Akira S, Murakami M (2006a) Toll-like receptor-mediated regulation of zinc homeostasis influences dendritic cell function. Nat Immunol 7(9):971
Kitamura H, Morikawa H, Kamon H, Iguchi M, Hojyo S, Fukada T, Yamashita S, Kaisho T, Akira S, Murakami M, Hirano T (2006b) Toll-like receptor-mediated regulation of zinc homeostasis influences dendritic cell function. Nat Immunol 7(9):971–977. https://doi.org/10.1038/ni1373
Klein C, Sunahara RK, Hudson TY, Heyduk T, Howlett AC (2002) Zinc inhibition of cAMP signaling. J Biol Chem 277(14):11859–11865. https://doi.org/10.1074/jbc.M108808200
Klein C, Heyduk T, Sunahara RK (2004) Zinc inhibition of adenylyl cyclase correlates with conformational changes in the enzyme. Cell Signal 16(10):1177–1185. https://doi.org/10.1016/j.cellsig.2004.03.008
Knapp LT, Klann E (2000) Superoxide-induced stimulation of protein kinase C via thiol modification and modulation of zinc content. J Biol Chem 275(31):24136–24145. https://doi.org/10.1074/jbc.M002043200
Kojima C, Kawakami A, Takei T, Nitta K, Yoshida M (2007) Angiotensin-converting enzyme inhibitor attenuates monocyte adhesion to vascular endothelium through modulation of intracellular zinc. J Pharmacol Exp Ther 323(3):855–860. https://doi.org/10.1124/jpet.107.127944
Kopf M, Le Gros G, Bachmann M, Lamers MC, Bluethmann H, Kohler G (1993) Disruption of the murine IL-4 gene blocks Th2 cytokine responses. Nature 362(6417):245–248. https://doi.org/10.1038/362245a0
Korichneva I, Hoyos B, Chua R, Levi E, Hammerling U (2002) Zinc release from protein kinase C as the common event during activation by lipid second messenger or reactive oxygen. J Biol Chem 277(46):44327–44331. https://doi.org/10.1074/jbc.M205634200
Kown MH, Van der Steenhoven T, Blankenberg FG, Hoyt G, Berry GJ, Tait JF, Strauss HW, Robbins RC (2000) Zinc-mediated reduction of apoptosis in cardiac allografts. Circulation 102(19. Suppl 3):Iii228–Iii232
Kown MH, van der Steenhoven TJ, Jahncke CL, Mari C, Lijkwan MA, Koransky ML, Blankenberg FG, Strauss HW, Robbins RC (2002) Zinc chloride-mediated reduction of apoptosis as an adjunct immunosuppressive modality in cardiac transplantation. J Heart Lung Transplant 21(3):360–365
Krezel A, Maret W (2007) Dual nanomolar and picomolar Zn(II) binding properties of metallothionein. J Am Chem Soc 129(35):10911–10921. https://doi.org/10.1021/ja071979s
Kury S, Dreno B, Bezieau S, Giraudet S, Kharfi M, Kamoun R, Moisan JP (2002) Identification of SLC39A4, a gene involved in acrodermatitis enteropathica. Nat Genet 31(3):239–240. https://doi.org/10.1038/ng913
Laity JH, Andrews GK (2007) Understanding the mechanisms of zinc-sensing by metal-response element binding transcription factor-1 (MTF-1). Arch Biochem Biophys 463(2):201–210. https://doi.org/10.1016/j.abb.2007.03.019

LaRochelle O, Gagne V, Charron J, Soh JW, Seguin C (2001) Phosphorylation is involved in the activation of metal-regulatory transcription factor 1 in response to metal ions. J Biol Chem 276(45):41879–41888. https://doi.org/10.1074/jbc.M108313200
Lee WW, Cui D, Czesnikiewicz-Guzik M, Vencio RZ, Shmulevich I, Aderem A, Weyand CM, Goronzy JJ (2008) Age-dependent signature of metallothionein expression in primary CD4 T cell responses is due to sustained zinc signaling. Rejuvenation Res 11(6):1001–1011. https://doi.org/10.1089/rej.2008.0747
Lepage LM, Giesbrecht JA, Taylor CG (1999) Expression of T lymphocyte p56(lck), a zinc-finger signal transduction protein, is elevated by dietary zinc deficiency and diet restriction in mice. J Nutr 129(3):620–627
Lienau S, Rink L, Weßels I (2018) The role of zinc in calprotectin expression in human monocytic cells. J Trace Elem Med Biol 49:106–112. https://doi.org/10.1016/j.jtemb.2018.04.022
Lin RS, Rodriguez C, Veillette A, Lodish HF (1998) Zinc is essential for binding of p56(lck) to CD4 and CD8alpha. J Biol Chem 273(49):32878–32882
Lowe NM et al (2013) Eurreca-estimating zinc requirements for deriving dietary reference values. Crit Rev Food Sci Nutr 53:1110–1123
Macian F (2005) NFAT proteins: key regulators of T-cell development and function. Nat Rev Immunol 5(6):472–484. https://doi.org/10.1038/nri1632
Maret W (2006) Zinc coordination environments in proteins as redox sensors and signal transducers. Antioxid Redox Signal 8(9–10):1419–1441. https://doi.org/10.1089/ars.2006.8.1419
Maret W (2017) Zinc in pancreatic islet biology, insulin sensitivity, and diabetes. Prev Nutr Food Sci 22(1):1–8. https://doi.org/10.3746/pnf.2017.22.1.1
Maret W, Jacob C, Vallee BL, Fischer EH (1999) Inhibitory sites in enzymes: zinc removal and reactivation by thionein. Proc Natl Acad Sci U S A 96(5):1936–1940
Maywald M, Rink L (2016) Zinc supplementation induces CD4+CD25+Foxp3+ antigen-specific regulatory T cells and suppresses IFN-gamma production by upregulation of Foxp3 and KLF-10 and downregulation of IRF-1. Eur J Nutr. https://doi.org/10.1007/s00394-016-1228-7
Maywald M, Meurer SK, Weiskirchen R, Rink L (2017) Zinc supplementation augments TGF-beta1-dependent regulatory T cell induction. Mol Nutr Food Res 61 (3). https://doi.org/10.1002/mnfr.201600493
Maywald M, Wang F, Rink L (2018) Zinc supplementation plays a crucial role in T helper 9 differentiation in allogeneic immune reactions and non-activated T cells. J Trace Elem Med Biol 50:482–488. https://doi.org/10.1016/j.jtemb.2018.02.004
Mills CF (1989) Physiology of zinc: general aspects
Mingari MC, Moretta A, Moretta L (1998) Regulation of KIR expression in human T cells: a safety mechanism that may impair protective T-cell responses. Immunol Today 19(4):153–157
Miyai T, Hojyo S, Ikawa T, Kawamura M, Irie T, Ogura H, Hijikata A, Bin BH, Yasuda T, Kitamura H, Nakayama M, Ohara O, Yoshida H, Koseki H, Mishima K, Fukada T (2014) Zinc transporter SLC39A10/ZIP10 facilitates antiapoptotic signaling during early B-cell development. Proc Natl Acad Sci U S A 111(32):11780–11785. https://doi.org/10.1073/pnas.1323549111
Mocchegiani E, Santarelli L, Muzzioli M, Fabris N (1995) Reversibility of the thymic involution and of age-related peripheral immune dysfunctions by zinc supplementation in old mice. Int J Immunopharmacol 17(9):703–718
Moore JB, Blanchard RK, McCormack WT, Cousins RJ (2001) cDNA array analysis identifies thymic LCK as upregulated in moderate murine zinc deficiency before T-lymphocyte population changes. J Nutr 131(12):3189–3196
Moulder K, Steward MW (1989) Experimental zinc deficiency: effects on cellular responses and the affinity of humoral antibody. Clin Exp Immunol 77(2):269–274
Mustelin T, Tasken K (2003) Positive and negative regulation of T-cell activation through kinases and phosphatases. Biochem J 371(Pt 1):15–27. https://doi.org/10.1042/bj20021637
Muzzioli M, Stecconi R, Donnini A, Re F, Provinciali M (2007) Zinc improves the development of human CD34+ cell progenitors towards natural killer cells and induces the expression of

GATA-3 transcription factor. Int J Biochem Cell Biol 39(5):955–965. https://doi.org/10.1016/j.biocel.2007.01.011
Osati-Ashtiani F, King LE, Fraker PJ (1998) Variance in the resistance of murine early bone marrow B cells to a deficiency in zinc. Immunology 94(1):94–100
Palacios EH, Weiss A (2004) Function of the Src-family kinases, Lck and Fyn, in T-cell development and activation. Oncogene 23(48):7990–8000. https://doi.org/10.1038/sj.onc.1208074
Park JA, Koh JY (1999) Induction of an immediate early gene egr-1 by zinc through extracellular signal-regulated kinase activation in cortical culture: its role in zinc-induced neuronal death. J Neurochem 73(2):450–456
Plum LM et al (2014) PTEN-inhibition by zinc ions augments interleukin-2-mediated Akt phosphorylation. Metallomics 6(7):1277–1287. https://doi.org/10.1039/c3mt00197k
Prasad AS (1981) Zinc deficiency and effects of zinc supplementation on sickle cell anemia subjects. Prog Clin Biol Res 55:99–122
Prasad AS (1985) Clinical manifestations of zinc deficiency. Annu Rev Nutr 5:341–363. https://doi.org/10.1146/annurev.nu.05.070185.002013
Prasad AS (2000) Effects of zinc deficiency on Th1 and Th2 cytokine shifts. J Infect Dis 182(Suppl 1):S62–S68. https://doi.org/10.1086/315916
Prasad AS (2002) Zinc deficiency in patients with sickle cell disease. Am J Clin Nutr 75(2):181–182
Prasad AS (2009) Impact of the discovery of human zinc deficiency on health. J Am Coll Nutr 28(3):257–265
Prasad AS, Halsted JA, Nadimi M (1961) Syndrome of iron deficiency anemia, hepatosplenomegaly, hypogonadism, dwarfism and geophagia. Am J Med 31:532–546
Prasad AS, Miale A Jr, Farid Z, Sandstead HH, Schulert AR (1963) Zinc metabolism in patients with the syndrome of iron deficiency anemia, hepatosplenomegaly, dwarfism, and hypogonadism. J Lab Clin Med 61:537–549
Prasad AS, Bao B, Beck FW, Kucuk O, Sarkar FH (2004) Antioxidant effect of zinc in humans. Free Radic Biol Med 37(8):1182–1190. https://doi.org/10.1016/j.freeradbiomed.2004.07.007
Radtke F, Heuchel R, Georgiev O, Hergersberg M, Gariglio M, Dembic Z, Schaffner W (1993) Cloned transcription factor MTF-1 activates the mouse metallothionein I promoter. EMBO J 12(4):1355–1362
Ragozzino D, Giovannelli A, Degasperi V, Eusebi F, Grassi F (2000) Zinc permeates mouse muscle ACh receptor channels expressed in BOSC 23 cells and affects channel function. J Physiol 529(Pt 1):83–91
Rajagopalan S, Long EO (1998) Zinc bound to the killer cell-inhibitory receptor modulates the negative signal in human NK cells. J Immunol 161(3):1299–1305
Rajagopalan S, Winter CC, Wagtmann N, Long EO (1995) The Ig-related killer cell inhibitory receptor binds zinc and requires zinc for recognition of HLA-C on target cells. J Immunol 155(9):4143–4146
Ressnerova A, Raudenska M, Holubova M, Svobodova M, Polanska H, Babula P, Masarik M, Gumulec J (2016) Zinc and copper homeostasis in head and neck Cancer: review and meta-analysis. Curr Med Chem 23(13):1304–1330
Rink L, Gabriel P (2000) Zinc and the immune system. Proc Nutr Soc 59(4):541–552
Rolles B, Maywald M, Rink L (2018) Influence of zinc deficiency and supplementation on NK cell cytotoxicity. J Funct Foods 48:322–328. https://doi.org/10.1016/j.jff.2018.07.027
Rolles B, Maywald M, Rink L (2019) Intracellular zinc homeostasis during cell activation and zinc deficiency (in preparation)
Romir J, Lilie H, Egerer-Sieber C, Bauer F, Sticht H, Muller YA (2007) Crystal structure analysis and solution studies of human Lck-SH3; zinc-induced homodimerization competes with the binding of proline-rich motifs. J Mol Biol 365(5):1417–1428. https://doi.org/10.1016/j.jmb.2006.10.058
Rosenkranz E, Hilgers RD, Uciechowski P, Petersen A, Plumakers B, Rink L (2015) Zinc enhances the number of regulatory T cells in allergen-stimulated cells from atopic subjects. Eur J Nutr. https://doi.org/10.1007/s00394-015-1100-1

Rosenkranz E, Maywald M, Hilgers RD, Brieger A, Clarner T, Kipp M, Plumakers B, Meyer S, Schwerdtle T, Rink L (2016a) Induction of regulatory T cells in Th1-/Th17-driven experimental autoimmune encephalomyelitis by zinc administration. J Nutr Biochem 29:116–123. https://doi.org/10.1016/j.jnutbio.2015.11.010

Rosenkranz E, Metz CH, Maywald M, Hilgers RD, Wessels I, Senff T, Haase H, Jager M, Ott M, Aspinall R, Plumakers B, Rink L (2016b) Zinc supplementation induces regulatory T cells by inhibition of Sirt-1 deacetylase in mixed lymphocyte cultures. Mol Nutr Food Res 60(3):661–671. https://doi.org/10.1002/mnfr.201500524

Saha AR, Hadden EM, Hadden JW (1995) Zinc induces thymulin secretion from human thymic epithelial cells in vitro and augments splenocyte and thymocyte responses in vivo. Int J Immunopharmacol 17(9):729–733

Saito Y, Saito K, Hirano Y, Ikeya K, Suzuki H, Shishikura K, Manno S, Takakuwa Y, Nakagawa K, Iwasa A, Fujikawa S, Moriya M, Mizoguchi N, Golden BE, Osawa M (2002) Hyperzincemia with systemic inflammation: a heritable disorder of calprotectin metabolism with rheumatic manifestations? J Pediatr 140(2):267–269. https://doi.org/10.1067/mpd.2002.121699

Sampson B, Kovar IZ, Rauscher A, Fairweather-Tait S, Beattie J, McArdle HJ, Ahmed R, Green C (1997) A case of hyperzincemia with functional zinc depletion: a new disorder? Pediatr Res 42(2):219–225. https://doi.org/10.1203/00006450-199708000-00015

Sandstead HH (1991) Zinc deficiency. A public health problem? Am J Dis Child. (1960 145(8):853–859

Shankar AH, Prasad AS (1998) Zinc and immune function: the biological basis of altered resistance to infection. Am J Clin Nutr 68(2 Suppl):447s–463s

Sheikh A, Shamsuzzaman S, Ahmad SM, Nasrin D, Nahar S, Alam MM, Al Tarique A, Begum YA, Qadri SS, Chowdhury MI, Saha A, Larson CP, Qadri F (2010) Zinc influences innate immune responses in children with enterotoxigenic *Escherichia coli*-induced diarrhea. J Nutr 140(5):1049–1056. https://doi.org/10.3945/jn.109.111492

Smith JC, Zeller JA, Brown ED, Ong SC (1976) Elevated plasmz zinc: a heritable anomaly. Science (New York, NY) 193(4252):496–498

Sood SM, Wu MX, Hill KA, Slattery CW (1999) Characterization of zinc-depleted alanyl-tRNA synthetase from *Escherichia coli*: role of zinc. Arch Biochem Biophys 368(2):380–384. https://doi.org/10.1006/abbi.1999.1314

Stelmashook EV, Isaev NK, Genrikhs EE, Amelkina GA, Khaspekov LG, Skrebitsky VG, Illarioshkin SN (2014) Role of zinc and copper ions in the pathogenetic mechanisms of Alzheimer's and Parkinson's diseases. Biochem Biokhimiia 79(5):391–396. https://doi.org/10.1134/s0006297914050022

Stennicke HR, Salvesen GS (1997) Biochemical characteristics of caspases-3, −6, −7, and −8. J Biol Chem 272(41):25719–25723

Suzuki N, Suzuki S, Millar DG, Unno M, Hara H, Calzascia T, Yamasaki S, Yokosuka T, Chen NJ, Elford AR, Suzuki J, Takeuchi A, Mirtsos C, Bouchard D, Ohashi PS, Yeh WC, Saito T (2006) A critical role for the innate immune signaling molecule IRAK-4 in T cell activation. Science 311(5769):1927–1932. https://doi.org/10.1126/science.1124256

Szewczyk B (2013) Zinc homeostasis and neurodegenerative disorders. Front Aging Neurosci 5:33. https://doi.org/10.3389/fnagi.2013.00033

Takahashi K, Akaishi E, Abe Y, Ishikawa R, Tanaka S, Hosaka K, Kubohara Y (2003) Zinc inhibits calcineurin activity in vitro by competing with nickel. Biochem Biophys Res Commun 307(1):64–68

Takaki H, Ichiyama K, Koga K, Chinen T, Takaesu G, Sugiyama Y, Kato S, Yoshimura A, Kobayashi T (2008) STAT6 inhibits TGF-beta1-mediated Foxp3 induction through direct binding to the Foxp3 promoter, which is reverted by retinoic acid receptor. J Biol Chem 283(22):14955–14962. https://doi.org/10.1074/jbc.M801123200

Tan SL, Parker PJ (2003) Emerging and diverse roles of protein kinase C in immune cell signalling. Biochem J 376(Pt 3):545–552. https://doi.org/10.1042/bj20031406

Tanaka Y (1989) Zinc inhibits pokeweed mitogen-induced development of immunoglobulin-secreting cells through augmentation of both CD4 and CD8 cells. Int J Immunopharmacol 11(6):673–679

Tang X, Shay NF (2001) Zinc has an insulin-like effect on glucose transport mediated by phosphoinositol-3-kinase and Akt in 3T3-L1 fibroblasts and adipocytes. J Nutr 131(5):1414–1420

Tillmann N, Engelhardt G, Rink L, Weßels I (2020, in preparation) Zinc in granulopoiesis

Trame S, Wessels I, Haase H, Rink L (2018) A short 18 items food frequency questionnaire biochemically validated to estimate zinc status in humans. J Trace Elem Med Biol 49:285–295. https://doi.org/10.1016/j.jtemb.2018.02.020

Truong-Tran AQ, Carter J, Ruffin RE, Zalewski PD (2001) The role of zinc in caspase activation and apoptotic cell death. Biometals 14(3–4):315–330

Ulges A, Klein M, Reuter S, Gerlitzki B, Hoffmann M, Grebe N, Staudt V, Stergiou N, Bohn T, Bruhl TJ, Muth S, Yurugi H, Rajalingam K, Bellinghausen I, Tuettenberg A, Hahn S, Reissig S, Haben I, Zipp F, Waisman A, Probst HC, Beilhack A, Buchou T, Filhol-Cochet O, Boldyreff B, Breloer M, Jonuleit H, Schild H, Schmitt E, Bopp T (2015) Protein kinase CK2 enables regulatory T cells to suppress excessive TH2 responses in vivo. Nat Immunol 16(3):267–275. https://doi.org/10.1038/ni.3083

Vallee BL, Falchuk KH (1993) The biochemical basis of zinc physiology. Physiol Rev 73(1):79–118

Vang T, Torgersen KM, Sundvold V, Saxena M, Levy FO, Skalhegg BS, Hansson V, Mustelin T, Tasken K (2001) Activation of the COOH-terminal Src kinase (Csk) by cAMP-dependent protein kinase inhibits signaling through the T cell receptor. J Exp Med 193(4):497–507

von Bulow V, Rink L, Haase H (2005) Zinc-mediated inhibition of cyclic nucleotide phosphodiesterase activity and expression suppresses TNF-alpha and IL-1 beta production in monocytes by elevation of guanosine 3′,5′-cyclic monophosphate. J Immunol. (Baltimore, Md : 1950 175(7):4697–4705

von Bulow V, Dubben S, Engelhardt G, Hebel S, Plumakers B, Heine H, Rink L, Haase H (2007) Zinc-dependent suppression of TNF-alpha production is mediated by protein kinase A-induced inhibition of Raf-1, I kappa B kinase beta, and NF-kappa B. J Immunol Baltimore, MD 1950 179(6):4180–4186

Wang C, Deng L, Hong M, Akkaraju GR, Inoue J, Chen ZJ (2001) TAK1 is a ubiquitin-dependent kinase of MKK and IKK. Nature 412(6844):346–351. https://doi.org/10.1038/35085597

Wang K, Zhou B, Kuo YM, Zemansky J, Gitschier J (2002) A novel member of a zinc transporter family is defective in acrodermatitis enteropathica. Am J Hum Genet 71(1):66–73. https://doi.org/10.1086/341125

Wellinghausen N, Rink L (1998) The significance of zinc for leukocyte biology. J Leukoc Biol 64(5):571–577

Wellinghausen N, Fischer A, Kirchner H, Rink L (1996) Interaction of zinc ions with human peripheral blood mononuclear cells. Cell Immunol 171(2):255–261. https://doi.org/10.1006/cimm.1996.0201

Wellinghausen N, Kirchner H, Rink L (1997a) The immunobiology of zinc. Immunol Today 18(11):519–521

Wellinghausen N, Martin M, Rink L (1997b) Zinc inhibits interleukin-1-dependent T cell stimulation. Eur J Immunol 27(10):2529–2535. https://doi.org/10.1002/eji.1830271010

Wessels I (2014) Epigenetics and metal deficiencies. Curr Nutr Rep 3(3):196–203. https://doi.org/10.1007/s13668-014-0091-5

Wessels I, Haase H, Engelhardt G, Rink L, Uciechowski P (2013) Zinc deficiency induces production of the proinflammatory cytokines IL-1β and TNFα in promyeloid cells via epigenetic and redox-dependent mechanisms. J Nutr Biochem 24(1):289–297

WHO (2002) The world health report 2002. World Health Organisation, Geneva

Williams RJ (1984) Zinc: what is its role in biology? Endeavour 8(2):65–70

Wu W, Jaspers I, Zhang W, Graves LM, Samet JM (2002) Role of Ras in metal-induced EGF receptor signaling and NF-kappaB activation in human airway epithelial cells. Am J Physiol Lung Cell Mol Physiol 282(5):L1040–L1048. https://doi.org/10.1152/ajplung.00390.2001

Wu W, Wang X, Zhang W, Reed W, Samet JM, Whang YE, Ghio AJ (2003) Zinc-induced PTEN protein degradation through the proteasome pathway in human airway epithelial cells. J Biol Chem 278(30):28258–28263. https://doi.org/10.1074/jbc.M303318200

Wuehler SE, Peerson JM, Brown KH (2005) Use of national food balance data to estimate the adequacy of zinc in national food supplies: methodology and regional estimates. Public Health Nutr 8(7):812–819

Xin L, Yang X, Cai G, Fan D, Xia Q, Liu L, Hu Y, Ding N, Xu S, Wang L, Li X, Zou Y, Pan F (2015) Serum levels of copper and zinc in patients with rheumatoid arthritis: a meta-analysis. Biol Trace Elem Res 168(1):1–10. https://doi.org/10.1007/s12011-015-0325-4

Yamasaki S, Sakata-Sogawa K, Hasegawa A, Suzuki T, Kabu K, Sato E, Kurosaki T, Yamashita S, Tokunaga M, Nishida K, Hirano T (2007) Zinc is a novel intracellular second messenger. J Cell Biol 177(4):637–645. https://doi.org/10.1083/jcb.200702081

Yamasaki S, Hasegawa A, Hojyo S, Ohashi W, Fukada T, Nishida K, Hirano T (2012) A novel role of the L-type calcium channel alpha1D subunit as a gatekeeper for intracellular zinc signaling: zinc wave. PLoS One 7(6):e39654. https://doi.org/10.1371/journal.pone.0039654

Yoder JH, Chong H, Guan KL, Han M (2004) Modulation of KSR activity in *Caenorhabditis elegans* by Zn ions, PAR-1 kinase and PP2A phosphatase. EMBO J 23(1):111–119. https://doi.org/10.1038/sj.emboj.7600025

Yokoyama WM, Plougastel BF (2003) Immune functions encoded by the natural killer gene complex. Nat Rev Immunol 3(4):304–316. https://doi.org/10.1038/nri1055

Yu M, Lee WW, Tomar D, Pryshchep S, Czesnikiewicz-Guzik M, Lamar DL, Li G, Singh K, Tian L, Weyand CM, Goronzy JJ (2011) Regulation of T cell receptor signaling by activation-induced zinc influx. J Exp Med 208(4):775–785. https://doi.org/10.1084/jem.20100031

Chapter 14
Zinc Signals in Inflammation

Sannette C. Hall and Daren L. Knoell

Abstract Zinc is an essential micronutrient for cell growth, differentiation, and survival and when deficient, is associated with increased susceptibility to infections and inflammatory disorders. Zinc homeostasis is critical for proper immune cell function and is therefore tightly regulated by zinc transporters. Recent evidence has highlighted zinc as an intracellular signaling molecule capable of modulating immune cell signaling. Slight changes in intracellular zinc, either by zinc deficiency or by excess zinc, can alter cellular signaling and immune cell function often resulting in increased inflammation. In this chapter, we discuss zinc signals in inflammation with a focus on zinc dependent modulation of select signaling pathways and the effects on immune cell function in response to potentially damaging challenges.

Keywords Zinc · Zinc deficiency · Inflammation · Zinc signal · Host defense · Innate and adaptive immunity

14.1 Introduction

Zinc (Zn) is an essential micronutrient for human health. Since the discovery of its importance in the 1960s, several studies have highlighted the role of zinc in immune cell function [extensively reviewed in (Maywald et al. 2017; Hojyo and Fukada 2016)]. A zinc imbalance, whether through zinc deficiency or excess zinc, can result in altered immune responses. Despite its known importance, it is estimated that nearly two billion people globally suffer from zinc deficiency (Prasad 2012; Wessells and Brown 2012), which is associated with a number of pathological conditions including increased susceptibility to infections and inflammation because of impaired immunological responses (Bonaventura et al. 2015; Gammoh and Rink 2017; Sapkota and Knoell 2018).

S. C. Hall · D. L. Knoell (✉)
College of Pharmacy, Pharmacy Practice and Science, University of Nebraska Medical Center, Omaha, NE, USA
e-mail: daren.knoell@unmc.edu

T. Fukada, T. Kambe (eds.), *Zinc Signaling*,
https://doi.org/10.1007/978-981-15-0557-7_14

Zinc is ubiquitous in cells and can be found in the cytoplasm, as well as in most organelles. It is an important structural component of many proteins, including transcription factors and enzymes, therefore intracellular zinc levels are tightly regulated by zinc sensors and zinc transporters to ensure homeostasis (Bonaventura et al. 2015; Hojyo and Fukada 2016; Maywald et al. 2017). Within the cell, Zn has structural, catalytic, and co-catalytic functions. In its structural role, Zn atoms maintain the tertiary structure of enzymes (analogous to disulfide bridges) involved in signaling pathways and may influence the catalytic activity of enzymes (Bonaventura et al. 2015; Murakami and Hirano 2008). Zinc also serves as a co-factor in Zn-finger domains to stabilize protein folding and regulate protein-protein or protein-nucleic acid interactions (Bonaventura et al. 2015; Hirano et al. 2008; Murakami and Hirano 2008). Zinc finger proteins (ZNFs) are one of the most abundant groups of proteins. Zinc finger domains are present in 3% of the human genome and the zinc proteome is estimated to have over 3000 proteins that have a wide array of functions including transcriptional regulation, DNA repair, cell migration, and signal transduction (Cassandri et al. 2017; Bonaventura et al. 2015).

Free intracellular zinc, also referred to as labile or available zinc, is regulated by Zn transporters that uptake and redistribute Zn to mediate signaling (Haase and Rink 2014). Transport of zinc, from either the extracellular space or organelles to the cytosol, is mediated by 14 ZIP (Zrt/Irt like) proteins, resulting in a net increase in cytosolic Zn levels. Zinc export or sequestering is mediated by 10 ZnT transporters, resulting in an overall decrease in cytosolic zinc. Other proteins, namely metallothioneins (MTs) and a zinc-sensing transcription factor, metal-responsive transcription factor (MTF)-1, also play a role in maintaining zinc homeostasis (Bonaventura et al. 2015; Haase and Rink 2014; Hojyo and Fukada 2016; Maret 2017). Free intracellular zinc has been shown to act like a signaling molecule, either as a neurotransmitter or as a secondary messenger (Bonaventura et al. 2015; Maywald et al. 2017; Murakami and Hirano 2008). In the brain, free or loosely bound zinc can be found in forebrain glutamatergic neurons sequestered into synaptic vesicles. When released into the synaptic cleft after presynaptic stimulation, Zn has the capacity to interact with and modulate glutamate receptors and transporters, among other synaptic targets (Paoletti et al. 2009; Toth 2011). Over the last decade, zinc has been established as a second messenger, comparable to calcium and cyclic adenosine monophosphate (cAMP). Relative to inflammation, Zn can act through Zn transporters and channels to regulate signal cascades mediated by a number of receptors including growth factor, cytokine, and patter recognition receptors (Bonaventura et al. 2015; Hojyo and Fukada 2016; Hirano et al. 2008; Maywald et al. 2017).

Although the pool of free intracellular zinc is small, slight changes in Zn levels can alter cellular signaling by affecting protein stability and structure of hormones, growth factors, and cytokines (Hojyo and Fukada 2016; Maywald et al. 2017). In this chapter, we discuss zinc signals in inflammation, particularly zinc dependent modulation of signaling pathways that modulate immune cell function and the corresponding consequences of altered Zn concentrations in the context of host defense and the balancing act of inflammation.

14.2 Zinc Signals and Inflammation

14.2.1 Zinc Signals

Intracellular zinc signals can be broadly characterized as either a zinc flux (early Zn signal or EZS) or a homeostatic zinc signal (late zinc signal or LZS). The EZS occurs within seconds to minutes and is triggered by external stimuli such as activation of pattern-recognition receptors (PRRs), cytokines, or growth factors. This results in a rapid change in intracellular Zn levels, through the release of Zn, typically from a Zn store, that directly influences signaling cascades independent of Zn transporter synthesis (Bonaventura et al. 2015; Fukada et al. 2011; Maywald et al. 2017). The altered zinc levels can then activate select Zn transporters to further facilitate sharp increases in cytosolic Zn (Maywald et al. 2017).

The LZS occurs within several hours and can lasts for several days. Unlike the EZS, the LZS is dependent on transcriptional change and altered expression of proteins involved in Zn homeostasis (Fukada et al. 2011; Maywald et al. 2017; Maret 2017). Both zinc signals have been shown to modulate immune cell signaling in monocytes/macrophages, antigen presenting cells, T-cells, and mast cells [reviewed in (Haase and Rink 2009; Hirano et al. 2008; Fukada et al. 2011; Maywald et al. 2017; Maret 2017; Nishida and Uchida 2018) and discussed further in this chapter].

14.2.2 Zinc Homeostasis and Signaling Pathways

As an intracellular signaling molecule, Zn has been implicated in various inflammatory signaling pathways. Signal transduction mediated by Zn often involves post-translational modification by changes in phosphorylation (Maywald et al. 2017; Maret 2017). Therefore, signaling pathways that utilize phosphorylation for downstream signal activation can be affected by altered Zn homeostasis (Fig. 14.1).

14.2.2.1 Zinc and the MAPK Pathway

As previously mentioned, early zinc signals (zinc flux or zinc wave), result in the release of stored Zn. Yamasaki and colleagues demonstrated that cross linking of high affinity IgE receptors (FcεRI) on mast cells induced release of Zn in the perinuclear area; a phenomenon they termed "zinc wave". This release was dependent on calcium influx and the activation of mitogen-activated protein kinases (MAPKs), thus establishing Zn as an important second messenger (Yamasaki et al. 2007). Since the discovery of this phenomenon in mast cells, other studies have shown that intracellular Zn can alter MAPK signaling. Zinc has been found to differentially regulate MAPKs in T-cells with increased activation of p38. Further investigation revealed that this zinc-induced activation was mediated by activation of the cAMP

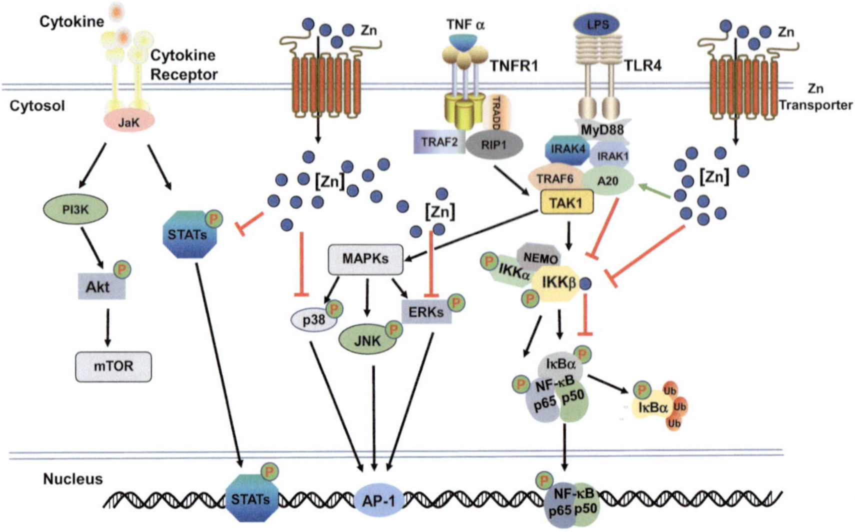

Fig. 14.1 Zinc signals and the inflammatory response. Zinc homeostasis is critical for proper immune cell function and has been shown to alter signal cascade in several inflammatory signaling pathways (discussed in this chapter). Increase in intracellular zinc, via Zn transporters, can modulate signal cascade directly or indirectly by inhibiting phosphorylation or activating pathway inhibitors that prevent nuclear translocation and gene expression. Green dots with P – phosphorylation; blue dots – intracellular zinc; black arrows – signal transduction, green arrow – activating effect of Zn; red lines – inhibition. LPS – lipopolysaccharide; TLR – toll-like receptor; MyD88 – myeloid differentiation primary response 88; IRAK- interleukin-1 receptor-associated kinase; TRAF – tumor necrosis factor receptor-associated factor; TAK – transforming growth factor beta activated kinase; IKK – I kappa B kinase; NFκB – nuclear factor kappa B; TRADD – tumor necrosis factor receptor type 1-associated death domain; TNF – tumor necrosis factor; TNFR – tumor necrosis factor receptor; MAPKs – mitogen activated protein kinase; ERK – extra-cellular signal regulated kinases; JNK – c-Jun N-terminal kinases; AP-1 – activator protein 1; STAT – signal transducer and activation of transcription; JaK – Janus kinase; PI3K – phosphatidyl-inositol-3-kinase

response element binding (CREB) protein and increased mRNA expression of Th1 cytokines (Honscheid et al. 2012). In the brain, increased intracellular Zn was found to inhibit extra-cellular signal regulated kinases (ERK)-1/2 activation and cell death in neurons (Ho et al. 2008). Other studies have suggested that Zn and ERK1/2 are important for normal brain development and can modulate neural progenitor cell proliferation. Zinc deficiency decreases ERK1/2 activity and suppresses neuronal progenitor proliferation, which has been shown to contribute to depression. Conversely, zinc supplementation has been shown to increase ERK1/2 phosphorylation, which has been shown to mitigate these deleterious effects [reviewed in (Nuttall and Oteiza 2012)].

Zinc homeostasis has been shown to affect signaling of other protein tyrosine kinases including phosphoinositide 3-kinase (PI3K) and CD45 (found on all leukocytes and regulates Src kinase family mediated signaling), and cell cycle controller CK2 (Maywald et al. 2017; Nimmanon et al. 2017; Taylor et al. 2012). Collectively, these observations indicate that Zn mediated phosphorylation of protein kinases can activate or inhibit inflammation depending upon the stimulus, cell type, and environment.

14.2.2.2 Zinc and the $NF_{\kappa}B$ Pathway

Like MAPKs, nuclear factor kappa B ($NF_{\kappa}B$) signaling is critical for proper immune cell function. $NF_{\kappa}B$ is a family of inducible transcription factors that regulate multiple aspects of the inflammatory response. The family is made up of five structurally related proteins namely, $NF_{\kappa}B1$ (p50), $NF_{\kappa}B2$ (p52), RelA (p65), RelC, and c-Rel, that form homo and heterodimers. These proteins are not synthesized de novo, but instead are present in an inactive form in the cytoplasm and regulated by another family of proteins known as inhibitors of $NF_{\kappa}B$ (IKB). $NF_{\kappa}B$ signaling can be divided into the classical/canonical and alternative/non-canonical pathways (triggered by activation of a subset of receptors). Activation of either pathway induces expression of multiple pro-inflammatory genes that are critical for the inflammatory process, cell adhesion, innate and adaptive immune cell function (Liu. et al. 2017; Jarosz et al. 2017).

Zinc has been shown to modulate the $NF_{\kappa}B$ pathway in various immune cells; however, results to date have been contradictory. In T-cells, zinc was found to be a positive regulator of $NF_{\kappa}B$ signaling, increasing expression of interleukin (IL)-2 (which plays a central role in T-helper cell function), and its receptor in HUT-78 cells. Further investigation revealed that this response was mediated by increased phosphorylation on IKB-α and subsequent translocation and activation of $NF_{\kappa}B$ by the IKB pathway (Bao et al. 2007). In monocytes, stimulation with toll-like receptor (TLR) ligands, LPS (TLR4) and Pam_3CSK_4 (TLR1/2), induced an increase in free zinc. In the case of LPS stimulation, this was associated with increased production of pro-inflammatory cytokines, suggesting a potential role of Zn signals in LPS-induced activation of $NF_{\kappa}B$ (and MAPK) pathways (Haase et al. 2008).

Conversely, the majority of literature suggests that zinc is a negative regulator of $NF_{\kappa}B$. In cancer cells, zinc was found to inhibit $NF_{\kappa}B$ and activator protein (AP)-1 activation. This was associated with reduced expression of several pro-angiogenic and pro-metastatic mediators, namely, vascular endothelial growth factor (VEGF), IL-6, IL-8, and matrix metalloproteinase (MMP)-9 (Uzzo et al. 2006). More recently, the ZIP8 transporter has been shown to be a negative feedback regulator of $NF_{\kappa}B$ through zinc-mediated inhibition of IKK. In lung epithelia and monocytes/macrophages, bacterial exposure significantly increased expression of ZIP8, which resulted in net increase of cytosolic Zn. This was found to inhibit $NF_{\kappa}B$-mediated signaling by blocking the IKK complex (Liu et al. 2013).

Zinc can also mediate its effects on the $NF_{\kappa}B$ signaling by modulating inhibitors of the pathway. Perhaps the most studied of these is the effects of zinc on the A20 protein, the main negative regulator of $NF_{\kappa}B$ activation via the tumor necrosis factor (TNF)-α receptor (TNFR)- and TLR-mediated pathways (Jarosz et al. 2017). Prasad and colleagues showed that zinc increased mRNA and protein expression of

A20 while decreasing oxidative stress and expression of IKK-α/ $NF_{K}B$ signaling in vitro (Prasad et al. 2011). Using the mouse macrophage cell line RAW 246.7, treatment with zinc oxide nanoparticles (ZO-NP) inhibited LPS-induced activation of $NF_{K}B$ by upregulating expression of A20 (Kim and Jeong 2015). In a rodent model of abdominal aorta aneurysm (AAA), in which chronic inflammation and degradation of elastin are the major contributors to pathogenesis, intraperitoneal injection with zinc sulphate ($ZnSO_4$) attenuated calcium chloride induced expansion of aortic diameter. Zinc supplementation also decreased infiltrating macrophages and lymphocytes in the aortas and induced expression of A20 that inhibited $NF_{K}B$ signaling pathway (Yan et al. 2016).

The evidence presented in these studies demonstrate that zinc alters $NF_{K}B$ signaling in immune cells. This zinc-mediated modulation of $NF_{K}B$ signaling can occur based on changes to key proteins involved in the signaling pathway and may vary depending on the activating signals and/or the type of immune cell involved in the response.

14.2.2.3 Zinc and GPCRs

There also exists a unique plasma membrane G-protein coupled receptor (ZnR/GPR39) that is sensitive to Zn and senses changes in extracellular Zn levels (Hershfinkel 2018). ZnR/GPR39 can be activated directly by endogenous release of zinc or indirectly by allosteric modulation of ZnR/GPR39 signaling (Sunuwar et al. 2017; Sato et al. 2016). Activation of ZnR/GPR39 induces the release of calcium via the inositol trisphosphate (IP3) pathway by signaling through Gαq, and has been shown to be important in cell signaling in neurons, keratinocytes, pancreatic cells, cancer cells, and bone (Hershfinkel 2018; Sunuwar et al. 2017).

ZnR/GPR39 has also been shown to be ubiquitously expressed throughout the gastrointestinal tract. A recent study demonstrated that engagement of the Zn sensing receptor enhanced tight junction assembly via activation of AMP activated protein kinase (AMPK) signaling, thus highlighting a potential therapeutic role of ZnR/GPR39 in intestinal barrier disruption (Pongkorpsakol et al. 2019). Zn has also been shown to upregulate protein kinase C (PKC)-ζ via activation of GPR39 and improve epithelial integrity by enhancing the abundance of tight junction protein zonula occludens (ZO)-1, following Salmonella challenge in vitro (Shao et al. 2017).

More recently, studies have identified an unexpected link between GPR-mediated signaling and Zn homeostasis. In mice, the ZIP14 transporter was shown to control GPR-mediated signaling by maintaining basal cAMP levels, through net increase in cytosolic zinc, and suppressing phosphodiesterase (PDE) activity (Hojyo et al. 2011). Another G-coupled protein receptor, GPR83, was also found to be activated by zinc. Under physiological conditions, Zn activated mouse GPR83 by binding to a cluster of ion-binding sensitive amino acids in an activation sensitive region of the receptor. This was not seen with calcium or magnesium, highlighting zinc as a potent agonist of GPR83. Further investigation revealed that the zinc-mediated activation was via the Gq/11 signaling pathway (Muller et al. 2013).

14.2.2.4 Zinc and Other Signaling Pathways

The Janus kinase/signal transducers and activators of transcription (JAK/STAT) signaling pathways are also affected by cellular zinc status. The mechanisms of JAK/STAT signaling have been extensively studied by several groups. Briefly, the signal cascade is initiated by extracellular binding of cytokine and/or growth factors to their respective receptors. This activates Jak, which then activates STATs, leading to dimerization, nuclear translocation, and DNA binding. The JAK/STAT signaling pathway regulates the expression of genes that are involved in cell proliferation, differentiation, survival, and apoptosis (Bafaro et al. 2017; Villarino et al. 2015).

Zinc deficiency was found to impair STAT1/3 signaling in fetal brain and neuroblastoma by inhibiting phosphorylation and nuclear translocation. Zn-mediated oxidative stress was also shown to impair STAT1/3 modulation, thus compounding the deleterious effects on fetal brain development (Supasai et al. 2017). In B-cells, ZIP10 was found to play a critical role in B-cell development and modulate early B-cell survival through activation of STAT3/5. ZIP10 was also highly expressed in human B-cell lymphoma samples and is believed to potentially exacerbate malignancy via Jak/STAT signaling, in coordination with Bcl2 (Miyai et al. 2014).

14.2.2.5 Zinc and ROS

Oxidative stress is essentially an imbalance between the production of reactive oxygen species (ROS) and the ability of cellular anti-oxidant systems to counteract their harmful effects. This leads to the oxidation of biomolecules that contribute to cellular damage and facilitate the development of several inflammatory diseases including cancer, diabetes, atherosclerosis, and neurodegeneration (Jarosz et al. 2017; Marreiro et al. 2017). The role of zinc as an anti-oxidant has been extensively studied. Zinc exhibits its anti-oxidant properties through catalytic activity of superoxide dismutase, upregulation of metallothionein (MT) expression, and protection of thiol groups from oxidation [discussed in (Lee 2018; Marreiro et al. 2017; Olechnowicz et al. 2018; Jarosz et al. 2017)].

As discussed in the studies outlined in the above reviews, zinc is a structural component of the superoxide dismutase (SOD) enzyme; the enzyme responsible for promoting conversion of harmful superoxide radicals to less harmful metabolites such as oxygen and hydrogen peroxide. Zinc, bound to MTs under physiological conditions, can be released from this complex during oxidative stress and intracellularly redistributed to exert anti-oxidant properties (Lee 2018; Marreiro et al. 2017; Olechnowicz et al. 2018; Jarosz et al. 2017).

It is important to note that the anti-oxidant properties of zinc are only afforded at physiological concentrations. Zinc deficiency has been associated with increased oxidative stress and inflammation in several in vitro and in vivo models of cardiovascular disease (Choi et al. 2018; Lee 2018). Chronic zinc deficiency also impairs the activity of MTs, by decreasing MT expression, and as such increases susceptibility to injury and inflammation mediated by oxidative stress (Jarosz et al. 2017;

Lee 2018; Olechnowicz et al. 2018). Other studies have demonstrated that at high concentrations, zinc can exert pro-oxidant effects. In the brain, elevated Zn levels have been shown to be toxic to astrocytes and neurons by inhibiting the activity of glutathione reductase, increasing oxidative stress (Bishop et al. 2007). In vascular smooth muscle cells (VSMCs), Zn overload (50 μM and higher) induced senescence and activated nicotinamide adenine dinucleotide phosphate hydrogen (NADPH) oxidase (Nox1) by increasing the production of mitochondrial ROS. This in part mediated the activation of $NF_{\kappa}B$, which when translocated to the nucleus, induced production of Nox1, further increasing cytoplasmic ROS levels (Salazar et al. 2017).

The above studies highlight multifunctional roles of zinc in redox regulation and its ability to act as an anti-oxidant or pro-oxidant based on zinc status. This further demonstrates the importance of zinc homeostasis in cell signaling and inflammation.

14.2.3 Zinc Signals in Acute Inflammation

Inflammation is a natural mechanism employed by the host to protect itself against infections, invading pathogens or injury. The primary goal of the inflammatory response is to detect and eliminate factors that interfere with homeostasis, thus restoring balance in the host. In some cases, however, resolution of inflammation is aberrant, and the inflammatory response persists leading to chronic inflammation that drives tissue damage (Gammoh and Rink 2017).

Nutritional inadequacies in early life contribute to poor lung function and impaired immunity resulting in more frequent and severe infections. Many children in developing countries suffer from zinc deficiency, which is associated with growth retardation, GI tract disorders, and increased susceptibility to diarrhea and respiratory infections (Prasad 2012; Basnet et al. 2015; Karim et al. 2017). Zinc supplementation has been shown to be associated with lower risk of acute lower respiratory infections in children and overall decreased mortality and morbidity (Karim et al. 2017; Malik et al. 2014; Roth et al. 2010).

14.2.3.1 Zinc and the Common Cold

The common cold is an acute infection of the upper respiratory tract that is spread by direct or indirect contact with secretions from an infected person or aerosol secretions. The cold lasts for about 7–10 days and symptoms include sore throat, rhinitis, cough, and fatigue (Allan and Arroll 2014). Zinc was suggested to be an effective treatment for the cold in the 1980s, however since then, there has been inconclusive evidence provided by clinical trials. Since the 2000s, several meta-analyses have been aimed at determining whether zinc supplementation can reduce the severity and duration of symptoms associated with the common cold.

A comparison of zinc lozenges and zinc intranasal gel versus placebo showed inconclusive results regarding zinc's ability to reduce duration of common cold symptoms (Arroll 2008). More recently, two meta-analyses have shown that zinc supplementation can reduce the duration and severity of a cold. In particular, evidence from 13 randomized placebo-controlled trials showed that if taken soon after the onset of cold symptoms, zinc could reduce both the duration and severity of these symptoms (Rao and Rowland 2011). Another study further examined these effects and reported that the reduction in symptoms were restricted to the dosage and formulation of zinc administered. Specifically, higher doses and administration of zinc acetate reduced the duration of symptoms, which was not seen with lower doses or other formulations (Science et al. 2012).

14.2.3.2 Zinc and Sepsis

Sepsis is a form of the systemic inflammatory response that is initiated by an infection and triggers an initial over-reaction of the immune system resulting in damage to the host, often in the form of organ failure. The response is characterized by an acute inflammatory phase, mediated by the innate immune system, followed by an immune suppressive phase, characterized by impaired innate and adaptive immune responses (Alker and Haase 2018; Bosmann and Ward 2013; Souffriau and Libert 2018). As previously mentioned, zinc homeostasis is critical for proper immune cell function and zinc deficiency is associated with increased risk of infection. Given that sepsis is caused by an exaggerated immune response to an infection, it should come as no surprise that zinc status plays a role in the pathology of sepsis.

A relationship between zinc status and sepsis has consistently been demonstrated by animal and human studies that have been the focus of recent reviews (Alker and Haase 2018; Bosmann and Ward 2013; Knoell and Liu 2010; Souffriau and Libert 2018). For the most part, zinc has been shown to provide protection against sepsis with a direct link with low zinc levels and increased sensitivity to develop systemic inflammatory responses and sepsis. The mechanism of protection seems to be mediated by the effects of zinc on the inflammatory response, chemotaxis, oxidative stress, and alteration of immune cell function.

14.2.4 Zinc Signals in Chronic Inflammation

14.2.4.1 Zinc and Allergic Airway Inflammation

Zinc deficiency is also associated with many chronic inflammatory diseases including allergic airway inflammation, rheumatoid arthritis, atherosclerosis, colitis, and dermatitis (Bonaventura et al. 2015; Fukada et al. 2011).

In humans, zinc deficiency has been linked to increased risk of asthma and its supplementation has been shown to decrease acute exacerbations in children

(Rerksuppaphol and Rerksuppaphol 2016). The underlying anti-inflammatory mechanisms of Zn supplementation in allergic airway inflammation have been further investigated using animal models. Early animal studies showed that Zn supplementation decreased infiltrating leukocytes, particularly eosinophils and lymphocytes, to the airways in acute and chronic models of allergic airway inflammation (Lang et al. 2007). Rats maintained on zinc restricted diets, which were sensitized and challenged with the experimental allergen ovalbumin (OVA), showed decreased numbers of eosinophils, neutrophils, and monocytes in broncho-alveolar lavage fluid. Zn supplementation resulted in suppressed expression of chemokines, eotaxin, and monocyte chemoattractant protein (MCP)-1, and increased mRNA expression of interferon (IFN)-γ (which inhibits Th2 cytokines) in the lung (Lu et al. 2012). In a similar study, Zn supplementation decreased airway hyperresponsiveness (AHR) and IgE levels; both of which are hallmark features of allergic airway inflammation (Morgan et al. 2011).

Zinc also plays a role in directly regulating the inflammatory response mediated by several cell types (namely, dendritic cells, T-cells, mast cells, and monocytes/macrophages) involved in the initiation and progression of allergic inflammation (Bonaventura et al. 2015; Fukada et al. 2011; Hojyo and Fukada 2016). The role of Zn homeostasis in immune cell signaling will be discussed in further details in the next section.

14.2.4.2 Zinc and Gastrointestinal (GI) Disorders

In the context of zinc and GI disorders, zinc deficiency can arise from either decreased nutritional intake, genetic defects that lead to malabsorption, or from disease mechanisms that causes malabsorption (biochemical deficiency) (Skrovanek et al. 2014). The ultimate effect of these factors is a disruption in zinc homeostasis that in turn alters zinc signaling and perpetuates inflammation.

The association between zinc and GI disorders, namely, Crohn's disease and ulcerative colitis, have been examined by many groups (Ananthakrishnan et al. 2015; Skrovanek et al. 2014). Results from meta-analysis of human studies have highlighted a link between zinc intake and digestive tract cancers. Higher Zn intake was associated with reduced colorectal cancer risk and with lower esophageal and gastric cancer risks in certain populations (Li et al. 2014). A recent systematic review identified several Zn transporters (ZIP 1, 2, 4–7, 10, and 14; ZnT 1, 8, and 19) as having some functional role in the GI system, potentially providing amenable targets for therapeutic intervention (Myers et al. 2017).

The zinc signals implicated in GI disorders (previously discussed) include: the modulation of ROS via SOD (Gilca-Blanariu et al. 2018; Skrovanek et al. 2014); modulation of MTs and zinc transporter expression (Skrovanek et al. 2014; Myers et al. 2017); as well as regulating cell signaling pathways that drive inflammation (Triner et al. 2018).

14.2.4.3 Zinc and Inflamm-Aging

Aging is an inevitable process associated with progressive pathological features including oxidative stress, nucleic acid damage, and altered immune cell function. Immune function decreases with age starting at about 60–65 years. The term "inflamm-aging" was coined in the early 2000s and is described as a chronic progressive increase in pro-inflammatory status associated with the aging process. It is associated with increased risk of development of inflammation related disorders including Alzheimer's disease, Parkinson's disease, autoimmunity, cancers, type II diabetes, osteoporosis, atherosclerosis, among others (Fulop et al. 2018; Maywald and Rink 2015; Xia et al. 2016). Inflamm-aging is a consequence of immunosenescence where the conserved innate immune system overtakes the more highly specialized adaptive immune system as we age (Fulop et al. 2018).

Many cells in the innate and adaptive immune response are affected by aging. Of these, T-cells seem to be the most affected by age-induced changes. The effects of aging on T-cells have been extensively reviewed by several groups (Chen et al. 2013; Nikolich-Zugich 2014; Salam et al. 2013). Age associated alterations in T-cell function include increased cytotoxicity, enhanced production of pro-inflammatory cytokines, reduced antigen-induced proliferation, and decreased cell division. These changes occur both in the thymus, the site of T-cell development and maturation, as well as in the periphery. Thymus involution, a gradual decrease in the structural integrity of the thymus with age, along with changes in genes that promote cell growth, proliferation and cell death, ultimately result in a decline in the production of naïve T-cells (Salam et al. 2013; Nikolich-Zugich 2014). Thymocytes also show age related changes in the expression of genes associated with T-cell receptor (TCR) signaling, antigen presentation, and lymphocyte development and function (Chen et al. 2013). Similar age-related effects are seen in mature peripheral T-cells. Studies show increased gene expression of apoptotic related genes and decreased expression of genes involved in cell growth and energy metabolism (Chen et al. 2013). Diversity of the TCR decreases with age and there is a shift in T-cell phenotype. Overall, there are increased numbers of $CD8^+$ cytotoxic T-cells and a decrease in $CD4^+$ T-helper cells. The $CD4^+$ repertoire also undergoes a shift in the Th1/Th2 paradigm in favor of the Th2 cytokines IL-4 and IL-10 (Xia et al. 2016; Chen et al. 2013; Nikolich-Zugich 2014; Salam et al. 2013).

There are striking similarities between immunosenescence and zinc deficiency including the effects on innate and adaptive immune cell function. Zinc deficiency is common in the elderly and when coupled with immunosenescence, increases susceptibility to infections, cancers, neurodegenerative, and autoimmune diseases (Maywald and Rink 2015; Szewczyk 2013). A recent study conducted in Japan found that a considerable portion (20% or greater) of the elderly had marginal to severe zinc deficiency, with the prevalence of zinc deficiency increasing with aging (Yasuda and Tsutsui 2016).

Inflamm-aging is also associated with changes in expression of zinc transporters and cytokine profiles in immune cells. Analysis of gene expression of zinc transporters and pro-inflammatory cytokines in lymphocytes isolated from young and old donors revealed that expression of ZIP1,2,3 were elevated in cells from young donors. Conversely, the pro-inflammatory cytokines IL-6 and TNF-α had higher expression in cells from elderly donors (Giacconi et al. 2012). In aged mice, zinc deficiency was associated with increased inflammation due to the reduction of intracellular zinc in immune cells. The pro-inflammatory response was enhanced by reduced expression of Zip6 and age-specific Zip6 dysregulation correlated with increased methylation of the Zip6 promoter (Wong et al. 2013). A follow up study by the same investigators showed that zinc deficiency in aged mice upregulated expression of intercellular adhesion molecule (ICAM)-1 and IL-6 gene expression with decreased IL-6 promoter methylation (Wong et al. 2015).

Zinc is involved in several neuronal processes. Evidence from molecular, genetic, and clinical studies have elucidated a role of zinc homeostasis in depression and psychosis in the elderly. While influx of zinc can drive neurotoxicity under clinical conditions such as traumatic brain injury, stroke, or epilepsy, there is increasing evidence to suggest that zinc deficiency leads to increased risk of developing neurological disorders, as well as learning and memory defects. In the brain, zinc can also positively modulate synaptic function and neurotrophic signaling, suggesting both pro- and anti-inflammatory roles of the metal ion in brain function (Cabrera 2015; Sensi et al. 2018; Szewczyk 2013).

Knowing that zinc deficiency potentiates immunosenescence and inflamm-aging in the elderly, zinc supplementation has been studied to determine whether it is beneficial. In an observational study, nursing home subjects with normal zinc concentration had lower incidences of pneumonia, shorter duration of pneumonia, and fewer new antibiotic prescription and days of antibiotic use than subjects with low zinc concentrations (Meydani et al. 2007). A 2016 randomized double-blind placebo-controlled trial to determine the effects of zinc supplementation (30 mg of Zn/day) on serum zinc concentration in elderly deficient subjects showed that supplementation increased serum concentration of zinc. Zinc supplementation also increased T-cell proliferation, denoted by increased anti-CD3/CD28, and the number of peripheral T-cells (Barnett et al. 2016). In healthy elderly subjects, moderate zinc supplementation was found to increase liable zinc that was associated with a reduction in the numbers of unspecific preactivated T-cells and led to improved T-cell responses upon stimulation (Kahmann et al. 2008). Collectively, the existing data highlights a correlation between zinc deficiency and inflamm-aging in the elderly and a role of zinc supplementation in reducing the extent of systemic inflammation which may reduce the existence of comorbid disease conditions, although this remains to be proven.

14.3 Altered Zinc Homeostasis and Immune Cell Signaling in Inflammation

Zinc homeostasis is critical for proper immune cell function. A growing body of evidence, which includes animal and human studies, highlights the effects of Zn on activation and function of innate and adaptive immune cells. The effect of Zn on immune cell function is multifaceted and can affect gene expression, cytokine production, maturation, and polarization of subsets of immune cells (Bonaventura et al. 2015; Haase and Rink 2014; Hojyo and Fukada 2016).

14.3.1 Zinc Signals and the Innate Immune System

The innate immune system is the first line of defense against invading pathogens. It is made up of monocytes/macrophages, dendritic cells (DCs), natural killer (NK) cells, and polymorphonuclear leukocytes (PMNs), among others. Cells of the innate immune system play a major role in recognizing and destroying or restricting pathogens until the more specific adaptive immune response can be mounted by the host (Elliott et al. 2014).

14.3.1.1 Monocyte/Macrophages

Macrophages are key components of the innate immune response to pathogens. They play a role in phagocytosis, antigen presentation, and immunomodulation, with the goal of maintaining homeostasis after pathogen exposure (Gao et al. 2018). Several studies have highlighted a relationship between zinc homeostasis and macrophage function. Macrophages express a number of zinc transporters under physiological conditions that can then be upregulated or downregulated during pathological conditions suggesting that Zn plays a critical role in macrophage function (Gao et al. 2018; Stafford et al. 2013).

The impact of zinc homeostasis upon monocytes/macrophage function is most highlighted in studies relating to innate immune signaling via LPS activation of TLR-4. In 2008, Haase and colleagues demonstrated that zinc signals were essential for LPS-induced signal transduction in monocytes. Their studies showed that LPS stimulation increased the concentration of free intracellular zinc that mediated activation of downstream signaling pathways, namely, MAPKs and $NF_{K}B$, as well as the release of pro-inflammatory cytokines. This effect was found to be abrogated by chelation of zinc, suggesting a zinc specific mechanism (Haase et al. 2008). Since then, several groups have focused on investigating the role of zinc signals, specifically in the context of zinc transporters and downstream signaling pathways in monocyte/macrophage function.

As discussed in previous sections, ZIP8 has been shown to be a negative regulator of $NF_{\kappa}B$ signaling in monocytes (Liu et al. 2013). Studies by the same group have revealed that ZIP8 is constitutively expressed on resting macrophages and significantly upregulated after LPS exposure. This was associated with reduced mRNA and protein expression of IL-10 but increased mRNA expression of pro-inflammatory cytokines (IL-6, IL-8 and TNF-α). Further investigation revealed that knockdown of ZIP8 inhibited LPS-mediated accumulation of intracellular zinc and prevented zinc-dependent reduction in IL-10. Supplementation with zinc was also found to reduce nuclear localization and activity of the transcription factor CCAAT/enhancer binding protein beta (C/EBPβ), which drives expression of IL-10 (Pyle et al. 2017). Zinc homeostasis was also found to be critical in monocyte adhesion and recruitment to inflamed arteries. ZIP8 was shown to be upregulated on monocytes adhering to aortas ex vivo, suggesting a role of the zinc transporter in regulating Zn influx and recruitment of monocytes to atherosclerotic lesions (Cheng et al. 2018).

Other zinc transporters have been implicated in macrophage function. In primary macrophages, ZIP14 was upregulated upon activation, which was dependent on calcium signaling, GC-rich DNA binding and downregulation of NFκB. Knockdown of ZIP14 attenuated expression of IL-6 and TNF-α, highlighting another zinc transporter that mediates LPS signaling (Sayadi et al. 2013). More recently, IL-4 has been shown to trigger a MT-3 and ZnT4 dependent increase in liable zinc stores in macrophages. This increase in zinc in turn was exploited by pathogens to promote their survival and drive inflammation in both human and animals alike (Subramanian Vignesh et al. 2016).

The impact of zinc deficiency and supplementation have been investigated in monocytes/macrophages given that they are at the forefront in host defense and therefore may be affected by changes in zinc status. Zinc deficiency induced changes in the number and distribution of leukocytes in rats. Rats on a zinc deficient diet (0.7 mg zinc/kg) had increased numbers of monocytes and granulocytes compared to animals on the control diet (Someya et al. 2009). Human peripheral blood mononuclear cells (PMBCs) cultured under zinc deficient conditions, either by using TPEN or CHELEX 100 resin, and exposed to different bacteria (*Streptococcus pneumoniae*, *Escherichia coli*, and *Staphylococcus aureus*) showed differential effects from zinc depletion. Phagocytosis of bacteria and oxidative burst were elevated whereas production of TNF-α and IL-6 were reduced (Mayer et al. 2014). In patients with type 2 diabetes mellitus, where immunity is suppressed, thus increasing susceptibility to several inflammatory disorders and cancers, supplementation with 30 mg of zinc per day significantly increased the proportion of monocytes expressing TNF-α, a key component in the host defense against cancers. It is worth mentioning that although there was a higher proportion of TNF-α bound to monocytes, there were no significant changes in plasma TNF-α and TNF-α expressing lymphocytes between the treated and placebo groups (Meksawan et al. 2014). Collectively, the data suggests that monocytes/macrophages can be affected by both short-term zinc signals, as well as changes in zinc status over time.

14.3.1.2 Dendritic Cells

Dendritic cells (DCs) are professional antigen presenting cells that bridge the gap between the innate and adaptive immune systems. Like macrophages, DCs express several PRRs that allow them to sense pathogens and then process and present antigens to T-cells, thus mediating the adaptive immune response (Bonaventura et al. 2015; Haase et al. 2008; Hojyo and Fukada 2016). Unlike macrophages, much less is known about zinc signals and dendritic cell function. In the early 2000s, it was shown that stimulation of murine DCs with LPS led to maturation of DCs, denoted by surface upregulation of major histocompatibility complex (MHC)-II and co-stimulatory molecules, mediated by overall decrease in intracellular free Zn levels. This decrease in intracellular zinc was the net result of changes in the differential expression of select zinc transporters. Expression of ZIP6 and ZIP10 was downregulated whereas ZnT1, 4, and 6 were upregulated. Zinc depletion by TPEN was shown to mimic the LPS-induced effects and increase the ability of DCs to prime naïve T-cells the effect of which was abrogated by zinc supplementation (Kitamura et al. 2006).

Since these findings were reported, other studies have investigated the effects of zinc homeostasis on DC function. In a model of *Histoplasma capsulatum* fungal infection, a global fungal pathogen, zinc treatment induced a tolerogenic phenotype in bone morrow derived dendritic cells (BMDCs) marked by diminished cell surface expression of the maturation marker MHC-II and upregulation of tolerogenic markers, namely, programmed death-ligand (PD-L)1 and PD-L2. These results were reproducible in vivo and skewed the T-helper cell response to the development of T-regulatory cells (Tregs) while decreasing Th17 responses. The results highlight a potential role of Zn in modulating DC function during fungal infection (George et al. 2016). Langerhans cells, the dendritic cells of the skin, were found to be decreased in mice and samples from human patients suffering from zinc deficiency (Kawamura et al. 2012). Conversely, another study showed that zinc induced apoptotic cell death in mouse DCs. In BMDCs, zinc was shown not to have any distinguishable effects on DC maturation but led to an increase in the number of cells in sub G-1 phase, a hallmark feature of apoptosis, stimulated the formation of ceramide, a known trigger of suicidal cell death, and resulted in apoptotic death that was absent in sphingomyelinase knockout DCs. This is one of the first studies to highlight a link between zinc-induced sphingomyelin activation and cell death in DCs, which in turn could affect the overall immune response (Shumilina et al. 2010).

14.3.1.3 Neutrophils

Neutrophils play a prominent role in acute inflammation and are rapidly recruited to the site of injury or infection soon after pathogen exposure, invasion, or inflammation. They are the largest leukocyte population in the blood and provide the first line of defense against pathogens by phagocytosis and oxidative burst, as well as secretion of multiple cytokines and chemokines (Maares and Haase 2016;

Subramanian Vignesh and Deepe 2016). Zinc deficiency has been shown to be associated with impaired neutrophil function in both animal models and human subjects. In zinc deficient rats, analysis of white blood cells revealed that there were marked increases in total number of neutrophils (Someya et al. 2009). Increased migration of neutrophils seen during zinc deficiency has been shown to be caused by membrane-induced damage and increased production of IL-8 (Finamore et al. 2008). Chelation of free Zn was found to impair chemotaxis, phagocytosis, oxidative burst, and cytokine production (the mechanisms of which have been previously discussed) in human neutrophils from healthy donors (Hasan et al. 2016). Conversely, zinc supplementation resulted in improved survival in a juvenile model of sepsis by enhancing neutrophil recruitment in the peritoneal cavity, enhanced phagocytosis, and neutrophil extracellular trap (NET) formation. This was associated with decreased serum cytokine levels and myeloperoxidase activity in the lung tissue, suggesting attenuation of the systemic inflammatory response (Ganatra et al. 2017).

14.3.1.4 Mast Cells

Tissue resident mast cells are central in orchestrating the immediate response to allergens by secreting mediators that act on the vasculature, connective tissue, structural and immune cells (Nishida and Uchida 2018). Mast cell activation, degranulation, and the production of pro-inflammatory cytokines are influenced by zinc signals. The discovery of the zinc wave phenomenon and establishment of zinc as a second messenger was first identified in mast cells (Yamasaki et al. 2007). The ZnT5 transporter is highly expressed on mast cells and was found to mediate delayed-type allergic responses. Studies using bone marrow-derived mast cells (BMMCs) demonstrated that ZnT5 is required for FcεRI- mediated activation of $NF_{K}B$ and the translocation of PKC to the plasma membrane, mediating cytokine production in delayed-type allergic reactions (Nishida et al. 2009).

14.3.2 Zinc Signals and the Adaptive Immune System

The adaptive immune response is mediated primarily by B- and T-lymphocytes. B-cells mature in the bone marrow and when activated mediate the humoral response. T-cells leave the bone marrow, mature in the thymus, and are more directly involved in the cellular-mediated responses (Blewett and Taylor 2012).

14.3.2.1 T-Lymphocytes

Zinc plays a role in T-cell maturation and development, as well as several T-cell signaling cascades including TCR and cytokine signaling pathways. The roles of zinc status in T-cell development and phenotype have been extensively discussed in two recent reviews (Blewett and Taylor 2012; Nishida and Uchida 2018). Broadly, studies examined in these reviews highlighted a role for zinc in lymphopenia (zinc deficiency was associated with decreased numbers and aberrant T-cell functions); skewing the Th1/Th2 polarization in favor of a Th1 phenotype; and decreasing the autoimmune response mediated by Th17 cell activation leading to modulation of T-helper cell responses that ameliorate inflammation (Blewett and Taylor 2012; Nishida and Uchida 2018).

More specifically, T-cell receptor stimulation induced zinc influx via ZIP6. The accumulation of free zinc ions resulted in reduced recruitment of Src homology region 2 domain-containing phosphatase (SHP)-1 to the TCR complex, increased phosphorylation of zeta chain of T-cell receptor associated protein kinase (ZAP)-70 and lowered threshold to TCR stimulation, providing a negative feedback mechanism in TCR signaling (Yu et al. 2011). More recently, zinc was shown to modulate several signaling pathways in T-cells, including NFAT, NFκB, and AP-1 signaling cascades. Stimulation of the TCR in human T-cells resulted in upregulation of ZIP1, ZIP4, ZIP6–9, and ZIP14. Of these transporters, ZIP6 was found to be a key component of the T-cell activation machinery (Colomar-Carando et al. 2019).

Zinc supplementation, at physiological doses (50 μM of $ZnSO_4$), significantly decreased cell proliferation and pro-inflammatory cytokine production from antigen-specific T-cells after reactivation. Mixed leukocyte cultures were shown to have an increased induction and stabilization of T-regs based on zinc induced upregulation of forkhead box (Fox)-P3 and Kruppel like factor (KLF)-10, essential transcription factors for T-reg development and function respectively, and downregulation of interferon regulatory factor (IRF)-1, a negative regulator of FoxP3 (Maywald and Rink 2017). In a similar study, supplementation with 50 μM of $ZnSO_4$ was shown to enhance cytosolic zinc concentrations in $CD3^+$ T-cells, increase Th1 cytokine responses, demonstrated by increased IL-10, TNF-α, and IFN-γ production after allergen challenge. Zinc also enhanced T-regs population and upregulated mRNA expression of cytotoxic T-lymphocyte antigen (CTLA)-4 (Rosenkranz et al. 2017). In another supplementation study, zinc aspartate suppressed expression of Th1 (IFN-γ), Th2 (IL-5), and Th17 (IL-17) cytokine production in pre-activated T-cells (Guttek et al. 2018). Collectively, studies suggest that zinc homeostasis is critical for proper T-cell function and its supplementation could potentially alleviate the inflammatory response mediated by Th2 and Th17 cells in allergic and autoimmune disorders.

14.3.2.2 B-Lymphocytes

Many of the signaling pathways used by T-cells already discussed, namely MAPKs, protein tyrosine kinases, nuclear factor of activated T-cells (NFAT), and NFκB, are also important to B-cell development, maturation, and function. However, unlike T-cells and other immune cells, B-cell development and function are less affected by an organism's zinc status. Zinc signals do not appear to directly affect B-cell function but overall zinc deficiency can alter lymphopoiesis and B-cell development (Bonaventura et al. 2015; Haase and Rink 2009; Hojyo and Fukada 2016; Maywald et al. 2017).

As an example, the ZIP10 transporter was found to play an important role in the B-cell receptor (BCR) signal transduction. ZIP10 deficiency reduced the mature B-cell population and shortened lifespan. Cells had dysregulated BCR signaling, which was triggered by reduction in CD45 receptor phosphatase activity resulting in hyperactivation of LYN, a key protein kinase that mediates BCR signaling (Hojyo et al. 2014). More recently, two studies have suggested a role of ZIP7 in B-cell activation and development. In the first study, human B-cells, infected with Epstein-Barr virus (EBV), showed differentially altered expression of zinc transporters, specifically ZIP7, which had increased mRNA and protein expression. Comparison of these findings to freshly isolated B-cells suggested that elevation of free intracellular zinc was a consequence of B-cell activation and proliferation (Ollig et al. 2019). In the second study, CRISPR-Cas9 mutagenesis was used to precisely model ZIP7 deficiency in mice and was found to reproduce the block in B-cell development previously observed in patients with agammaglobulinemia and absent B-cells. This was seen in hypomorphic alleles as homozygosity for null alleles caused embryonic death. B-cells from mutant mice had intrinsic failure of development seen in development arrest at the late pre-B-cell to immature B-cell transition. Cells also had reduced cytoplasmic zinc and defects in BCR signaling characterized by increased phosphatase activity and decreased phosphorylation downstream of the pre-BCR and BCR. Developmental blockade from pre-B-cells to B-cells is a characteristic feature seen in patients with agammaglobulinemia and these results highlight a potential role of zinc homeostasis in immunodeficiency (Anzilotti et al. 2019).

Taken together, these findings highlight the effects of zinc signals mediated by Zn transporters on both the innate and adaptive immune responses. Zinc influences proliferation, differentiation, and survival of cells involved in the first line of host defense and can subsequently influence the nature of the immune response generated by adaptive immune cells.

14.4 Zinc Toxicity

Our focus so far has been in the context of zinc signals associated with zinc deficiency and supplementation at physiological concentrations. The potential effects of over consumption of zinc and/or zinc toxicity are largely unknown. As discussed in this review, zinc has a broad impact upon the human proteome, playing both structural and catalytic roles. Additionally, there are multiple proteins that are largely responsible for maintaining zinc balance within and outside of cells, namely, zinc transporters and MTs. This suggests that the body is also designed to tolerate or manage excessive zinc content, unlike other metal ions that share similar properties.

Unlike zinc deficiency, which is widespread and has a number of implications on immune cell function and inflammation, zinc toxicity is relatively rare and based on human case studies, mainly arises from short-term exposure to high zinc concentrations (Plum et al. 2010). Perhaps the most studied toxic reaction to zinc in humans is inhalation of industrial fumes containing zinc: the “metal fume fever”

(MFF). Metal fume fever is mainly caused by inhalation of zinc oxide resulting in localized inflammation with symptoms that include fever, chills, muscle soreness, nausea, chest pain, and cough. The symptoms are usually acute and resolve within a few days following exposure (Plum et al. 2010). A recent analysis of reported cases of metal fume fever showed inhalation of excessively high concentrations of zinc oxide fumes, 77–600 mg Zn/m^3, were required to induce metal fume fever. Although the exact mechanism of MFF has yet to be elucidated, pathophysilogical mechanisms reported included production of pro-inflammatory cytokines, neutrophil activation, and the formation of oxygen radicals (Greenberg and Vearrier 2015). While metal fume fever is typically benign, chronic exposure can lead to more severe cases and longer recovery periods following cessation of exposure (Greenberg and Vearrier 2015; Plum et al. 2010).

Oral intake of large doses of supplemental zinc over prolonged periods can result in copper deficiency caused by competitive absorption of zinc and copper mediated by MTs in red blood cells. Dietary zinc upregulates the expression of MTs and MTs binds copper with a higher affinity than zinc resulting in decreased available copper ions (Plum et al. 2010). A 2005 case report identified three patients with zinc-induced copper deficiencies exhibiting clinical symptoms of severe neutropenia, anemia, and peripheral neutropathy. Copper deficiency was identified by bone marrow biopsy and identification of high zinc levels (Willis et al. 2005). In a more recent case, a patient was diagnosed with pancytopenia, low levels of all blood cells including red and white blood cells and platelets, secondary to hypocuprenia from zinc toxicity (Johnsrud et al. 2017).

Zinc has also been shown to be toxic on a cellular level specifically in the brain. Excess accumulation of zinc in the central nervous system (CNS) can lead to excitotoxicity, oxidative stress, impaired energy production, and neural and glial cell death. Zinc-induced neurotoxicity has been implicated in a number of neurodegenerative disorders including epilepsy, Alzheimer's, and Parkinson's diseases, all of which are comprehensively described in (Morris and Levenson 2012; Sensi et al. 2011).

Zinc is essential for normal immune cell function and has been shown to have relatively low toxicity, given the rare incidences of zinc toxicity in humans. Nevertheless, when zinc toxicity does occur, it can be damaging to the host both at a cellular and systemic level.

14.5 Conclusions

The studies in this chapter have highlighted the many roles of zinc signals in immune cell function and inflammation. Despite the known and emerging roles of zinc homeostasis in proper immune cell function, zinc deficiency remains a global problem that contributes to severe immunodeficiency. Results from studies conducted in the last two decades have established zinc as a bona fide intracellular signaling molecule that modulates multiple signaling cascades within multiple cell types involved in innate and adaptive immune function. By doing so, zinc plays an essential role in mediating the host response to infection and danger signals, thereby balancing the extent of inflammation observed both during the acute and later stages of damage and recovery. A prevailing theme to most studies conducted so far is that zinc must be present in sufficient quantities at the onset of danger to catalyze a robust response to control inflammation and set the stage for effective resolution. If this is not the case, often observed is exuberant inflammation, excessive damage, and poorer prognosis both in the context of acute and chronic inflammatory-based disease.

A great deal remains unknown regarding the actions of zinc signals in immune cell function and inflammation. A more in-depth understanding of cellular zinc homeostasis, zinc signals, and the molecular mechanisms that mediate dysfunction will further establish zinc as a critical micronutrient in health and disease, potentially providing novel therapeutic options for patient care.

References

Alker W, Haase H (2018) Zinc and Sepsis. Nutrients 10(8). https://doi.org/10.3390/nu10080976. doi: E976 [pii]

Allan GM, Arroll B (2014) Prevention and treatment of the common cold: making sense of the evidence. CMAJ 186(3):190–199. https://doi.org/10.1503/cmaj.121442

Ananthakrishnan AN, Khalili H, Song M et al (2015) Zinc intake and risk of Crohn's disease and ulcerative colitis: a prospective cohort study. Int J Epidemiol 44(6):1995–2005. https://doi.org/10.1093/ije/dyv301

Anzilotti C, Swan DJ, Boisson B et al (2019) An essential role for the Zn(2+) transporter ZIP7 in B cell development. Nat Immunol. https://doi.org/10.1038/s41590-018-0295-8

Arroll B (2008) Common cold. BMJ Clin Evid 2008:1510. doi:1510 [pii]

Bafaro E, Liu Y, Xu Y et al (2017) The emerging role of zinc transporters in cellular homeostasis and cancer. Signal Transduct Target Ther 2. https://doi.org/10.1038/sigtrans.2017.29. Epub 2017 July 28. doi:17029 [pii]

Bao B, Prasad AS, Beck FW et al (2007) Zinc up-regulates NF-kappaB activation via phosphorylation of IkappaB in HUT-78 (Th0) cells. FEBS Lett 581(23):4507–4511. doi:S0014-5793(07)00906-4 [pii]

Barnett JB, Dao MC, Hamer DH et al (2016) Effect of zinc supplementation on serum zinc concentration and T cell proliferation in nursing home elderly: a randomized, double-blind, placebo-controlled trial. Am J Clin Nutr 103(3):942–951. https://doi.org/10.3945/ajcn.115.115188

Basnet S, Mathisen M, Strand TA (2015) Oral zinc and common childhood infections--an update. J Trace Elem Med Biol 31:163–166. https://doi.org/10.1016/j.jtemb.2014.05.006

Bishop GM, Dringen R, Robinson SR (2007) Zinc stimulates the production of toxic reactive oxygen species (ROS) and inhibits glutathione reductase in astrocytes. Free Radic Biol Med 42(8):1222–1230. doi:S0891-5849(07)00049-4 [pii]

Blewett HJ, Taylor CG (2012) Dietary zinc deficiency in rodents: effects on T-cell development, maturation and phenotypes. Nutrients 4(6):449–466. https://doi.org/10.3390/nu4060449

Bonaventura P, Benedetti G, Albarede F et al (2015) Zinc and its role in immunity and inflammation. Autoimmun Rev 14(4):277–285. https://doi.org/10.1016/j.autrev.2014.11.008

Bosmann M, Ward PA (2013) The inflammatory response in sepsis. Trends Immunol 34(3):129–136. https://doi.org/10.1016/j.it.2012.09.004

Cabrera AJ (2015) Zinc, aging, and immunosenescence: an overview. Pathobiol Aging Age Relat Dis 5:25592. https://doi.org/10.3402/pba.v5.25592

Cassandri M, Smirnov A, Novelli F et al (2017) Zinc-finger proteins in health and disease. Cell Death Discov 3:17071. https://doi.org/10.1038/cddiscovery.2017.71
Chen G, Lustig A, Weng NP (2013) T cell aging: a review of the transcriptional changes determined from genome-wide analysis. Front Immunol 4:121. https://doi.org/10.3389/fimmu.2013.00121
Cheng G, Chang FJ, You PH et al (2018) ZIP8 induces monocyte adhesion to the aortas ex-vivo by regulating zinc influx. Int Immunopharmacol 62:203–211. doi:S1567-5769(18)30269-8 [pii]
Choi S, Liu X, Pan Z (2018) Zinc deficiency and cellular oxidative stress: prognostic implications in cardiovascular diseases. Acta Pharmacol Sin 39(7):1120–1132. https://doi.org/10.1038/aps.2018.25
Colomar-Carando N, Meseguer A, Company-Garrido I et al (2019) Zip6 transporter is an essential component of the lymphocyte activation machinery. J Immunol 202(2):441–450. https://doi.org/10.4049/jimmunol.1800689
Elliott DE, Siddique SS, Weinstock JV (2014) Innate immunity in disease. Clin Gastroenterol Hepatol 12(5):749–755. https://doi.org/10.1016/j.cgh.2014.03.007
Finamore A, Massimi M, Conti Devirgiliis L et al (2008) Zinc deficiency induces membrane barrier damage and increases neutrophil transmigration in Caco-2 cells. J Nutr 138(9):1664–1670. doi:138/9/1664 [pii]
Fukada T, Yamasaki S, Nishida K et al (2011) Zinc homeostasis and signaling in health and diseases: zinc signaling. J Biol Inorg Chem 16(7):1123–1134. https://doi.org/10.1007/s00775-011-0797-4
Fulop T, Larbi A, Dupuis G et al (2018) Immunosenescence and Inflamm-aging as two sides of the same coin: friends or foes? Front Immunol 8:1960. https://doi.org/10.3389/fimmu.2017.01960
Gammoh NZ, Rink L (2017) Zinc in infection and inflammation. Nutrients 9(6):624. https://doi.org/10.3390/nu9060624. doi:E624 [pii]
Ganatra HA, Varisco BM, Harmon K et al (2017) Zinc supplementation leads to immune modulation and improved survival in a juvenile model of murine sepsis. Innate Immun 23(1):67–76. https://doi.org/10.1177/1753425916677073
Gao H, Dai W, Zhao L et al (2018) The role of zinc and zinc homeostasis in macrophage function. J Immunol Res 2018:6872621. https://doi.org/10.1155/2018/6872621
George MM, Subramanian Vignesh K, Landero Figueroa JA et al (2016) Zinc induces dendritic cell Tolerogenic phenotype and skews regulatory T cell-Th17 balance. J Immunol 197(5):1864–1876. https://doi.org/10.4049/jimmunol.1600410
Giacconi R, Malavolta M, Costarelli L et al (2012) Comparison of intracellular zinc signals in nonadherent lymphocytes from young-adult and elderly donors: role of zinc transporters (zip family) and proinflammatory cytokines. J Nutr Biochem 23(10):1256–1263. https://doi.org/10.1016/j.jnutbio.2011.07.005
Gilca-Blanariu GE, Diaconescu S, Ciocoiu M et al (2018) New insights into the role of trace elements in IBD. Biomed Res Int 2018:1813047. https://doi.org/10.1155/2018/1813047
Greenberg MI, Vearrier D (2015) Metal fume fever and polymer fume fever. Clin Toxicol (Phila) 53(4):195–203. https://doi.org/10.3109/15563650.2015.1013548
Guttek K, Wagenbrett L, Reinhold A et al (2018) Zinc aspartate suppresses proliferation and Th1/Th2/Th17 cytokine production of pre-activated human T cells in vitro. J Trace Elem Med Biol 49:86–90. doi:S0946-672X(18)30126-3 [pii]
Haase H, Ober-Blobaum JL, Engelhardt G et al (2008) Zinc signals are essential for lipopolysaccharide-induced signal transduction in monocytes. J Immunol 181(9):6491–6502. doi:181/9/6491 [pii]
Haase H, Rink L (2009) Functional significance of zinc-related signaling pathways in immune cells. Annu Rev Nutr 29:133–152. https://doi.org/10.1146/annurev-nutr-080508-141119
Haase H, Rink L (2014) Zinc signals and immune function. Biofactors 40(1):27–40. https://doi.org/10.1002/biof.1114
Hasan R, Rink L, Haase H (2016) Chelation of free Zn(2)(+) impairs chemotaxis, phagocytosis, oxidative burst, degranulation, and cytokine production by neutrophil granulocytes. Biol Trace Elem Res 171(1):79–88. https://doi.org/10.1007/s12011-015-0515-0

Hershfinkel M (2018) The zinc sensing receptor, ZnR/GPR39, in health and disease. Int J Mol Sci 19(2). https://doi.org/10.3390/ijms19020439. doi:E439 [pii]

Hirano T, Murakami M, Fukada T et al (2008) Roles of zinc and zinc signaling in immunity: zinc as an intracellular signaling molecule. Adv Immunol 97:149–176. https://doi.org/10.1016/S0065-2776(08)00003-5

Ho Y, Samarasinghe R, Knoch ME et al (2008) Selective inhibition of mitogen-activated protein kinase phosphatases by zinc accounts for extracellular signal-regulated kinase 1/2-dependent oxidative neuronal cell death. Mol Pharmacol 74(4):1141–1151. https://doi.org/10.1124/mol.108.049064

Hojyo S, Fukada T (2016) Roles of zinc signaling in the immune system. J Immunol Res 2016:6762343. https://doi.org/10.1155/2016/6762343

Hojyo S, Fukada T, Shimoda S et al (2011) The zinc transporter SLC39A14/ZIP14 controls G-protein coupled receptor-mediated signaling required for systemic growth. PLoS One 6(3):e18059. https://doi.org/10.1371/journal.pone.0018059

Hojyo S, Miyai T, Fujishiro H et al (2014) Zinc transporter SLC39A10/ZIP10 controls humoral immunity by modulating B-cell receptor signal strength. Proc Natl Acad Sci U S A 111(32):11786–11791. https://doi.org/10.1073/pnas.1323557111

Honscheid A, Dubben S, Rink L et al (2012) Zinc differentially regulates mitogen-activated protein kinases in human T cells. J Nutr Biochem 23(1):18–26. https://doi.org/10.1016/j.jnutbio.2010.10.007

Jarosz M, Olbert M, Wyszogrodzka G et al (2017) Antioxidant and anti-inflammatory effects of zinc. Zinc-dependent NF-kappaB signaling. Inflammopharmacology 25(1):11–24. https://doi.org/10.1007/s10787-017-0309-4

Johnsrud J, Abdallah AO, Schichman SA et al (2017) Image diagnosis: zinc-induced copper deficiency causing pancytopenia recognized on bone marrow examination. Perm J 21. https://doi.org/10.7812/TPP/16-077. doi:10.7812/TPP/16-077 [doi]

Kahmann L, Uciechowski P, Warmuth S et al (2008) Zinc supplementation in the elderly reduces spontaneous inflammatory cytokine release and restores T cell functions. Rejuvenation Res 11(1):227–237. https://doi.org/10.1089/rej.2007.0613

Karim T, Muhit M, Khandaker G (2017) Interventions to prevent respiratory diseases - nutrition and the developing world. Paediatr Respir Rev 22:31–37. doi:S1526-0542(16)30112-9 [pii]

Kawamura T, Ogawa Y, Nakamura Y et al (2012) Severe dermatitis with loss of epidermal Langerhans cells in human and mouse zinc deficiency. J Clin Invest 122(2):722–732. https://doi.org/10.1172/JCI58618

Kim MH, Jeong HJ (2015) Zinc oxide nanoparticles suppress LPS-induced NF-kappaB activation by inducing A20, a negative regulator of NF-kappaB, in RAW 264.7 macrophages. J Nanosci Nanotechnol 15(9):6509–6515

Kitamura H, Morikawa H, Kamon H et al (2006) Toll-like receptor-mediated regulation of zinc homeostasis influences dendritic cell function. Nat Immunol 7(9):971–977. doi:ni1373 [pii]

Knoell DL, Liu MJ (2010) Impact of zinc metabolism on innate immune function in the setting of sepsis. Int J Vitam Nutr Res 80(4–5):271–277. https://doi.org/10.1024/0300-9831/a000034

Lang C, Murgia C, Leong M et al (2007) Anti-inflammatory effects of zinc and alterations in zinc transporter mRNA in mouse models of allergic inflammation. Am J Physiol Lung Cell Mol Physiol 292(2):L577–L584. doi:00280.2006 [pii]

Lee SR (2018) Critical role of zinc as either an antioxidant or a Prooxidant in cellular systems. Oxidative Med Cell Longev 2018:9156285. https://doi.org/10.1155/2018/9156285

Li P, Xu J, Shi Y et al (2014) Association between zinc intake and risk of digestive tract cancers: a systematic review and meta-analysis. Clin Nutr 33(3):415–420. https://doi.org/10.1016/j.clnu.2013.10.001

Liu MJ, Bao S, Galvez-Peralta M et al (2013) ZIP8 regulates host defense through zinc-mediated inhibition of NF-kappaB. Cell Rep 3(2):386–400. https://doi.org/10.1016/j.celrep.2013.01.009

Liu T, Zhang L, Joo D et al (2017) NF-kappaB signaling in inflammation. Signal Transduct Target Ther 2. https://doi.org/10.1038/sigtrans.2017.23. Epub 2017 Jul 14. doi:17023 [pii]

Lu H, Xin Y, Tang Y et al (2012) Zinc suppressed the airway inflammation in asthmatic rats: effects of zinc on generation of eotaxin, MCP-1, IL-8, IL-4, and IFN-gamma. Biol Trace Elem Res 150(1–3):314–321. https://doi.org/10.1007/s12011-012-9493-7

Maares M, Haase H (2016) Zinc and immunity: an essential interrelation. Arch Biochem Biophys 611:58–65. doi:S0003-9861(16)30074-1 [pii]

Malik A, Taneja DK, Devasenapathy N et al (2014) Zinc supplementation for prevention of acute respiratory infections in infants: a randomized controlled trial. Indian Pediatr 51(10):780–784

Maret W (2017) Zinc in cellular regulation: the nature and significance of "zinc signals". Int J Mol Sci 18(11). https://doi.org/10.3390/ijms18112285. doi:E2285 [pii]

Marreiro DD, Cruz KJ, Morais JB et al (2017) Zinc and oxidative stress: current mechanisms. Antioxidants (Basel) 6(2):https://doi.org/10.3390/antiox6020024. doi:E24 [pii]

Mayer LS, Uciechowski P, Meyer S et al (2014) Differential impact of zinc deficiency on phagocytosis, oxidative burst, and production of pro-inflammatory cytokines by human monocytes. Metallomics 6(7):1288–1295. https://doi.org/10.1039/c4mt00051j

Maywald M, Rink L (2015) Zinc homeostasis and immunosenescence. J Trace Elem Med Biol 29:24–30. https://doi.org/10.1016/j.jtemb.2014.06.003

Maywald M, Rink L (2017) Zinc supplementation induces CD4(+)CD25(+)Foxp3(+) antigen-specific regulatory T cells and suppresses IFN-gamma production by upregulation of Foxp3 and KLF-10 and downregulation of IRF-1. Eur J Nutr 56(5):1859–1869. https://doi.org/10.1007/s00394-016-1228-7

Maywald M, Wessels I, Rink L (2017) Zinc signals and immunity. Int J Mol Sci 18(10):https://doi.org/10.3390/ijms18102222. doi:E2222 [pii]

Meksawan K, Sermsri U, Chanvorachote P (2014) Zinc supplementation improves anticancer activity of monocytes in type-2 diabetic patients with metabolic syndrome. Anticancer Res 34(1):295–299. doi:34/1/295 [pii]

Meydani SN, Barnett JB, Dallal GE et al (2007) Serum zinc and pneumonia in nursing home elderly. Am J Clin Nutr 86(4):1167–1173. doi:86/4/1167 [pii]

Miyai T, Hojyo S, Ikawa T et al (2014) Zinc transporter SLC39A10/ZIP10 facilitates antiapoptotic signaling during early B-cell development. Proc Natl Acad Sci U S A 111(32):11780–11785. https://doi.org/10.1073/pnas.1323549111

Morgan CI, Ledford JR, Zhou P et al (2011) Zinc supplementation alters airway inflammation and airway hyperresponsiveness to a common allergen. J Inflamm (Lond) 8:36–9255-8-36. https://doi.org/10.1186/1476-9255-8-36

Morris DR, Levenson CW (2012) Ion channels and zinc: mechanisms of neurotoxicity and neurodegeneration. J Toxicol 2012:785647. https://doi.org/10.1155/2012/785647

Muller A, Kleinau G, Piechowski CL et al (2013) G-protein coupled receptor 83 (GPR83) signaling determined by constitutive and zinc(II)-induced activity. PLoS One 8(1):e53347. https://doi.org/10.1371/journal.pone.0053347

Murakami M, Hirano T (2008) Intracellular zinc homeostasis and zinc signaling. Cancer Sci 99(8):1515–1522. https://doi.org/10.1111/j.1349-7006.2008.00854.x

Myers S, Shastri MD, Adulcikas J et al (2017) Zinc and gastrointestinal disorders: a role for the zinc transporters zips and ZnTs. Curr Pharm Des 23(16):2328–2332. https://doi.org/10.2174/1381612823666170124115850

Nikolich-Zugich J (2014) Aging of the T cell compartment in mice and humans: from no naive expectations to foggy memories. J Immunol 193(6):2622–2629. https://doi.org/10.4049/jimmunol.1401174

Nimmanon T, Ziliotto S, Morris S et al (2017) Phosphorylation of zinc channel ZIP7 drives MAPK, PI3K and mTOR growth and proliferation signalling. Metallomics 9(5):471–481. https://doi.org/10.1039/c6mt00286b

Nishida K, Hasegawa A, Nakae S et al (2009) Zinc transporter Znt5/Slc30a5 is required for the mast cell-mediated delayed-type allergic reaction but not the immediate-type reaction. J Exp Med 206(6):1351–1364. https://doi.org/10.1084/jem.20082533

Nishida K, Uchida R (2018) Role of zinc signaling in the regulation of mast cell-, basophil-, and T cell-mediated allergic responses. J Immunol Res 2018:5749120. https://doi.org/10.1155/2018/5749120

Nuttall JR, Oteiza PI (2012) Zinc and the ERK kinases in the developing brain. Neurotox Res 21(1):128–141. https://doi.org/10.1007/s12640-011-9291-6

Olechnowicz J, Tinkov A, Skalny A et al (2018) Zinc status is associated with inflammation, oxidative stress, lipid, and glucose metabolism. J Physiol Sci 68(1):19–31. https://doi.org/10.1007/s12576-017-0571-7

Ollig J, Kloubert V, Taylor KM et al (2019) B cell activation and proliferation increase intracellular zinc levels. J Nutr Biochem 64:72–79. doi:S0955-2863(18)30724-1 [pii]

Paoletti P, Vergnano AM, Barbour B et al (2009) Zinc at glutamatergic synapses. Neuroscience 158(1):126–136. https://doi.org/10.1016/j.neuroscience.2008.01.061

Plum LM, Rink L, Haase H (2010) The essential toxin: impact of zinc on human health. Int J Environ Res Public Health 7(4):1342–1365. https://doi.org/10.3390/ijerph7041342

Pongkorpsakol P, Buasakdi C, Chantivas T et al (2019) An agonist of a zinc-sensing receptor GPR39 enhances tight junction assembly in intestinal epithelial cells via an AMPK-dependent mechanism. Eur J Pharmacol 842:306–313. doi:S0014-2999(18)30623-X [pii]

Prasad AS (2012) Discovery of human zinc deficiency: 50 years later. J Trace Elem Med Biol 26(2–3):66–69. https://doi.org/10.1016/j.jtemb.2012.04.004

Prasad AS, Bao B, Beck FW et al (2011) Zinc-suppressed inflammatory cytokines by induction of A20-mediated inhibition of nuclear factor-kappaB. Nutrition 27(7–8):816–823. https://doi.org/10.1016/j.nut.2010.08.010

Pyle CJ, Akhter S, Bao S et al (2017) Zinc modulates endotoxin-induced human macrophage inflammation through ZIP8 induction and C/EBPbeta inhibition. PLoS One 12(1):e0169531. https://doi.org/10.1371/journal.pone.0169531

Rao G, Rowland K (2011) PURLs: zinc for the common cold – not if, but when. J Fam Pract 60(11):669–671. jfp_6011g [pii]

Rerksuppaphol S, Rerksuppaphol L (2016) Zinc supplementation in children with asthma exacerbation. Pediatr Rep 8(4):6685. https://doi.org/10.4081/pr.2016.6685

Rosenkranz E, Hilgers RD, Uciechowski P et al (2017) Zinc enhances the number of regulatory T cells in allergen-stimulated cells from atopic subjects. Eur J Nutr 56(2):557–567. https://doi.org/10.1007/s00394-015-1100-1

Roth DE, Richard SA, Black RE (2010) Zinc supplementation for the prevention of acute lower respiratory infection in children in developing countries: meta-analysis and meta-regression of randomized trials. Int J Epidemiol 39(3):795–808. https://doi.org/10.1093/ije/dyp391

Salam N, Rane S, Das R et al (2013) T cell ageing: effects of age on development, survival & function. Indian J Med Res 138(5):595–608. doi:IndianJMedRes_2013_138_5_595_124628 [pii]

Salazar G, Huang J, Feresin RG et al (2017) Zinc regulates Nox1 expression through a NF-kappaB and mitochondrial ROS dependent mechanism to induce senescence of vascular smooth muscle cells. Free Radic Biol Med 108:225–235. S0891-5849(17)30186-7 [pii]

Sapkota M, Knoell DL (2018) Essential role of zinc and zinc transporters in myeloid cell function and host defense against infection. J Immunol Res 2018:4315140. https://doi.org/10.1155/2018/4315140

Sato S, Huang XP, Kroeze WK et al (2016) Discovery and characterization of novel GPR39 agonists allosterically modulated by zinc. Mol Pharmacol 90(6):726–737. doi:mol.116.106112 [pii]

Sayadi A, Nguyen AT, Bard FA et al (2013) Zip14 expression induced by lipopolysaccharides in macrophages attenuates inflammatory response. Inflamm Res 62(2):133–143. https://doi.org/10.1007/s00011-012-0559-y

Science M, Johnstone J, Roth DE et al (2012) Zinc for the treatment of the common cold: a systematic review and meta-analysis of randomized controlled trials. CMAJ 184(10):E551–E561. https://doi.org/10.1503/cmaj.111990

Sensi SL, Granzotto A, Siotto M et al (2018) Copper and zinc dysregulation in Alzheimer's disease. Trends Pharmacol Sci 39(12):1049–1063. doi:S0165-6147(18)30179-2 [pii]
Sensi SL, Paoletti P, Koh JY et al (2011) The neurophysiology and pathology of brain zinc. J Neurosci 31(45):16076–16085. https://doi.org/10.1523/JNEUROSCI.3454-11.2011
Shao YX, Lei Z, Wolf PG et al (2017) Zinc supplementation, via GPR39, upregulates PKCzeta to protect intestinal barrier integrity in Caco-2 cells challenged by Salmonella enterica Serovar typhimurium. J Nutr 147(7):1282–1289. https://doi.org/10.3945/jn.116.243238
Shumilina E, Xuan NT, Schmid E et al (2010) Zinc induced apoptotic death of mouse dendritic cells. Apoptosis 15(10):1177–1186. https://doi.org/10.1007/s10495-010-0520-x
Skrovanek S, DiGuilio K, Bailey R et al (2014) Zinc and gastrointestinal disease. World J Gastrointest Pathophysiol 5(4):496–513. https://doi.org/10.4291/wjgp.v5.i4.496
Someya Y, Tanihata J, Sato S et al (2009) Zinc-deficiency induced changes in the distribution of rat white blood cells. J Nutr Sci Vitaminol (Tokyo) 55(2):162–169. doi:JST.JSTAGE/jnsv/55.162 [pii]
Souffriau J, Libert C (2018) Mechanistic insights into the protective impact of zinc on sepsis. Cytokine Growth Factor Rev 39:92–101. doi:S1359-6101(17)30188-0 [pii]
Stafford SL, Bokil NJ, Achard ME et al (2013) Metal ions in macrophage antimicrobial pathways: emerging roles for zinc and copper. Biosci Rep 33(4). https://doi.org/10.1042/BSR20130014
Subramanian Vignesh K, Deepe GS Jr (2016) Immunological orchestration of zinc homeostasis: the battle between host mechanisms and pathogen defenses. Arch Biochem Biophys 611:66–78. doi:S0003-9861(16)30039-X [pii]
Subramanian Vignesh K, Landero Figueroa JA, Porollo A et al (2016) IL-4 induces Metallothionein 3- and SLC30A4-dependent increase in intracellular Zn(2+) that promotes pathogen persistence in macrophages. Cell Rep 16(12):3232–3246. doi:S2211-1247(16)31138-X [pii]
Sunuwar L, Gilad D, Hershfinkel M (2017) The zinc sensing receptor, ZnR/GPR39, in health and disease. Front Biosci (Landmark Ed) 22:1469–1492. doi:4554 [pii]
Supasai S, Aimo L, Adamo AM et al (2017) Zinc deficiency affects the STAT1/3 signaling pathways in part through redox-mediated mechanisms. Redox Biol 11:469–481. doi:S2213-2317(16)30353-6 [pii]
Szewczyk B (2013) Zinc homeostasis and neurodegenerative disorders. Front Aging Neurosci 5:33. https://doi.org/10.3389/fnagi.2013.00033
Taylor KM, Hiscox S, Nicholson RI et al (2012) Protein kinase CK2 triggers cytosolic zinc signaling pathways by phosphorylation of zinc channel ZIP7. Sci Signal 5(210):ra11. https://doi.org/10.1126/scisignal.2002585
Toth K (2011) Zinc in neurotransmission. Annu Rev Nutr 31:139–153. https://doi.org/10.1146/annurev-nutr-072610-145218
Triner D, Castillo C, Hakim JB et al (2018) Myc-associated zinc finger protein regulates the proinflammatory response in colitis and colon cancer via STAT3 Signaling. Mol Cell Biol 38(22):https://doi.org/10.1128/MCB.00386-18. Print 2018 Nov 15. doi:e00386–18 [pii]
Uzzo RG, Crispen PL, Golovine K et al (2006) Diverse effects of zinc on NF-kappaB and AP-1 transcription factors: implications for prostate cancer progression. Carcinogenesis 27(10):1980–1990. doi:bgl034 [pii]
Villarino AV, Kanno Y, Ferdinand JR et al (2015) Mechanisms of Jak/STAT signaling in immunity and disease. J Immunol 194(1):21–27. https://doi.org/10.4049/jimmunol.1401867
Wessells KR, Brown KH (2012) Estimating the global prevalence of zinc deficiency: results based on zinc availability in national food supplies and the prevalence of stunting. PLoS One 7(11):e50568. https://doi.org/10.1371/journal.pone.0050568
Willis MS, Monaghan SA, Miller ML et al (2005) Zinc-induced copper deficiency: a report of three cases initially recognized on bone marrow examination. Am J Clin Pathol 123(1):125–131
Wong CP, Magnusson KR, Ho E (2013) Increased inflammatory response in aged mice is associated with age-related zinc deficiency and zinc transporter dysregulation. J Nutr Biochem 24(1):353–359. https://doi.org/10.1016/j.jnutbio.2012.07.005

Wong CP, Rinaldi NA, Ho E (2015) Zinc deficiency enhanced inflammatory response by increasing immune cell activation and inducing IL6 promoter demethylation. Mol Nutr Food Res 59(5):991–999. https://doi.org/10.1002/mnfr.201400761
Xia S, Zhang X, Zheng S et al (2016) An update on Inflamm-aging: mechanisms, prevention, and treatment. J Immunol Res 2016:8426874. https://doi.org/10.1155/2016/8426874
Yamasaki S, Sakata-Sogawa K, Hasegawa A et al (2007) Zinc is a novel intracellular second messenger. J Cell Biol 177(4):637–645. doi:jcb.200702081 [pii]
Yan YW, Fan J, Bai SL et al (2016) Zinc prevents abdominal aortic aneurysm formation by induction of A20-mediated suppression of NF-kappaB pathway. PLoS One 11(2):e0148536. https://doi.org/10.1371/journal.pone.0148536
Yasuda H, Tsutsui T (2016) Infants and elderlies are susceptible to zinc deficiency. Sci Rep 6:21850. https://doi.org/10.1038/srep21850
Yu M, Lee WW, Tomar D et al (2011) Regulation of T cell receptor signaling by activation-induced zinc influx. J Exp Med 208(4):775–785. https://doi.org/10.1084/jem.20100031

Chapter 15
Zinc Transporters and Zinc Signaling in Skin Formation and Diseases

Bum-Ho Bin, Mi-Gi Lee, Takafumi Hara, Teruhisa Takagishi, and Toshiyuki Fukada

Abstract The skin is the first body region to manifest zinc deficiency. Recent studies have revealed that the zinc transporters, especially zinc importers belonging to the Zrt-Irt-like (ZIP) family, play crucial roles in skin homeostasis. Fourteen ZIP members have been identified in humans, with at least 6 members being related to skin development and maintenance. ZIP1, ZIP2, ZIP4, and ZIP10 are associated with epidermal morphogenesis and disorders, whereas ZIP7 and ZIP13 are essential for dermis formation and collagen metabolism. Mouse models in combination with clinical data have shown the molecular pathogenic mechanisms involving ZIP members. Although other family members have not been well studied with respect to their role in the skin, their direct or indirect associations are also considerable and they are believed to be drug targets for skin diseases. Therefore, precise analysis and understanding of ZIP family members are indispensable for the care and treatment of skin disorders.

Keywords Zinc transporter · Zinc signal · Skin · Epidermis · Dermis

15.1 Zinc Transporters and Zinc Signaling

Zinc is an essential mineral in the skin (Ogawa et al. 2018; Prasad 2014; Tasman-Jones and Kay 1975), and the skin is the first body region to manifest zinc deficiency. In humans, the skin has the third highest abundance of zinc among tissues and zinc

B.-H. Bin
Applied Biotechnology, Ajou University, Suwon, South Korea

M.-G. Lee
Bio-Center, Gyeonggido Business and Science Accelerator, Suwon, South Korea

T. Hara · T. Takagishi
Faculty of Pharmaceutical Sciences, Tokushima Bunri University, Tokushima, Japan

T. Fukada (✉)
Molecular and Cellular Physiology, Faculty of Pharmaceutical Sciences,
Tokushima Bunri University, Tokushima, Japan
e-mail: fukada@ph.bunri-u.ac.jp

T. Fukada, T. Kambe (eds.), *Zinc Signaling*,
https://doi.org/10.1007/978-981-15-0557-7_15

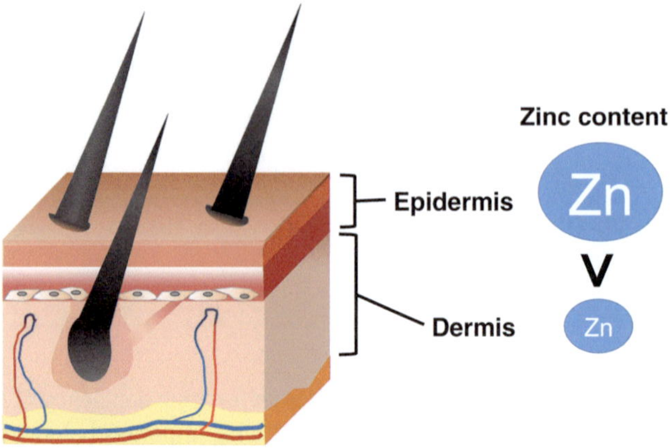

Fig. 15.1 Zinc content in the skin
Zinc is more abundant in the epidermis than in the dermis, which indicates that the proliferation of keratinocytes in the epidermis requires zinc-mediated cellular processes

is the second most abundant trace element after iron (Ogawa et al. 2016, 2018). An adult body with an average weight of 70 kg contains between 2 and 3 g of zinc. Zinc is present in all tissues and liquids of the body, and its concentration is higher in the epidermis than in the dermis (Fig. 15.1) because it is required for active cell proliferation and differentiation of epidermal keratinocytes (Bin et al. 2017c, 2018a, c; Ogawa et al. 2019). In fact, about 10% of proteins encoded in the human genome including transcription factors and enzymes may have zinc-binding motifs (Andreini et al. 2006). In addition to these molecules, certain enzymes such as caspase and protein tyrosine phosphatase are sensitive to the zinc ion, which binds to the active center and inhibits their enzymatic activity. Moreover, CD4 and Lck cross-link via the zinc atom in T cells (Huse et al. 1998). Although initially considered as a component of proteins, recent attention has been drawn to the importance of the zinc ion, mediated via zinc transporters, as a signaling factor (called the zinc signal), highlighting the importance of zinc as an integral component of the cellular response to environmental cues (Fukada and Kambe 2014).

As discussed in the other chapters, zinc in cells is transported by 2 distinguished types of transporters: the Zrt-Irt-like (ZIP) family and the ZnT family of proteins (Fig. 15.2) (Bin et al. 2018c; Fukada and Kambe 2011; Hara et al. 2017). ZnT, which is a zinc-exporting protein belonging to solute carrier 30A (SLC30A), carries zinc from the inside of the cell to the outside, or from the cytoplasm to the organelle lumen. ZnT proteins generally have 6 transmembrane domains, and the N- and C-terminal domains are located in the cytoplasm. To date, 10 proteins belonging to this family, ZnT1 to ZnT10, have been identified. Like ZnT2, some allow the passage of zinc from the mother to the embryo or the newborn via milk; thus, loss of function of mutations in ZnT2 affects the zinc level in maternal breast milk, resulting in severe epidermal dermatitis (Golan et al. 2017; Itsumura et al. 2013; Lee et al. 2015). In contrast, the import of zinc into cells in humans is mainly maintained by solute carrier 39A (SLC39A), which belongs to the ZIP family

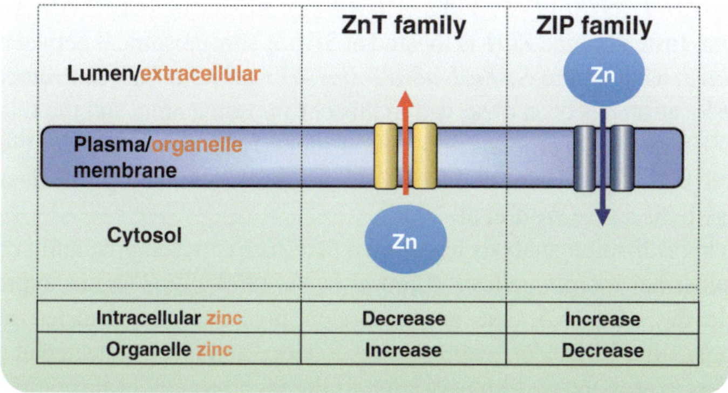

Fig. 15.2 Zinc transporter families and their contribution to zinc levels
There are 2 major zinc transporter families: SLC30A/ZnT transports zinc to the lumen or organelle, whereas the SLC39A/ZIP family transports zinc to the cytosol

(Fukada and Kambe 2011; Hara et al. 2017). In humans, there are 14 SLC39A proteins (SLC39A [or ZIP] 1–14), which have 8 transmembrane domains and their C- and N-terminal domains are on the extracellular side. ZIP proteins are localized in the plasma membrane, where they mediate the transport of zinc from the outside to the inside of the cells or to the membranes of intracellular organelles (where the transport takes place from the lumen of the organelle to the cytoplasm), resulting in elevated intracellular zinc levels.

The involvement of ZIP proteins in human diseases has been recently uncovered (Takagishi et al. 2017). For example, a nonsense mutation at the level of the gene coding for ZIP12 is responsible for a form of schizophrenia (Chowanadisai 2014; Chowanadisai et al. 2013), and hypomorphic mutations of ZIP7 are associated with a novel type of human immunodeficiency with agammaglobulinemia (Anzilotti et al. 2019). ZIP6 and ZIP10 are reported to be involved in the progression of breast cancer (Hashemi et al. 2007; Hwang et al. 2010; Kagara et al. 2007; Matsui et al. 2017). In addition, recent data have suggested that several ZIP members are associated with skin disorders. In this chapter, we outline the role of ZIP family members in skin formation and diseases.

15.2 ZIP1

ZIP1 was isolated as a member of the IRT1 (iron regulated transporter) family of transporters (Lioumi et al. 1999). *ZIP1* is mapped at 1q21.3 of the human chromosome at the telomeric end of the epidermal differentiation complex (EDC). The EDC contains many keratinocyte differentiation- and stratification-associated precursors such as loricrin, involucrin, and small proline-rich proteins, implying the involvement of *ZIP1* in epidermis formation. More precisely, *ZIP1* is located as telomere-*ZIP1-S100A1-A13-A2-A3-A4-A5-A6-A7-A8-A12-A9-A11-A10*-centromere. Instead, mice *Zip1* is located at 3F1 of chromosome 3 between *S100a9* and *S100a13*, as telomere-*S100a1-a8-a9-Zip1-a13-a6-a3-a4-a5*-centromere. Thus, gene translocation has been observed in this chromosome area, and the relationship between ZIP1 and S100 proteins may be considerably involved in epidermal homeostasis, as S100A7 (psoriasin) and S100A15 are associated with psoriasis and autoimmune skin diseases (Wolf et al. 2008).

In situ hybridization analysis has shown that *Zip1* is developmentally controlled during epidermal morphogenesis (Lioumi et al. 1999). *Zip1* is not expressed by embryonic day (E) 15.5, but its expression is apparently detected at E17.5. Interestingly, after birth, *Zip1* expression is reduced and rarely detected at postnatal day 21. This expression pattern may support the vigorous cell proliferation and differentiation during embryonic epidermal morphogenesis, as many proteins that are essential for DNA synthesis and cell proliferation need zinc for their functions. In fact, when proliferation of fibroblasts is stimulated by growth factors such as platelet-derived growth factor, epidermal growth factor, and insulin-like growth factor I, the labile intracellular pool of zinc is dramatically increased with the elevation of *ZIP1* (Simpson and Xu 2006). Taken together, ZIP1 might play a crucial role in skin morphogenesis by regulating zinc homeostasis in the epidermis.

15.3 ZIP2

Based on a synchrotron radiation high-energy x-ray fluorescence (SR-XRF) study and a neutron activation analysis, the zinc level in the epidermis was reported to be about 6 times greater than that in the dermis (Inoue et al. 2014). To identify the important ZIP proteins in this event, several ZIP proteins were analyzed and ZIP2 was suggested to be one of the main zinc transporters in the epidermis. ZIP2 was not expressed in the basal layer, but accumulated in all other parts of the epidermis. After the induction of keratinocytes by calcium for differentiation, the *ZIP2* expression was increased, but the expression of *ZIP1* and *ZIP3* was comparable. These expression patterns well correlated with the results of SR-XRF analysis of zinc. In addition, when extracellular zinc was chelated by Chelex-100, the *ZIP2* expression was significantly increased for transporting zinc to modulate the zinc balance in keratinocytes. Although ZIP2 is not essential in mice, *Zip2* knockout mice showed epidermal blistering under zinc deficiency during early embryogenesis (Peters et al. 2007).

15.4 ZIP4

ZIP4 is produced by the *slc39A4* gene located on chromosome 8 in 8q24 (Kury et al. 2002; Nakano et al. 2003). ZIP4 has been shown to be a key importer of zinc in intestinal cells such as enterocytes of the intestine and visceral endoderm

(Kambe and Andrews 2009; Wang et al. 2004). ZIP4, as other ZIP family members, contains 8 transmembrane domains organized into 2 blocks of 3 and 5, which are separated by a histidine-rich cytoplasmic metal-binding site. ZIP4 possesses a long N-terminus facing the opposite side of the cytoplasm, with about 340 amino acids. During zinc deficiency, the N-terminus is cleaved by an unknown mechanism, and this seems to contribute to the plasma localization of ZIP4 in intestinal cells to increase the zinc uptake. ZIP4 was discovered when mutations were linked to the human genetic disease acrodermatitis enteropathica (AE) (Andrews 2008; Hurley et al. 1979; Maverakis et al. 2007; Michaelsson 1990; Tucker et al. 1976; Walravens et al. 1978). First described by Brandt in 1936 and named by Danbolt, AE is also known as Danbolt-Closs syndrome. This disease is due to an autosomal recessive mutation in ZIP4, causing a disruption of its transport function and a reduction in zinc absorption. AE usually occurs during maternal weaning. Dozens of pathogenic mutations have been identified in ZIP4. Some of them prohibit the proper cleavage of the N-terminus, leading to abnormal cellular distribution. Others lead to low expression, causing defects in zinc uptake. As a result, patients with AE require daily zinc supplements throughout their lives. As zinc supplements can address the problem of these patients, it is obvious that there are other transport mechanisms, although less effective, in the enterocytes. In fact, enterocytes increase the expression of ZIP4 when food intake is low, with ZIP4 more localized in the apical membrane with the cleavage of the N-terminus. This increases the zinc acquisition from the diet.

In mice fed a zinc-deficient diet, ZIP4 is overexpressed in the small intestine; however, its expression decreases after zinc supplementation (Andrews 2008; Dufner-Beattie et al. 2003, 2007; Geiser et al. 2012). These results suggest that the regulation of ZIP4 is essential for zinc uptake and plays a role in zinc homeostasis by increasing the ZIP4 expression when the zinc intake is limited or decreasing. As both ZIP2 and ZIP4 are overexpressed in the case of zinc deficiency, a common regulatory mechanism is likely responsible for this response.

Recent studies have demonstrated that human ZIP4 is also expressed in the epidermal keratinocytes to transport zinc into the cytoplasm (Bin et al. 2017a). ZIP4 expression is enriched in proliferative and undifferentiated epidermal keratinocytes in the basal layer of the epidermis, like the expression of zinc-dependent protein metallothionein (MT), implying that ZIP4 is involved in the zinc homeostasis of the basal layer. Zinc-binding master transcription factor ΔNp63 is also expressed in the basal layer for epidermal homeostasis. When ZIP4 is depleted, the ΔNp63 activity is decreased, resulting in epidermal malformation (Fig. 15.3). ΔNp63 is one of the variants of p63 that lacks the N-terminal transactivation domain, and it possesses Cys205 as an essential residue for zinc-dependent regulation. Mutations in Cys205 lead to abnormal nuclear localization of ΔNp63, and to loss of its functions. Taken together, ZIP4-mediated zinc homeostasis is essential for epidermal formation by supporting the ΔNp63 activity.

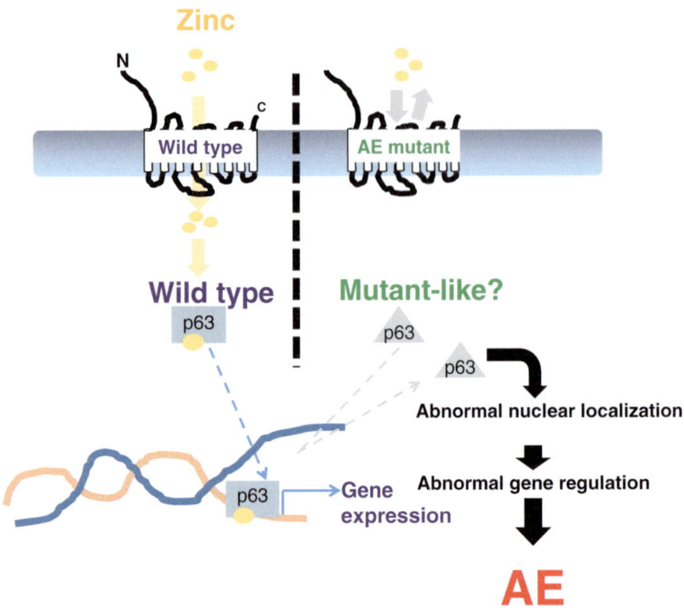

Fig. 15.3 ZIP4 supports p63 activity for epidermal homeostasis
ZIP4 supplies zinc for p63 activity. ZIP4 mutation leads to zinc deficiency, resulting in improper functioning of p63 and thus inducing epidermal disorders such as acrodermatitis enteropathica (AE). Zinc is depicted as yellow circles. (Modified from Bin et al. 2017a)

15.5 ZIP7

ZIP7 is especially localized in the endoplasmic reticulum (ER) of almost all tissues and is believed to play a role as a zinc transporter (Hogstrand et al. 2009; Taylor et al. 2012). Zinc is released by ZIP7 through the phosphorylation of ZIP7 by an important enzyme called casein kinase 2. This enzyme stimulates the discharge of zinc ions from the ER, which provides a signal transduction pathway. The result of this pathway contributes to the activation of the receptors present in the cell surface, such as epidermal growth factor.

ZIP7 transporters have important roles in the growth of the dermis (Bin et al. 2017b). ZIP7 deletion under the control of the collagen 1 promoter in mice leads to dysgenesis of the dermis due to loss of collagen 1-producing cells. ZIP7 is expressed in cells of mesenchymal origin, such as fibroblasts. When ZIP7 is depleted, the ER stress is greatly increased, leading to induction of cell death (Fig. 15.4). In the ER, protein disulfide isomerases (PDIs) are highly accumulated to mediate proper protein folding by facilitating disulfide bonding via a cysteine-rich motif. ZIP7 normally exports zinc from the luminal side of the ER to the cytoplasm. However, when ZIP7 loses its function, zinc accumulates within the luminal side of the ER, leading to an increase in the luminal zinc level within the ER. This then results in zinc-dependent aggregation of PDIs, probably via cysteine residues. Subsequently, ER

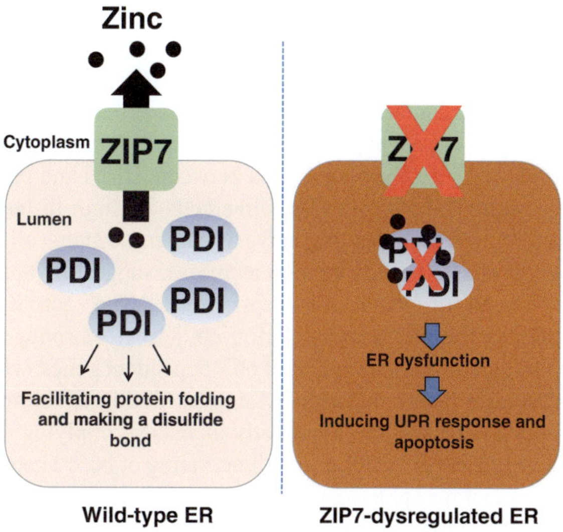

Fig. 15.4 Schematic model for the role of ZIP7 in mesenchymal stem cells
When ZIP7 is dysregulated (right), the endoplasmic reticulum (ER) luminal zinc level is increased, which accelerates the zinc-dependent aggregation of protein disulfide isomerases (PDIs), causing ER dysfunction and finally inducing unfolded protein response (UPR) and apoptosis. (Modified from Bin et al. 2017b)

dysfunction occurs, resulting in unfolded protein responses including apoptosis. Therefore, the source of cell numbers for collagen production is reduced in a ZIP7-deficient dermis. Most recently, a novel type of immune-deficient disorder caused by hypomorphic mutations of ZIP7 was identified, which affected not only B-cell-mediated immunity but also skin formation (Anzilotti et al. 2019; Fukada et al. 2019). Interestingly, the hypomorphic ZIP7 mutants impaired the B-cell antigen receptor signaling pathways, but did not increase ER stress, indicating that ZIP7 is involved in multiple regulatory systems that participate in physiological events.

15.6 ZIP10

ZIP10 is widely expressed but particularly highly in lymphocytes such as B-cells, as ZIP10 plays a role in B-cell development and functions (Hojyo et al. 2014; Miyai et al. 2014). In addition, ZIP10 is also expressed in epithelial tissues, predominantly in the hair follicles, and is necessary for epidermis formation (Bin et al. 2017c).

During mouse embryo development, the ZIP10 expression patterns are similar to those of MT. Its expression is accumulated within hair follicles, which contain diverse epidermal stem cells, indicating the involvement of ZIP10 in epidermal morphogenesis. ZIP10 depletion in mice epidermis causes significant defects in epidermis formation. The mice show a partially distinctly formed epidermis, in that the dorsal epidermis progresses to stratification, but is very thin compared with normal epidermis, whereas the ventral epidermis has an entirely undifferentiated epithelium. This might be the result of the relatively different speeds of epidermis formation in each body part. ZIP10 expression is detected in the embryo from E15, and is extremely elevated until just before birth. At E15, almost every basal epithelium is ready for differentiation. Thus, ZIP10 is not essential for basal epithelium formation. Instead, from E15, vigorous cell proliferation and stratification need to progress for epidermis formation. During this period, ZIP10 seems to supply zinc to those cells. In fact, ZIP10-positive cells in mouse epidermis show a high association with p63-depedent genes that are crucial for epidermis differentiation and stratification, implying the cooperation between ZIP10 and p63 for epidermal morphogenesis. Interestingly, ZIP10-positive cells are high in *lgr6* expression and low in *gli1* expression such that they serve as epidermal stem cell markers. When the epidermis is injured, ZIP10 expression increases, suggesting that ZIP10 may mark epidermal stem cell subsets. As ZIP4 supports the ΔNp63 activity, ZIP10 is also a crucial partner for the p63 activity during epidermal morphogenesis. ZIP10 overexpression upregulates the expression of p63-dependent genes, whereas ZIP10 depletion downregulates their expression. Moreover, during epidermal morphogenesis, from E14, both ZIP10 and p63 are greatly increased, implying that more zinc will be needed during this period for proper functioning of p63. Therefore, the zinc-ZIP10-p63 axis is crucial for epidermal morphogenesis in the embryo.

ZIP10 may support diverse protein functions. Recent data have demonstrated that ZIP10 is involved in the activities of epigenetic enzymes (Fig. 15.5). ZIP10 depletion in reconstituted human skin equivalent reconstructed from keratinocytes results in the downregulation of the activity of the histone acetyltransferases (HATs) (Bin et al. 2018b). HATs induce the acetylation of histones to open the chromosome to initiate the transcription of genes such as *FLG* and *MT* for epidermal homeostasis. Therefore, ZIP10 is essential for the coordinated actions of transcription factors and epigenetic enzymes for proper epidermal morphogenesis and homeostasis.

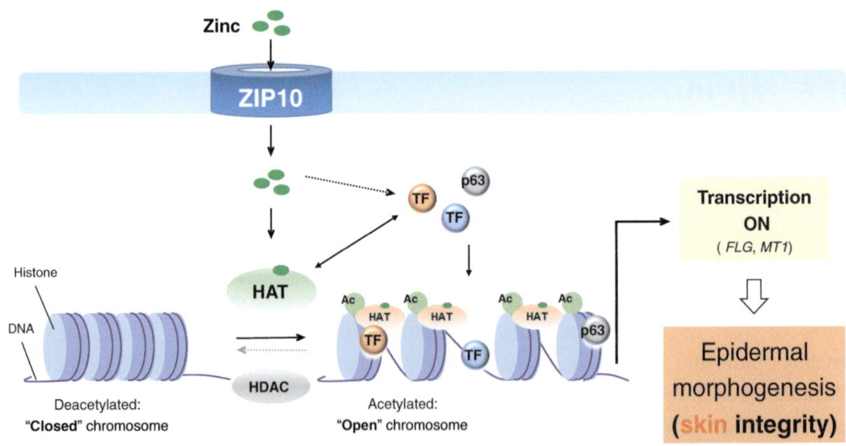

Fig. 15.5 Schematic model for the role of ZIP10 in human epidermal homeostasis
ZIP10 mediates zinc influx into the cytoplasm, affecting the activity of transcription factors (TFs) including p63 and epigenetic enzymes such as histone acetyltransferases (HATs). The cooperation of ZIP10 with HATs and TFs is essentially involved in epidermal morphogenesis and homeostasis. *HDAC* histone deacetylase

15.7 ZIP13

ZIP13 is located in the perinuclear zone of cells. ZIP13 is mainly located in the Golgi apparatus in fibroblasts and osteoblasts, suggesting that this protein functions as an intracellular zinc transporter similar to another intracellular ZIP member, like ZIP7 (Bin et al. 2011; Fukada et al. 2008). ZIP13 is associated with Ehlers-Danlos syndrome spondylodysplastic type 3 (EDSSPD3; OMIM # 612350) (Fukada et al. 2008; Giunta et al. 2008). Patients with general Ehlers-Danlos syndrome (EDS) show disorders in connective tissues with hyperelastic and translucent skin and hypermobile fingers, as pathogenic mutations occur in diverse fibrous proteins and enzymes involved in collagen production. Although patients with EDSSPD3 show similar signs and symptoms to those with general EDS, no apparent mutations in those proteins have been reported. Instead, to date, 2 mutations have been identified from patients with EDSSPD3: G64D and ΔFLA in the *SLC39A13* gene, which encodes the ZIP13 protein (Fukada et al. 2008; Giunta et al. 2008). These 2 mutations reduce the protein stability and lead to rapid protein degradation via an ER-associated degradation pathway, resulting in loss of function (Bin et al. 2011, 2014a, b). Two molecules (valosine-containing protein and heat shock protein 90) that bind to the ZIP13 mutant have been identified, and they are involved in protein degradation for the transfer of ZIP13 to the proteasome machinery. Loss of function of ZIP13 reduces the cytoplasmic zinc level and causes dysgenesis of the dermis due to defects in collagen production. ZIP13-null mice also show identical phenotypes to those of patients with EDSSPD3, including thin and weak dermis and growth retardation. Further analysis has demonstrated that the transforming growth factor-β signaling pathway, the major collagen production pathway in the dermis, is blocked by the irregularity of the nuclear translocation of SMAD transcription factors (Fukada et al. 2008). SMAD is a zinc finger transcription factor that needs zinc to bind to the promoter region of the *collagen* gene (*Col*). Reduced cytoplasmic zinc level in ZIP13-mutated or ZIP13-depleted fibroblasts may contribute to the dysregulation of SMAD by inhibiting the proper zinc distribution to SMAD; however, the precise molecular mechanism of how ZIP13 controls the nuclear translocation of SMAD is still unclear.

As discussed in Chap. 12, ZIP13 is also involved in adipocyte browning in the hypodermis, which is the lowest layer of the skin (Fukunaka et al. 2017). Patients with EDSSPD3 show lipoatrophy (Fukada et al. 2008), and in an animal model analysis using ZIP13-null mice, the stability of the CCAAT/enhancer binding protein-β (C/EBP-β), which is associated with the dominant transcriptional co-regulator PR domain containing 16 in directing adipocyte lineage processing, is regulated by ZIP13-mediated zinc. Although the precise mechanism underlying the association between ZIP13 and C/EBP-β is still unclear, there is no doubt that ZIP13 is a molecule that regulates the zinc homeostasis of both the dermis and the hypodermis.

15.8 Other ZIP Family Members

ZIP8 deletion reduces the MT (*Mt1*) expression as a consequence of decreased intracellular zinc levels, and increases the responses of pro-inflammatory cytokines such as tumor necrosis factor-α and interleukin-1β by inhibiting the nuclear factor-κB pathway through zinc-mediated downmodulation of IκB kinase activity (Liu et al. 2013). Because immune responses can affect skin development and homeostasis, it is possible that ZIP8 might indirectly influence skin development and homeostasis. As ZIP14 expression is changed by inflammatory signaling pathways (Aburto-Luna et al. 2017; Galvez-Peralta et al. 2014; Sayadi et al. 2013) and its expression is altered by ER stress in mesenchymal stem cells (Bin et al. 2017b; Homma et al. 2013), ZIP14 may be involved in human skin pathophysiology under certain stress signaling pathways.

15.9 Conclusion

The ZIP members mentioned in this chapter have clear involvements in skin health and diseases. Therefore, further investigations will be needed to gain a precise understanding of zinc regulation in the skin, as well as to sufficiently elucidate the association of the remaining members with skin development and homeostasis. Because zinc transporters are related to diverse cellular events including immune responses, organ homeostasis, and aging, which can also affect the skin condition, the indirect involvements of zinc transporters could be considerable.

Such observations and speculations emphasize the importance of considering zinc transporters as drug targets in skin-related diseases and of understanding the mechanisms by which zinc transporters control the skin homeostasis, so as to fully understand the roles of zinc transporter-mediated zinc signals in the regulation of normal skin and disease conditions.

References

Aburto-Luna V, Trevino S, Santos-Lopez G et al (2017) Hepatic mobilization of zinc after an experimental surgery, and its relationship with inflammatory cytokines release, and expression of metallothionein and Zip14 transporter. Inflamm Res 66(2):167–175

Andreini C, Banci L, Bertini I, Rosato A (2006) Counting the zinc-proteins encoded in the human genome. J Proteome Res 5(1):196–201

Andrews GK (2008) Regulation and function of Zip4, the acrodermatitis enteropathica gene. Biochem Soc Trans 36(Pt 6):1242–1246

Anzilotti C, Swan DJ, Boisson B et al (2019) An essential role for the Zn(2+) transporter ZIP7 in B cell development. Nat Immunol 20(3):350–361

Bin BH, Fukada T, Hosaka T et al (2011) Biochemical characterization of human ZIP13 protein: a homo-dimerized zinc transporter involved in the spondylocheiro dysplastic Ehlers-Danlos syndrome. J Biol Chem 286(46):40255–40265

Bin BH, Hojyo S, Hosaka T et al (2014a) Molecular pathogenesis of spondylocheirodysplastic Ehlers-Danlos syndrome caused by mutant ZIP13 proteins. EMBO Mol Med 6(8):1028–1042

Bin BH, Hojyo S, Ryong Lee T, Fukada T (2014b) Spondylocheirodysplastic Ehlers-Danlos syndrome (SCD-EDS) and the mutant zinc transporter ZIP13. Rare Dis 2(1):e974982

Bin BH, Bhin J, Kim NH et al (2017a) An acrodermatitis enteropathica-associated Zn transporter, ZIP4, regulates human epidermal homeostasis. J Invest Dermatol 137(4):874–883

Bin BH, Bhin J, Seo J et al (2017b) Requirement of zinc transporter SLC39A7/ZIP7 for dermal development to fine-tune endoplasmic reticulum function by regulating protein disulfide isomerase. J Invest Dermatol 137(8):1682–1691

Bin BH, Bhin J, Takaishi M et al (2017c) Requirement of zinc transporter ZIP10 for epidermal development: implication of the ZIP10-p63 axis in epithelial homeostasis. Proc Natl Acad Sci U S A 114(46):12243–12248

Bin BH, Hojyo S, Seo J et al (2018a) The role of the Slc39a family of zinc transporters in zinc homeostasis in skin. Nutrients 10(2):219

Bin BH, Lee SH, Bhin J et al (2018b) The epithelial zinc transporter ZIP10 epigenetically regulates human epidermal homeostasis by modulating histone acetyltransferase activity. Br J Dermatol 180(4):869–880

Bin BH, Seo J, Kim ST (2018c) Function, structure, and transport aspects of ZIP and ZnT zinc transporters in immune cells. J Immunol Res 2018:9365747

Chowanadisai W (2014) Comparative genomic analysis of slc39a12/ZIP12: insight into a zinc transporter required for vertebrate nervous system development. PLoS One 9(11):e111535

Chowanadisai W, Graham DM, Keen CL, Rucker RB, Messerli MA (2013) Neurulation and neurite extension require the zinc transporter ZIP12 (slc39a12). Proc Natl Acad Sci U S A 110(24):9903–9908

Dufner-Beattie J, Wang F, Kuo YM, Gitschier J, Eide D, Andrews GK (2003) The acrodermatitis enteropathica gene ZIP4 encodes a tissue-specific, zinc-regulated zinc transporter in mice. J Biol Chem 278(35):33474–33481

Dufner-Beattie J, Weaver BP, Geiser J et al (2007) The mouse acrodermatitis enteropathica gene Slc39a4 (Zip4) is essential for early development and heterozygosity causes hypersensitivity to zinc deficiency. Hum Mol Genet 16(12):1391–1399

Fukada T, Kambe T (2011) Molecular and genetic features of zinc transporters in physiology and pathogenesis. Metallomics 3(7):662–674

Fukada T, Kambe T (eds) (2014) Zinc signals in cellular functions and disorders. Springer, Tokyo

Fukada T, Civic N, Furuichi T et al (2008) The zinc transporter SLC39A13/ZIP13 is required for connective tissue development; its involvement in BMP/TGF-beta signaling pathways. PLoS One 3(11):e3642

Fukada T, Hojyo S, Hara T, Takagishi T (2019) Revisiting the old and learning the new of zinc in immunity. Nat Immunol 20(3):248–250

Fukunaka A, Fukada T, Bhin J et al (2017) Zinc transporter ZIP13 suppresses beige adipocyte biogenesis and energy expenditure by regulating C/EBP-β expression. PLoS Genet 13(8):e1006950

Galvez-Peralta M, Wang Z, Bao S, Knoell DL, Nebert DW (2014) Tissue-specific induction of mouse ZIP8 and ZIP14 divalent cation/bicarbonate symporters by, and cytokine response to, inflammatory signals. Int J Toxicol 33(3):246–258

Geiser J, Venken KJ, De Lisle RC, Andrews GK (2012) A mouse model of acrodermatitis enteropathica: loss of intestine zinc transporter ZIP4 (Slc39a4) disrupts the stem cell niche and intestine integrity. PLoS Genet 8(6):e1002766

Giunta C, Elcioglu NH, Albrecht B et al (2008) Spondylocheiro dysplastic form of the Ehlers-Danlos syndrome--an autosomal-recessive entity caused by mutations in the zinc transporter gene SLC39A13. Am J Hum Genet 82(6):1290–1305

Golan Y, Kambe T, Assaraf YG (2017) The role of the zinc transporter SLC30A2/ZnT2 in transient neonatal zinc deficiency. Metallomics 9(10):1352–1366

Hara T, Takeda TA, Takagishi T, Fukue K, Kambe T, Fukada T (2017) Physiological roles of zinc transporters: molecular and genetic importance in zinc homeostasis. J Physiol Sci 67(2):283–301

Hashemi M, Ghavami S, Eshraghi M, Booy EP, Los M (2007) Cytotoxic effects of intra and extracellular zinc chelation on human breast cancer cells. Eur J Pharmacol 557(1):9–19
Hogstrand C, Kille P, Nicholson RI, Taylor KM (2009) Zinc transporters and cancer: a potential role for ZIP7 as a hub for tyrosine kinase activation. Trends Mol Med 15(3):101–111
Hojyo S, Miyai T, Fujishiro H et al (2014) Zinc transporter SLC39A10/ZIP10 controls humoral immunity by modulating B-cell receptor signal strength. Proc Natl Acad Sci U S A 111(32):11786–11791
Homma K, Fujisawa T, Tsuburaya N et al (2013) SOD1 as a molecular switch for initiating the homeostatic ER stress response under zinc deficiency. Mol Cell 52(1):75–86
Hurley LS, Lonnerdal B, Stanislowski AG (1979) Zinc citrate, human milk, and acrodermatitis enteropathica. Lancet 1(8117):677–678
Huse M, Eck MJ, Harrison SC (1998) A Zn2+ ion links the cytoplasmic tail of CD4 and the N-terminal region of Lck. J Biol Chem 273(30):18729–18733
Hwang JJ, Kim HN, Kim J et al (2010) Zinc(II) ion mediates tamoxifen-induced autophagy and cell death in MCF-7 breast cancer cell line. Biometals 23(6):997–1013
Inoue Y, Hasegawa S, Ban S et al (2014) ZIP2 protein, a zinc transporter, is associated with keratinocyte differentiation. J Biol Chem 289(31):21451–21462
Itsumura N, Inamo Y, Okazaki F et al (2013) Compound heterozygous mutations in SLC30A2/ZnT2 results in low milk zinc concentrations: a novel mechanism for zinc deficiency in a breast-fed infant. PLoS One 8(5):e64045
Kagara N, Tanaka N, Noguchi S, Hirano T (2007) Zinc and its transporter ZIP10 are involved in invasive behavior of breast cancer cells. Cancer Sci 98(5):692–697
Kambe T, Andrews GK (2009) Novel proteolytic processing of the ectodomain of the zinc transporter ZIP4 (SLC39A4) during zinc deficiency is inhibited by acrodermatitis enteropathica mutations. Mol Cell Biol 29(1):129–139
Kury S, Dreno B, Bezieau S et al (2002) Identification of SLC39A4, a gene involved in acrodermatitis enteropathica. Nat Genet 31(3):239–240
Lee S, Hennigar SR, Alam S, Nishida K, Kelleher SL (2015) Essential role for zinc transporter 2 (ZnT2)-mediated zinc transport in mammary gland development and function during lactation. J Biol Chem 290(21):13064–13078
Lioumi M, Ferguson CA, Sharpe PT et al (1999) Isolation and characterization of human and mouse ZIRTL, a member of the IRT1 family of transporters, mapping within the epidermal differentiation complex. Genomics 62(2):272–280
Liu MJ, Bao S, Galvez-Peralta M et al (2013) ZIP8 regulates host defense through zinc-mediated inhibition of NF-kappaB. Cell Rep 3(2):386–400
Matsui C, Takatani-Nakase T, Hatano Y, Kawahara S, Nakase I, Takahashi K (2017) Zinc and its transporter ZIP6 are key mediators of breast cancer cell survival under high glucose conditions. FEBS Lett 591(20):3348–3359
Maverakis E, Fung MA, Lynch PJ et al (2007) Acrodermatitis enteropathica and an overview of zinc metabolism. J Am Acad Dermatol 56(1):116–124
Michaelsson G (1990) Acrodermatitis enteropathica: zinc in epidermis in relation to changes in dosage of zinc. Acta Derm Venereol 70(1):90–91
Miyai T, Hojyo S, Ikawa T et al (2014) Zinc transporter SLC39A10/ZIP10 facilitates antiapoptotic signaling during early B-cell development. Proc Natl Acad Sci U S A 111(32):11780–11785
Nakano A, Nakano H, Nomura K, Toyomaki Y, Hanada K (2003) Novel SLC39A4 mutations in acrodermatitis enteropathica. J Invest Dermatol 120(6):963–966
Ogawa Y, Kawamura T, Shimada S (2016) Zinc and skin biology. Arch Biochem Biophys 611:113–119
Ogawa Y, Kinoshita M, Shimada S, Kawamura T (2018) Zinc and skin disorders. Nutrients 10(2):199
Ogawa Y, Kinoshita M, Shimada S, Kambe T, Kawamura T (2019) Zinc transporters in the epidermis. J Dermatol 46(7):e243–e245

Peters JL, Dufner-Beattie J, Xu W et al (2007) Targeting of the mouse Slc39a2 (Zip2) gene reveals highly cell-specific patterns of expression, and unique functions in zinc, iron, and calcium homeostasis. Genesis 45(6):339–352

Prasad AS (2014) Impact of the discovery of human zinc deficiency on health. J Trace Elem Med Biol 28(4):357–363

Sayadi A, Nguyen AT, Bard FA, Bard-Chapeau EA (2013) Zip14 expression induced by lipopolysaccharides in macrophages attenuates inflammatory response. Inflamm Res 62(2):133–143

Simpson M, Xu Z (2006) Increased abundance of labile intracellular zinc during cell proliferation was due to increased retention of extracellular zinc in 3T3 cells. J Nutr Biochem 17(8):541–547

Takagishi T, Hara T, Fukada T (2017) Recent advances in the role of SLC39A/ZIP zinc transporters in vivo. Int J Mol Sci 18(12)

Tasman-Jones C, Kay RG (1975) Letter: zinc deficiency and skin lesions. N Engl J Med 293(16):830

Taylor KM, Hiscox S, Nicholson RI, Hogstrand C, Kille P (2012) Protein kinase CK2 triggers cytosolic zinc signaling pathways by phosphorylation of zinc channel ZIP7. Sci Signal 5(210):ra11

Tucker SB, Schroeter AL, Brown PW Jr, McCall JT (1976) Acquired zinc deficiency. Cutaneous manifestations typical of acrodermatitis enteropathica. JAMA 235(22):2399–2402

Walravens PA, Hambidge KM, Neldner KH et al (1978) Zinc metabolism in acrodermatitis enteropathica. J Pediatr 93(1):71–73

Wang F, Kim BE, Dufner-Beattie J, Petris MJ, Andrews G, Eide DJ (2004) Acrodermatitis enteropathica mutations affect transport activity, localization and zinc-responsive trafficking of the mouse ZIP4 zinc transporter. Hum Mol Genet 13(5):563–571

Wolf R, Howard OM, Dong HF et al (2008) Chemotactic activity of S100A7 (Psoriasin) is mediated by the receptor for advanced glycation end products and potentiates inflammation with highly homologous but functionally distinct S100A15. J Immunol 181(2):1499–1506

Chapter 16
Post-translational Mechanisms of Zinc Signalling in Cancer

Thirayost Nimmanon and Kathryn M. Taylor

Abstract Three groups of proteins are actively involved in the control of intracellular zinc, consisting of ZIP channels (SLC39A), ZnT transporters (SLC30A), and metallothioneins. Malfunctions of many zinc transport proteins, especially those belonging to the ZIP family which increase cytosolic zinc availability, have been associated with cancer. Importantly, post-translational modifications have been reported to play an increasing role in the functional control of ZIP channels. In this chapter, we therefore detail the established role of zinc signalling in cancer, with an emphasis on breast cancer, as well as demonstrate effects of post-translational modifications by phosphorylation and proteolytic cleavage.

Keywords Cancer · SLC39A · SLC30A · SLC39A7 · SLC39A6 · SLC39A10 · ZIP channels · ZnT transporters · Phosphorylation · Proteolytic cleavage

16.1 Introduction

The importance of zinc in human health and diseases cannot be overstated, as has previously and only partially been described in this book. The human body has approximately 2 g or 30 mmol of zinc, with 60% of total body zinc contained in the skeletal muscle (King et al. 2000), and the highest zinc concentration in the choroid and the prostatic secretion (FAO/WHO 2002). Amongst all the metals, only zinc associates with all the 6 enzyme classes (Vallee and Galdes 1984). Furthermore, 10% of human protein has been designated as zinc protein, according to a bioinfor-

T. Nimmanon
Department of Pathology, Phramongkutklao College of Medicine, Bangkok, Thailand
e-mail: thirayost@pcm.ac.th

K. M. Taylor (✉)
Cardiff School of Pharmacy and Pharmaceutical Sciences, Cardiff University, Cardiff, UK
e-mail: TaylorKM@cardiff.ac.uk

T. Fukada, T. Kambe (eds.), *Zinc Signaling*,
https://doi.org/10.1007/978-981-15-0557-7_16

matics approach (Andreini et al. 2006, Passerini et al. 2007). Not surprisingly, zinc is involved in numerous physiological processes (Kambe et al. 2015), as evidently supported by disturbances of multiple human organ systems in people with severe zinc deficiency (Hambidge 2000), the severest end of the spectrum of which is called acrodermatitis enteropathica (Moynahan 1974). Despite much advancement of the insight into pathogenesis of zinc deficiency, this condition has still been reported to have relatively high prevalence worldwide, with 17.3% of the population estimated to be at risk (Wessells and Brown 2012). Given the high impact on the public health globally of either subclinical or clinical zinc deficiency and the lack of a dependable diagnostic techniques, zinc is reasonably referred to as a micronutrient of remarkable biological and epidemiological significance (Hambidge and Krebs 2007).

Zinc helps to maintain human health in various ways. Zinc is dispensable for the immune system, both innate and adaptive (Haase and Rink 2014, Bonaventura et al. 2015), the antioxidant mechanism (Oteiza 2012) and the wound-healing process (Lansdown et al. 2007), with intrinsic anti-bacterial (David 2012) and anti-fungal (Reeder et al. 2011) properties. As such, different forms of zinc are included in dietary supplements (Schwingshackl et al. 2015), topical therapeutic preparations (Sadeghian et al. 2011), and even anti-dandruff shampoos (Schwartz et al. 2013). Zinc supplementation has also been proved to prevent underweight and growth stunting in children prone to have zinc deficiency (Brown et al. 2002), quicken recovery time of acute diarrhoea (Walker et al. 2015) and common cold (Fashner et al. 2012), and halt progression of early stages of age-related macular degeneration (Lawrenson and Grzybowski 2015). It is therefore apparent that zinc plays substantial roles in both the homeostasis of the human body and the battle against various diseases.

Zinc actively participates in multiple biochemical mechanisms (Hambidge and Krebs 2007; FAO/WHO 2002). Furthermore, zinc helps the body to fight against invaders through its involvement in the immune system (Haase and Rink 2014), protects cells from oxidative damage through its pro-antioxidant properties (Maret 2008), and modulates human brain functions through its action as a neuromodulator in zincergic neurons (Takeda et al. 2013; Kay and Tóth 2008). Additionally, it has been demonstrated in mast cells (Yamasaki et al. 2007), Chinese hamster ovary cells (Pandey et al. 2010), and breast cancer cells (Taylor et al. 2012) that zinc also functions as a second messenger, satisfying at least four out of five criteria for a second messenger proposed by Sutherland (Sutherland et al. 1968; Aley et al. 2013). In breast cancer cells, zinc is released from the stores after cellular exposure to an extracellular first messenger, either zinc plus the zinc ionophore pyrithione or epidermal growth factor (EGF) plus the calcium ionophore ionomycin, producing a zinc wave (Taylor et al. 2012). Consequentially, the released zinc acts as a second messenger, mediating phosphorylation on ERK1/2 and AKT (Taylor et al. 2012) through its inhibitory action on protein tyrosine phosphatases (Wilson et al. 2012). Since zinc is a second messenger, zinc-mediated cellular effects can now be observed in the time scale of minutes, independently of its transcriptional roles, which may take hours or days (Hirano et al. 2008). Following this remarkable role of zinc, the importance of zinc signalling in pathogenesis of human diseases, including carcinogenesis, has increasingly been unravelled.

16.2 Control of Intracellular Zinc Level

Intracellular zinc homeostasis is stringently regulated, given that too much zinc results in apoptosis or necrosis (Kim et al. 1999), whereas too little zinc results in cell growth arrest (MacDonald 2000) or apoptosis (Fraker 2005). This regulation is carried out by three families of proteins: ZIP channels or zinc importers (SLC39A), ZnT transporters or zinc exporters (SLC30A), and zinc-binding proteins, particularly metallothioneins (Lichten and Cousins 2009). ZIP channels increase zinc bioavailability by importing zinc from the extracellular space or the intracellular stores into the cytosol (Kambe et al. 2015) through an as-yet-unknown mechanism, although phosphorylation (Taylor et al. 2012) and heterodimerisation (Hogstrand et al. 2013) have been reported for some ZIP channels. In contrast, ZnT transporters, functioning as Zn^{2+}/H^{+} exchangers (Ohana et al. 2009), counterbalance ZIP channels by mobilising zinc in the opposite direction (Kambe et al. 2015). Metallothioneins substantially participate in the control of intracytosolic free zinc levels through a zinc-buffering reaction under a steady state and a zinc-muffling reaction upon zinc flux into the cytosol (Colvin et al. 2010). Metallothioneins are so efficient that cytosolic free zinc levels are maintained within a narrow range of high pM to low nM (Krezel and Maret 2006).

Experimental data performed in cortical neuronal cells have proposed a "muffler" model, which incorporated into the equations a protein that has high affinity for zinc ions and deep cellular stores where zinc is sequestrated and released upon activation (Colvin et al. 2008). This model strongly supports the capability of zinc to act as a second messenger, given that zinc is stored intracellularly, released upon activation, and initiates intracytosolic phosphorylation signalling cascades (Taylor et al. 2012). This model also underlines the key role of ZIP channels located in the intracellular zinc stores in zinc homeostasis. In light of the muffler model (Colvin et al. 2010), when zinc ions are transported in to a cell, they are immediately muffled by the metallothioneins and shuttled to the stores such as the ER. When the cell is exposed to an extracellular stimulus, organellar ZIP channels are activated, mobilising zinc ions from the stores into the cytosol where phosphorylation cascades commence.

16.3 Classification and Predicted Membrane Topology of Zinc Transport Proteins

Much scientific data have been reported concerning the association of ZIP channels and their post-translational modifications with cancer. These channels, particularly those belonging to the LIV-1 subfamily, will therefore be the main focus of this chapter.

16.3.1 *ZIP Channels*

A variety of evidence supports that ZIP proteins can act as ion channels. Firstly, they are responsible for bringing zinc down the transmembrane concentration gradient to the cytosol, which contains a negligible amount of zinc ions (Alam and Kelleher 2012), though it is never really negligible with regard to their impact on cellular physiology (Maret 2013). Secondly, ZIPB, a ZIP homolog from the bacteria *Bordetella bronchiseptica*, is molecularly characterised as a non-saturable ion-selective electro-diffusional channel responsible for passive cellular uptake of zinc (Lin et al. 2010), supporting the biochemical properties of an ion channel (Nelson and Cox 2013). However, the exact mechanism of ZIP channel-mediated zinc transport is relatively obscure with no supporting data from investigations using electrophysiological methods such as a patch clamp technique (Kornreich 2007). Because of these collective data, ZIP transport proteins are referred to as ZIP channels in this chapter.

Fourteen human proteins have been identified as ZIP channels, which are phylogenetically divided into four subfamilies of ZIP channels: gufA (ZIP11), subfamily I (ZIP9), subfamily II (ZIP1-3), and the LIV-1 subfamily (ZIP4-8, ZIP8, ZIP10, and ZIP12-14, also previously called LZT) (Taylor et al. 2007; Taylor and Nicholson 2003). ZIP7 is exceptionally located in the ER, and ZIP9, ZIP11, and ZIP13 are found in the Golgi, whilst all the other ZIP channels are known to reside on the plasma membrane (Fukada and Kambe 2011). Structurally, computational analysis has predicted that all ZIP channels have 8 transmembrane domains (TM) with a long extracytosolic N-terminus, a short extracytosolic C-terminus, and a long variable intracytosolic loop between TM3 and TM4, which is histidine-rich and thereby supposedly zinc-binding (Guerinot 2000) (Fig. 16.1).

Many members of the LIV-1 subfamily have two extra histidine-rich regions at the long extracytosolic N-terminus and the extracytosolic loop between TM2 and TM3 (Taylor and Nicholson 2003). Importantly, all the LIV-1 subfamily members contain within TM5 a HEXPHEXGD sequence (H, histidine; E, glutamate; P, proline; D, aspartate; G, glycine; and X, any amino acid), a consensus motif that is detected only in this subfamily and predicted to be a catalytic zinc-binding site (Taylor and Nicholson 2003) because of its sequence similarity to a zinc-binding region in zinc metallopeptidases (zincins) (Hooper 1994). In fact, this region has long been thought to contribute to selective zinc transport (Guerinot 2000), and this was further confirmed by the capability of ZIP8 and ZIP14, which have the initial histidine residue in this motif changed to a glutamate (E) residue (Taylor et al. 2007), to also transport cadmium, manganese (Girijashanker et al. 2008), and iron (Wang et al. 2012).

The unique localisation of ZIP7 and ZIP13 on intracellular membranes has been firmly established by co-localisation of recombinant ZIP7 with an ER marker, calreticulin (Taylor et al. 2004; Valentine et al. 2007), and recombinant ZIP13 with a Golgi marker, GM130 (Bin et al. 2011), using an immunofluorescence technique. In contrast to the majority of the LIV-1 subfamily members which mainly function as zinc importers (Hogstrand et al. 2009), these intracellular ZIP channels instead

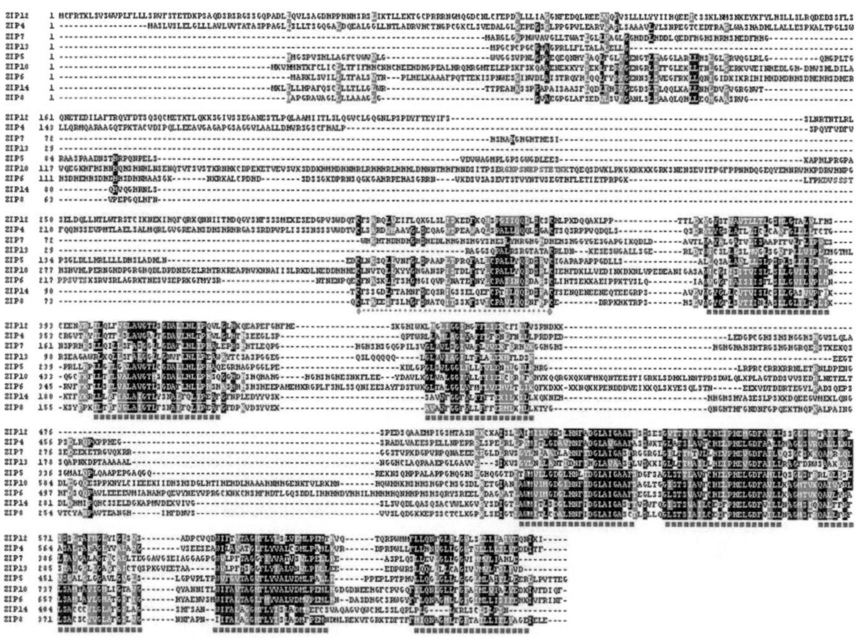

Fig. 16.1 Amino acid sequences of all human ZIP channels were aligned. Residues that were at least 50% identical (black) and complementary (grey) were shaded. The green stars highlight the predicted N-terminal CPALLY motif, which is not present in ZIP7 and ZIP13. The red squares mark the predicted transmembrane domains (TM). The red residues in ZIP6 and ZIP10 are the N-terminal predicted PEST sites

function as gatekeepers for zinc release from intracellular stores. In addition, because of the ubiquitous expression of ZIP7 in the human body (Taylor et al. 2004) and the demonstration of CK2-induced ZIP7 phosphorylation on residues S275 and S276 (Taylor et al. 2012) which results in activation of growth-promoting tyrosine phosphorylation cascades (Taylor et al. 2008), ZIP7 is reasonably referred to as "a hub for tyrosine kinase activation" (Hogstrand et al. 2009). It is noteworthy that ZIP9, which is located on the Golgi membrane, may function as a gatekeeper for zinc release from intracellular stores where endogenous ZIP7 is not present, such as in chicken B cells (DT40) (Kambe et al. 2006). In these cells, ZIP9-mediated zinc release from cellular stores results in protein tyrosine phosphatase inhibition and subsequent activation of downstream effectors (Taniguchi et al. 2013). This signifies that intracellularly located ZIP channels can be functionally interchangeable, and when ZIP7 is chronically deficient in mammalian cells, another ZIP channel, such as ZIP9, may effectively compensate the gatekeeper function of ZIP7.

ZIP6 and ZIP10 are the closest paralogues grouped in the same sub-branch in a phylogenetic tree and are both strongly associated with breast cancer migration and metastasis (Nimmanon and Taylor 2014). A unique characteristic for ZIP6 and ZIP10 is the presence within the N terminus of a PEST motif (Figs. 16.1 and 16.2), which is a potential cleavage site for ubiquitination-mediated proteolysis (Meyer

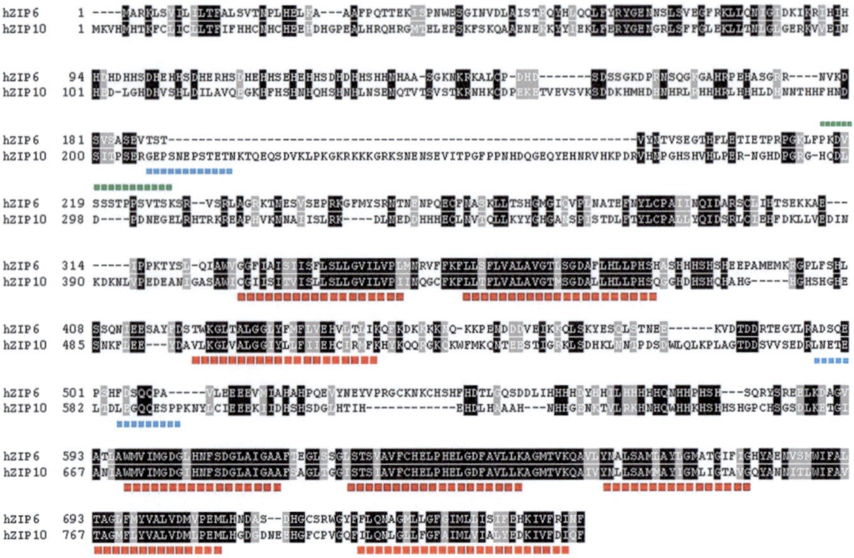

Fig. 16.2 Amino acid sequences of ZIP6 and ZIP10 were aligned. Residues that were at least 70% identical (black) and complementary (grey) across all the LIV − 1 subfamily members were shaded. The predicted TM regions (Taylor and Nicholson 2003) (red squares), the CPALLY motif (green stars), and the PEST motifs (the ZIP6 motif beneath red squares and the ZIP10 motifs above blue squares) are indicated

et al. 2011, Ramakrishna et al. 2011). This motif is hydrophilic, containing 12 negatively charged amino acids, especially proline (P), glutamate (E), serine (S), and threonine (T) (Rogers et al. 1986, Rechsteiner and Rogers 1996). ZIP10 has an additional PEST motif within the cytosolic loop between TM3 and TM4 (Fig. 16.2). The presence of a PEST motif may generally imply a shortened half-life of the proteins because it is a site for ubiquitination-mediated degradation (Meyer et al. 2011; Ramakrishna et al. 2011). Nevertheless, it may instead lead to proteolytic cleavage which can be required for channel activation, a phenomenon previously reported in the CaVβ3 subunit of voltage-gated calcium channels (Sandoval et al. 2006). This cleavage has also been observed for both ZIP6 (Hogstrand et al. 2013) and ZIP10 (Ehsani et al. 2012), zinc transport function of which is activated by cleavage in the N-terminus or the ectodomain. This will be discussed further in the following sections.

16.3.2 ZnT Transporters

Ten human proteins have been recognised as ZnT transporters, which can be classified into 4 subfamilies (Kambe 2012). ZnT transporters catalyse H^+/Zn^{2+} exchange as the means of zinc transport (Ohana et al. 2009). However, ZnT9 is now believed to be a nuclear coactivator instead of a zinc transporter (Kambe 2012). Most ZnT transporters are estimated to have 6 TMs, but ZnT5 is shown to have as many as 12 TMs (Cousins et al. 2006). In contrast to ZIP channels, both the N-terminus and C-terminus are on the cytosolic side of the membrane (Palmiter and Huang 2004). There is a long histidine-rich intracytosolic loop located between TM4 and TM5 (Palmiter and Huang 2004), potentially responsible for zinc binding and selectivity. This histidine-rich region is replaced by a serine-rich region for ZnT6 (Seve et al. 2004), suggesting that this ZnT may also transport other metals. Amongst all the ZnT transporters, only ZnT1 is located on the cell membrane and therefore crucially needed for zinc excretion from the cells (Palmiter and Findley 1995).

16.4 Zinc and Cancer

Zinc is widely believed to be a cancer-preventative substance, thanks to its immunity-enhancing and antioxidant properties (Dhawan and Chadha 2010). Although without any well-established evidence from a clinical trial, this presumption is firmly supported by high prevalence of low serum zinc levels amongst patients with cancers of the breast (Arinola and Charles-Davies 2008; Gumulec et al. 2014), the head and neck (Buntzel et al. 2007; Gumulec et al. 2014), the lung (Gumulec et al. 2014), the gastrointestinal tract (Boz et al. 2005; Gumulec et al. 2014), the liver (Gumulec et al. 2014), the gallbladder (Gupta et al. 2005), the female genital organs (Martin-Lagos et al. 1997, Naidu et al. 2007), and the prostate gland (Gumulec et al. 2014). Measurement of hair zinc levels, which is thought to be as reliable as serum zinc levels in indicating body zinc status (Lowe et al. 2009), also confirmed the association between low body zinc status and breast cancer (Memon et al. 2007). Notwithstanding, conflicting data do exist with regard to the association of serum zinc levels with cancer. One meta-analysis observed decreased serum zinc levels in most cancer types (Gumulec et al. 2014), whereas another smaller meta-analysis reported no change in serum zinc levels, but low hair zinc levels, in breast cancer (Wu et al. 2015). This discrepancy can be at least partly due to various factors, either physiological or pathological, that may interfere with measurement of serum zinc levels (King 2011).

Cellular zinc deficiency, experimentally induced by either zinc supply limitation or zinc chelation, results in upregulation of genes involved in DNA repair with an increase in oxidative DNA damage (Ho et al. 2003; Ho and Ames 2002). Despite the increase in P53 gene expression (Ho et al. 2003), the function of this DNA repair protein, which requires zinc for the site specificity of DNA binding (Loh 2010), is severely impaired unless zinc is sufficiently supplied (Bruinsma et al. 2002). This is experimental evidence of cellular zinc deficiency as a predisposing factor for DNA damage, which is a known mechanism for carcinogenesis.

In vivo studies have also been conducted to demonstrate the role of zinc in cancer prevention. Studies in mice and rats have demonstrated cancer-preventative effects of zinc supplementation in lung cancer (Satoh et al. 1993), colonic cancer (Dani et al. 2007), and oesophageal cancer (Fong et al. 2001). Nevertheless, conflicting data have been reported in clinical studies, indicating zinc intake as a risk factor (Kolonel et al. 1988), a preventative factor (Kristal et al. 1999), and also a non-associated factor (Chang et al. 2004; Leitzmann et al. 2003; Gonzalez et al. 2009) for prostatic cancer. A large clinical trial revealed that zinc could prevent prostatic cancer, although other vitamins and minerals that were also supplied with zinc in this study might have contributed to the effect (Meyer et al. 2005). It is also noteworthy that because of the lack of a reliable body zinc status indicator (Wieringa et al. 2015), determining any relationship of body zinc status with cancer can be extremely challenging. Altogether, these data imply that the role of zinc in either cancer-promoting or cancer-preventing mechanisms still requires further clinical investigation.

Zinc is generally perceived as a preventative factor for carcinogenesis and therefore reasonably expected to be present in cancer cells or tissues at a lower level than their non-cancerous counterparts. This expectation is proved to be true for many types of cancer, including prostatic cancer, lung cancer, and thyroid cancer (Huang et al. 2006). However, breast cancer tissue surprisingly contains higher zinc levels than in normal breast tissue (Taylor et al. 2011). Compared to non-neoplastic breast tissue, breast cancer tissue was shown to have a seven-fold mean zinc level increase when using emission spectrography (Mulay et al. 1971) and an approximately two-fold zinc level increase when using atomic absorption spectrophotometry (Margalioth et al. 1983; Jin et al. 1999). A more recent study using synchrotron radiation microprobe x−ray fluorescence revealed a 1.5-fold increase in tissue zinc levels, specifically in oestrogen receptor-positive breast cancer when compared to surrounding non-neoplastic breast tissue, but this increase was not seen in oestrogen receptor-negative breast cancer cases (Farquharson et al. 2009). Not only does the zinc level change in breast cancer, but zinc isotopic composition has also been reported to be altered in breast cancer, with a lighter zinc isotope being detected in breast cancer tissue than that found in blood and non-neoplastic breast tissue (Larner et al. 2015). These data indicate that the change in cancer tissue zinc levels does not necessarily reflect the total body zinc status, but may instead be attributed to abnormal expression of zinc transport proteins.

16.5 ZIP Channels in Breast Cancer

The neoplasm that has the strongest link to abnormal functions of ZIP channels is breast cancer (Taylor et al. 2011). ZIP6 (LIV-1), the first zinc transport protein associated with breast cancer, was shown to have increased gene expression in oestrogen receptor-positive cases (Manning et al. 1993) and cases associated with regional lymph node metastasis (Manning et al. 1994). After having been extensively investigated, ZIP6 is now known as a reliable indicator of oestrogen receptor-positive breast cancer (Tozlu et al. 2006; Schneider et al. 2006). Similar associations have also been reported for ZIP10, another ZIP channel that is grouped together in the same sub-branch with ZIP6 (figure ref) in a phylogenetic tree (Taylor and Nicholson 2003). ZIP10 gene expression is also enhanced in oestrogen-positive cases (Taylor et al. 2007) and cases with regional lymph node metastasis (Kagara et al. 2007). In contrast to ZIP6 and ZIP10, the ZIP7 gene and protein expression has been linked specifically to tamoxifen-resistant breast cancer cells (Taylor et al. 2007) and aggressiveness of these cells (Taylor et al. 2008). More details are going to be discussed in the following sections.

16.5.1 ZIP7 in Breast Cancer

ZIP7, which has a ubiquitous gene expression in the human body (Taylor et al. 2004), plays an essential role in the regulation of zinc release from intracellular stores and thereby activation of tyrosine kinases, including those required for cell growth and migration (Hogstrand et al. 2009). ZIP7 is mapped to the region for the major histocompatibility complex II (MHC II) on chromosome 6p21.3 (Ando et al. 1996), consistent with the recently proved role of its signalling in the immune response, specifically B-cell functions (Ollig et al. 2019). Its unique localisation in the ER has been confirmed by immunofluorescence that co-localised it with calreticulin, an ER marker, in MCF-7 breast cancer cells (Taylor et al. 2004) and with ZnT5 variant B, which is also known to be located in the ER, in Hela cervical cancer cells (Thornton et al. 2011).

ZIP7 overexpression has been observed in tamoxifen-resistant breast cancer cells (TAMR cells) (Knowlden et al. 2003), consistent with the discovery of a cellular zinc level in these cells twice as much as in the MCF-7 hormone-responsive breast cancer cells (Taylor et al. 2008). TAMR cells not only survive and grow in the presence of the selective oestrogen receptor modulator, but also acquire aggressiveness (Hiscox et al. 2004), attributed to increased activation of phosphorylation of well-known oncogenic signalling molecules including EGFR (Knowlden et al. 2003), IGF-1R (Jones et al. 2004), and Src (Hiscox et al. 2006). Importantly, ZIP7-mediated zinc release from intracellular stores has been demonstrated to be responsible for the activation of EGFR, IGF-1R, and Src (Taylor et al. 2008), probably due to the strong inhibitory effect of zinc on protein tyrosine phosphatase 1B (Haase and Maret 2005; Bellomo et al. 2014), an enzyme that phosphorylates all these signalling molecules (Bourdeau et al. 2005) (Fig. 16.3). The activation of these tyrosine kinases results in a phosphorylating activation of ERK1/2 and AKT, promoting cell growth and migration (Taylor et al. 2008). Inhibiting ZIP7 can therefore become a strategy to prevent development of drug resistance and decrease aggressiveness of the hormone-resistant cells.

ZIP7 functional control came to light when it was shown to involve protein kinase CK2-mediated phosphorylation on residues S275 and S276, which are situated in the cytosolic loop between TM3 and TM4 (Taylor et al. 2012) (Fig. 16.3). These two residues are contained within a region that matches a consensus motif for CK2 binding, **S/T**-X-X-E (S, serine; T, threonine; E, glutamine; X, any amino acid)

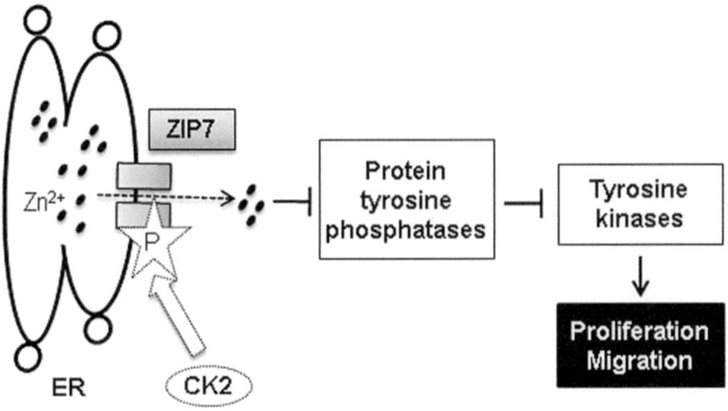

Fig. 16.3 The schematic illustrates a simplified ZIP7-mediated zinc-signalling pathway. ZIP7 in the ER is functionally activated by CK2-mediated phosphorylation on residues S275 and S276 (Taylor et al. 2012). This ZIP7 activation results in zinc release from cellular stores, inhibition of protein tyrosine phosphatases, and activation of tyrosine kinases that are involved in cell proliferation and migration

(Franchin et al. 2015). This post-translational modification results in prompt channel opening within 2 min, followed by zinc release from the ER and activation of ERK1/2 and AKT, with consequential growth and migration of breast cancer cells (Taylor et al. 2012) (Fig. 16.3). CK2 is a known oncoprotein contributing to carcinogenesis through activation of numerous predicted signalling molecules (Tawfic et al. 2001). It is therefore plausible that ZIP7 is one of the molecules responsible for oncogenic effects driven by CK2 (Nimmanon and Taylor 2015). Further investigations revealed that maximal ZIP7 activation requires phosphorylation not only on residues S275 and S276, but also on residues S293 and T294 (Nimmanon et al. 2017), suggesting that hierarchical phosphorylation (Lodish et al. 2000), a phenomenon common for CK2-mediated phosphorylation (St-Denis et al. 2015), may be involved in ZIP7 activation control. Moreover, using antibody arrays, several protein kinases were shown to be phosphorylated after phosphorylation-mediated ZIP7 activation, including those involved in MAPK, PI3K-AKT, and MTOR pathways (Nimmanon et al. 2017) (Table 16.1). Amongst these protein kinases, zinc has been experimentally confirmed to either directly or indirectly induce phosphorylation of CREB (Nuttall and Oteiza 2012), p70 S6 Kinase (Nuttall and Oteiza 2012, Nimmanon et al. 2017), and GSK-3β (Nimmanon et al. 2017), all of which are known to promote cell growth and survival (Sakamoto and Frank 2009; Soda et al. 2008; Piedfer et al. 2013). This therefore links ZIP7 to multiple carcinogenetic pathways and explains the role of ZIP7 as an independent predictor of poor clinical outcome in breast cancer (Taylor et al. 2007) according to the Oncomine cancer microarray database (Rhodes et al. 2004).

The discovery of phosphorylation as a mechanism for ZIP7 activation gives rise to a twofold implication. First, because of the lack of an effective biomarker for body zinc status (Wieringa et al. 2015), the antibody that specifically binds the

Table 16.1 Downstream effectors of ZIP7-mediated zinc release

Receptor-tyrosine kinases (RTK) arrays		
ALK	ErbB3	EphA7
EGFR	ErbB4	FGFR3
ErbB2	EphA1	RYK
Phospho-kinase and mitogen-activated protein kinases (MAPK) arrays		
AKT	HSP27	PRAS40
AMPK	HSP60	STAT2
c – Jun	JNK	STAT5
CREB	p38α MAPK	STAT6
ERK1/2	p53	WNK1
GSK-3α/β	p70 S6	

This table details proteins that were activated downstream of ZIP7-mediated zinc release when tested using arrays of receptor tyrosine kinases, the MAPK pathway or multiple phospho-kinases (Nimmanon et al. 2017)

phosphorylated form of ZIP7 might be used to determine ZIP7 activity and thus the zinc bioavailability within the cells. Second, given the role of activated ZIP7 in breast cancer aggressive behaviour particularly in tamoxifen-resistant breast cases, targeting activated ZIP7 might benefit these patients. The antibody that recognises ZIP7 when residues S275 and S276 are phosphorylated has been developed in house (Nimmanon et al. 2017). Using this antibody with an immunohistochemistry technique, breast cancer samples are shown to have a significant increase in pZIP7 compared to normal breast samples with a clearer result than that seen when using a total ZIP7 antibody (Ziliotto et al. 2019).

16.5.2 ZIP6 in Breast Cancer

ZIP6 (SLC39A6), previously recognised as LIV-1 (Taylor et al. 2003), is a 4.4-kb-long mRNA (Manning et al. 1988) with its gene being mapped to chromosome 18q12.2. Unlike ZIP7, ZIP6 is preferentially expressed in some tissues, particularly brain and hormonally regulated tissues (Taylor et al. 2003), although the relevant GenBank page suggests that it is expressed in all tissues with some variability. ZIP6 gene is oestrogen-regulated, highly expressed in hormonally responsive breast cancer cells (Manning et al. 1993, Manning et al. 1995), especially after being treated with oestrogen (Manning et al. 1988). The positive relationship between ZIP6 expression and oestrogen receptor status has already been confirmed and established by larger scale studies utilising real-time reverse-transcription PCR (Tozlu et al. 2006; Taylor et al. 2007) and microarray analysis (Schneider et al. 2006). In addition, ZIP6 gene expression is prevalent in regional lymph node metastasis (Manning et al. 1994). Because of these relationships, ZIP6 is now a potential biomarker for the luminal A molecular subtype of breast cancer (Schnitt 2010). Given that this molecular subtype of breast cancer has good prognosis, high ZIP6 expression indicates favourable outcome (Kim et al. 2012).

The involvement of ZIP6 in carcinogenesis was first related to effects on the process of epithelial–mesenchymal transition (EMT) observed during embryonic development (Yamashita et al. 2004), whereby epithelial cells acquire a mesenchymal phenotype allowing them to migrate (Thiery et al. 2009). In zebrafish during gastrulation, ZIP6 was shown to be transcriptionally activated by signal transducer and activator of transcription 3 (STAT3) (Yamashita et al. 2004). This increase in ZIP6 expression resulted in nuclear retention of Snail, decreased expression of E-cadherin, and cell migration (Yamashita et al. 2004). This phenomenon has also been observed in MCF-7 breast cancer cells (Taylor et al. 2007), signifying that ZIP6 may also be involved in the EMT process in breast cancer, the process required for neoplastic cell invasion and metastasis (Thiery et al. 2009). In breast cancer cells, STAT3 induces ZIP6 gene expression and consequential production of a ZIP6 pro-protein, which is stored in the ER (Hogstrand et al. 2013) (Fig. 16.4). Western blotting analysis using ZIP6 antibodies to different epitopes detected different bands of ZIP6, indicating that it is N-terminally cleaved before being moved to the plasma membrane, the location where it can function as a cellular zinc importer (Hogstrand et al. 2013) (Fig. 16.4). Zinc ions that flux into the cell directly phosphorylate

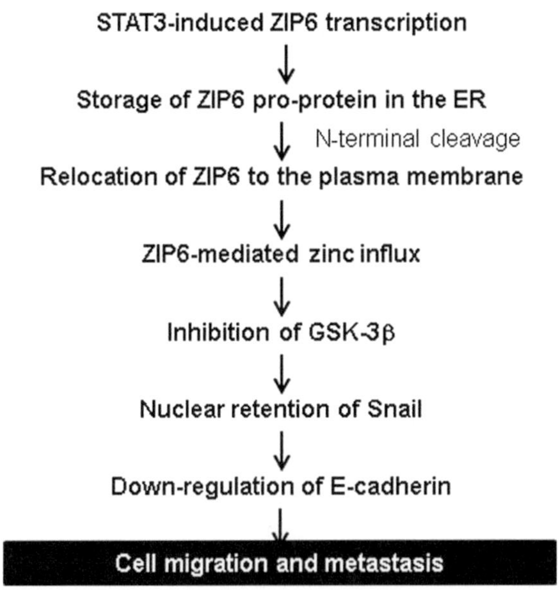

Fig. 16.4 The schematic illustrates ZIP6 post-translational modification in breast cancer EMT. STAT3 induces ZIP6 gene expression and production of a ZIP6 pro-protein, which is stored in the ER and N-terminally cleaved before being moved to the plasma membrane, the location where it imports zinc into the cells (Hogstrand et al. 2013). Zinc ions that flux into the cell phosphorylate GSK-3β, resulting in retention of Snail within the nucleus, in E-cadherin downregulation, cell detachment, migration, and metastasis (Hogstrand et al. 2013)

GSK-3β (Ilouz et al. 2002), and AKT (Lee et al. 2009, Ohashi et al. 2015), the latter of which also phosphorylates GSK-3β (Moore et al. 2013) (Fig. 16.4). GSK-3β in a normal situation phosphorylates the nuclear repressor Snail (Zhou et al. 2004), relocating it to the cytoplasm where it is degraded in the proteosomes (Bauer et al. 2009). Zinc-induced inhibitory phosphorylation of GSK-3β therefore retains Snail within the nucleus, resulting in E-cadherin downregulation, cell detachment, migration, and metastasis (Hogstrand et al. 2013) (Fig. 16.4). These data link ZIP6 to cancer cell invasion and metastasis, explaining the previous observation of its relationship with regional lymph node involvement in breast cancer.

16.5.3 ZIP10 in Breast Cancer

ZIP10 and ZIP6 are the closest paralogues of the LIV-1 subfamily of ZIP channels and are both involved in breast carcinogenesis. The ZIP10 gene is mapped to chromosome 2q32.3, and widely expressed throughout the body, preferentially in the central nervous system, the lungs, the kidneys, the bowels, and the genital organs (Lizio et al. 2015). The same gastrulation effects attributed to ZIP6 have also been demonstrated for ZIP10 (Taylor et al. 2016). Like ZIP6, ZIP10 gene expression is positively correlated with oestrogen receptor status (Taylor et al. 2007) and regional lymph node involvement (Kagara et al. 2007). High ZIP10 gene expression was observed in highly aggressive cell lines (MDA-MB-231 and MDAMB-435S) compared to less aggressive ones, and knocking down ZIP10 and zinc chelation successfully inhibited cell migration (Kagara et al. 2007), establishing the essential role of ZIP10 in breast cancer EMT. Importantly, a recent study demonstrated that ZIP6 forms a heterodimer with ZIP10 in the ER (Taylor et al. 2016), confirming the close companionship between the two ZIP channels in functioning as cellular zinc importers.

16.6 Zinc Signalling in Prostate Cancer

The prostate gland contains the highest concentration of zinc in the body (FAO/WHO 2002). It is essential for sperm release and motility, keeping fertility in males (Yoshida et al. 2008). It inhibits m-aconitase, keeping a high amount of citric acid in prostatic secretion (Costello et al. 2005). However, zinc is decreased in prostatic cancer tissue compared to non-neoplastic tissue (Ho and Song 2009), supporting the possible role of zinc supplementation in prostate cancer prevention. This preventative effect can be attributed to the role of zinc in protecting the cells from oxidative DNA damage (Yan et al. 2008) and suppressing cell growth by promoting pro-apoptotic conditions (Feng et al. 2002; Feng et al. 2008).

At least six members of the ZIP family of zinc transport proteins have been reported to have altered expression in prostate cancer. ZIP1-4 are reduced in prostate cancer, consistent with the decrease in the cellular zinc level (Huang et al. 2006; Desouki et al. 2007; Chen et al. 2012). ZIP1 is the most extensively characterised, being shown to have depressed gene expression at an early stage of prostate cancer (Franklin et al. 2005) and thereby indicated as a tumour suppressor for prostate cancer (Costello and Franklin 2006; Franklin and Costello 2007). In contrast to ZIP1-4, ZIP6 (Lue et al. 2011) and ZIP9 (Thomas et al. 2014) have been reported to increase, with the former being linked to prostate cancer EMT. ZnT1 (Hasumi et al. 2003) and ZnT4 (Beck et al. 2004) are decreased in prostate cancer tissue, with as-yet-unclear mechanism in carcinogenesis.

16.7 Post-translational Modifications of ZIP Channels

The transcriptional control of ZIP channels has been well characterised (Cousins et al. 2006). However, relatively little is known regarding their post-translational controls. Discovering the mechanisms might allow development of new strategies for cancer treatment. Two processes in particular have been shown to play important roles in ZIP channel controls, including phosphorylation and proteolytic cleavage (Nimmanon and Taylor 2017).

16.7.1 Phosphorylation

Phosphorylation is a well-known post-translational process whereby a phosphoryl group from ATP is attached to a serine (84%), threonine (15%), or tyrosine residue (less than 1%) (Humphrey et al. 2015). This process is mediated by cellular kinases and reversed by phosphatases (Cheng et al. 2011). It is involved in virtually all cellular aspects (Cohen 2002), coordinating with other post-translational modifications (Humphrey et al. 2015). Abnormal control of this mechanism therefore results in aberrant cell growth and the development of cancer. Its importance in carcinogenesis is suggested by a high-throughput bioinformatic approach in cancer exhibiting high prevalence of mutations affecting phosphorylatable residues, which is twice as high as the control data set (Radivojac et al. 2008). Moreover, a high proportion of eukaryotic kinases were mapped to the amplicons frequently involved by oncogenic mutations (Manning et al. 2002). Importantly, tyrosine kinases have successfully been targeted by inhibitors and monoclonal antibodies. For example, Imatinib mesylate (Gleevec), a Bcr–Abl tyrosine kinase inhibitor, has become the first-line treatment for chronic myeloid leukaemia (Henkes et al. 2008), and Trastuzumab (Herceptin), a monoclonal antibody targeting erbB2 or HER2/neu, has enormously benefitted patients with breast cancer that overexpresses HER2 (Pinto et al. 2013). Given the importance of this post-translational modification in various cellular processes including cancer, it is reasonable to speculate that ZIP channels may be phosphorylated as a means of functional control. The discovery of CK2-mediated phosphorylation of ZIP7 on residues S275 and S276 as a key mechanism for activating ZIP7 (Taylor et al. 2012) therefore validates this speculation.

Since ZIP7 is activated by phosphorylation, it is possible that other ZIP channels may also be activated in a similar fashion. Online databases, including PhosphoNet (Kinexus Bioinformatics Corporation), PhosphoSitePlus (Hornbeck et al. 2012), and NetPhos (Blom et al. 2004), list multiple sites in the intracytosolic loop between TM3 and TM4 that have been either predicted from amino acid sequences or experimentally confirmed by mass spectrometry to be phosphorylated in many other ZIP channels (Table 16.2). Remarkably, some ZIP channels other than ZIP7 also contain serine residues predicted to be phosphorylated by CK2, including ZIP6 and ZIP10 (Table 16.2). It is therefore possible that CK2 might simultaneously activate both intracellularly located ZIP7 to release zinc from the cellular stores and plasma membrane-located ZIP channels, such as ZIP6 and ZIP10, to refill the stores with zinc once depleted, allowing cells to have adequate cytosolic free zinc to exert its oncogenic action. Noteworthy, phosphorylation sites are present not only in ZIP channels, but also in ZnT transporters (Hogstrand et al. 2009), some of which are even predicted to be phosphorylated by CK2, such as those in ZnT6 (Huang et al. 2002) and ZnT8 (Chimienti et al. 2004). The relevance of phosphorylation sites to the control of zinc transport proteins need to be properly and individually investigated, given the simple consensus sequences and thus the high probability of false positives.

16.7.2 Proteolytic Cleavage of ZIP Channels

Proteolytic cleavage has been reported as a mechanism required for relocation and zinc ion transport function of ZIP channels as well as ZIP channels. ZIP4 has been demonstrated during zinc deficiency to be proteolytically cleaved at the extracellular N-terminus, leaving its 37-kDa ZIP4 peptide portion without its ectodomain to import zinc into the cells and thus enhancing zinc intestinal absorption (Kambe and Andrews 2009). Following this discovery, another pair of LIV-1 subfamily members, ZIP10 (Schmitt-Ulms et al. 2009) and ZIP6 (Hogstrand et al. 2013) were shown to be also functionally activated by N-terminal cleavage. It is noteworthy that interactome analyses have indicated these two ZIP channels as candidate interactors of cellular prion protein (Watts et al. 2009). To compensate for cellular starvation of zinc or manganese, ZIP10 undergoes ecto-domain shedding at its N-terminus (Schmitt-Ulms et al. 2009), reminiscent of the prion protein activation mechanism (Altmeppen et al. 2012), resulting in ZIP10 activation and cellular import of these trace metals (Schmitt-Ulms et al. 2009). In cancer, ZIP6 activation involves a complex process of post-translational modifications, including N-terminal proteolytic cleavage (Hogstrand et al. 2013), as has previously been detailed (Fig. 16.4). Altogether, these findings implicate that N-terminal proteolytic cleavage is a key mechanism for activation of ZIP channels, particularly ZIP4, ZIP10, and ZIP6.

Table 16.2 Phosphorylation sites in the intracytosolic loop between TM3 and TM4 of ZIP channels belonging to the LIV-1 subfamily with associated and predicted kinases

ZIP	Site	Sequence	Predicted kinases
4	S439	**PCGHSSHSH**	mTOR
	S440	**CGHSSHSHG**	mTOR
	S442	**HSSHSHGGH**	mTOR
	S447	**HGGHSHGVS**	TBK1
	S451	**SHGVSLQLA**	TBK1
	S457	**QLAPSELRQ**	GSK3 family
	S469	**PHEGSRADL**	PKA
	S478	**VAEESPELL**	JNK family
	S490	**PRRLSPELR**	p38 MAPK
5	S336	**PENGSGMAL**	
	S359	**REKNSQHPP**	DNAPK, PKG
	S374	**HQGHSHGHQ**	PKC
	S381	**HQGGTDITW**	cdc2
6	S471	**KKQLSKYES**	CK2, GSK3
	Y473	**QLSKYESQL**	
	S475	**SKYESQLST**	GSK3, DNAPK, PIKK
	S478	**ESQLSTNEE**	CK2, CK1
	T479	**SQLSTNEEK**	CK2, PKC
	T486	**EKVDTDDRT**	CK2, GSK3
	T490	**TDDRTEGYL**	
	Y493	**RTEGYLRAD**	(EGFR)
	S498	**LRADSQEPS**	GSK3, DNAPK, ATM
	S502	**SQEPSHFDS**	GSK3
	S506	**SHFDSQQPA**	DNAPK, ATM, PIKK
	Y528	**PQEVYNEYV**	
	Y531	**VYNEYVPRG**	
	S542	**NKCHSHFHD**	PKA
	S547	**HFHDTLGQS**	GSK3, NEK
	S551	**TLGQSDDLI**	CK2
	S577	**HHPHSHSQR**	PKG1, GSK3
	S579	**PHSHSQRYS**	DNAPK, ATM, GSK3
	S583	**SQRYSREEL**	CK2
7	S247	**GHGHSHGHG**	cdc2
	S255	**GHAHSHTRG**	mTOR/FRAP, PIM1
	S260	**HTRGSHGHG**	mTOR/FRAPm, PKA
	S269	**RQERSTKEK**	AKT1/2, CK2, PKC
	T270	**QERSTKEKQ**	PIM1, PIM3, PKC
	S275	**KEKQSSEEE**	CK2a
	S276	**EKQSSEEEE**	CK2a, CK1
	T283	**EEKETRGVQ**	BARK1
	S293	**RRGGSTVPK**	MAPKAPK, PKA, PKC

(continued)

Table 16.2 (continued)

ZIP	Site	Sequence	Predicted kinases
	T294	RGGSTVPKD	PIM1, PKC
8	S275	FDNVSVVSL	NEK10, CK1, mTOR
	S278	VSVVSLQDG	CK1
	S287	KKEPSSCTC	mTOR, CK1
	S288	KEPSSCTCL	mTOR, CK1, PKA
	S298	GPKLSEIGT	PKA
10	Y521	MFKHYKQQR	CSK, (INSR)
	T536	MKQNTEEST	ATR, (GSK3)
	S539	NTEESTIGR	PIM1, BARK1, PLK1
	T540	TEESTIGRK	LRRK2, PLK1
	S546	GRKLSDHKL	PIM3/1, PKA, RSK
	T553	KLNNTPDSD	JNK1/3, MAPK
	S556	NTPDSDWLQ	IKKb, CK2, (GSK3)
	T567	PLAGTDDSV	CK2, (cdc2, GSK3)
	S570	GTDDSVVSE	PIM2/1, PLK1
	S573	DSVVSEDRL	CK2, PIM2, CK1, CK2
	T580	RLNETELTD	Trb3, CK2
	T583	ETELTDLEG	CK2, cdc2
	S591	GQQESPPKN	CDK2/1, MAPK
	Y596	PPKNYLCIE	SYK, (INSR)
	S610	DHSHSDGLH	NEK10, CK2
	T634	GENKTVLRK	PIM1, PLK1, PKC
	S648	HHKHSHHSH	PKC
	S651	HSHHSHGPC	PKC, CK1
	S657	GPCHSGSDL	CDK1, PIM1
	S659	CHSGSDLKE	mTOR, cdc2
12	S470	ILLVSPNDK	(GSK3)
	S478	KQGLSLVNG	mTOR, PKA
	S494	LALNSELSD	JNK, mTOR, CK1
	S497	NSELSDQAG	mTOR, (CK1)
	S505	GRGKSASTI	PIM, PKA, RSK
	S507	GKSASTIQL	PIM, PKC
	T508	KSASTIQLK	PKC, CK1
	S513	IQLKSPEDS	ERK5, JNK, p38MAPK
	S517	SPEDSQAAE	ATR, PIKK, DNAPK
	S526	MPIGSMTAS	RHOK, ERK5, GSK3
	T528	IGSMTASNR	CDK2, ERK5, (GSK3)
	S530	SMTASNRKC	RHOK, CDK, PKC
	S538	CKAISLLAI	PKA
13	S172	MFLDSKEEG	CK2
	S178	EEGTSQAPN	ATR, ATM, DNAPK
	S213	AVVRSIKVS	PKC, GSK3

(continued)

Table 16.2 (continued)

ZIP	Site	Sequence	Predicted kinases
	S217	**SIKVSGYLN**	PKC, PKA
14	S256	**HHGHSHYAS**	IKKE, GSK3
	Y258	**GHSHYASES**	
	S262	**YASESLPSK**	CK1, IKKE, PKC
	S265	**ESLPSKKDQ**	CK2, PKC
	S292	**QHCSSELDG**	RAF family
	S309	**VIVGSLSVQ**	PKA, DNAPK
	S311	**VGSLSVQDL**	(cdc2)
	S318	**DLQASQSAC**	ATR, NEK2, PIKK
	Y330	**KGVRYSDIG**	TRK family, ABL
	S331	**GVRYSDIGT**	GSK3, PKA
	T335	**SDIGTLAWM**	DNAPK

Table showing the predicted phosphorylation sites in the intracellular loop of the LIV-1 family of ZIP channels between TM3 and 4. These predictions are an amalgamation of results from using PhosphoNet (Kinexus Bioinformatics Corporation), PhosphoSitePlus (Hornbeck et al. 2012), and NetPhos (Blom et al. 2004), and ELM (Gouw et al. 2018). Values in brackets were sites predicted by Netphos which met the threshold value of 0.5, but the actual kinase prediction was less than 0.5. Blank spaces represent no kinase prediction

16.8 Conclusions

It is becoming clear that ZIP channels may play an important role in multiple cancers. This is perhaps not surprising, as zinc is central to the growth of cells and when zinc is present in increased amounts in diseases such as cancer, it can easily activate many growth signalling pathways. Since ZIP7 was demonstrated to require phosphorylation in order to release zinc from stores (Taylor et al. 2012), it is a likely prospect that other ZIP channels can also be activated in this way. Analyses of the potential of these protein sequences for phosphorylation reveal numerous opportunities (Table 16.2), increasing the likelihood that phosphorylation offers some acceptable functional control. This is an exciting area of research that needs much more experimental examination and testing before valid conclusions can be drawn and the full extent of relevance to disease can be discovered. Additionally, N-terminal cleavage of ZIP channels has been demonstrated for at least three family members to date and in all cases, this cleavage has influenced the activity of the proteins concerned. The fact that prions have descended from the LIV-1 family (Schmitt-Ulms et al. 2009) adds credence to the belief that the N-terminal cleavage is crucial to controlling protein functional activity. Prions are known to be cleaved in functionally relevant ways and therefore discovery of the proteases involved with each ZIP channel should provide valuable information of functional control. Legends.

References

Alam S, Kelleher SL (2012) Cellular mechanisms of zinc dysregulation: a perspective on zinc homeostasis as an etiological factor in the development and progression of breast cancer. Nutrients 4(8):875–903

Aley PK, Singh N, Brailoiu GC, Brailoiu E, Churchill GC (2013) Nicotinic acid adenine dinucleotide phosphate (NAADP) is a second messenger in muscarinic receptor-induced contraction of guinea pig trachea. J Biol Chem 288(16):10986–10993

Altmeppen HC, Puig B, Dohler F, Thurm DK, Falker C, Krasemann S, Glatzel M (2012) Proteolytic processing of the prion protein in health and disease. Am J Neurodegener Dis 1(1):15–31

Ando A, Kikuti YY, Shigenari A, Kawata H, Okamoto N, Shiina T, Chen L, Ikemura T, Abe K, Kimura M, Inoko H (1996) cDNA cloning of the human homologues of the mouse Ke4 and Ke6 genes at the centromeric end of the human MHC region. Genomics 35(3):600–602

Andreini C, Banci L, Bertini I, Rosato A (2006) Counting the zinc-proteins encoded in the human genome. J Proteome Res 5(1):196–201

Arinola OG, Charles-Davies MA (2008) Micronutrient levels in the plasma of Nigerian females with breast cancer. Afr J Biotechnol 7(11):1620–1623

Bauer K, Dowejko A, Bosserhoff AK, Reichert TE, Bauer RJ (2009) P-cadherin induces an epithelial-like phenotype in oral squamous cell carcinoma by GSK-3beta-mediated snail phosphorylation. Carcinogenesis 30(10):1781–1788

Beck FW, Prasad AS, Butler CE, Sakr WA, Kucuk O, Sarkar FH (2004) Differential expression of hZnT-4 in human prostate tissues. Prostate 58(4):374–381

Bellomo E, Massarotti A, Hogstrand C, Maret W (2014) Zinc ions modulate protein tyrosine phosphatase 1B activity. Metallomics 6(7):1229–1239

Bin BH, Fukada T, Hosaka T, Yamasaki S, Ohashi W, Hojyo S, Miyai T, Nishida K, Yokoyama S, Hirano T (2011) Biochemical characterization of human ZIP13 protein: a homo-dimerized zinc transporter involved in the spondylocheiro dysplastic Ehlers-Danlos syndrome. J Biol Chem 286(46):40255–40265

Blom N, Sicheritz-Pontén T, Gupta R, Gammeltoft S, Brunak S (2004) Prediction of post-translational glycosylation and phosphorylation of proteins from the amino acid sequence. Proteomics 4(6):1633–1649

Bonaventura P, Benedetti G, Albarède F, Miossec P (2015) Zinc and its role in immunity and inflammation. Autoimmun Rev 14(4):277–285

Bourdeau A, Dube N, Tremblay ML (2005) Cytoplasmic protein tyrosine phosphatases, regulation and function: the roles of PTP1B and TC-PTP. Curr Opin Cell Biol 17(2):203–209

Boz A, Evliyaoglu O, Yildirim M, Erkan N, Karaca B (2005) The value of serum zinc, copper, ceruloplasmin levels in patients with gastrointestinal tract cancers. Turk J Gastroenterol 16(2):81–84

Brown KH, Peerson JM, Rivera J, Allen LH (2002) Effect of supplemental zinc on the growth and serum zinc concentrations of prepubertal children: a meta-analysis of randomized controlled trials. Am J Clin Nutr 75(6):1062–1071

Bruinsma JJ, Jirakulaporn T, Muslin AJ, Kornfeld K (2002) Zinc ions and cation diffusion facilitator proteins regulate Ras-mediated signaling. Dev Cell 2(5):567–578

Buntzel J, Bruns F, Glatzel M, Garayev A, Mucke R, Kisters K, Schafer U, Schonekaes K, Micke O (2007) Zinc concentrations in serum during head and neck cancer progression. Anticancer Res 27(4A):1941–1943

Chang ET, Hedelin M, Adami HO, Gronberg H, Balter KA (2004) Re: zinc supplement use and risk of prostate cancer. J Natl Cancer Inst 96(14):1108; author reply 1108-9

Chen QG, Zhang Z, Yang Q, Shan GY, Yu XY, Kong CZ (2012) The role of zinc transporter ZIP4 in prostate carcinoma. Urol Oncol 30(6):906–911

Cheng HC, Qi RZ, Paudel H, Zhu HJ (2011) Regulation and function of protein kinases and phosphatases. Enzyme Res 2011:794089

Chimienti F, Devergnas S, Favier A, Seve M (2004) Identification and cloning of a beta-cell-specific zinc transporter, ZnT-8, localized into insulin secretory granules. Diabetes 53(9):2330–2337

Cohen P (2002) The origins of protein phosphorylation. Nat Cell Biol 4(5):E127–E130

Colvin RA, Bush AI, Volitakis I, Fontaine CP, Thomas D, Kikuchi K, Holmes WR (2008) Insights into Zn2+ homeostasis in neurons from experimental and modeling studies. Am J Physiol Cell Physiol 294(3):C726–C742

Colvin RA, Holmes WR, Fontaine CP, Maret W (2010) Cytosolic zinc buffering and muffling: their role in intracellular zinc homeostasis. Metallomics 2(5):306–317

Costello LC, Franklin RB (2006) The clinical relevance of the metabolism of prostate cancer; zinc and tumor suppression: connecting the dots. Mol Cancer 5:17

Costello LC, Franklin RB, Feng P, Tan M, Bagasra O (2005) Zinc and prostate cancer: a critical scientific, medical, and public interest issue (United States). Cancer Causes Control 16:901–915

Cousins RJ, Liuzzi JP, Lichten LA (2006) Mammalian zinc transport, trafficking, and signals. J Biol Chem 281(34):24085–24089

Dani V, Goel A, Vaiphei K, Dhawan DK (2007) Chemopreventive potential of zinc in experimentally induced colon carcinogenesis. Toxicol Lett 171(1-2):10–18

David R (2012) Antimicrobials: why zinc is bad for bacteria. Nat Rev Microbiol 10(1):4–4

Desouki MM, Geradts J, Milon B, Franklin RB, Costello LC (2007) hZip2 and hZip3 zinc transporters are down regulated in human prostate adenocarcinomatous glands. Mol Cancer 6:37

Dhawan DK, Chadha VD (2010) Zinc: a promising agent in dietary chemoprevention of cancer. Indian J Med Res 132(December):676–682

Ehsani S, Salehzadeh A, Huo H, Reginold W, Pocanschi CL, Ren H, Wang H, So K, Sato C, Mehrabian M, Strome R, Trimble WS, Hazrati LN, Rogaeva E, Westaway D, Carlson GA, Schmitt-Ulms G (2012) LIV-1 ZIP ectodomain shedding in prion-infected mice resembles cellular response to transition metal starvation. J Mol Biol 422(4):556–574

FAO/WHO (2002) 'Zinc', FAO/WHO expert consultation on human vitamin and mineral requirements. Rome, pp 257–270

Farquharson MJ, Al-Ebraheem A, Geraki K, Leek R, Jubb A, Harris AL (2009) Zinc presence in invasive ductal carcinoma of the breast and its correlation with oestrogen receptor status. Phys Med Biol 54(13):4213

Fashner J, Ericson K, Werner S (2012) Treatment of the common cold in children and adults. Am Fam Physician 86(2):153–159

Feng P, Li TL, Guan ZX, Franklin RB, Costello LC (2002) Direct effect of zinc on mitochondrial apoptogenesis in prostate cells. Prostate 52:311–318

Feng P, Li T, Guan Z, Franklin RB, Costello LC (2008) The involvement of Bax in zinc-induced mitochondrial apoptogenesis in malignant prostate cells. Mol Cancer 7:25

Fong LY, Nguyen VT, Farber JL (2001) Esophageal cancer prevention in zinc-deficient rats: rapid induction of apoptosis by replenishing zinc. J Natl Cancer Inst 93(20):1525–1533

Fraker PJ (2005) Roles for cell death in zinc deficiency. J Nutr 135(3):359–362

Franchin C, Cesaro L, Salvi M, Millioni R, Iori E, Cifani P, James P, Arrigoni G, Pinna L (2015) Quantitative analysis of a phosphoproteome readily altered by the protein kinase CK2 inhibitor quinalizarin in HEK-293T cells. Biochim Biophys Acta 1854(6):609–623

Franklin RB, Costello LC (2007) Zinc as an anti-tumor agent in prostate cancer and in other cancers. Arch Biochem Biophys 463(2):211–217

Franklin RB, Feng P, Milon B, Desouki MM, Singh KK, Kajdacsy-Balla A, Bagasra O, Costello LC (2005) hZIP1 zinc uptake transporter down regulation and zinc depletion in prostate cancer. Mol Cancer 4:32

Fukada T, Kambe T (2011) Molecular and genetic features of zinc transporters in physiology and pathogenesis. Metallomics 3(7):662–674

Girijashanker K, He L, Soleimani M, Reed JM, Li H, Liu Z, Wang B, Dalton TP, Nebert DW (2008) Slc39a14 gene encodes ZIP14, a metal/bicarbonate symporter: similarities to the ZIP8 transporter. Mol Pharmacol 73(5):1413–1423

Gonzalez A, Peters U, Lampe JW, White E (2009) Zinc intake from supplements and diet and prostate cancer. Nutr Cancer 61(2):206–215

Gouw M, Michael S, Samano-Sanchez H, Kumar M, Zeke A, Lang B, Bely B, Chemes LB, Davey NE, Deng Z, Diella F, Gurth CM, Huber AK, Kleinsorg S, Schlegel LS, Palopoli N, Roey KV, Altenberg B, Remenyi A, Dinkel H, Gibson TJ (2018) The eukaryotic linear motif resource – 2018 update. Nucleic Acids Res 46(D1):D428–D434

Guerinot ML (2000) The ZIP family of metal transporters. Biochim Biophys Acta 1465(1-2):190–198

Gumulec J, Masarik M, Adam V, Eckschlager T, Provaznik I, Kizek R (2014) Serum and tissue zinc in epithelial malignancies: a meta-analysis. PLoS One 9(6):e99790

Gupta SK, Singh SP, Shukla VK (2005) Copper, zinc, and Cu/Zn ratio in carcinoma of the gallbladder. J Surg Oncol 91(3):204–208

Haase H, Maret W (2005) Protein tyrosine phosphatases as targets of the combined insulinomimetic effects of zinc and oxidants. Biometals 18(4):333–338

Haase H, Rink L (2014) Zinc signals and immune function. Biofactors 40(1):27–40

Hambidge M (2000) Human zinc deficiency. J Nutr 130(5S Suppl):1344S–1349S

Hambidge KM, Krebs NF (2007) Zinc deficiency: a special challenge. J Nutr 137(4):1101–1105

Hasumi M, Suzuki K, Matsui H, Koike H, Ito K, Yamanaka H (2003) Regulation of metallothionein and zinc transporter expression in human prostate cancer cells and tissues. Cancer Lett 200(2):187–195

Henkes M, van der Kuip H, Aulitzky WE (2008) Therapeutic options for chronic myeloid leukemia: focus on imatinib (Glivec, Gleevectrade mark). Ther Clin Risk Manag 4(1):163–187

Hirano T, Murakami M, Fukada T, Nishida K, Yamasaki S, Suzuki T (2008) Roles of zinc and zinc signaling in immunity: zinc as an intracellular signaling molecule. Adv Immunol 97:149–176

Hiscox S, Morgan L, Barrow D, Dutkowski C, Wakeling A, Nicholson RI (2004) Tamoxifen resistance in breast cancer cells is accompanied by an enhanced motile and invasive phenotype: inhibition by gefitinib ('Iressa', ZD1839). Clin Exp Metastasis 21(3):201–212

Hiscox S, Morgan L, Green TP, Barrow D, Gee J, Nicholson RI (2006) Elevated Src activity promotes cellular invasion and motility in tamoxifen resistant breast cancer cells. Breast Cancer Res Treat 97(3):263–274

Ho E, Ames BN (2002) Low intracellular zinc induces oxidative DNA damage, disrupts p53, NFkappa B, and AP1 DNA binding, and affects DNA repair in a rat glioma cell line. Proc Natl Acad Sci U S A 99(26):16770–16775

Ho E, Song Y (2009) Zinc and prostatic cancer. Curr Opin Clin Nutr Metab Care 12(6):640–645

Ho E, Courtemanche C, Ames BN (2003) Zinc deficiency induces oxidative DNA damage and increases p53 expression in human lung fibroblasts. J Nutr 133(8):2543–2548

Hogstrand C, Kille P, Nicholson RI, Taylor KM (2009) Zinc transporters and cancer: a potential role for ZIP7 as a hub for tyrosine kinase activation. Trends Mol Med 15(3):101–111

Hogstrand C, Kille P, Ackland ML, Hiscox S, Taylor KM (2013) A mechanism for epithelial-mesenchymal transition and anoikis resistance in breast cancer triggered by zinc channel ZIP6 and STAT3 (signal transducer and activator of transcription 3). Biochem J 455(2):229–237

Hooper NM (1994) Families of zinc metalloproteases. FEBS Lett 354(1):1–6

Hornbeck PV, Kornhauser JM, Tkachev S, Zhang B, Skrzypek E, Murray B, Latham V, Sullivan M (2012) PhosphoSitePlus: a comprehensive resource for investigating the structure and function of experimentally determined post-translational modifications in man and mouse. Nucleic Acids Res 40(Database issue):D261–D270

Huang L, Kirschke CP, Gitschier J (2002) Functional characterization of a novel mammalian zinc transporter, ZnT6. J Biol Chem 277(29):26389–26395

Huang L, Kirschke CP, Zhang Y (2006) Decreased intracellular zinc in human tumorigenic prostate epithelial cells: a possible role in prostate cancer progression. Cancer Cell Int 6:10

Humphrey SJ, James DE, Mann M (2015) Protein phosphorylation: a major switch mechanism for metabolic regulation. Trends Endocrinol Metab 26(12):676–687

Ilouz R, Kaidanovich O, Gurwitz D, Eldar-Finkelman H (2002) Inhibition of glycogen synthase kinase-3β by bivalent zinc ions: insight into the insulin-mimetic action of zinc. Biochem Biophys Res Commun 295(1):102–106

Jin R, Bay B, Tan P, Tan BK (1999) Metallothionein expression and zinc levels in invasive ductal breast carcinoma. Oncol Rep 6(4):871–875

Jones HE, Goddard L, Gee JM, Hiscox S, Rubini M, Barrow D, Knowlden JM, Williams S, Wakeling AE, Nicholson RI (2004) Insulin-like growth factor-I receptor signalling and acquired resistance to gefitinib (ZD1839; Iressa) in human breast and prostate cancer cells. Endocr Relat Cancer 11(4):793–814

Kagara N, Tanaka N, Noguchi S, Hirano T (2007) Zinc and its transporter ZIP10 are involved in invasive behavior of breast cancer cells. Cancer Sci 98(5):692–697

Kambe T (2012) Molecular architecture and function of ZnT transporters. Curr Top Membr 69:199–220

Kambe T, Andrews GK (2009) Novel proteolytic processing of the ectodomain of the zinc transporter ZIP4 (SLC39A4) during zinc deficiency is inhibited by acrodermatitis enteropathica mutations. Mol Cell Biol 29(1):129–139

Kambe T, Suzuki T, Nagao M, Yamaguchi-Iwai Y (2006) Sequence similarity and functional relationship among eukaryotic ZIP and CDF transporters. Genomics Proteom Bioinform 4(1):1–9

Kambe T, Tsuji T, Hashimoto A, Itsumura N (2015) The physiological, biochemical, and molecular roles of zinc transporters in zinc homeostasis and metabolism. Physiol Rev 95(3):749–784

Kay AR, Tóth K (2008) Is zinc a neuromodulator? Sci Signal 1(19):re3

Kim YH, Kim EY, Gwag BJ, Sohn S, Koh JY (1999) Zinc-induced cortical neuronal death with features of apoptosis and necrosis: mediation by free radicals. Neuroscience 89(1):175–182

Kim HS, Park I, Cho HJ, Gwak G, Yang K, Bae BN, Kim KW, Han S, Kim HJ, Kim YD (2012) Analysis of the potent prognostic factors in luminal-type breast cancer. J Breast Cancer 15(4):401–406

King JC (2011) Zinc: an essential but elusive nutrient. Am J Clin Nutr 94(2):679S–684S

King JC, Shames DM, Woodhouse LR (2000) Zinc homeostasis in humans. J Nutr 130(5S Suppl):1360S–1366S

Knowlden JM, Hutcheson IR, Jones HE, Madden T, Gee JM, Harper ME, Barrow D, Wakeling AE, Nicholson RI (2003) Elevated levels of epidermal growth factor receptor/c-erbB2 heterodimers mediate an autocrine growth regulatory pathway in tamoxifen-resistant MCF-7 cells. Endocrinology 144(3):1032–1044

Kolonel LN, Yoshizawa CN, Hankin JH (1988) Diet and prostatic cancer: a case-control study in Hawaii. Am J Epidemiol 127(5):999–1012

Kornreich BG (2007) The patch clamp technique: principles and technical considerations. J Vet Cardiol 9(1):25–37

Krezel A, Maret W (2006) Zinc-buffering capacity of a eukaryotic cell at physiological pZn. J Biol Inorg Chem 11(8):1049–1062

Kristal AR, Stanford JL, Cohen JH, Wicklund K, Patterson RE (1999) Vitamin and mineral supplement use is associated with reduced risk of prostate cancer. Cancer Epidemiol Prev Biomarkers 8(10):887–892

Lansdown AB, Mirastschijski U, Stubbs N, Scanlon E, Agren MS (2007) Zinc in wound healing: theoretical, experimental, and clinical aspects. Wound Repair Regen 15(1):2–16

Larner F, Woodley LN, Shousha S, Moyes A, Humphreys-Williams E, Strekopytov S, Halliday AN, Rehkamper M, Coombes RC (2015) Zinc isotopic compositions of breast cancer tissue. Metallomics 7(1):112–117

Lawrenson JG, Grzybowski A (2015) Controversies in the use of nutritional supplements in ophthalmology. Curr Pharm Des *21*(32):4667–4672

Lee S, Chanoit G, McIntosh R, Zvara DA, Xu Z (2009) Molecular mechanism underlying Akt activation in zinc-induced cardioprotection. Am J Physiol Heart Circ Physiol 297(2):H569–H575

Leitzmann MF, Stampfer MJ, Wu KN, Colditz GA, Willett WC, Giovannucci EL (2003) Zinc supplement use and risk of prostate cancer. J Natl Cancer I 95(13):1004–1007

Lichten LA, Cousins RJ (2009) Mammalian zinc transporters: nutritional and physiologic regulation. Annu Rev Nutr 29:153–176

Lin W, Chai J, Love J, Fu D (2010) Selective electrodiffusion of zinc ions in a Zrt-, Irt-like protein, ZIPB. J Biol Chem 285(50):39013–39020

Lizio M, Harshbarger J, Shimoji H, Severin J, Kasukawa T, Sahin S, Abugessaisa I, Fukuda S, Hori F, Ishikawa-Kato S, Mungall CJ, Arner E, Baillie JK, Bertin N, Bono H, de Hoon M, Diehl AD, Dimont E, Freeman TC, Fujieda K, Hide W, Kaliyaperumal R, Katayama T, Lassmann T, Meehan TF, Nishikata K, Ono H, Rehli M, Sandelin A, Schultes EA, 't Hoen PA, Tatum Z, Thompson M, Toyoda T, Wright DW, Daub CO, Itoh M, Carninci P, Hayashizaki Y, Forrest AR, Kawaji H, Consortium F (2015) Gateways to the FANTOM5 promoter level mammalian expression atlas. Genome Biol 16. pp. 22

Lodish H, Berk A, Zipursky SL, Matsudaira P, Baltimore D, Darnell J (2000) Hierarchical structure of proteins. Molecular cell biology. W. H. Freeman, New York. Available at: http://www.ncbi.nlm.nih.gov/books/NBK21581. Accessed 12 Nov 2015

Loh SN (2010) The missing zinc: p53 misfolding and cancer. Metallomics 2(7):442–449

Lowe NM, Fekete K, Decsi T (2009) Methods of assessment of zinc status in humans: a systematic review. Am J Clin Nutr 89(6):2040S–2051S

Lue HW, Yang X, Wang R, Qian W, Xu RZ, Lyles R, Osunkoya AO, Zhou BP, Vessella RL, Zayzafoon M, Liu ZR, Zhau HE, Chung LW (2011) LIV-1 promotes prostate cancer epithelial-to-mesenchymal transition and metastasis through HB-EGF shedding and EGFR-mediated ERK signaling. PLoS One 6(11):e27720

MacDonald RS (2000) The role of zinc in growth and cell proliferation. J Nutr 130(5S Suppl):1500S–1508S

Manning DL, Daly RJ, Lord PG, Kelly KF, Green CD (1988) Effects of oestrogen on the expression of a 4.4 kb mRNA in the ZR-75-1 human breast cancer cell line. Mol Cell Endocrinol 59(3):205–212

Manning DL, McClelland RA, Gee JM, Chan CM, Green CD, Blamey RW, Nicholson RI (1993) The role of four oestrogen-responsive genes, pLIV1, pS2, pSYD3 and pSYD8, in predicting responsiveness to endocrine therapy in primary breast cancer. Eur J Cancer 29A(10):1462–1468

Manning DL, Robertson JF, Ellis IO, Elston CW, McClelland RA, Gee JM, Jones RJ, Green CD, Cannon P, Blamey RW et al (1994) Oestrogen-regulated genes in breast cancer: association of pLIV1 with lymph node involvement. Eur J Cancer 30A(5):675–678

Manning DL, McClelland RA, Knowlden JM, Bryant S, Gee JM, Green CD, Robertson JF, Blamey RW, Sutherland RL, Ormandy CJ (1995) Differential expression of oestrogen regulated genes in breast cancer. Acta Oncol 34(5):641–646

Manning G, Whyte DB, Martinez R, Hunter T, Sudarsanam S (2002) The protein kinase complement of the human genome. Science 298(5600):1912–1934

Maret W (2008) Metallothionein redox biology in the cytoprotective and cytotoxic functions of zinc. Exp Gerontol 43(5):363–369

Maret W (2013) Zinc biochemistry: from a single zinc enzyme to a key element of life. Adv Nutr 4(1):82–91

Margalioth EJ, Schenker JG, Chevion M (1983) Copper and zinc levels in normal and malignant tissues. Cancer 52(5):868–872

Martin-Lagos F, Navarro-Alarcon M, Terres-Martos C, Lopez GD 1 SH, Lopez-Martinez MC (1997) Serum copper and zinc concentrations in serum from patients with cancer and cardiovascular disease. Sci Total Environ 204(1):27–35

Memon AU, Kazi TG, Afridi HI, Jamali MK, Arain MB, Jalbani N, Syed N (2007) Evaluation of zinc status in whole blood and scalp hair of female cancer patients. Clin Chim Acta 379(1-2):66–70

Meyer F, Galan P, Douville P, Bairati I, Kegle P, Bertrais S, Estaquio C, Hercberg S (2005) Antioxidant vitamin and mineral supplementation and prostate cancer prevention in the SU.VI.MAX trial. Int J Cancer 116(2):182–186

Meyer RD, Srinivasan S, Singh AJ, Mahoney JE, Gharahassanlou KR, Rahimi N (2011) PEST motif serine and tyrosine phosphorylation controls vascular endothelial growth factor receptor 2 stability and downregulation. Mol Cell Biol 31(10):2010–2025

Moore SF, van den Bosch MT, Hunter RW, Sakamoto K, Poole AW, Hers I (2013) Dual regulation of glycogen synthase kinase 3 (GSK3)α/β by protein kinase C (PKC)α and Akt promotes thrombin-mediated integrin αIIbβ3 activation and granule secretion in platelets. J Biol Chem 288(6):3918–3928

Moynahan EJ (1974) Letter: Acrodermatitis enteropathica: a lethal inherited human zinc-deficiency disorder. Lancet 2(7877):399–400

Mulay IL, Roy R, Knox BE, Suhr NH, Delaney WE (1971) Trace-metal analysis of cancerous and noncancerous human tissues. J Natl Cancer Inst 47(1):1–13

Naidu MS, Suryakar AN, Swami SC, Katkam RV, Kumbar KM (2007) Oxidative stress and antioxidant status in cervical cancer patients. Indian J Clin Biochem 22(2):140–144

Nelson DL, Cox MM (2013) Biological membranes and transport. In: CourseSmart international e-book for principles of biochemistry, 6th edn. Palgrave Macmillan, New York, pp 385–432

Nimmanon T, Taylor KM (2014) Zinc signalling and cancer. In: Fukada T, Kambe T (eds) Zinc signals in cellular functions and disorders. Springer, Tokyo, pp 285–313

Nimmanon T, Taylor K (2015) Cellular zinc signalling is triggered by CK2. In: Ahmed K, Issinger O-G, Szyszka R (eds) Protein kinase CK2 cellular function in Normal and disease states, Advances in biochemistry in health and disease, vol 12. Springer, Cham, pp 141–157

Nimmanon T, Taylor KM (2017) Posttranslational mechanisms of zinc Signaling. In: Collins JF (ed) Molecular, genetic, and nutritional aspects of major and trace minerals. Academic Press, Amsterdam, pp 273–281

Nimmanon T, Ziliotto S, Morris S, Flanagan L, Taylor KM (2017) Phosphorylation of zinc channel ZIP7 drives MAPK, PI3K and mTOR growth and proliferation signalling. Metallomics 9(5):471–481

Nuttall JR, Oteiza PI (2012) Zinc and the ERK kinases in the developing brain. Neurotox Res 21(1):128–141

Ohana E, Hoch E, Keasar C, Kambe T, Yifrach O, Hershfinkel M, Sekler I (2009) Identification of the Zn2+ binding site and mode of operation of a mammalian Zn2+ transporter. J Biol Chem 284(26):17677–17686

Ohashi K, Nagata Y, Wada E, Zammit PS, Shiozuka M, Matsuda R (2015) Zinc promotes proliferation and activation of myogenic cells via the PI3K/Akt and ERK signaling cascade. Exp Cell Res 333(2):228–237

Ollig J, Kloubert V, Taylor KM, Rink L (2019) B cell activation and proliferation increase intracellular zinc levels. J Nutr Biochem 64:72–79

Oteiza PI (2012) Zinc and the modulation of redox homeostasis. Free Radic Biol Med 53(9):1748–1759

Palmiter RD, Findley SD (1995) Cloning and functional characterization of a mammalian zinc transporter that confers resistance to zinc. EMBO J 14(4):639–649

Palmiter RD, Huang L (2004) Efflux and compartmentalization of zinc by members of the SLC30 family of solute carriers. Pflugers Arch 447(5):744–751

Pandey NR, Vardatsikos G, Mehdi MZ, Srivastava AK (2010) Cell-type-specific roles of IGF-1R and EGFR in mediating Zn2+−induced ERK1/2 and PKB phosphorylation. J Biol Inorg Chem 15(3):399–407

Passerini A, Andreini C, Menchetti S, Rosato A, Frasconi P (2007) Predicting zinc binding at the proteome level. BMC Bioinform 8:39

Piedfer M, Bouchet S, Tang RP, Billard C, Dauzonne D, Bauvois B (2013) p70S6 kinase is a target of the novel proteasome inhibitor 3,3 '-diamino-4 '-methoxyflavone during apoptosis in human myeloid tumor cells. BBA-Mol Cell Res 1833(6):1316–1328

Pinto AC, Ades F, de Azambuja E, Piccart-Gebhart M (2013) Trastuzumab for patients with HER2 positive breast cancer: delivery, duration and combination therapies. Breast 22(Suppl 2):S152–S155

Radivojac P, Baenziger PH, Kann MG, Mort ME, Hahn MW, Mooney SD (2008) Gain and loss of phosphorylation sites in human cancer. Bioinformatics 24(16):i241–i247
Ramakrishna S, Suresh B, Lim KH, Cha BH, Lee SH, Kim KS, Baek KH (2011) PEST motif sequence regulating human NANOG for proteasomal degradation. Stem Cells Dev 20(9):1511–1519
Rechsteiner M, Rogers SW (1996) PEST sequences and regulation by proteolysis. Trends Biochem Sci 21(7):267–271
Reeder NL, Kaplan J, Xu J, Youngquist RS, Wallace J, Hu P, Juhlin KD, Schwartz JR, Grant RA, Fieno A, Nemeth S, Reichling T, Tiesman JP, Mills T, Steinke M, Wang SL, Saunders CW (2011) Zinc pyrithione inhibits yeast growth through copper influx and inactivation of iron-sulfur proteins. Antimicrob Agents Chemother 55(12):5753–5760
Rhodes DR, Yu J, Shanker K, Deshpande N, Varambally R, Ghosh D, Barrette T, Pandey A, Chinnaiyan AM (2004) ONCOMINE: a cancer microarray database and integrated data-mining platform. Neoplasia 6(1):1–6
Rogers S, Wells R, Rechsteiner M (1986) Amino acid sequences common to rapidly degraded proteins: the PEST hypothesis. Science 234(4774):364–368
Sadeghian G, Ziaei H, Nilforoushzadeh MA (2011) Treatment of localized psoriasis with a topical formulation of zinc pyrithione. Acta Dermatovenerol Alp Pannonica Adriat 20(4):187–190
Sakamoto KM, Frank DA (2009) CREB in the pathophysiology of cancer: implications for targeting transcription factors for cancer therapy. Clin Cancer Res 15(8):2583–2587
Sandoval A, Oviedo N, Tadmouri A, Avila T, De Waard M, Felix R (2006) Two PEST-like motifs regulate Ca^{2+}/calpain-mediated cleavage of the CaVbeta3 subunit and provide important determinants for neuronal Ca^{2+} channel activity. Eur J Neurosci 23(9):2311–2320
Satoh M, Kondo Y, Mita M, Nakagawa I, Naganuma A, Imura N (1993) Prevention of carcinogenicity of anticancer drugs by metallothionein induction. Cancer Res 53(20):4767–4768
Schmitt-Ulms G, Ehsani S, Watts JC, Westaway D, Wille H (2009) Evolutionary descent of prion genes from the ZIP family of metal ion transporters. PLoS One 4(9):e7208
Schneider J, Ruschhaupt M, Buness A, Asslaber M, Regitnig P, Zatloukal K, Schippinger W, Ploner F, Poustka A, Sultmann H (2006) Identification and meta-analysis of a small gene expression signature for the diagnosis of estrogen receptor status in invasive ductal breast cancer. Int J Cancer 119(12):2974–2979
Schnitt SJ (2010) Classification and prognosis of invasive breast cancer: from morphology to molecular taxonomy. Mod Pathol 23(Suppl 2):S60–S64
Schwartz JR, Bacon RA, Shah R, Mizoguchi H, Tosti A (2013) Therapeutic efficacy of anti-dandruff shampoos: a randomized clinical trial comparing products based on potentiated zinc pyrithione and zinc pyrithione/climbazole. Int J Cosmet Sci 35(4):381–387
Schwingshackl L, Hoffmann G, Buijsse B, Mittag T, Stelmach-Mardas M, Boeing H, Gottschald M, Dietrich S, Arregui M, Dias S (2015) Dietary supplements and risk of cause-specific death, cardiovascular disease, and cancer: a protocol for a systematic review and network meta-analysis of primary prevention trials. Syst Rev 4:34
Seve M, Chimienti F, Devergnas S, Favier A (2004) 'In silico identification and expression of SLC30 family genes: an expressed sequence tag data mining strategy for the characterization of zinc transporters' tissue expression'. BMC Genomics 5(1):32
Soda M, Willert K, Kaushansky K, Geddis AE (2008) Inhibition of GSK-3beta promotes survival and proliferation of megakaryocytic cells through a beta-catenin-independent pathway. Cell Signal 20(12):2317–2323
St-Denis N, Gabriel M, Turowec JP, Gloor GB, Li SS, Gingras AC, Litchfield DW (2015) Systematic investigation of hierarchical phosphorylation by protein kinase CK2. J Proteome 118:49–62
Sutherland E, Robison G, Butcher R (1968) Some aspects of the biological role of adenosine 3′, 5′-monophosphate (cyclic AMP). Circulation 37(2):279–306
Takeda A, Nakamura M, Fujii H, Tamano H (2013) Synaptic Zn(2+) homeostasis and its significance. Metallomics 5(5):417–423

Taniguchi M, Fukunaka A, Hagihara M, Watanabe K, Kamino S, Kambe T, Enomoto S, Hiromura M (2013) Essential role of the zinc transporter ZIP9/SLC39A9 in regulating the activations of Akt and Erk in B-cell receptor signaling pathway in DT40 cells. PLoS One 8(3):e58022
Tawfic S, Yu S, Wang H, Faust R, Davis A, Ahmed K (2001) Protein kinase CK2 signal in neoplasia. Histol Histopathol 16(2):573–582
Taylor KM, Nicholson RI (2003) The LZT proteins; the LIV-1 subfamily of zinc transporters. Biochim Biophys Acta 1611(1-2):16–30
Taylor KM, Morgan HE, Johnson A, Hadley LJ, Nicholson RI (2003) Structure-function analysis of LIV-1, the breast cancer-associated protein that belongs to a new subfamily of zinc transporters. Biochem J 375(Pt 1):51–59
Taylor KM, Morgan HE, Johnson A, Nicholson RI (2004) Structure-function analysis of HKE4, a member of the new LIV-1 subfamily of zinc transporters. Biochem J 377(Pt 1):131–139
Taylor KM, Morgan HE, Smart K, Zahari NM, Pumford S, Ellis IO, Robertson JF, Nicholson RI (2007) The emerging role of the LIV-1 subfamily of zinc transporters in breast cancer. Mol Med 13(7–8):396–406
Taylor KM, Vichova P, Jordan N, Hiscox S, Hendley R, Nicholson RI (2008) ZIP7-mediated intracellular zinc transport contributes to aberrant growth factor signaling in antihormone-resistant breast cancer cells. Endocrinology 149(10):4912–4920
Taylor KM, Gee JMW, Kille P (2011) Zinc and Cancer. In: Rink L (ed) Zinc in human health. IOS Press, Amsterdam, pp 283–304
Taylor KM, Hiscox S, Nicholson RI, Hogstrand C, Kille P (2012) Protein kinase CK2 triggers cytosolic zinc signaling pathways by phosphorylation of zinc channel ZIP7. Sci Signal 5(210):ra11
Taylor KM, Muraina I, Brethour D, Schmitt-Ulms G, Nimmanon T, Ziliotto S, Kille P, Hogstrand C (2016) Zinc transporter ZIP10 forms a heteromer with ZIP6 which regulates embryonic development and cell migration. Biochem J *473*(16):2531–2544
Thiery JP, Acloque H, Huang RY, Nieto MA (2009) Epithelial-mesenchymal transitions in development and disease. Cell 139(5):871–890
Thomas P, Pang Y, Dong J, Berg AH (2014) Identification and characterization of membrane androgen receptors in the ZIP9 zinc transporter subfamily: II. Role of human ZIP9 in testosterone-induced prostate and breast cancer cell apoptosis. Endocrinology 155(11):4250–4265
Thornton JK, Taylor KM, Ford D, Valentine RA (2011) Differential subcellular localization of the splice variants of the zinc transporter ZnT5 is dictated by the different C-terminal regions. PLoS One 6(8):e23878
Tozlu S, Girault I, Vacher S, Vendrell J, Andrieu C, Spyratos F, Cohen P, Lidereau R, Bieche I (2006) Identification of novel genes that co-cluster with estrogen receptor alpha in breast tumor biopsy specimens, using a large-scale real-time reverse transcription-PCR approach. Endocr Relat Cancer 13(4):1109–1120
Valentine RA, Jackson KA, Christie GR, Mathers JC, Taylor PM, Ford D (2007) ZnT5 variant B is a bidirectional zinc transporter and mediates zinc uptake in human intestinal Caco-2 cells. J Biol Chem 282(19):14389–14393
Vallee BL, Galdes A (1984) The metallobiochemistry of zinc enzymes. Adv Enzymol Relat Areas Mol Biol 56:283–430
Walker CL, Taneja S, LeFevre A, Black RE, Mazumder S (2015) Appropriate management of acute diarrhea in children among public and private providers in Gujarat, India: a cross-sectional survey. Glob Health Sci Pract 3(2):230–241
Wang CY, Jenkitkasemwong S, Duarte S, Sparkman BK, Shawki A, Mackenzie B, Knutson MD (2012) ZIP8 is an Iron and zinc transporter whose cell-surface expression is up-regulated by cellular Iron loading. J Biol Chem 287(41):34032–34043
Watts JC, Huo H, Bai Y, Ehsani S, Jeon AH, Won AH, Shi T, Daude N, Lau A, Young R, Xu L, Carlson GA, Williams D, Westaway D, Schmitt-Ulms G (2009) Interactome analyses identify ties of PrP and its mammalian paralogs to oligomannosidic N-glycans and endoplasmic reticulum-derived chaperones. PLoS Pathog 5(10):e1000608

Wessells KR, Brown KH (2012) Estimating the global prevalence of zinc deficiency: results based on zinc availability in national food supplies and the prevalence of stunting. PLoS One 7(11):e50568

Wieringa FT, Dijkhuizen MA, Fiorentino M, Laillou A, Berger J (2015) Determination of zinc status in humans: which indicator should we use? Nutrients 7(5):3252–3263

Wilson M, Hogstrand C, Maret W (2012) Picomolar concentrations of free zinc(II) ions regulate receptor protein-tyrosine phosphatase β activity. J Biol Chem 287(12):9322–9326

Wu X, Tang J, Xie M (2015) Serum and hair zinc levels in breast cancer: a meta-analysis. Sci Rep 5:12249

Yamasaki S, Sakata-Sogawa K, Hasegawa A, Suzuki T, Kabu K, Sato E, Kurosaki T, Yamashita S, Tokunaga M, Nishida K, Hirano T (2007) Zinc is a novel intracellular second messenger. J Cell Biol 177(4):637–645

Yamashita S, Miyagi C, Fukada T, Kagara N, Che YS, Hirano T (2004) Zinc transporter LIVI controls epithelial-mesenchymal transition in zebrafish gastrula organizer. Nature 429(6989):298–302

Yan M, Song Y, Wong CP, Hardin K, Ho E (2008) Zinc deficiency alters DNA damage response genes in normal human prostate epithelial cells. J Nutr 138(4):667–673

Yoshida K, Kawano N, Yoshiike M, Yoshida M, Iwamoto T, Morisawa M (2008) Physiological roles of semenogelin I and zinc in sperm motility and semen coagulation on ejaculation in humans. Mol Hum Reprod 14(3):151–156

Zhou BP, Deng J, Xia W, Xu J, Li YM, Gunduz M, Hung MC (2004) Dual regulation of snail by GSK-3beta-mediated phosphorylation in control of epithelial-mesenchymal transition. Nat Cell Biol 6(10):931–940

Ziliotto S, Gee JMW, Ellis IO, Green AR, Finlay P, Gobbato A, Taylor KM (2019) Activated zinc transporter ZIP7 as an indicator of anti-hormone resistance in breast cancer. Metallomics 11(9):1579–1592

Chapter 17
Zinc Signaling (Zinc'ing) in Intestinal Function

Michal Hershfinkel, Johnny Iv, and Shannon L. Kelleher

Abstract Zinc plays a well-documented role in intestinal function, perturbations in zinc homeostasis are associated with impaired barrier function and disease. A large repertoire of zinc transport proteins (ZnTs, ZIPs), TRP channels, and a distinct zinc receptor, ZnR/GPR39, are expressed along the intestinal epithelial cells. We will discuss the cellular roles of zinc ions, Zn^{2+}, in activating signaling and physiological function of multiple cells along the intestinal epithelium. We will further describe the network of Zn^{2+}-homeostatic proteins that are responsible for maintaining Zn^{2+} concentrations within the cytosol and their role in maintaining cellular functions. This review will reveal the tip of the iceberg that is currently known regarding the role of intestinal Zn^{2+} and will elucidate the need for extensive studies that will expound the complete map and physiological roles of Zn^{2+}-homeostatic proteins.

Keywords Zinc · ZnT · SLC30A · ZIP · SLC39A · ZnR/GPR39 · Intestine · Zinc absorption · Barrier function

17.1 Introduction

Zinc is an essential nutrient that plays a role in growth and proper function of all physiological systems, among them immune, epithelial, and digestive. Zinc ions (Zn^{2+}) are critical for >300 different biological pathways, therefore aberrations in cellular zinc concentrations will inherently influence numerous signaling systems

M. Hershfinkel (✉)
Department of Physiology and Cell Biology, Faculty of Health Sciences, Ben-Gurion University of the Negev, Beer-Sheva, Israel
e-mail: hmichal@bgu.ac.il

J. Iv · S. L. Kelleher
Department of Biomedical and Nutritional Sciences, University of Massachusetts Lowell, Lowell, MA, USA
e-mail: shannon_kelleher@uml.edu

T. Fukada, T. Kambe (eds.), *Zinc Signaling*,
https://doi.org/10.1007/978-981-15-0557-7_17

and intracellular second messengers that coordinate cell renewal, differentiation, function, and death (Levaot and Hershfinkel 2018; Hennigar and Kelleher 2012). Zinc is absorbed from the diet by intestinal epithelial cells, mostly enterocytes in the upper part of the small intestine, after being released from dietary complexes into zinc ions (Zn^{2+}). Under circumstances when absorption is not sufficient in the small intestine, such as in cases of chronic intestinal inflammation, Zn^{2+} can also be absorbed in the colon. In addition to the zinc from the diet, endogenous sources of luminal Zn^{2+} include Zn^{2+} secreted from salivary gland vesicles and Paneth cell granules, and Zn^{2+} released from pancreatic digestive enzymes (Gopalsamy et al. 2015; Frederickson et al. 1987; Frederickson and Danscher 1990; Sondergaard et al. 2003; Ishii et al. 1999). High turnover of intestinal epithelial cells involves shedding of mature apical cells into the lumen and constant migration of newly differentiated cells from the crypts to the apical surfaces in the small intestine and colon (Watson et al. 2009). Apoptosis of the shed cells can result in release of endogenous Zn^{2+} from cellular proteins (Sharir et al. 2010), and this may account for free Zn^{2+} in the intestinal lumen. Absorption of the luminal Zn^{2+} by the epithelial cells plays a role in maintaining whole body zinc homeostasis. Importantly, the changes in luminal and cellular levels of Zn^{2+} regulate physiological function of intestinal epithelial cells.

The luminal-facing cells of the intestinal epithelium are derived from intestinal stem cells (iSCs) and include enterocytes, goblet cells, enteroendocrine cells, tuft cells, and Paneth cells. These cells express Zn^{2+} transporters, both from the ZIP (SLC39A/Zrt- and Irt-like protein) and ZnT (Solute carrier, SLC30A/Zinc transporter) families (Fig. 17.1), which govern the movement of Zn^{2+} into the cytosol and out of it, respectively. As such, Zn^{2+} efflux transporters expressed on the apical cells of the epithelium may transport Zn^{2+} from the cells into the intestinal lumen and may provide another source of luminal Zn^{2+}. Indeed, the Zn^{2+} transporters ZnT1, ZnT5, and ZnT6 are localized to the apical side of intestinal cells suggesting they may transport Zn^{2+} into the lumen (Yu et al. 2007; Jou et al. 2009). Changes in dietary zinc uptake were shown to influence the expression level and localization of the Zn^{2+} transporters, further attesting to their roles in zinc absorption, as well as Zn^{2+} signaling (Jou et al. 2009). Emerging evidence implicate specific Zn^{2+} transporters, not only in regulating intracellular Zn^{2+} levels, but also in playing key roles that govern intestinal epithelial cell development and physiological function (Wang and Zhou 2010; Myers et al. 2017).

Dyshomeostasis of zinc, resulting from changes in diet or Zn^{2+} transporter dysfunction, may lead to deficiency or excess of this ion but can also impair intestinal cell function. Indeed, zinc dyshomeostasis is associated with disease. Zinc deficiency is manifested by impaired function of the digestive system, most prominent of which are severe diarrhea and inflammatory bowel disease (IBD) (Walker and Black 2010; Luk et al. 2002; Maret and Sandstead 2006). The general role of Zn^{2+} in preserving intestinal epithelium in human disease is underscored by the effect of zinc supplementation on reducing diarrhea (Walker and Black 2010; Scrimgeour and Lukaski 2008) and permeability of the gut (Sturniolo et al. 2001), particularly in individuals who are zinc deficient. The World Health Organization (WHO) sug-

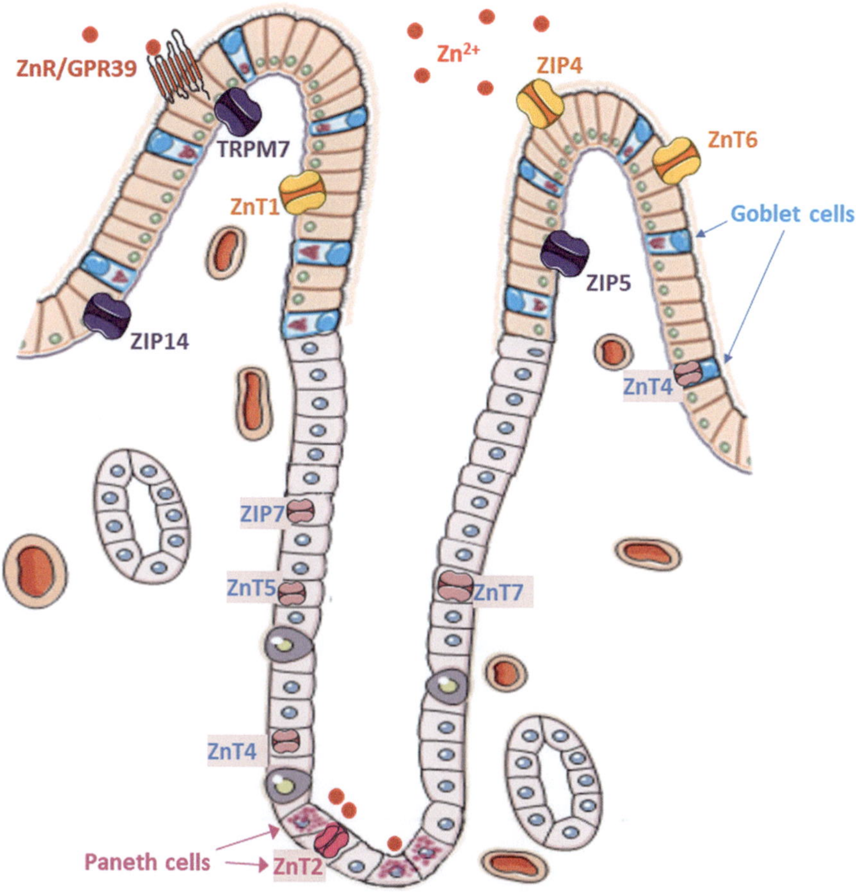

Fig. 17.1 Schematic representation of the localization of Zn^{2+} homeostatic proteins in intestinal epithelial cells. In light pink are crypt cells, which include the intestinal stem cells (iSC) and Paneth cells; while villus cells (or apical surface cells in the large intestine) that are facing the intestinal lumen are shown in orange and represent enterocytes (or colonocytes in the large intestine) or Goblet cells (blue cells). Apical or basolateral expression of Zn^{2+} homeostatic proteins is marked by the location of relevant transporters relative to the cell it is shown on; and transporters localized within cellular compartments are marked by a pink background to the protein name. Figure adjusted from modules of Servier Medical Art (http://www.servier.com)

gests zinc supplementation for effective treatment of diarrhea, and this is supported by multiple studies showing the role of this nutrient in shortening the disease phase and reducing mortality (Lazzerini and Wanzira 2016 21; Crisinel et al. 2015; Hoque and Binder 2006). Although zinc deficiency is associated with IBD, it is unclear if it predisposes one to IBD, or if IBD results in zinc deficiency. This is illustrated by the fact that IBD is more prominent in countries where nutritional zinc deficiency is less common. Notably, availability of Zn^{2+} in the intestine may also be impaired by phytate, and a high fiber diet may also result in zinc deficiency that is more common in developed countries (Gibson et al. 2018). In addition to a direct role on intestinal epithelial physiology, zinc regulates immune system function, and may thus be also involved in mucosal inflammation, such as seen in IBD. Finally, zinc supplementation is suggested to enhance appetite during chemotherapy treatment or in anorexia nervosa patients. In these disease states, it is not clear if zinc deficiency is a consequence of the low micronutrient uptake and zinc supplementation merely repletes the deficiency, or if zinc serves as a signaling ion that triggers cellular pathways to enhance recovery. More recently, excess zinc intake through supplementation has been modestly associated with IBD, and studies in animal models provide compelling evidence that excess zinc exposure in otherwise healthy and non-zinc deficient subjects, causes oxidative stress, inflammation, and negatively impacts intestine function (Podany et al. 2019). Thus, understanding the mechanisms underlying the roles of zinc in the digestive system is important to optimizing human health and eliminating the burden of disease.

17.2 The Role of Zinc in Intestinal Function

The intestinal epithelium must maintain a firm and selective barrier to prevent invasion of bacteria and pathogens into the body, but at the same time to enable selective uptake and removal of nutrients and solutes. This is achieved by: (i) continuous renewal of the intestinal epithelium, a process requiring rapid cell proliferation and iSC differentiation into multiple cell types found along the lumen; (ii) formation of a physical tight barrier mediated by expression of apical junctional proteins; (iii) regulation of ion (K^+, Cl^-, and Na^+) transport across the intestinal epithelium that establishes osmotic gradients and allows water and nutrient absorption (Barrett 2008; Zachos et al. 2009); and (iv) regulated secretion of antimicrobial peptides, cytokines, and other immunologically active factors that maintain mucosal immunity and the gut microbiome. Evidence supports a critical role for zinc in all of these functions.

The intestinal epithelium is composed of an array of differentiated epithelial cells that derive from the multipotent iSCs. Under homeostatic conditions, iSCs replenish the various cell types that line the lumen (Potten and Loeffler 1990). These remarkable cells feature the ability to self-renew while giving rise to intermediate transit-amplifying progenitor cells (TA cells) that continually differentiate to renew intestinal epithelial cells that comprise the crypt-villus axis (Rizk and Barker 2012; Potten and Loeffler 1990). The most predominate cell type is the enterocyte that primarily functions to absorb nutrients through various integral membrane transporters and channels, and large macromolecules through endocytosis (Overeem et al. 2016). Enterocytes are critical to maintaining active absorptive function of this tissue on one hand and on the other hand, these cells are responsible for forming the intestinal barrier that attenuates paracellular movement across the epithelium. The intestinal barrier is composed of a mucus layer and apically expressed transmembrane-junctional proteins, e.g., claudins, occluding, and zonula occludens-1, which seal the apical surface and block paracellular movement. Chronic inflammation and diarrhea associated with IBD have been linked to disruption in epithelial barrier functions (Barbara 2006; Catalioto et al. 2011). Breakdown of the apical junctional complexes during inflammatory conditions, induces uncontrolled paracellular transport that precedes actual damage of the tissue, and is therefore considered an important stage in the pathogenesis of these diseases (Bertiaux-Vandaele et al. 2011). In addition to serving as the absorption site for zinc, the physiological function of the enterocyte is regulated by this ion (Walker and Black 2010; Sazawal et al. 1995; Lindenmayer et al. 2014). Zinc deficiency impairs barrier function and increases permeability and cell death (Finamore et al. 2008; Geiser et al. 2012). Barrier function and expression of claudin proteins are impaired during disease, and zinc supplementation to Shigella-infected intestinal cells ameliorates claudin expression, and thereby restores barrier function (Sarkar et al. 2019). Studies in piglets show that zinc repletion enhances expression of the tight junction (TJ) proteins occludin and zonula occludens-1, and promotes increased villus height, width, crypt depth, and surface area (Wang et al. 2018). Studies in vitro also show that TJ proteins are regulated by zinc supplementation to the growth medium. At the cellular level, Zn^{2+} regulates barrier function by enhancing the cell membrane localization of TJ proteins claudin-2 and -4, and modulating phosphorylation state of TJ proteins through extracellular signal-regulated kinase (ERK)1/2 (Sarkar et al. 2019). Zn^{2+} also promotes barrier formation (Finamore et al. 2008; Geiser et al. 2012; Glover and Hogstrand 2003; Wang et al. 2013) in colon epithelial cell (colonocyte) cultures, and plays a major role in enhancing proliferation and survival of colonocytes (Cohen et al. 2012; Sunuwar et al. 2017b). The underlying signaling pathways associated with these processes are activated by an upstream regulator, a zinc-sensing G-protein coupled receptor (ZnR/GPR39, see Sect. 17.4). However, excess zinc in non-zinc deficient subjects can have a negative effect on barrier function as well, as zinc induces oxidative stress and intestinal permeability through decreasing e-cadherin expression (Podany et al. 2019).

Goblet cells are the second most-predominant cell type and are responsible for secreting mucus, which is quintessential to the integrity of the mucosal barrier, and ensures not only the hydration of the underlying intestinal layers but is critical for neutralizing toxins and infectious agents (Kim and Khan 2013; Kim and Ho 2010; Pelaseyed et al. 2014), and providing an energy source for gut microbiota. Effects of Zn^{2+} on mucus secretion have been described in fish and are also associated with regulation of absorption by the epithelium (Khan and McGeer 2013; Glover and Hogstrand 2003). Studies in vitro show that mucins provide a zinc source for absorption, as mucins contain multiple zinc-binding sites with biologically relevant affinity within one mucin molecule (Maares et al. 2018). Zinc deficiency increases mucus production (Bolick et al. 2014); however, effects on goblet cell number were not reported. Interestingly, excess zinc exposure increases mucus production and goblet cell number in mice (Podany et al. 2019), broilers (Shah et al. 2018), and in vitro in HT-29 cells (Arriaga et al. 2017). Although the mechanisms are not currently understood, one possibility is that excess zinc increases KLF-4 and/or BMP signaling in iSCs, which drives differentiation of iSCs into goblet cells (Chen et al. 2018).

Finally, protection of the intestinal system is supported also by secretion of antimicrobial agents and cytokines by the Paneth cells, and by immune cells from the lamina propria beneath the epithelial layer. The highly specialized and secretory Paneth cells are situated in the crypts of Lieberkühn and adjacent to the iSCs, and known for their large secretory granules through which they secrete antimicrobial peptides (e.g., a-defensin, lysozyme, and secretory phospholipase A2), cytokines, enzymes, growth factors, and a large amount of Zn^{2+} (Stappenbeck 2009; Giblin et al. 2006; Danscher and Stoltenberg 2005). These granule factors are required to neutralize bacterial pathogens and maintain microbial homeostasis, as well as provide critical growth factors such as Wnt to influence iSC proliferation and differentiation (Stappenbeck 2009). Loss of Paneth cells leads to reduction of iSCs, which underscore the Paneth cell's role as the "gatekeeper of the intestine". Severe zinc deficiency in rats does not affect granule Zn^{2+} content or Paneth cell ultrastructure, suggesting these specialized cells are relatively resistant to zinc depletion (Moran and Lewis 1985). However, evidence from human studies indicate that loss of Paneth cell granules is associated with low serum zinc levels, which may affect host susceptibility in these individuals (Kelly et al. 2004). The importance of zinc in host-microbe interactions is underscored by numerous studies, and changes to ZIP transporters induced by inflammatory cytokines have been described in the lung (Galvez-Peralta et al. 2014), similar mechanisms may affect the intestinal epithelium as well. Importantly, secretion of Zn^{2+} or Zn^{2+}-chelating agents may regulate the luminal microbial survival (Malavia et al. 2017; Shank et al. 2018). Indeed, excess zinc exposure in neonatal mice alters the gut microbiome such that several pathogenic taxa are enriched (Podany et al. 2019). Thus, zinc may also indirectly regulate barrier function of the intestinal epithelium via modulation of antimicrobial/antibacterial mechanisms.

17.3 Zinc Transporters in the Digestive Tract

17.3.1 Transporters Responsible for Uptake

The transporter that is most closely related to Zn^{2+} uptake is ZIP4, and loss of this transporter is embryonically lethal. ZIP4 is found on the apical membrane of intestinal cells where it imports zinc from the lumen, and its expression is modulated by proteolytic processes following changes in enterocyte Zn^{2+} concentrations (Hashimoto et al. 2016; Dufner-Beattie et al. 2003, 2007). A profound example of zinc deficiency is seen in the disease acrodermatitis enteropathica (AE) where mutations in ZIP4 underlie the disease mechanism (Wang et al. 2002). A primary defect in AE is reduced transport of Zn^{2+} across the apical membrane of enterocyte, which leads to diarrhea, impaired immune function, gastrointestinal lesions, and severe and potentially lethal zinc deficiency (Nistor et al. 2016). Importantly, supplementation with dietary zinc can rescue these patients. Knockdown of ZIP4 in the intestinal epithelium abrogates Zn^{2+} absorption, but also disrupts iSC differentiation and eventually results in loss of the columnar epithelium on the villi (Geiser et al. 2012). This is associated with profound loss of labile Zn^{2+} within Paneth cell granules (Geiser et al. 2012), which may impair Paneth cell-iSC interactions that are critical for maintaining the stem cell niche. Localization of ZIP5 on the basolateral membrane of enterocytes cells suggests this transporter is also involved in response to systemic zinc or zinc excretion. Indeed, knockdown of ZIP5 results in abrogated zinc concentrations in the liver, and specific knockdown of ZIP5 in enterocytes increases pancreatic zinc accumulation (Geiser et al. 2013). Expression of ZIP5 is also tightly regulated by dietary zinc deficiency when it is degraded, which is likely a protective mechanism to reduce zinc loss. Basolateral expression of ZIP14 on enterocytes suggested that this transporter is also involved in absorption of this ion (Guthrie et al. 2015). Indeed ZIP14 knockout mice showed increase in accumulation of Zn^{2+} in endosomes in the small intestinal epithelial cells and increased expression of the Zn^{2+} binding metallothioneins (Guthrie et al. 2015). Interestingly, ZIP14 deficient mice showed impaired TJ complexes and increased permeability, which may also result due to impaired Zn^{2+}-dependent activation of ZnR/GPR39 (see Sect. 17.4). The ubiquitous zinc transporter ZnT1 is predominantly expressed on the basolateral entercocyte membrane and is likely responsible for Zn^{2+} efflux and absorption into systemic circulation (Yu et al. 2007). Possible ZnT1 apical expression in enterocytes suggests it may also be responsible for Zn^{2+} efflux back into the lumen following excess zinc exposure (Jou et al. 2009). Dietary zinc deficiency affects expression of ZnT1 in the intestine, with even mild zinc deficiency reducing levels of ZnT1 in the jejunum (Jou et al. 2009), is due to the cellular role of Zn^{2+} regulating metal transcription factor 1 (MTF1) phosphorylation (Langmade et al. 2000).

The Transient Receptor Potential proteins are a family of non-selective ion channels that are widely expressed (Bouron et al. 2015). While Zn^{2+} modulates, and largely inhibits, TRP channels, several members of this family can transport Zn^{2+} ions, for example TRPA1 (Hu et al. 2009), TRPC6 (Gibon et al. 2011), TRPM3 (Wagner et al. 2010), TRPML1 (Kiselyov et al. 2011; Eichelsdoerfer et al. 2010), and TRPM7 (Inoue et al. 2010). A role for TRPM3 was suggested in pancreatic β-cell absorption of zinc, where Zn^{2+} is stored in insulin granules (Kristiansen et al. 2001), and may also potentiate Zn^{2+} influx through voltage-gated calcium channels (Wagner et al. 2010). Efflux of lysosomal Zn^{2+} is mediated by TRPML1, and orchestrated activity of this transporter with ZnT4 maintains cellular Zn^{2+} homeostasis (Kukic et al. 2013). Recently, specific intestinal knock down of TRPM7 was shown to modulate zinc status in mice, as well as downregulation of several Zn^{2+} transporters, including ZIP4 (Mittermeier et al. 2019). Thus, ZIP4 decrease may also have a role in reducing absorption of Zn^{2+}, a direct role for TRPM7 in Zn^{2+} absorption still needs to be verified.

17.3.2 Transporters Responsible for Intracellular Zn^{2+} Homeostasis

Expression studies indicate that ZnT4 is found in both the absorptive enterocytes and the goblet cells (Yu et al. 2007; Murgia et al. 2006, 1999). Previously reported on the basolateral side of enterocytes in rats (Murgia et al. 1999), ZnT4 is also localized to the Golgi apparatus and to intracellular vesicles other cell types (McCormick et al. 2016). Although the major phenotype in ZnT4-null mice is a reduction in milk zinc levels known as "lethal-milk" syndrome (McCormick et al. 2016), a spontaneous truncation mutation in ZnT4 also results in ZIP4 expression and low zinc absorption (Murgia et al. 2006). ZnT4-deletion leads to reduced activation of STAT5, which is also critical for homeostatic regulation of gut integrity and immunity (Gilbert et al. 2012). Attenuation of STAT5 in intestinal organoids fail to form crypts and are unable to expand, suggesting that ZnT4-mediated zinc transport plays a role in crypt expansion (Gilbert et al. 2015). As ZnT4 is required for epithelial differentiation via STAT5 in mammary cells (McCormick et al. 2016), it may play a similar role in enterocytes and goblets cells. ZnT5 and ZnT7 are primarily localized to the basal side of enterocytes (Yu et al. 2007), which is consistent with the localization of these proteins on the Golgi and vesicular compartments (Ishihara et al. 2006). Expression of ZnT5 and ZnT7 is higher in developing cells in the crypts, likely due to the high rates of protein production that requires Zn^{2+} (Suzuki et al. 2005a; Suzuki et al. 2005b). While ZnT4, 5, and 7 are found mostly in the duodenum and jejunum, ZnT7 alone is expressed in the ileum. ZnT6, which is largely associated with ZnT5 as a heterodimer found on intracellular organelles, is also expressed on the apical side of small intestinal cells (Yu et al. 2007); however, the role of ZnT6 is not understood (Fukunaka et al. 2009; Ohana et al. 2009).

Paneth cells have a unique requirement for zinc. Their large secretory granules contain a large amount of Zn^{2+} in addition to antimicrobial proteins and cytokines (Danscher and Stoltenberg 2005; Elmes and Jones 1981). Zinc depletion following genetic deletion of ZIP4 in enterocytes, results in Paneth cell death and their replacement by mucin-secreting goblet cells (Geiser et al. 2012). This may be due to changes in Wnt signaling as it plays a well-known role as the master regulator of the sophisticated network that modulates the iSC niche (Fevr et al. 2007). Wnt binds to Frizzled receptor, thereby stabilizing β-catenin transcription factor complex, which drives a variety of cell signaling pathways that control cell fate (Clevers 2006). Deletion of ZIP4 reduces SOX9 and dephosphorylates GSK3, thereby reducing mTOR1 activity, which is closely interconnected with Wnt signaling pathways. A zinc transporter critical for Paneth cell function is ZnT2. ZnT2 is localized to the large secretory granules and is responsible for the accumulation of granular Zn^{2+}. Deletion of ZnT2 in mice does not alter the gross morphology of the small intestine, but reduces Zn^{2+} content in Paneth cells and results in disorganization of their secretory granules altogether (Podany et al. 2016). In accordance with the major role of the Paneth cells in gut microbial homeostasis and immune susceptibility, ZnT2 knockout mice show diminished antibacterial response to lipopolysaccharide, increased gut permeability, and systemic inflammation (Podany et al. 2016). This suggests ZnT2 may protect against IBD; however, further studies are required to address the human health implications.

Epithelial regeneration is particularly rapid in the cells of the intestinal mucosa that interface with the lumen of the digestive tract. ZIP7 is detected in the intestinal crypt, and localized to subcellular compartments such as the endoplasmic reticulum (ER) and Golgi apparatus in TA cells and Paneth cells, where it exports Zn^{2+} into the cytoplasm (Huang et al. 2005; Ohashi et al. 2016). Targeted deletion of ZIP7 in the intestinal epithelium in mice reveals aberrations in Paneth cell granules and the loss of $Olfm4^{+}$ stem cells in the crypt, suggesting ZIP7 plays a key role in mediating iSC proliferation and differentiation (Ohashi et al. 2016) and rapid recovery from injury. Phosphorylation regulates ZIP7 function and it is required for activation of AKT signaling and myoblast differentiation (Mnatsakanyan et al. 2018; Taylor et al. 2012), suggesting a potential mechanism in the intestine. In addition, ZIP7 deletion increases susceptibility to irradiation-induced damage that requires rapid epithelial recovery (Ohashi et al. 2016). A similar mechanism was also associated with dermal stem cell survival (Bin et al. 2017). Moreover, deletion of ZIP7 in cultured organoids leads to degeneration and failure to undergo crypt fission, suggesting that ZIP7 is essential to regeneration of intestinal crypts (Ohashi et al. 2016). Interestingly, the inability to release Zn^{2+} from the ER leads to ER stress and hypodense granules, suggesting the ER may provide the pool of Zn^{2+} that accumulates in Paneth cell granules (Ohashi et al. 2016; Podany et al. 2016).

17.4 Intestinal Function of ZnR/GPR39

Transient changes in extracellular Zn^{2+} concentrations suggested that this ion might act as a first messenger, triggering cellular signaling via interaction with membrane proteins. Indeed, Zn^{2+} acts as a modulator of numerous membrane proteins, thereby affecting cellular signaling pathways. For example, Zn^{2+} modulates the function the store operated Ca^{2+} signals (Gore et al. 2004) or purinergic receptor responses (Mendez-Barredo et al. 2018; Jiang 2009). But, exogenous Zn^{2+} also acts as an essential activator of a distinct Zn^{2+} sensing receptor (Hershfinkel et al. 2001). A Zn^{2+}-triggered Gq-dependent mechanism for activation of Ca^{2+} cellular signaling is mediated by a functional metabotropic receptor, ZnR/GPR39, that senses changes in extracellular Zn^{2+} in colonocytes (Hershfinkel 2018). Initial studies indicated that ZnR/GPR39 is widely-expressed throughout the digestive system (Moechars et al. 2006; Depoortere 2012). In native colon tissue, luminal application of Zn^{2+} is sufficient to activate ZnR/GPR39, suggesting that the receptor is present on the apical side of colon epithelial cells (Cohen et al. 2012). The Zn^{2+}- triggered ZnR/GPR39-dependent Ca^{2+} signaling and downstream MAP kinase and AKT activation, play a key role in enhancing proliferation and differentiation of epithelial cells (Hershfinkel et al. 2001; Cohen et al. 2012, 2014; Azriel-Tamir et al. 2004). Colonocytes are exposed to numerous apoptotic signals, e.g., the short chain fatty acid butyrate present at high concentrations in the colon lumen, which imposes acidosis and cell death (Yu et al. 2010; Scharlau et al. 2009; Bordonaro et al. 2008). Prolonged exposure of colonocyte cell cultures to butyrate induces cell death that is attenuated by activation of ZnR/GPR39-dependent MAP kinase and AKT pathways and subsequent upregulation of the pro-survival glycoprotein, clusterin (Cohen et al. 2012). The role of ZnR/GPR39 is likely not limited to the colonocytes, as luminal Zn^{2+} accelerates healing of gastric ulcers (Opoka et al. 2010), which may involve ZnR/GPR39-dependent enhanced proliferation.

Activation of ZnR/GPR39 in colonocyte cell cultures also enhances expression of TJ proteins zonula occludens-1 and occludin (Cohen et al. 2014), critical elements of the intestinal barrier (Furuse et al. 1994; Chiba et al. 2008). While several studies showed that Zn^{2+} enhances TJ formation, only later was ZnR/GPR39 identified as the underlying mechanism by using inhibitors of the ZnR/GPR39 signaling pathway, as well as silencing of the receptor itself (Cohen et al. 2014). A significant reduction in TJ barrier function is also monitored in colonic epithelium of ZnR/GPR39 knockout mice (Cohen et al. 2012). Accordingly, treatment of ZnR/GPR39-expressing or knockout mice with dextran sodium sulfate (DSS) to induce colitis, resulted in increased inflammatory response in the knockout mice (Sunuwar et al. 2016). Using the DSS model permitted exploration of the recovery period, which showed that ZnR/GPR39 expression allows enhanced proliferation and recovery of the mucosal layer including TJ formation. In contrast, ZnR/GPR39 knockout mice had reduced survival and slower recovery of the mucosal layer after removal of DSS. Importantly, this study reveals a direct role for Zn^{2+}, via ZnR/GPR39, in activating epithelial cell signaling leading to a tightening of the epithelial barrier. Controversial results regarding the role of dietary zinc in IBD may stem from partial understanding of the mechanism, as such, supplementation with zinc during periods of ZnR/GPR39 desensitization or in patients who lack a functional receptor may yield negative results for the role of zinc. In addition, the interaction of supplemental Zn^{2+} with multiple Zn^{2+} transporters clearly affects multiple pathways. Thus, a specific agonist for ZnR/GPR39 may serve as a better entity to test the effect of ZnR/GPR39 in human patients with IBD.

Epithelial intestinal cells have an essential role in maintaining ion gradients across the epithelial barrier. In the colon, Zn^{2+} was suggested to modulate ion transport either under normal conditions or during experimentally induced diarrhea (Medani et al. 2012; Canani et al. 2005). Activation of ZnR/GPR39 in colon cell lines and native colonocytes promotes upregulation of Na^+/H^+ exchanger (NHE) activity (Azriel-Tamir et al. 2004; Cohen et al. 2012) and basolateral K^+/Cl^- cotransporter, KCC1, activity (Sunuwar et al. 2017a). Impaired ion transport balance results in diarrhea, and is seen in cholera toxin-treated mouse models. Indeed, ZnR/GPR39 upregulation of KCC1 activity in wild type mice results in a much milder diarrhea compared to that observed in ZnR/GPR39 knockout mice (Sunuwar et al. 2017a). In addition, ZnR/GPR39 upregulation of NHE activity may be mediated by the apical NHE3 that also has a role in attenuating diarrhea (Singh et al. 2012; Donowitz et al. 2013). Diarrhea itself may impose zinc deficiency, and zinc supplementation in patients with dietary zinc deficiency may improve general health and immune system function. However, the identification of ZnR/GPR39-dependent activation of ion transport suggests that Zn^{2+} acts as a specific signaling molecule. Moreover, ZnR/GPR39 serves as a direct target for intestinal Zn^{2+} that can be therapeutically modulated to alleviate diarrhea.

17.5 Conclusions and Future Directions

Zinc is essential for function of all the systems in the body, and its primary entry pathway is through gastrointestinal uptake. Multiple transporters are responsible for maintaining zinc homeostasis and critical cellular functions in the gastrointestinal system. Zn^{2+} in the enterocytes is responsible for the formation of a tight barrier between the intestinal lumen and the body; maintaining a stable barrier is critical to maintaining a healthy gut. Paneth and goblet cell functions are critically dependent upon Zn^{2+}, and as such, are essential for promoting host-microbe interactions. A key gap in knowledge is our profound lack of understanding of the role of Zn^{2+} in regulating iSC function and differentiation. Although the function of only few of the Zn^{2+} transporters has been elucidated to-date, it is clear that they serve an important role in development and maintenance of the healthy intestine. In addition, Zn^{2+} acts as a signaling ion and activates cellular signaling pathways either directly or via modulation of its transporters. As such, chelation of Zn^{2+} by compounds that can be found in the intestinal lumen, e.g., phytate or citrate, not only affects zinc absorption leading to deficiency but can also impair Zn^{2+} – dependent signaling. Loss of endogenous Zn^{2+} following its chelation by citrate or amyloid-β resulted in changes in ZnR/GPR39 signaling that were shown to modulate functions of prostate cells or neurons, respectively (Dubi et al. 2008; Abramovitch-Dahan et al. 2016). The complexity of the Zn^{2+} transporting network is far from being clear, as changes in each of the transporters, or levels or subcellular localization of Zn^{2+} itself, affect the expression level, localization, and function of other transporters. The requirement for multiple compensatory pathways for each piece of this puzzle only attests to the importance of the Zn^{2+} transporting network in maintaining gastrointestinal homeostasis. Revealing the multitude of relations between these Zn^{2+} transporters remains a task to be investigated. The Zn^{2+} sensing receptor, ZnR/GPR39, is also activated in the intestinal epithelial cells and regulates important functions including barrier recovery and ion transport. Its interaction with other Zn^{2+} homeostatic proteins has not been elucidated, but adds another layer of complexity to the Zn^{2+} transporting network. Future studies are expected to identify and link the dots of this physiologically critical network.

References

Abramovitch-Dahan C, Asraf H, Bogdanovic M, Sekler I, Bush AI, Hershfinkel M (2016) Amyloid beta attenuates metabotropic zinc sensing receptor, mZnR/GPR39, dependent Ca($^{2+}$), ERK1/2 and Clusterin signaling in neurons. J Neurochem 139(2):221–233. https://doi.org/10.1111/jnc.13760

Arriaga JM, Bravo AI, Mordoh J, Bianchini M (2017) Metallothionein 1G promotes the differentiation of HT-29 human colorectal cancer cells. Oncol Rep 37(5):2633–2651

Azriel-Tamir H, Sharir H, Schwartz B, Hershfinkel M (2004) Extracellular zinc triggers ERK-dependent activation of Na+/H+ exchange in colonocytes mediated by the zinc-sensing receptor. J Biol Chem 279(50):51804–51816

Barbara G (2006) Mucosal barrier defects in irritable bowel syndrome. Who left the door open? Am J Gastroenterol 101(6):1295–1298
Barrett KE (2008) New ways of thinking about (and teaching about) intestinal epithelial function. Adv Physiol Educ 32(1):25–34
Bertiaux-Vandaele N, Youmba SB, Belmonte L, Lecleire S, Antonietti M, Gourcerol G, Leroi AM, Dechelotte P, Menard JF, Ducrotte P, Coeffier M (2011) The expression and the cellular distribution of the tight junction proteins are altered in irritable bowel syndrome patients with differences according to the disease subtype. Am J Gastroenterol 106(12):2165–2173
Bin BH, Bhin J, Seo J, Kim SY, Lee E, Park K, Choi DH, Takagishi T, Hara T, Hwang D, Koseki H, Asada Y, Shimoda S, Mishima K, Fukada T (2017) Requirement of zinc transporter SLC39A7/ZIP7 for dermal development to fine-tune endoplasmic reticulum function by regulating protein disulfide isomerase. J Invest Dermatol 137(8):1682–1691
Bolick DT, Kolling GL, Moore JH 2nd, de Oliveira LA, Tung K, Philipson C, Viladomiu M, Hontecillas R, Bassaganya-Riera J, Guerrant RL (2014) Zinc deficiency alters host response and pathogen virulence in a mouse model of enteroaggregative Escherichia coli-induced diarrhea. Gut Microbes 5(5):618–627
Bordonaro M, Lazarova DL, Sartorelli AC (2008) Butyrate and Wnt signaling: a possible solution to the puzzle of dietary fiber and colon cancer risk? Cell Cycle 7(9):1178–1183
Bouron A, Kiselyov K, Oberwinkler J (2015) Permeation, regulation and control of expression of TRP channels by trace metal ions. Pflugers Arch 467(6):1143–1164
Canani RB, Cirillo P, Buccigrossi V, Ruotolo S, Passariello A, De Luca P, Porcaro F, De Marco G, Guarino A (2005) Zinc inhibits cholera toxin-induced, but not Escherichia coli heat-stable enterotoxin-induced, ion secretion in human enterocytes. J Infect Dis 191(7):1072–1077
Catalioto RM, Maggi CA, Giuliani S (2011) Intestinal epithelial barrier dysfunction in disease and possible therapeutical interventions. Curr Med Chem 18(3):398–426
Chen W, Zhuo M, Lu X, Xia X, Zhao Y, Huang Z, Xu J, Li W, Yu C (2018) SRC-3 protects intestine from DSS-induced colitis by inhibiting inflammation and promoting goblet cell differentiation through enhancement of KLF4 expression. Int J Biol Sci 14(14):2051–2064
Chiba H, Osanai M, Murata M, Kojima T, Sawada N (2008) Transmembrane proteins of tight junctions. Biochim Biophys Acta 1778(3):588–600
Clevers H (2006) Wnt/beta-catenin signaling in development and disease. Cell 127(3):469–480
Cohen L, Azriel-Tamir H, Arotsker N, Sekler I, Hershfinkel M (2012) Zinc sensing receptor signaling, mediated by GPR39, reduces butyrate-induced cell death in HT29 colonocytes via upregulation of clusterin. PLoS One 7(4):e35482
Cohen L, Sekler I, Hershfinkel M (2014) The zinc sensing receptor, ZnR/GPR39, controls proliferation and differentiation of colonocytes and thereby tight junction formation in the colon. Cell Death Dis 5(5):e1307
Crisinel PA, Verga ME, Kouame KS, Pittet A, Rey-Bellet CG, Fontaine O, Di Paolo ER, Gehri M (2015) Demonstration of the effectiveness of zinc in diarrhoea of children living in Switzerland. Eur J Pediatr 174(8):1061–1067
Danscher G, Stoltenberg M (2005) Zinc-specific autometallographic in vivo selenium methods: tracing of zinc-enriched (ZEN) terminals, ZEN pathways, and pools of zinc ions in a multitude of other ZEN cells. J Histochem Cytochem 53(2):141–153
Depoortere I (2012) GI functions of GPR39: novel biology. Curr Opin Pharmacol 12(6):647–652
Donowitz M, Ming Tse C, Fuster D (2013) SLC9/NHE gene family, a plasma membrane and organellar family of Na(+)/H(+) exchangers. Mol Asp Med 34(2–3):236–251
Dubi N, Gheber L, Fishman D, Sekler I, Hershfinkel M (2008) Extracellular zinc and zinc-citrate, acting through a putative zinc-sensing receptor, regulate growth and survival of prostate cancer cells. Carcinogenesis 29(9):1692–1700
Dufner-Beattie J, Wang F, Kuo YM, Gitschier J, Eide D, Andrews GK (2003) The acrodermatitis enteropathica gene ZIP4 encodes a tissue-specific, zinc-regulated zinc transporter in mice. J Biol Chem 278(35):33474–33481

Dufner-Beattie J, Weaver BP, Geiser J, Bilgen M, Larson M, Xu W, Andrews GK (2007) The mouse acrodermatitis enteropathica gene Slc39a4 (Zip4) is essential for early development and heterozygosity causes hypersensitivity to zinc deficiency. Hum Mol Genet 16(12):1391–1399

Eichelsdoerfer JL, Evans JA, Slaugenhaupt SA, Cuajungco MP (2010) Zinc dyshomeostasis is linked with the loss of mucolipidosis IV-associated TRPML1 ion channel. J Biol Chem 285(45):34304–34308. https://doi.org/10.1074/jbc.C110.165480

Elmes ME, Jones JG (1981) Paneth cell zinc: a comparison of histochemical and microanalytical techniques. Histochem J 13(2):335–337

Fevr T, Robine S, Louvard D, Huelsken J (2007) Wnt/beta-catenin is essential for intestinal homeostasis and maintenance of intestinal stem cells. Mol Cell Biol 27(21):7551–7559

Finamore A, Massimi M, Conti Devirgiliis L, Mengheri E (2008) Zinc deficiency induces membrane barrier damage and increases neutrophil transmigration in Caco-2 cells. J Nutr 138(9):1664–1670

Frederickson CJ, Danscher G (1990) Zinc-containing neurons in hippocampus and related CNS structures. Prog Brain Res 83:71–84

Frederickson CJ, Perez-Clausell J, Danscher G (1987) Zinc-containing 7S-NGF complex. Evidence from zinc histochemistry for localization in salivary secretory granules. J Histochem Cytochem 35(5):579–583

Fukunaka A, Suzuki T, Kurokawa Y, Yamazaki T, Fujiwara N, Ishihara K, Migaki H, Okumura K, Masuda S, Yamaguchi-Iwai Y, Nagao M, Kambe T (2009) Demonstration and characterization of the heterodimerization of ZnT5 and ZnT6 in the early secretory pathway. J Biol Chem 284(45):30798–30806

Furuse M, Itoh M, Hirase T, Nagafuchi A, Yonemura S, Tsukita S, Tsukita S (1994) Direct association of occludin with ZO-1 and its possible involvement in the localization of occludin at tight junctions. J Cell Biol 127(6. Pt 1):1617–1626

Galvez-Peralta M, Wang Z, Bao S, Knoell DL, Nebert DW (2014) Tissue-specific induction of mouse ZIP8 and ZIP14 divalent cation/bicarbonate symporters by, and cytokine response to, inflammatory signals. Int J Toxicol 33(3):246–258

Geiser J, Venken KJ, De Lisle RC, Andrews GK (2012) A mouse model of acrodermatitis enteropathica: loss of intestine zinc transporter ZIP4 (Slc39a4) disrupts the stem cell niche and intestine integrity. PLoS Genet 8(6):e1002766

Geiser J, De Lisle RC, Andrews GK (2013) The zinc transporter Zip5 (Slc39a5) regulates intestinal zinc excretion and protects the pancreas against zinc toxicity. PLoS One 8(11):e82149

Giblin LJ, Chang CJ, Bentley AF, Frederickson C, Lippard SJ, Frederickson CJ (2006) Zinc-secreting Paneth cells studied by ZP fluorescence. J Histochem Cytochem 54(3):311–316

Gibon J, Tu P, Bohic S, Richaud P, Arnaud J, Zhu M, Boulay G, Bouron A (2011) The over-expression of TRPC6 channels in HEK-293 cells favours the intracellular accumulation of zinc. Biochim Biophys Acta 1808(12):2807–2818

Gibson RS, Raboy V, King JC (2018) Implications of phytate in plant-based foods for iron and zinc bioavailability, setting dietary requirements, and formulating programs and policies. Nutr Rev 76(11):793–804

Gilbert S, Zhang R, Denson L, Moriggl R, Steinbrecher K, Shroyer N, Lin J, Han X (2012) Enterocyte STAT5 promotes mucosal wound healing via suppression of myosin light chain kinase-mediated loss of barrier function and inflammation. EMBO Mol Med 4(2):109–124

Gilbert S, Nivarthi H, Mayhew CN, Lo YH, Noah TK, Vallance J, Rulicke T, Muller M, Jegga AG, Tang W, Zhang D, Helmrath M, Shroyer N, Moriggl R, Han X (2015) Activated STAT5 confers resistance to intestinal injury by increasing intestinal stem cell proliferation and regeneration. Stem Cell Rep 4(2):209–225

Glover CN, Hogstrand C (2003) Effects of dissolved metals and other hydrominerals on in vivo intestinal zinc uptake in freshwater rainbow trout. Aquat Toxicol 62(4):281–293

Gopalsamy GL, Alpers DH, Binder HJ, Tran CD, Ramakrishna BS, Brown I, Manary M, Mortimer E, Young GP (2015) The relevance of the colon to zinc nutrition. Nutrients 7(1):572–583

Gore A, Moran A, Hershfinkel M, Sekler I (2004) Inhibitory mechanism of store-operated Ca^{2+} channels by zinc. J Biol Chem 279(12):11106–11111

Guthrie GJ, Aydemir TB, Troche C, Martin AB, Chang SM, Cousins RJ (2015) Influence of ZIP14 (slc39A14) on intestinal zinc processing and barrier function. Am J Physiol Gastrointest Liver Physiol 308(3):G171–G178. https://doi.org/10.1152/ajpgi.00021.2014

Hashimoto A, Nakagawa M, Tsujimura N, Miyazaki S, Kizu K, Goto T, Komatsu Y, Matsunaga A, Shirakawa H, Narita H, Kambe T, Komai M (2016) Properties of Zip4 accumulation during zinc deficiency and its usefulness to evaluate zinc status: a study of the effects of zinc deficiency during lactation. Am J Physiol Regul Integr Comp Physiol 310(5):R459–R468

Hennigar SR, Kelleher SL (2012) Zinc networks: the cell-specific compartmentalization of zinc for specialized functions. Biol Chem 393(7):565–578

Hershfinkel M (2018) The zinc sensing receptor, ZnR/GPR39, in health and disease. Int J Mol Sci 19(2):439

Hershfinkel M, Moran A, Grossman N, Sekler I (2001) A zinc-sensing receptor triggers the release of intracellular Ca^{2+} and regulates ion transport. Proc Natl Acad Sci U S A 98(20):11749–11754

Hoque KM, Binder HJ (2006) Zinc in the treatment of acute diarrhea: current status and assessment. Gastroenterology 130(7):2201–2205

Hu H, Bandell M, Petrus MJ, Zhu MX, Patapoutian A (2009) Zinc activates damage-sensing TRPA1 ion channels. Nat Chem Biol 5(3):183–190

Huang L, Kirschke CP, Zhang Y, Yu YY (2005) The ZIP7 gene (Slc39a7) encodes a zinc transporter involved in zinc homeostasis of the Golgi apparatus. J Biol Chem 280(15):15456–15463

Inoue K, Branigan D, Xiong ZG (2010) Zinc-induced neurotoxicity mediated by transient receptor potential melastatin 7 channels. J Biol Chem 285(10):7430–7439

Ishihara K, Yamazaki T, Ishida Y, Suzuki T, Oda K, Nagao M, Yamaguchi-Iwai Y, Kambe T (2006) Zinc transport complexes contribute to the homeostatic maintenance of secretory pathway function in vertebrate cells. J Biol Chem 281(26):17743–17750

Ishii K, Sato M, Akita M, Tomita H (1999) Localization of zinc in the rat submandibular gland and the effect of its deficiency on salivary secretion. Ann Otol Rhinol Laryngol 108(3):300–308

Jiang LH (2009) Inhibition of P2X(7) receptors by divalent cations: old action and new insight. Eur Biophys J 38(3):339–346

Jou MY, Hall AG, Philipps AF, Kelleher SL, Lonnerdal B (2009) Tissue-specific alterations in zinc transporter expression in intestine and liver reflect a threshold for homeostatic compensation during dietary zinc deficiency in weanling rats. J Nutr 139(5):835–841

Kelly P, Feakins R, Domizio P, Murphy J, Bevins C, Wilson J, McPhail G, Poulsom R, Dhaliwal W (2004) Paneth cell granule depletion in the human small intestine under infective and nutritional stress. Clin Exp Immunol 135(2):303–309

Khan FR, McGeer JC (2013) Zn-stimulated mucus secretion in the rainbow trout (Oncorhynchus mykiss) intestine inhibits Cd accumulation and Cd-induced lipid peroxidation. Aquat Toxicol 142-143:17–25

Kim YS, Ho SB (2010) Intestinal goblet cells and mucins in health and disease: recent insights and progress. Curr Gastroenterol Rep 12(5):319–330

Kim JJ, Khan WI (2013) Goblet cells and mucins: role in innate defense in enteric infections. Pathogens 2(1):55–70

Kiselyov K, Colletti GA, Terwilliger A, Ketchum K, Lyons CW, Quinn J, Muallem S (2011) TRPML: transporters of metals in lysosomes essential for cell survival? Cell Calcium 50(3):288–294. https://doi.org/10.1016/j.ceca.2011.04.009

Kristiansen LH, Rungby J, Sondergaard LG, Stoltenberg M, Danscher G (2001) Autometallography allows ultrastructural monitoring of zinc in the endocrine pancreas. Histochem Cell Biol 115(2):125–129

Kukic I, Lee JK, Coblentz J, Kelleher SL, Kiselyov K (2013) Zinc-dependent lysosomal enlargement in TRPML1-deficient cells involves MTF-1 transcription factor and ZnT4 (Slc30a4) transporter. Biochem J 451(2):155–163. https://doi.org/10.1042/BJ20121506

Langmade SJ, Ravindra R, Daniels PJ, Andrews GK (2000) The transcription factor MTF-1 mediates metal regulation of the mouse ZnT1 gene. J Biol Chem 275(44):34803–34809

Lazzerini M, Wanzira H (2016) Oral zinc for treating diarrhoea in children. Cochrane Database Syst Rev (12, 12):CD005436

Levaot N, Hershfinkel M (2018) How cellular Zn($^{2+}$) signaling drives physiological functions. Cell Calcium 75:53–63

Lindenmayer GW, Stoltzfus RJ, Prendergast AJ (2014) Interactions between zinc deficiency and environmental enteropathy in developing countries. Adv Nutr 5(1):1–6

Luk HH, Ko JK, Fung HS, Cho CH (2002) Delineation of the protective action of zinc sulfate on ulcerative colitis in rats. Eur J Pharmacol 443(1–3):197–204

Maares M, Keil C, Koza J, Straubing S, Schwerdtle T, Haase H (2018) In vitro studies on zinc binding and buffering by intestinal mucins. Int J Mol Sci 19(9)

Malavia D, Crawford A, Wilson D (2017) Nutritional immunity and fungal pathogenesis: the struggle for micronutrients at the host-pathogen Interface. Adv Microb Physiol 70:85–103

Maret W, Sandstead HH (2006) Zinc requirements and the risks and benefits of zinc supplementation. J Trace Elem Med Biol 20(1):3–18

McCormick NH, Lee S, Hennigar SR, Kelleher SL (2016) ZnT4 (SLC30A4)-null ("lethal milk") mice have defects in mammary gland secretion and hallmarks of precocious involution during lactation. Am J Physiol Regul Integr Comp Physiol 310(1):R33–R40

Medani M, Bzik VA, Rogers A, Collins D, Kennelly R, Winter DC, Brayden DJ, Baird AW (2012) Zinc sulphate attenuates chloride secretion in human colonic mucosae in vitro. Eur J Pharmacol 696(1–3):166–171

Mendez-Barredo LH, Rodriguez-Melendez JG, Gomez-Coronado KS, Guerrero-Alba R, Valdez-Morales EE, Espinosa-Luna R, Barajas-Espinosa A, Barajas-Lopez C (2018) Physiological concentrations of zinc have dual effects on P2X myenteric receptors of Guinea pig. Cell Mol Neurobiol 38(7):1439–1449

Mittermeier L, Demirkhanyan L, Stadlbauer B, Breit A, Recordati C, Hilgendorff A, Matsushita M, Braun A, Simmons DG, Zakharian E, Gudermann T, Chubanov V (2019) TRPM7 is the central gatekeeper of intestinal mineral absorption essential for postnatal survival. Proc Natl Acad Sci U S A 15 (1810633116):1810633116

Mnatsakanyan H, Serra RSI, Rico P, Salmeron-Sanchez M (2018) Zinc uptake promotes myoblast differentiation via Zip7 transporter and activation of Akt signalling transduction pathway. Sci Rep 8(1):13642

Moechars D, Depoortere I, Moreaux B, de Smet B, Goris I, Hoskens L, Daneels G, Kass S, Ver Donck L, Peeters T, Coulie B (2006) Altered gastrointestinal and metabolic function in the GPR39-obestatin receptor-knockout mouse. Gastroenterology 131(4):1131–1141

Moran JR, Lewis JC (1985) The effects of severe zinc deficiency on intestinal permeability: an ultrastructural study. Pediatr Res 19(9):968–973

Murgia C, Vespignani I, Cerase J, Nobili F, Perozzi G (1999) Cloning, expression, and vesicular localization of zinc transporter Dri 27/ZnT4 in intestinal tissue and cells. Am J Phys 277(6):G1231–G1239

Murgia C, Vespignani I, Rami R, Perozzi G (2006) The Znt4 mutation inlethal milk mice affects intestinal zinc homeostasis through the expression of other Zn transporters. Genes Nutr 1(1):61–70

Myers S, Shastri MD, Adulcikas J, Sohal SS, Norouzi S (2017) Zinc and gastrointestinal disorders: a role for the zinc transporters Zips and ZnTs. Curr Pharm Des 23(16):2328–2332. https://doi.org/10.2174/1381612823666170124115850

Nistor N, Ciontu L, Frasinariu OE, Lupu VV, Ignat A, Streanga V (2016) Acrodermatitis Enteropathica: a case report. Medicine (Baltimore) 95(20):e3553

Ohana E, Hoch E, Keasar C, Kambe T, Yifrach O, Hershfinkel M, Sekler I (2009) Identification of the Zn^{2+} binding site and mode of operation of a mammalian Zn^{2+} transporter. J Biol Chem 284(26):17677–17686

Ohashi W, Kimura S, Iwanaga T, Furusawa Y, Irie T, Izumi H, Watanabe T, Hijikata A, Hara T, Ohara O, Koseki H, Sato T, Robine S, Mori H, Hattori Y, Watarai H, Mishima K, Ohno H, Hase K, Fukada T (2016) Zinc transporter SLC39A7/ZIP7 promotes intestinal epithelial self-renewal by resolving ER stress. PLoS Genet 12(10):e1006349

Opoka W, Adamek D, Plonka M, Reczynski W, Bas B, Drozdowicz D, Jagielski P, Sliwowski Z, Adamski P, Brzozowski T (2010) Importance of luminal and mucosal zinc in the mechanism of experimental gastric ulcer healing. J Physiol Pharmacol 61(5):581–591

Overeem AW, Posovszky C, Rings EH, Giepmans BN, van ISC (2016) The role of enterocyte defects in the pathogenesis of congenital diarrheal disorders. Dis Model Mech 9(1):1–12

Pelaseyed T, Bergstrom JH, Gustafsson JK, Ermund A, Birchenough GM, Schutte A, van der Post S, Svensson F, Rodriguez-Pineiro AM, Nystrom EE, Wising C, Johansson ME, Hansson GC (2014) The mucus and mucins of the goblet cells and enterocytes provide the first defense line of the gastrointestinal tract and interact with the immune system. Immunol Rev 260(1):8–20

Podany AB, Wright J, Lamendella R, Soybel DI, Kelleher SL (2016) ZnT2-mediated zinc import into Paneth cell granules is necessary for coordinated secretion and Paneth cell function in mice. Cell Mol Gastroenterol Hepatol 2(3):369–383

Podany A, Rauchut J, Wu T, Kawasawa YI, Wright J, Lamendella R, Soybel DI, Kelleher SL (2019) Excess dietary zinc intake in neonatal mice causes oxidative stress and alters intestinal host-microbe interactions. Mol Nutr Food Res 63(3):e1800947

Potten CS, Loeffler M (1990) Stem cells: attributes, cycles, spirals, pitfalls and uncertainties. Lessons for and from the crypt. Development 110(4):1001–1020

Rizk P, Barker N (2012) Gut stem cells in tissue renewal and disease: methods, markers, and myths. Wiley Interdiscip Rev Syst Biol Med 4(5):475–496

Sarkar P, Saha T, Sheikh IA, Chakraborty S, Aoun J, Chakrabarti MK, Rajendran VM, Ameen NA, Dutta S, Hoque KM (2019) Zinc ameliorates intestinal barrier dysfunctions in shigellosis by reinstating claudin-2 and -4 on the membranes. Am J Physiol Gastrointest Liver Physiol 316(2):G229–G246

Sazawal S, Black RE, Bhan MK, Bhandari N, Sinha A, Jalla S (1995) Zinc supplementation in young children with acute diarrhea in India. N Engl J Med 333(13):839–844

Scharlau D, Borowicki A, Habermann N, Hofmann T, Klenow S, Miene C, Munjal U, Stein K, Glei M (2009) Mechanisms of primary cancer prevention by butyrate and other products formed during gut flora-mediated fermentation of dietary fibre. Mutat Res 682(1):39–53

Scrimgeour AG, Lukaski HC (2008) Zinc and diarrheal disease: current status and future perspectives. Curr Opin Clin Nutr Metab Care 11(6):711–717

Shah M, Zaneb H, Masood S, Khan RU, Ashraf S, Sikandar A, Rehman HFU, Rehman HU (2018) Effect of dietary supplementation of zinc and multi-microbe probiotic on growth traits and alteration of intestinal architecture in broiler. Probiotics Antimicrob Proteins 22(10):018–9424

Shank JM, Kelley BR, Jackson JW, Tweedie JL, Franklin D, Damo SM, Gaddy JA, Murphy CN, Johnson JG (2018) The host antimicrobial protein Calgranulin C participates in the control of campylobacter jejuni growth via zinc sequestration. Infect Immun 86(6):00234–00218

Sharir H, Zinger A, Nevo A, Sekler I, Hershfinkel M (2010) Zinc released from injured cells is acting via the Zn^{2+}–sensing receptor, ZnR, to trigger signaling leading to epithelial repair. J Biol Chem 285(34):26097–26106

Singh V, Raheja G, Borthakur A, Kumar A, Gill RK, Alakkam A, Malakooti J, Dudeja PK (2012) Lactobacillus acidophilus upregulates intestinal NHE3 expression and function. Am J Physiol Gastrointest Liver Physiol 303(12):G1393–G1401

Sondergaard LG, Stoltenberg M, Flyvbjerg A, Brock B, Schmitz O, Danscher G, Rungby J (2003) Zinc ions in beta-cells of obese, insulin-resistant, and type 2 diabetic rats traced by autometallography. APMIS 111(12):1147–1154

Stappenbeck TS (2009) Paneth cell development, differentiation, and function: new molecular cues. Gastroenterology 137(1):30–33

Sturniolo GC, Di Leo V, Ferronato A, D'Odorico A, D'Inca R (2001) Zinc supplementation tightens "leaky gut" in Crohn's disease. Inflamm Bowel Dis 7(2):94–98

Sunuwar L, Medini M, Cohen L, Sekler I, Hershfinkel M (2016) The zinc sensing receptor, ZnR/GPR39, triggers metabotropic calcium signalling in colonocytes and regulates occludin recovery in experimental colitis. Philos Trans R Soc Lond Ser B Biol Sci 371(1700)

Sunuwar L, Asraf H, Donowitz M, Sekler I, Hershfinkel M (2017a) The Zn(2+)-sensing receptor, ZnR/GPR39, upregulates colonocytic Cl(−) absorption, via basolateral KCC1, and reduces fluid loss. Biochim Biophys Acta Mol basis Dis 1863(4):947–960

Sunuwar L, Gilad D, Hershfinkel M (2017b) The zinc sensing receptor, ZnR/GPR39, in health and disease. Front Biosci (Landmark Ed) 22:1469–1492

Suzuki T, Ishihara K, Migaki H, Ishihara K, Nagao M, Yamaguchi-Iwai Y, Kambe T (2005a) Two different zinc transport complexes of cation diffusion facilitator proteins localized in the secretory pathway operate to activate alkaline phosphatases in vertebrate cells. J Biol Chem 280(35):30956–30962

Suzuki T, Ishihara K, Migaki H, Matsuura W, Kohda A, Okumura K, Nagao M, Yamaguchi-Iwai Y, Kambe T (2005b) Zinc transporters, ZnT5 and ZnT7, are required for the activation of alkaline phosphatases, zinc-requiring enzymes that are glycosylphosphatidylinositol-anchored to the cytoplasmic membrane. J Biol Chem 280(1):637–643

Taylor KM, Hiscox S, Nicholson RI, Hogstrand C, Kille P (2012) Protein kinase CK2 triggers cytosolic zinc signaling pathways by phosphorylation of zinc channel ZIP7. Sci Signal 5(210):ra11

Wagner TF, Drews A, Loch S, Mohr F, Philipp SE, Lambert S, Oberwinkler J (2010) TRPM3 channels provide a regulated influx pathway for zinc in pancreatic beta cells. Pflugers Arch 460(4):755–765

Walker CL, Black RE (2010) Zinc for the treatment of diarrhoea: effect on diarrhoea morbidity, mortality and incidence of future episodes. Int J Epidemiol 39(Suppl 1 (1)):i63–i69

Wang X, Zhou B (2010) Dietary zinc absorption: a play of Zips and ZnTs in the gut. IUBMB Life 62(3):176–182. https://doi.org/10.1002/iub.291. [doi]

Wang K, Zhou B, Kuo YM, Zemansky J, Gitschier J (2002) A novel member of a zinc transporter family is defective in acrodermatitis enteropathica. Am J Hum Genet 71(1):66–73

Wang X, Valenzano MC, Mercado JM, Zurbach EP, Mullin JM (2013) Zinc supplementation modifies tight junctions and alters barrier function of CACO-2 human intestinal epithelial layers. Dig Dis Sci 58(1):77–87

Wang C, Zhang L, Ying Z, He J, Zhou L, Zhang L, Zhong X, Wang T (2018) Effects of dietary zinc oxide nanoparticles on growth, diarrhea, mineral deposition, intestinal morphology, and barrier of weaned piglets. Biol Trace Elem Res 185(2):364–374

Watson AJ, Duckworth CA, Guan Y, Montrose MH (2009) Mechanisms of epithelial cell shedding in the mammalian intestine and maintenance of barrier function. Ann N Y Acad Sci 1165:135–142

Yu YY, Kirschke CP, Huang L (2007) Immunohistochemical analysis of ZnT1, 4, 5, 6, and 7 in the mouse gastrointestinal tract. J Histochem Cytochem 55(3):223–234

Yu DC, Waby JS, Chirakkal H, Staton CA, Corfe BM (2010) Butyrate suppresses expression of neuropilin I in colorectal cell lines through inhibition of Sp1 transactivation. Mol Cancer 9(276):276

Zachos NC, Kovbasnjuk O, Donowitz M (2009) Regulation of intestinal electroneutral sodium absorption and the brush border Na+/H+ exchanger by intracellular calcium. Ann N Y Acad Sci 1165:240–248

Chapter 18
In Situ Imaging of Zinc with Synthetic Fluorescent Probes

Jiyao Yu and Christoph J. Fahrni

Abstract Small-molecule synthetic fluorescent probes represent powerful tools for interrogating labile zinc pools in biology. Transported into cells through passive diffusion, they are particularly well suited for visualizing dynamic changes of zinc in living systems. When employing fluorescent probes for visualizing biological zinc, there are important limitations that differentiate this approach from quantitative microanalytical techniques. Notably, fluorescent probes engage in a competitive exchange equilibrium with endogenous ligands and proteins and thus detect changes of labile Zn(II) pools rather than total cellular zinc levels. Focusing on design approaches and photophysical concepts, this chapter offers an overview of the most widely employed fluorogenic and ratiometric probes for the detection of Zn(II) in a biological environment, discusses concepts relevant for the design and application of Zn(II)-responsive probes for two-photon excitation microscopy, and outlines current challenges and limitations of their application in biological systems.

Keywords Exchangeable zinc · Fluorescence imaging · Fluorogenic probes · Emission-ratiometric probes · Two-photon excitation microscopy

18.1 Introduction

Cellular zinc levels are regulated through a complex network of subcellular storage sites such as metallothioneins and membrane-localized transporters that control influx and efflux pathways (Kambe et al. 2015; Hara et al. 2017). While a significant fraction of cellular zinc is tightly bound to proteins where it serves as a catalytic

J. Yu · C. J. Fahrni (✉)
School of Chemistry and Biochemistry, Petit Institute for Bioengineering and Bioscience, Georgia Institute of Technology, Atlanta, GA, USA
e-mail: fahrni@chemistry.gatech.edu

T. Fukada, T. Kambe (eds.), *Zinc Signaling*,
https://doi.org/10.1007/978-981-15-0557-7_18

cofactor or structural support, there is increasing evidence that cells also maintain a labile pool of zinc that might be involved in signaling processes (Khan et al. 2014; Vergnano et al. 2014). Synthetic fluorescent probes offer unique opportunities to explore the nature of this exchangeable zinc pool and to advance our understanding of zinc-dependent signaling and regulatory processes in biology (Carter et al. 2014; Fahrni et al. 2017). Due to their lipophilicity and low molecular weight, synthetic probes can passively diffuse across biological membranes to reach cellular targets in a noninvasive manner, thus enabling the visualization of dynamic changes in live cells. A number of Zn(II)-selective fluorescent probes are commercially available and can be employed with conventional fluorescence microscopes available in most cell biology laboratories. As no additional equipment is required, the detection of Zn(II) with fluorescent probes is cost-effective and often the method of choice for assessing cellular zinc levels. However, fluorescent probes do not report on total cellular zinc concentrations, rather, they reflect on the fraction of labile or exchangeable Zn(II) that is buffered by an ensemble of endogenous proteins and bioligands (Maret 2015). Because fluorescent probes inherently engage in a competitive metal exchange equilibrium, the observed response is dependent on the probe interaction, notably the relative affinities and exchange kinetics. For this reason, fluorescence imaging with Zn(II)-selective probes cannot substitute for quantitative imaging techniques such as x-ray fluorescence microscopy or secondary ion mass spectrometry imaging (McRae et al. 2009). Instead, fluorescent probes offer complementary insights into the dynamics of labile Zn(II).

18.2 Fluorogenic Probes

18.2.1 Design Principles and Applications

At a molecular level, fluorogenic probes are composed of an organic fluorophore tethered to a chelator, which triggers a fluorescence increase upon metal ion binding. The photophysical mechanisms that couple the fluorescence output with the metal binding event are diverse but typically involve controlling the photoinduced movement of electrons or protons in a metal ion-dependent fashion. For example, 8-hydroxy-quinoline and 6-methoxy-8-*p*-toluenesulfonamido-quinoline (TSQ), two of the earliest renditions for probing biological zinc, employ a heterocyclic quinoline ring, which serves both as metal coordination site and fluorophore moiety. The donor attached in the 8-position turns the quinoline ring into a bidentate chelator, which forms with Zn(II) a complex with 2:1 stoichiometry (Fig. 18.1) (Fahrni and O'Halloran 1999; Nasir et al. 1999). While the Zn(II) complex is brightly fluorescent, a non-radiative excited state intramolecular proton transfer pathway is likely responsible for quenching of the quinoline emission in the free probe (Lan and Liu 2015). The underpinning photophysical mechanism is best illustrated with a simplified Jablonski diagram as shown in Fig. 18.1b. Upon photoexcitation, the

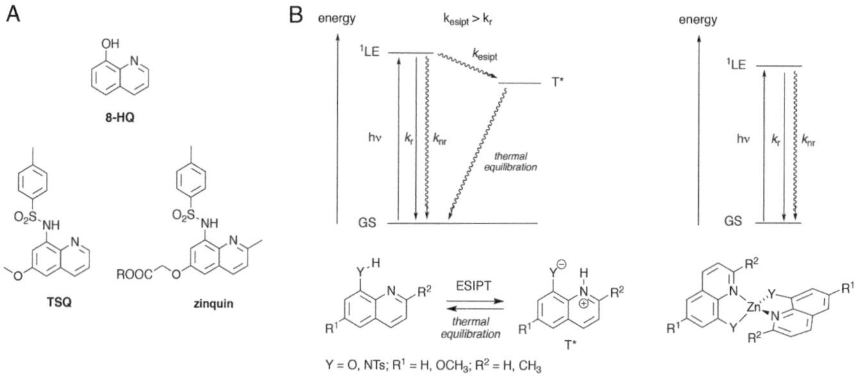

Fig. 18.1 Quinoline-based Zn(II)-selective fluorescent probes. (**a**) Molecular structures of 8-hydroxy-quinoline (8HQ), 6-methoxy-8-*p*-toluenesulfonamido-quinoline (TSQ), and zinquin. (**b**) Simplified Jablonski diagram illustrating the photophysical mechanism for Zn(II)-induced emission enhancement. Excitation of the quinoline fluorophore promotes an exited state intramolecular proton transfer (ESIPT) to yield the zwitterionic tautomer T∗, which thermally equilibrates back to the ground state (GS). In the presence of Zn(II), the bidentate quinoline ligand forms a tetrahedral complex, where ESIPT quenching is no longer possible and the fluorescence is switched on

basicity of the quinoline nitrogen increases concomitantly with the acidity of the neighboring hydroxy or sulfonamide group, thus promoting a proton transfer to yield a zwitterionic intermediate, which then thermally equilibrates back to the original ground state protomer without photon emission. In the complexed form, Zn(II) occupies both the proton donor and acceptor sites, thus inhibiting the original proton transfer quenching pathway and restoring fluorescence of the quinoline fluorophore. Because the quinoline heterocycle acts as a strong photobase in its own right (Hunt and Dawlaty 2018), the fluorescence remains quenched even under basic conditions where the hydroxy or sulfonamide groups are deprotonated. Although other divalent transition metal ions such as Mn(II), Fe(II), or Co(II) can form analogous complexes, their open d-shell electron configuration results in efficient fluorescence quenching.

The selectivity of TSQ for the detection of biological zinc within the chemically complex environment of tissues was corroborated with the neo-Timm stain (Frederickson et al. 1987; Frederickson 1989), a histochemical method based on autometallographic enhancement of zinc selenide nanoparticles formed from histochemically reactive zinc upon perfusion with sodium selenide (Danscher 1981). Such correlative imaging studies were not only critical to validate the observed fluorescence staining pattern but also to promote overall confidence when employing synthetic probes in other biological systems. To increase retention in live cells, Zalewski et al. developed the quinoline derivative zinquin carrying an ester group (Zalewski et al. 1993; Mahadevan et al. 1996). Upon cleavage by cellular esterases, the ester group is converted into a negatively charged carboxylate anion with has a much-decreased membrane permeability.

The early success of quinoline-based probes for the in situ detection of biological zinc stimulated further developments, which evolved into a vibrant area of research that has produced a large body of literature as summarized in several comprehensive reviews (Carter et al. 2014; Jiang and Guo 2004; Kikuchi et al. 2004; Lim et al. 2004; Nolan and Lippard 2009; Xu et al. 2010; Carter et al. 2014; Chen et al. 2015; Chabosseau et al. 2018; Sfrazzetto et al. 2016). While the probes reported thus far cover a broad range of binding affinities and exhibit diverse photophysical characteristics, both in terms of fluorescence contrast and available emission wavelengths, the literature on their biological applications remains comparatively sparse. Because most probes require elaborate multistep synthetic procedures, the majority of biological studies have been conducted with reagents that are available from commercial sources. With these limitations in mind, the following discussion focuses on general guidelines for the application of these reagents rather than reviewing the extensive literature on Zn(II)-responsive fluorogenic probes that are less readily accessible. As a newer development, the design and application of fluorescent probes for two-photon excitation microscopy (TPEM) will also be reviewed, including some of the latest development concerning ratiometric Zn(II)-selective fluorescent probes.

The majority of commercially available Zn(II)-selective fluorescent probes respond with an increase in fluorescence intensity upon saturation with Zn(II) (Table 18.1). With the exception of TSQ and zinquin, all probes employ a xanthene derivative as fluorophore moiety, which is linked to various chelators acting as Zn(II) recognition sites (Fig. 18.2). In contrast to the quinoline-based probes discussed above, the Zn(II)-induced fluorescence switching of xanthene derivatives relies on a photoinduced electron transfer (PET) mechanism rather than excited state proton transfer (Fig. 18.3). Specifically, excitation of the fluorophore promotes an electron into an empty antibonding orbital, and the resulting vacancy in the bonding orbital is filled by accepting an electron from the tethered chelator acting as an electron-rich donor. The resulting radical ion pair undergoes charge recombination to furnish the original ground state, a process that usually occurs through thermal equilibration without emission of a photon. Because Zn(II) coordination lowers the energy of the electron donor, formation of the radical ion pair becomes less favorable, and the driving force for the electron transfer is diminished. As a consequence, radiative deactivation can now effectively compete with PET, which would otherwise quench the fluorescence. Thus, the observed Zn(II)-induced fluorescence enhancement can be rationalized based on the efficiency of PET quenching in the free probe and its inhibition upon Zn(II) binding.

As evident from Table 18.1, different combinations of chelator and fluorophore moieties produce quite different fluorescence contrasts, which directly affect the limit of detection and overall sensitivity. In addition to the magnitude of fluorescence enhancement upon saturation with Zn(II), the limit of detection also depends on the molar brightness, which corresponds to the product of the absorption cross section (molar extinction coefficient) at the wavelength of excitation and the quantum yield of the Zn(II)-bound probe. Probes with higher fluorescence contrast and brightness can be employed at a lower concentration and thus are capable of detect-

Table 18.1 Photophysical and thermodynamic properties of commercially available Zn(II)-selective fluorogenic probes

	Free probe			Zn(II)-bound probe							
Probe	λ_{abs} [nm]	λ_{em} [nm]	Φ_F [a]	λ_{abs} [nm]	ε [b] [10^4 $M^{-1}cm^{-1}$]	λ_{em} [nm]	Φ_F [a]	log*K* [c]	Φ ε [d]	f_e [e]	References
TSQ	369	535	0.002	369	0.61	535	0.34	7.7	0.2	170	Pearce et al. (2001), Nowakowski et al. (2015), and Radford et al. (2013)
Zinquin acid	336	n.d.	n.d.	361	0.79	484	0.13	13.5[f]	0.1	n.d.	Fahrni and O'Halloran (1999), Zalewski et al. (1993), and Mahadevan et al. (1996)
FluoZin-1	496	515	0.005	496	n.d.	515	n.d.	5.1	n.d.	200	Royzen et al. (2005), and Gee et al. (2002a)
FluoZin-2	495	495	n.d.	495	n.d.	525	n.d.	5.7	n.d.	12	Gee et al. (2002a)
FluoZin-3	491	517	0.005	495	7.1	516	0.43	7.8	3.1	86	Gee et al. (2002b), and Marszalek et al. (2016)
RhodZin-3	550	575	n.d.	550	n.d.	575	n.d.	7.2	n.d.	75	Gee et al. (2002a) and Sensi et al. (2003)
ZnAF-1	492	514	0.022	492	7.4	514	0.23	9.1	1.7	10	Hirano et al. (2000)
ZnAF-2	490	514	0.023	492	7.8	514	0.36	8.6	2.8	16	Hirano et al. (2000)
ZnAF-1F	489	514	0.004	492	7.7	514	0.17	8.7	1.3	43	Hirano et al. (2002)
ZnAF-2F	490	514	0.006	492	7.4	514	0.24	8.3	1.8	40	Hirano et al. (2002))
Zinpyr-1	515	529	0.39	507	8.0	529	0.87	9.2	7.0	1.8	Walkup et al. (2000)
Newport Green DCF	505	535	n.d.	505	8.2	535	n.d.	6.0	n.d.	5	Carter et al. (2014), and Iain Johnson (2010)
Phen Green SK	507	532	n.d.	507	8.6	532	n.d.	n.d.	n.d.	n.d.	Carter et al. (2014) and Iain Johnson (2010)

[a]Fluorescence quantum yield
[b]Molar absorption coefficient at absorption maximum (λ_{abs})
[c]Apparent complex stability constant of the probe-Zn(II) complex
[d]Molar brightness (product of molar absorption coefficient and quantum yield, unit [10^4 $M^{-1}cm^{-1}$])
[e]Fluorescence enhancement factor upon Zn(II) saturation [f]logβ_2 at pH 7.20

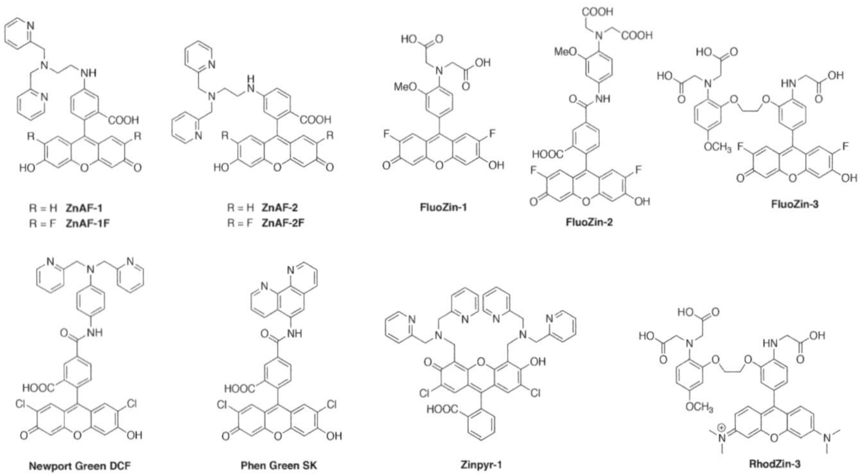

Fig. 18.2 Molecular structures of commercially available Zn(II)-selective fluorogenic probes

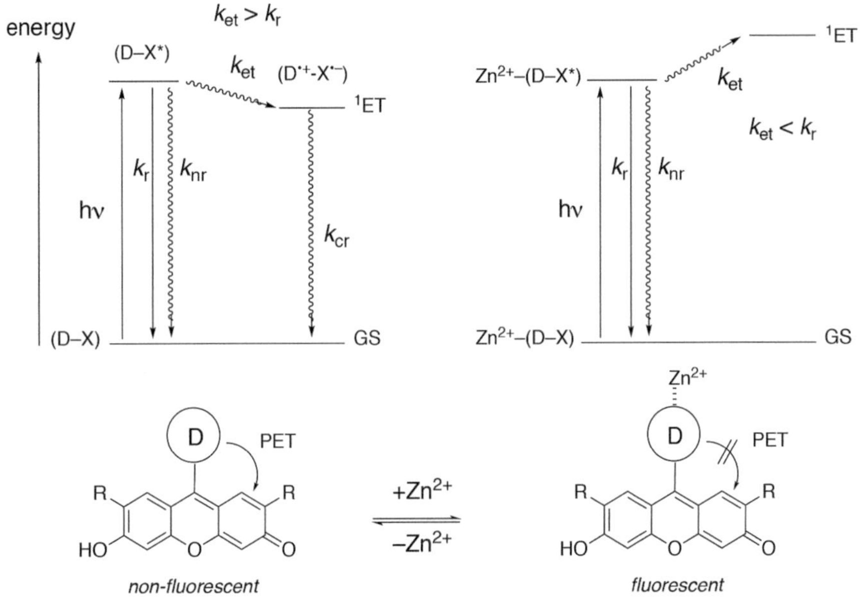

Fig. 18.3 Principle of photoinduced electron transfer (PET) fluorescence switching in xanthene-based Zn(II)-responsive fluorescent probes. Photoexcitation of the xanthene moiety promotes an electron transfer from the nearby donor (D) to yield a radical ion pair ($D^{\bullet+}$-$X^{\bullet-}$), which undergoes non-radiative charge recombination back to the ground state (GS). Because the electron transfer rate constant (k_{et}) is faster than that of radiative deactivation (k_r), the fluorescence is quenched. Upon metal binding to the donor, the electron transfer is rendered less favorable, and radiative deactivation is now the preferred deactivation pathway

ing lower Zn(II) concentrations. Even for the brightest probes, fluorescence intensity measurements with conventional fluorimeters still require probe concentrations in the high nanomolar range to achieve a robust signal-to-noise ratio.

The Zn(II)-selective response of fluorogenic probes can be harnessed to determine the total Zn(II) concentration in analytical samples by fluorimetry. If the Zn(II) dissociation constants of the probe is significantly lower than the total probe concentration, the fluorescence intensity increases with a steady slope upon addition of Zn(II) up to the saturation point. Given this linear relationship, a simple calibration curve can be established and utilized to determine the total Zn(II) concentration of an unknown sample by fluorimetry, provided no other ligands are present that bind Zn(II) with similar or higher affinity than the probe. To exclude such interference, the sample is therefore best digested by boiling with concentrated acid followed by neutralization with base. Following this approach, fluorescent probes are well suited to measure total Zn(II) concentrations in a wide range of samples, including bulk cells or tissues, thus offering a low-cost alternative to microanalytical methods that require expensive instruments such as inductively coupled plasma mass spectrometry (ICP-MS), atomic emission spectroscopy (AES), or total reflection x-ray fluorescence elemental analysis (TXRF) (McRae et al. 2009).

18.2.2 Visualizing Zinc in Cells and Tissues

Due to their lipophilicity and small molecular size, many fluorescent probes can diffuse across lipid bilayers to reach subcellular compartments and organelles in a noninvasive fashion. Although hydrophilic charged probes are usually not membrane-permeant, they are useful for detecting extracellular analytes as demonstrated, for example, with the dynamic release of Zn(II) ions from pancreatic β-cells (Gee et al. 2002b). To enable cellular uptake and increased retention, fluorescent probes containing carboxylic acids can be masked as ester derivatives, an approach that was pioneered by Tsien and coworkers in the early 1980s (Tsien 1981). Upon cellular uptake, the ester group is cleaved by nonspecific cytosolic esterases to release the negatively charged carboxylate anion, thus rendering the probe membrane impermeant and increasing its overall cellular retention. Due to more favorable hydrolysis kinetics, acetoxymethyl (AM) esters are preferred over methyl or ethyl ester derivatives, and most of the carboxylic acid-containing probes listed in Table 18.1 are also commercially available as AM ester derivatives. When using AM esters for live cell imaging studies, the cell culture medium can be supplemented with millimolar concentrations of pyruvate to counteract potential toxicity of formaldehyde, which is a by-product of the AM ester hydrolysis (Tsien and Pozzan 1989). While AM ester derivatives have been widely used for cellular in vitro studies, high levels of extracellular esterases in the vascular space may cleave the probe prior to cellular uptake, thus rendering AM derivatives ineffective for monitoring intracellular processes in vivo (Jobsis et al. 2007). Altogether, the ability to mask charged probes and deliver them into the cytosol of live cells permitted for the first time the noninvasive observation of various metal ions, a key development that revolutionized cell biology research.

Analogous to the fluorimetric analysis of bulk samples, the intracellular concentration of a fluorescent probe must be sufficiently high to achieve a robust signal-to-noise ratio for fluorescence imaging. The following consideration provides a rough estimate for the lower limit of the cellular fluorophore concentration. Assuming an average focal volume of around 1 fL for a confocal imaging setup, a single probe molecule placed within this volume would yield a nominal concentration of 1.7 nM. Because the emission of a single fluorophore would be obscured by the autofluorescent background within a cellular environment, the probe must be present at significantly higher concentration, thus placing the recommended loading into the high nano- to low micromolar concentration window.

The first fluorescent probes for imaging metal ions in live cells were geared toward monitoring intracellular Ca(II) and Mg(II) fluxes, which occur at concentrations that are significantly higher than the typical cellular loading levels of the probe. Under these conditions, the analyte concentration is not limiting, and the fluorescence response toward increasing metal ion concentrations follows a sigmoidal binding isotherm centered around the probe K_d (Fig. 18.4). As these probes engage with the metal ion in a 1:1 binding equilibrium, the fractional saturation f, defined as the ratio of the metal-bound to total probe concentration, is a direct measure of the free metal ion concentration according to Eq. (18.1):

$$f = \frac{[M]}{K_d + [M]} \tag{18.1}$$

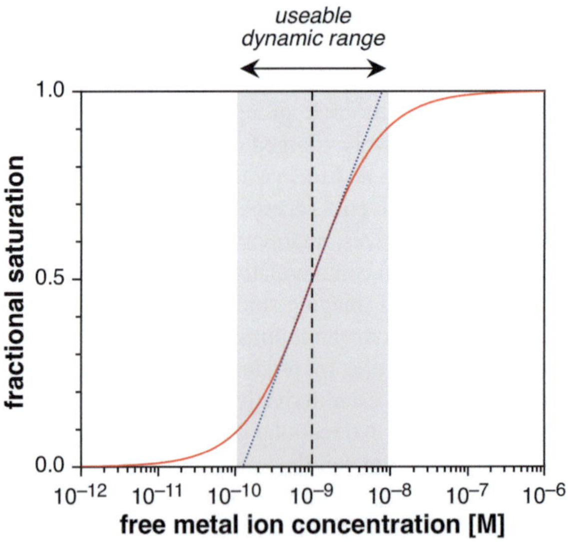

Fig. 18.4 Binding isotherm for a probe-metal complexation equilibrium with 1:1 stoichiometry assuming a complex dissociation constant (K_d) of 1 nM. The shaded area indicates the free metal ion concentration range where the fractional saturation of the probe resides between 10 and 90%. As illustrated with the dotted blue trace, a linear approximation of the probe response is only valid for metal ion concentrations close to the probe K_d

Conversely, the above relationship can be used to calculate the free metal ion concentration based on the fractional saturation of the probe after solving for [M]:

$$[\mathrm{M}] = \frac{K_d f}{f - 1} \tag{18.2}$$

Based on Eq. (18.1), it is apparent that the fractional saturation of the probe is 50% if the free metal ion concentration is equal to the K_d. Because the slope of the binding isotherm is the steepest within one logarithmic unit below and above the K_d, the dynamic range window of the probe is limited to approximately two decades centered around the K_d (shaded area in Fig. 18.4). Because the concentration of the probe-metal complex is significantly lower than the total metal concentration, the fractional probe saturation represents a direct measure of the cellular metal status.

The success of these early studies stimulated the development of fluorescence probes for other biologically relevant metal ions, including zinc (Maret 2015). However, there is an important difference that must be taken into consideration when employing Zn(II)-selective fluorescent probes. Although total cellular zinc levels may approach low millimolar concentrations, a significant portion of zinc remains buried within the binding pockets of proteins and is therefore not accessible, even for probes with much tighter binding affinities. Unlike Ca(II) and Mg(II), which rapidly equilibrate within the cytosol, only a fraction of the total cellular Zn(II) is exchangeable and can be detected with a fluorescent probe. Increasing evidence suggests that this labile, exchangeable zinc pool is buffered at high pico- to low nanomolar concentrations (Qin et al. 2011; Vinkenborg et al. 2009; Hessels and Merkx 2014), presumably involving an ensemble of proteins such as metallothioneins and low-molecular-weight bioligands (Colvin et al. 2010). Although the actual buffer depth, which corresponds to the total concentration of exchangeable Zn(II), is not known, it likely resides in the tens of micromolar range and thus should be sufficient to saturate a probe that is present at low micromolar concentrations. Provided the probe K_d matches the buffer window of the exchangeable Zn(II) pool, the fractional saturation can still be used as an indicator of the cellular zinc status as discussed above. However, it is important to note that there is no linear relationship between the probe response and *total* cellular Zn(II) concentration; instead, the probe only reports on changes in buffered levels. For this reason, in-cell analysis of the fluorescence response of a probe cannot be used as a substitute for measuring total Zn(II) levels by bulk analytical methods such as ICP-MS or AES.

The tight buffering of the exchangeable Zn(II) pool has another important implication when employing fluorescent probes. Because the buffered Zn(II) levels are several orders of magnitude lower than the probe concentration, the Zn(II) exchange kinetics would be exceedingly slow if the probe could only access free aquated Zn(II) ions. In view of the short experimental equilibration times, the exchange of Zn(II) between endogenous ligands and the probe likely occurs through an associate mechanism involving a transient ternary complex. Because the kinetics for formation of the ternary complex is expected to depend on both the electrostatic and steric nature of the probe and the coordination environment of the endogenous Zn(II) sites, it is conceivable that structurally different fluorescent probes might produce different results, even if their Zn(II)-binding affinities are identical. Such differences might be especially pronounced when acquiring data with protein-based genetically encoded vs. small synthetic fluorescent probes (Qin et al. 2013).

18.2.3 Limitations and Artifacts

When employing fluorogenic probes for monitoring cellular Zn(II) levels, it is important to keep in mind that the observed fluorescence signal is only an indirect measure that might be influenced by factors other than Zn(II). For example, changes in environmental polarity might trigger a fluorescence increase without binding of the target analyte. This is especially of concern for probes that employ a PET switching mechanism because the charge-separated radical intermediate is destabilized in a low polarity environment such as lipid bilayers (Fahrni et al. 2003). Even without polarity-induced fluorescence increases, accumulation of a fluorogenic probe within a compartment leads to a fluorescence enhancement due to the background fluorescence of the unbound probe. It is quite common that synthetic fluorescent probes exhibit an uneven cellular distribution. Depending on the molecular size and polarity, they often localize to specific subcellular structures and compartments. For example, the fluorescence staining pattern of zinquin in untreated HeLa cells appeared uniform with a mostly cytosolic distribution (Devergnas et al. 2004), while FluoZin-3 localized to the Golgi apparatus (Qin et al. 2013), and Zinpyr-1 was sequestered mostly into mitochondria (Lu et al. 2016). Thus, accumulation within subcellular compartments, partitioning into lipid bilayers, or association with lipid droplets might produce bright staining by the free probe alone, thus leading to potential imaging artifacts. To avoid misinterpretations, the reversibility of the fluorescence signal is best tested with a membrane-permeant high-affinity chelator such as TPEN. Ternary complex formation by association with proteins (Meeusen et al. 2012; Karim and Petering 2016) or low-molecular ligands (Staszewska et al. 2013; Marszalek et al. 2018) might present another source of potential artifacts. For example, detailed subcellular fractionation studies of cell lysate revealed that TSQ and zinquin (Fig. 18.1) form both ternary adducts with zinc proteins (Nowakowski et al. 2015). Rather to competitively chelate Zn(II) from proteins, the probes form ternary adducts, which in fact appear to be the major source of the observed fluorescence staining in cells.

18.3 Ratiometric Fluorescent Probes

Developed more than 30 years ago for the visualization of calcium ion fluxes (Grynkiewicz et al. 1985), ratiometric probes respond with a spectral shift rather than a large intensity increase upon binding of the analyte. Due to this spectral shift, the ratio of fluorescence intensities at two different wavelengths is different for the free and metal-bound probe but independent of the probe concentration, thus rendering ratiometric microscopy less prone to artifacts than intensity-based imaging with fluorogenic probes. Furthermore, the intensity ratio can be directly related to the actual buffered metal ion concentration [M] according to Eq. (18.3) (Tsien 1988):

$$[\mathrm{M}] = K_d \left(\frac{R - R_{\mathrm{min}}}{R_{\mathrm{max}} - R} \right) \left(\frac{S_{\mathrm{f}}}{S_{\mathrm{b}}} \right) \quad (18.3)$$

where K_{d} is the dissociation constant of the metal-bound indicator, R_{min} and R_{max} are the minimum and maximum intensity ratios for the free and metal-bound indicator, and S_{f} and S_{b} are instrument-dependent calibration constants. The instrument-dependent term $S_{\mathrm{f}}/S_{\mathrm{b}}$ becomes obsolete if one of the acquisition wavelengths is chosen at the crossover point of the calibration spectra. In this case, the limiting ratio values R_{min} and R_{max} can also be used to determine the fractional probe saturation f for a given intensity ratio R according to Eq. (18.4):

$$f = \frac{R - R_{\mathrm{min}}}{R_{\mathrm{max}} - R_{\mathrm{min}}} \quad (18.4)$$

where f is defined as the ratio of the metal-bound and total probe concentrations. The dissociation constant K_{d} is best determined from an independent calibration using fluorescence intensity data at a single wavelength. It is also important to note that Eq. (18.3) is only valid if the probe engages with the metal ion in a simple 1:1 complexation equilibrium, both in calibration media and within the chemically complex environment of cells.

As the ratiometric probe must be present at low micromolar concentrations to produce an adequate fluorescence signal over cellular background, it raises the question of how this approach can be used to detect Zn(II) at nanomolar levels and below. Instead of capturing free aquazinc(II) ions, the probe engages in a competitive exchange equilibrium with endogenous Zn(II) ligands, likely through an associative mechanism as described in Sect. 18.2. Thus, the probe affinity dictates to what degree Zn(II) can be chelated from this pool. Upon equilibration, the fractional saturation of the probe, which can be derived from the intensity ratio, reflects the thermodynamic availability of cellular Zn(II) based on the isotherm shown in Fig. 18.4. As long as the buffer depth, corresponding to the total concentration of exchangeable Zn(II), is significantly larger than the probe concentration, the ratiometric probe response is expected to accurately reflect the cellular zinc status. As the buffer depth might vary significantly with the cellular zinc status, ratiometric probes should be employed at the lowest possible concentration where a satisfactory signal over background ratio can be achieved. If the probe concentration exceeds the buffer capacity, the resulting fractional saturation no longer reflects the buffered levels. Such probe overloading is particularly of concern when using acetoxymethyl (AM) ester derivatives, which yield increased cellular retention but might produce higher than desired intracellular probe concentrations (Thompson et al. 2012).

Compared to fluorogenic probes, the body of literature on Zn(II)-responsive ratiometric probes is much smaller. An overview of various designs is shown in Fig. 18.5 and the corresponding photophysical properties are compiled in Table 18.2. FuraZin and IndoZin were both derived from the established Ca(II)-selective indicators fura-2 and indo-1 (Gee et al. 2002a). By truncating the original BAPTA ligand, the affinity for Ca(II) was significantly reduced; however, the Zn(II)-binding affinities are too weak for monitoring cytosolic Zn(II) levels, which are buffered at in the low nanomolar range (vide infra). Similarly, ZnAF-R2 was derived from the molecular framework of the Ca(II) probe fura-2 (Maruyama et al. 2002). Although the *N*,*N*-di(2-picolyl)ethylenediamine (DPEN) moiety increases the Zn(II) affinity by approximately tenfold over FuraZin, it still remains too low for accessing endogenous Zn(II) pools. In contrast, the Zn(II) affinity of Zinbo-5 is well matched for biological applications (Taki et al. 2004). The probe responds with a 36 nm red-shift of the emission as Zn(II) coordination displaces the phenolic proton. The coumarin-based probe ZnIC revealed a significant shift of the emission maximum upon saturation with Zn(II) (Komatsu et al. 2007). Employed in live HEK293 cells as well as rat hippocampal slices, the probe revealed dynamic changes of labile Zn(II) availability. The structurally related probe DPA-COUM-4 showed a Zn(II)-induced shift of the absorption and emission maximum; however, the probe was only character-

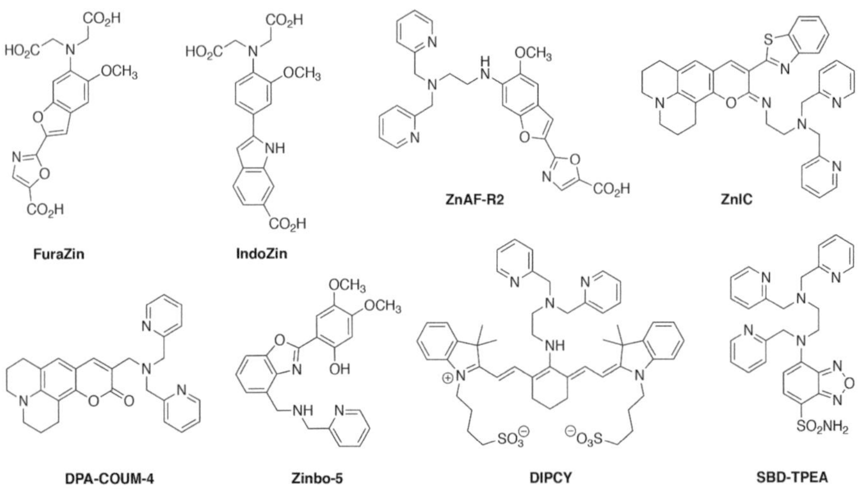

Fig. 18.5 Molecular structures of Zn(II)-selective ratiometric fluorescent probes

Table 18.2 Photophysical and thermodynamic properties of ratiometric Zn(II)-responsive fluorescent probes

	Free probe		Zn(II)-bound probe				
Probe	λ_{abs} [nm]	λ_{em} [nm]	λ_{abs} [nm]	λ_{em} [nm]	log$K$	R_{max}/R_{min}	References
FuraZin	378	510	339	510	5.7	9.0	Gee et al. (2002a)
IndoZin	350	480	350	395	5.5	n.d.	Gee et al. (2002a)
ZnAF-R2	365	495	335	495	6.6	7.0	Maruyama et al. (2002)
ZnIC	513	543	524	558	11.9	2.4	Komatsu et al. (2007)
DPA-COUM-4	400	484	431	505	6.3	n.d.	Lim and Bruckner (2004)
Zinbo-5	337	407	376	443	8.7	33	Taki et al. (2004)
DIPCY	627	758	671	765	7.6	1.5	Kiyose et al. (2006)
SBD-TPEA	456	585	386	545	8.7	1.7	Liu et al. (2014)

ized in methanol and not further tested within a biological system (Lim and Bruckner 2004). Specifically designed for NIR imaging, the cyanine derivative DIPCY responded with a strong shift of the maximum excitation wavelength upon saturation with Zn(II). With a K_d around 2.1 nM (log$K = 8.7$), the tris-picolylamine derivative SBD-TPEA offers an affinity that is well matched for biological applications as demonstrated for ratiometric imaging of labile Zn(II) in HepG2 cells as well as zebrafish embryos (Liu et al. 2014).

18.4 Fluorescent Probes for Two-Photon Excitation Microscopy

18.4.1 Design Principles and Applications

Originally developed by Webb and coworkers (Denk et al. 1990), two-photon excitation microscopy (TPEM) employs an ultrafast-pulsed laser as excitation source to promote simultaneous absorption of two photons by the fluorophore. Compared to conventional fluorescence imaging, only half of the photon energy is required to reach the same excited state manifold of the fluorophore, thus shifting the excitation window toward the near-infrared region. The shift to longer-wavelength excitation is particularly advantageous for intravital imaging as it results in increased penetration depth, decreased cellular autofluorescence, and reduced phototoxicity (Schießl and Castrop 2016). Because the two-photon absorption process scales with the square of the incident light intensity, the focal volume is significantly smaller compared to one-photon excitation. As a result, only fluorophore molecules that reside near the focal plane are excited, thus yielding intrinsic 3D imaging capabilities and enabling live cell imaging studies over extended periods by minimizing photobleaching (Schießl and Castrop 2016).

While most established fluorescent probes can be used for TPEM, their relative brightness might be significantly different. In analogy to one-photon excitation, the fluorophore brightness in TPEM depends on the quantum yield and absorption cross section; however, as the quantum mechanical selection rules that govern the two-photon absorption process are different from those for linear absorption, not all fluorophores are equally well-suited for TPEM. A compilation of the two-photon absorption properties for a selection of commonly employed fluorophores and fluorescent probes is shown in Table 18.3. An extensive list of two-photon probes and their 2PA maxima, molecular brightness, and emission ranges has been recently reported by Fiole and coworkers (Ricard et al. 2018). To better compare the one- and two-photon excited fluorescence properties across the different fluorophores, their relative brightness has been normalized to fluorescein as a reference. Among the tabulated fluorophores, rhodamine 6G provides the greatest detection sensitivity in both excitation modes. While the brightness of Alexa Fluor 488 and BODIPY 492 is similar compared to fluorescein, the two-photon excited fluorescence output is significantly attenuated due to the lower 2PA cross sections. In contrast, the cyanine dye Cy5.5 is 35% brighter compared to fluorescein in two-photon excitation mode. Genetically encoded fluorescent proteins such as EGFP (enhanced green fluorescent protein) are particularly attractive for labeling proteins in live cells and whole organisms. A wide variety of fluorescent proteins with emission maxima across the entire visible spectrum are available, and most of them offer a two-photon excited brightness comparable or better than fluorescein (Drobizhev et al. 2011). In general, the two-photon absorption cross section of fluorophores with absorption maxima in the UV range is significantly lower. For example, the brightness of the Ca(II)-responsive fluorescent probes fura-2 or indo-1 is less than 10% compared to fluorescein and decreases even further upon saturation metal binding (Table 18.3). The low two-photon absorption cross section is due to the smaller size of the conjugated π-system, which limits both the oscillator strength and the magnitude of the excited state dipole moment. Nevertheless, the two-photon excited brightness of such dyes is often significantly increased at wavelengths below the 680 nm limit of pulsed Ti:sapphire lasers commonly used in TPEM. For example, the UV-excitable nuclear stain DAPI is approximately tenfold brighter using two-photon excitation at 590 nm compared to excitation above 700 nm using the same laser power (Tragardh et al. 2015).

The theoretical principles that govern the two-photon absorption cross section of a fluorophore are well established (Albota et al. 1998; Rumi et al. 2000), and several comprehensive reviews have been published that summarize various design approaches for fluorophores with high two-photon absorption cross sections and their application in biological systems (Sumalekshmy and Fahrni 2011; Huang et al. 2010; Kim and Cho 2011; Yao and Belfield 2012; Sarkar et al. 2014; Kim and Cho 2015; Liu et al. 2017). At a fundamental level, the two-photon absorption cross section is proportional to the imaginary part of the second-order hyperpolarizability, which increases with increasing excited state polarization. In the simplest case, the

Table 18.3 Two-photon absorption properties of fluorophores and fluorescent probes adopted from linear microscopy

	One-photon abs		Two-photon abs				Rel. brightness[a]			
Probe	λ_{abs} [nm]	ε [10^4 $M^{-1}cm^{-1}$]	λ_{abs} [nm]	δ [b] [GM]	λ_{em} [nm]	Φ_f [c]	1P	2P	Solvent	Ref.
Fluorescein	490	7.6	790	47	520	0.93	1.00	1.00	H_2O, pH 11	Makarov et al. (2008) and Sjoback et al. (1995)
Alexa Fluor 488	495	7.3	940	33	519	0.92	0.95	0.69	H_2O, pH 11	Mütze et al. (2012)
Rhodamine 6G	530	11.6	700	120	565	0.95	1.56	2.61	H_2O, pH 11	Albota et al. (1998) and Kubin and Fletcher (1982)
BODIPY 492	500	7.9	920	15	509	0.95	1.06	0.31	Ethanol	Mütze et al. (2012) and Loudet and Burgess (2007)
Cy3	550	15.0	700	140	565	0.04	0.08	0.13	PBS	Albota et al. (1998) and Mujumdar et al. (1993)
Cy5.5	675	25.0	1280	210	702	0.28	0.99	1.35	PBS	Kobat et al. (2009) and Umezawa et al. (2009)
Fura-2	362	2.7	750	12	512	0.23	0.09	0.06	HEPES, pH 7	Grynkiewicz et al. 1985 and Wokosin et al. (2004)
Fura-2 + Ca(II)	335	3.3	750	1	505	0.49	0.23	0.01	30 mM MOPS	Grynkiewicz et al. 1985 and Wokosin et al. (2004)
Indo-1	349	3.4	730	4.5	485	0.38	0.18	0.04	100 mM KCl	Grynkiewicz et al. 1985, Xu and Webb (1996), and Wang and Yeh (2012)
EGFP	492	4.6	927	39	510	0.76	0.49	0.68	H_2O, pH 8	Drobizhev et al. (2011)

[a]Brightness for one- (1P) and two-photon (2P) excited fluorescence emission relative to fluorescein, calculated as the product of absorption cross section and quantum yield. [b]Two-photon absorption cross section at the wavelength indicated in column to the left. [c]Fluorescence quantum yield

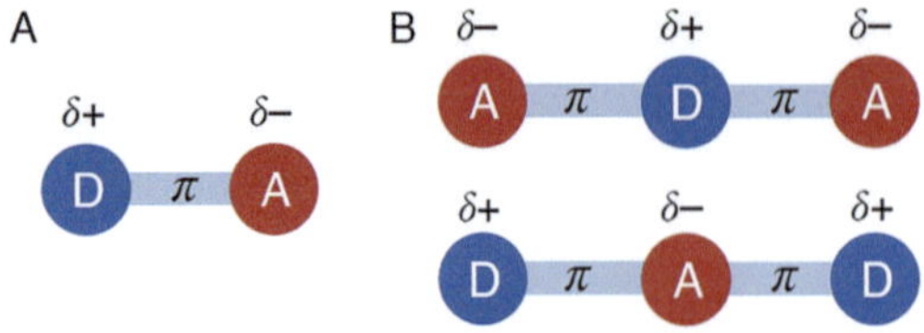

Fig. 18.6 Fluorophore architectures for achieving high two-photon absorption cross sections. (**a**) Simple non-centrosymmetric dipolar arrangement of an acceptor (A) and donor (D) moiety to a fluorophore π-system (π). (**b**) Centrosymmetric architectures with a central donor (top) or acceptor (bottom) moiety

degree of intramolecular charge transfer (ICT) or excited state polarization can be enhanced by modifying the π-system of a linear fluorophore with an electron-rich donor (D) and an electron-deficient acceptor (A) group (Fig. 18.6A). Although such fluorophores are already polarized in the ground state, photoexcitation results in further charge redistribution from the donor to acceptor end of the π-system, thus yielding a strong enhancement of the dipole moment in the excited state compared to the ground state. By extending the fluorophore π-system from a dipolar to a centrosymmetric quadrupolar structure (Fig. 18.6B), very large 2PA cross sections can be achieved (Rumi et al. 2000). While the increase in excited state polarization is beneficial for achieving a high 2PA cross section, it might also have an adverse effect on the quantum yield, especially in polar solvents due to enhanced non-radiative deactivation pathways promoted through stronger solvent-solute interactions. For this reason, enhancing the excited state polarization might in fact yield a fluorophore with lower overall brightness when employed in TPEM.

Although several Zn(II)-selective probes have been employed in TPEM (Khan et al. 2014; Lee et al. 2016; Chang et al. 2004), only few have been specifically designed to achieve a large 2PA cross section. For example, the fluorogenic probe AZn2 takes advantage of the favorable two-photon absorption properties of the donor-acceptor substituted naphthalene derivative acedan (2-acetyl-6-(dimethylamino)naphthalene) (Kim et al. 2008). The probe exhibits an emission maximum at 499 nm and responds with an approximately 50-fold increase in fluorescence intensity, likely mediated through a photoinduced electron transfer switching mechanism as described above for fluorescein-based probes (Fig. 18.7). With a two-photon absorption cross section of 140 GM at 780 nm and a quantum yield of 0.49 for Zn(II)-bound form, the brightness of AZn2 is more than twofold higher compared to fluorescein when used in TPEM. Functionalized with a *N*,*N*-di(2-picolyl)ethylenediamine (DPEN) moiety as Zn(II)-receptor site, the probe offers a Zn(II) dissociation constant of 1.1 nM (pH 7.2, 0.1 M KCl), which is well matched for imaging labile Zn(II) pools within biological systems. Upon incubation in acute hippocampal slices, AZn2 revealed elevated fluorescence in the stratum lucidum of the CA3 region and the hilus of the dentate gyrus, areas that are both known to contain elevated levels of mobile Zn(II) based on histochemical detection (Cole et al. 1999). As an extension of this design, structural modifications of the chelating moi-

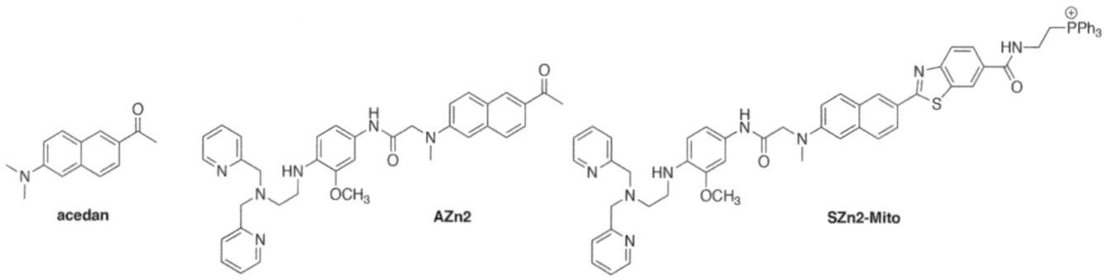

Fig. 18.7 Zn(II)-selective fluorogenic probes for two-photon microscopy employing donor-acceptor substituted naphthalene π-system as fluorophore

ety yielded a series of Zn(II)-responsive probes covering an extended range of dissociation constants from 8 nM to 12 μM while offering similar photophysical properties (Danish et al. 2011). Replacing the acetyl group with a benzothiazole as acceptor moiety, the fluorescence emission of the structurally related probe SZN2-Mito was red-shifted to 536 nm, and the two-photon absorption cross section increased to 155 GM at 750 nm. By attaching the lipophilic triphenylphosphonium group, SZn2-Mito and the related probe SZn-Mito selectively localized to mitochondria (Baek et al. 2012; Masanta et al. 2011). Interestingly, an analogous probe SZn-C containing a carboxylic acid in lieu of the amide substituent selectively localized to the Golgi apparatus (Singh et al. 2015).

18.4.2 Ratiometric Probes for Two-Photon Microscopy

With the ability to follow dynamic changes of analyte concentrations in a semiquantitative fashion, ratiometric imaging is the preferred choice over simple intensity-based approaches. However, not all ratiometric probes are suitable for TPEM. Although the excitation wavelength of femtosecond-pulsed Ti:sapphire lasers can be tuned over a broad range, it is not possible to switch dynamically between two different wavelengths. Ratiometric TPEM must therefore rely on fluorescent probes that respond with a chromatic shift of the emission rather than excitation maximum. As indicated by the data listed in Table 18.2, Zn(II)-induced shifts of the emission maxima are, however, small for most ratiometric probes. The observed small changes are an intrinsic property of the underlying probe architecture in which the metal receptor also functions as the electron donor of a push-pull fluorophore system. The intramolecular charge redistribution that occurs upon photoexcitation results in partial buildup of a positive charge on the donor moiety. Coulomb repulsion by the bound metal ion thus weakens its interaction with the fluorophore, which therefore adopts a polarized charge transfer state similar to that of the free probe. As weakening of the metal-fluorophore interaction proceeds on a faster timescale than radiative deactivation, emission occurs from the equilibrated excited state with only a small change in energy.

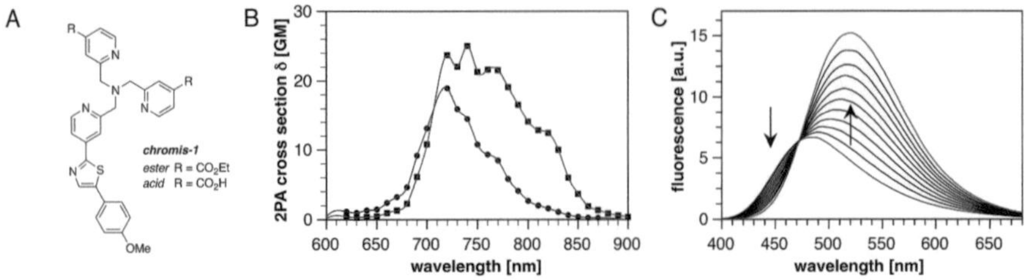

Fig. 18.8 Spectral properties of the ratiometric Zn(II)-selective fluorescent probe chromis-1 (pH 7.0, 10 mM PIPES, 0.1 M KCl, 25 °C). (**a**) Molecular structure of chromis-1 ester and its acid form. (**b**) Two-photon absorption cross section of chromis-1 acid in the free (blue trace) and Zn(II)-saturated (red trace) form. (**c**) Fluorescence emission response upon addition of Zn(II) with 0.1 molar eq aliquots. (Data adopted from ref. (Bourassa et al. 2018). Copyright American Chemical Society 2018)

To design a probe that offers both a large 2PA cross section and a shift of the emission maximum upon metal ion coordination, we devised a push-pull fluorophore architecture where the metal ion receptor is integrated within the acceptor rather than the donor site (Sumalekshmy et al. 2007). With this configuration, the metal-fluorophore interaction is strengthened upon photoexcitation, as it generates a partial negative charge on the acceptor site, thus leading to Coulomb attraction rather than repulsion. Recently implemented in chromis-1 (Fig. 18.8a) (Bourassa et al. 2018), this design approach yielded a large shift of the emission maximum from 483 to 520 nm upon saturation with Zn(II) and a balanced 2PA cross section for the free and Zn(II)-bound forms of 19 and 25 GM, respectively (Fig. 18.8b, c). Despite the small size of the fluorophore π-system, the 2PA action cross section of Zn(II)-bound chromis-1 still approaches 18 GM, corresponding to about 40% of the value for fluorescein.

The acid form of chromis-1 binds Zn(II) with a K_d of 49 ± 13 pM at pH 7.0 (0.1 M KCl, 25 °C). Although other biologically relevant transition metals such as Mn(II), Fe(II), and Co(II) also coordinate to chromis-1, their open d-shell electron configuration leads to fluorescence quenching, likely mediated through energetically low-lying metal-centered d-d states. While chromis-1 acid is not membrane-permeant, the corresponding ethyl ester can readily enter cells. As illustrated by TPEM imaging of live NIH 3 T3 mouse fibroblasts, incubation with chromis-1 ester for 15 minutes in serum-free medium yielded bright fluorescence staining throughout the cytoplasm (Fig. 18.9a). For ratiometric analysis, fluorescence micrographs were acquired by collecting the emission intensity between 425–462 nm (BP1) and 478–540 nm (BP2) with two separate detectors. The corresponding ratio image, which was derived by dividing BP2/BP1, revealed a rather uniform intensity ratio around 0.60 ± 0.16 across all cells (Fig. 18.9a, right). Exposure to a mixture of Zn(II) sulfate and the ionophore pyrithione (ZnPyr) resulted in an immediate increase of the intensity ratio to 2.5 ± 0.6, which was reversed to a slightly lower

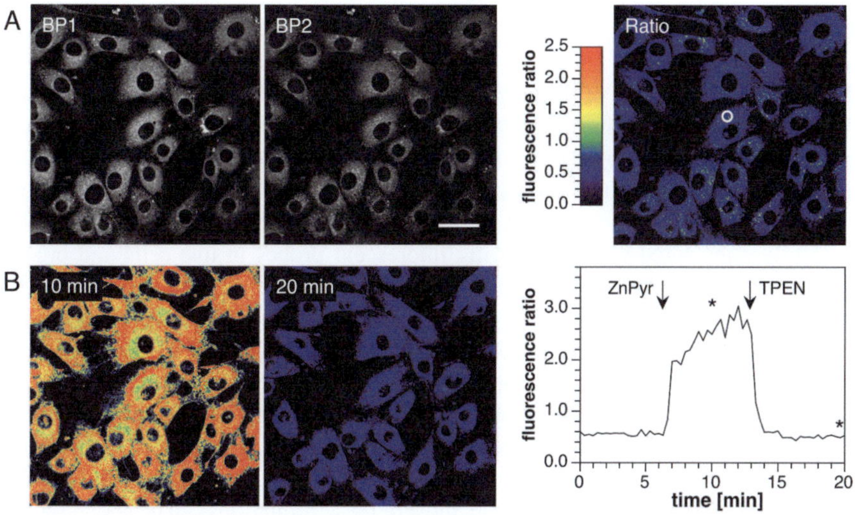

Fig. 18.9 Ratiometric imaging of labile Zn(II) pools in live NIH 3T3 mouse fibroblasts with chromis-1 ester by TPEM (excitation at 720 nm). (**a**) Left: Fluorescence intensity images acquired with 425–462 nm (BP1) and 478–540 nm (BP2) emission channels. Right: Intensity ratio image with R = BP2/BP1. (**b**) Left: Ratio images (BP2/BP1) after addition of 50 μM $ZnSO_4$ and 5 μM pyrithione (ZnPyr), followed by addition of 100 μM TPEN at 13 min. Right: Time course of the average intensity ratio change for the ROI indicated with a white circle in panel A. The asterisks indicate time points for the respective ratio images shown to the left. Scale bar: 40 μm. Data adopted from ref. (Bourassa et al. 2018). Copyright American Chemical Society 2018

value around 0.54 ± 0.13 upon addition of the membrane permeant high-affinity chelator TPEN. A plot of the intensity ratio change as a function of time illustrates the rapid dynamics of Zn(II) binding and competitive chelation by TPEN (Fig. 18.9b, right). Using the K_d of chromis-1 ester (2.4 ± 0.4 nM) and the limiting ratio values after exposure to Zn(II)-pyrithione and TPEN as R_{max} and R_{min}, the fractional saturation of chromis-1 can be calculated with Eq. (18.4) to yield 3–6% under basal conditions. Based on this value, the buffered Zn(II) concentrations reside between 50 and 100 pM, a range that agrees well with previous reports using genetically encoded ratiometric Zn(II) probes (Qin et al. 2011; Vinkenborg et al. 2009; Hessels and Merkx 2014). Chromis-1 was also suitable to visualize dynamic changes of endogenous zinc pools. For example, exposure to the thiol-selective oxidant 2,2′-dithiodipyridine (DTDP) induced a rapid release of Zn(II), presumably from proteins containing sulfhydryl binding sites. Similar to recent observations with a Cu(I)-selective ratiometric fluorescent probe (Morgan et al. 2019), the elevated levels gradually re-equilibrated back to basal conditions. As differences in cellular loading are cancelled out by ratiometric image analysis, this approach is more reliable than intensity-based imaging for monitoring subtle changes in buffered Zn(II) levels. For example, chromis-1 revealed a small but significant decrease in buffered Zn(II) levels upon differentiation of developing into mature oligodendrocytes (Bourassa et al. 2018).

18.5 Conclusions

Since the early success of visualizing histochemically reactive zinc in brain tissue with fluorogenic quinoline derivatives, an extensive range of Zn(II)-responsive fluorescent probes has been developed. The ability to visualize Zn(II) in a noninvasive fashion has provided unique opportunities to unravel the dynamics of biological signaling events in real time. However, there are important limitations that must be carefully considered when employing fluorescent probes in live cells and organisms. First and foremost, fluorescent probes do not report on total zinc levels but only on the exchangeable fraction. The probe dissociation constant should therefore match the buffered concentration of the exchangeable Zn(II) pool. Second, different probes might produce different staining patterns due to variations in subcellular distribution, binding affinity, or pH sensitivity. Third, the response of fluorogenic probes might be susceptible to potential imaging artifacts caused by subcellular sequestration, ternary complex formation with low-molecular-weight ligands, or the association with proteins. In principle, most of these limitations can be addressed by using ratiometric probes; however, only few probes have been developed that offer a dissociation constants within the biologically relevant concentration window, and none of these are commercially available. Moreover, in-cell protein labeling techniques such as HaloTag or SnapTag can readily be adapted to direct synthetic probes to specific subcellular localization (Li et al. 2015; Los et al. 2008; Tomat et al. 2008; Kamiya and Johnsson 2010), thus addressing a key shortcoming over genetically encoded protein-based probes. Regardless of the approach, it is important to keep in mind that fluorescent probes interact with a chemically complex environment and that the identity of the ligands and proteins that participate in buffering labile Zn(II) remains mostly elusive. As the labile Zn(II) pool likely equilibrates through an associate exchange involving ternary complexes (Heinz et al. 2005; Wommer et al. 2002), the underlying kinetic barriers add an additional level of complexity. At present, the kinetics of Zn(II) exchange reactions remains mostly unexplored but might prove to be critical for understanding cellular zinc regulation and signaling. Altogether, Zn(II)-selective fluorescent probes have already become an indispensable tool in Zn(II) biology research, and the most recent advances in the development of ratiometric probes offer new exciting opportunities for unraveling the complexity of cellular zinc homeostasis.

Acknowledgments Financial support from the National Science Foundation under grant CHE-1306943 and the National Institutes of Health under the award number R01GM067169 is gratefully acknowledged. We also thank Dr. M. Thomas Morgan for a critical review of the manuscript.

References

Albota MA et al (1998) Two-photon fluorescence excitation cross sections of biomolecular probes from 690 to 960 nm. Appl Opt 37:7352–7356

Baek NY et al (2012) A highly sensitive two-photon fluorescent probe for mitochondrial zinc ions in living tissue. Chem Commun 48:4546–4548

Bourassa D et al (2018) Chromis-1, a ratiometric fluorescent probe optimized for two-photon microscopy reveals dynamic changes in labile Zn(II) in differentiating oligodendrocytes. ACS Sens 3:458–467

Carter KP et al (2014) Fluorescent sensors for measuring metal ions in living systems. Chem Rev 114:4564–4601

Chabosseau P et al (2018) Sensors for measuring subcellular zinc pools. Metallomics 10:229–239

Chang CJ et al (2004) ZP8, a neuronal zinc sensor with improved dynamic range; imaging zinc in hippocampal slices with two-photon microscopy. Inorg Chem 43:6774–6779

Chen YC et al (2015) Photoluminescence imaging of Zn^{2+} in living systems. Chem Soc Rev 44:4517–4546

Cole TB et al (1999) Elimination of zinc from synaptic vesicles in the intact mouse brain by disruption of the ZnT3 gene. Proc Natl Acad Sci U S A 96:1716–1721

Colvin R et al (2010) Cytosolic zinc buffering and muffling: their role in intracellular zinc homeostasis. Metallomics 2:306–317

Danish IA et al (2011) Two-photon probes for Zn^{2+} ions with various dissociation constants. Detection of Zn^{2+} ions in live cells and tissues by two-photon microscopy. Chem Asian J 6:1234–1240

Danscher G (1981) Histochemical-demonstration of heavy-metals - a revised version of the sulfide silver method suitable for both light and electron-microscopy. Histochemistry 71:1–16

Denk W et al (1990) 2-photon laser scanning fluorescence microscopy. Science 248:73–76

Devergnas S et al (2004) Differential regulation of zinc efflux transporters ZnT-1, ZnT-5 and ZnT-7 gene expression by zinc levels: a real-time RT-PCR study. Biochem Pharmacol 68:699–709

Drobizhev M et al (2011) Two-photon absorption properties of fluorescent proteins. Nat Meth 8:393–399

Fahrni C, O'Halloran T (1999) Aqueous coordination chemistry of quinoline-based fluorescence probes for the biological chemistry of zinc. J Am Chem Soc 121:11448–11458

Fahrni C et al (2003) Tuning the photoinduced electron-transfer thermodynamics in 1,3,5-triaryl-2-pyrazoline fluorophores: X-ray structures, photophysical characterization, computational analysis, and in vivo evaluation. J Am Chem Soc 125:3799–3812

Fahrni CJ et al (2017) Probing biological trace metals with fluorescent indicators. In: White AR (ed) Metals in the brain: measurement and imaging, Neuromethods, vol 124. Springer, New York, pp 71–107

Frederickson CJ (1989) Neurobiology of zinc and zinc-containing neurons. Int Rev Neurobiol 31:145–238

Frederickson CJ et al (1987) A quinoline fluorescence method for visualizing and assaying the histochemically reactive zinc (bouton zinc) in the brain. J Neurosci Meth 20:91–103

Gee KR et al (2002a) Measuring zinc in living cells: a new generation of sensitive and selective fluorescent probes. Cell Calcium 31:245–251

Gee KR et al (2002b) Detection and imaging of zinc secretion from pancreatic beta-cells using a new fluorescent zinc indicator. J Am Chem Soc 124:776–778

Grynkiewicz G et al (1985) A new generation of Ca^{2+} indicators with greatly improved fluorescence properties. J Biol Chem 260:3440–3450

Hara T et al (2017) Physiological roles of zinc transporters: molecular and genetic importance in zinc homeostasis. J Physiol Sci 67:283–301

Heinz U et al (2005) On the competition for available zinc. J Biol Chem 280:3197–3207

Hessels AM, Merkx M (2014) Genetically-encoded FRET-based sensors for monitoring Zn^{2+} in living cells. Metallomics 7:258

Hirano T et al (2000) Highly zinc-selective fluorescent sensor molecules suitable for biological applications. J Am Chem Soc 122:12399–12400
Hirano T et al (2002) Improvement and biological applications of fluorescent probes for zinc, ZnAFs. J Am Chem Soc 124:6555–6562
Huang CB et al (2010) Two-photon fluorescence sensors. Prog Chem 22:2408–2419
Hunt JR, Dawlaty JM (2018) Photodriven deprotonation of alcohols by a quinoline photobase. J Phys Chem A 122:7931–7940
Jiang PJ, Guo ZJ (2004) Fluorescent detection of zinc in biological systems: recent development on the design of chemosensors and biosensors. Coord Chem Rev 248:205–229
Jobsis PD et al (2007) Limited utility of acetoxymethyl (AM)-based intracellular delivery systems, in vivo: interference by extracellular esterases. J Microsc 226:74–81
Johnson ID, Spence MTZ (2010) Molecular probes handbook, a guide to fluorescent probes and labeling technologies, 11th edn. Life Technologies
Kambe T et al (2015) The physiological, biochemical, and molecular roles of zinc transporters in zinc homeostasis and metabolism. Physiol Rev 95:749–784
Kamiya M, Johnsson K (2010) Localizable and highly sensitive calcium indicator based on a BODIPY fluorophore. Anal Chem 82:6472–6479
Karim M, Petering D (2016) Newport green, a fluorescent sensor of weakly bound cellular Zn^{2+}: competition with proteome for Zn^{2+}. Metallomics 8:201–210
Khan M et al (2014) Two-photon imaging of Zn^{2+} dynamics in mossy fiber boutons of adult hippocampal slices. Proc Natl Acad Sci USA 111:6786–6791
Kikuchi K et al (2004) Zinc sensing for cellular application. Curr Opin Chem Biol 8:182–191
Kim HM, Cho BR (2011) Two-photon fluorescent probes for metal ions. Chem Asian J 6:58–69
Kim HM, Cho BR (2015) Small-molecule two-photon probes for bioimaging applications. Chem Rev 115:5014–5055
Kim HM et al (2008) Two-photon fluorescent probes for intracellular free zinc ions in living tissue. Angew Chem Int Ed Engl 47:5167–5170
Kiyose K et al (2006) Development of a ratiometric fluorescent zinc ion probe in near-infrared region, based on tricarbocyanine chromophore. J Am Chem Soc 128:6548–6549
Kobat D et al (2009) Deep tissue multiphoton microscopy using longer wavelength excitation. Opt Express 17:13354–13364
Komatsu K et al (2007) Development of an iminocoumarin-based zinc sensor suitable for ratiometric fluorescence imaging of neuronal zinc. J Am Chem Soc 129:13447–13454
Kubin RF, Fletcher AN (1982) Fluorescence quantum yields of some rhodamine dyes. J Lumin 27:455–462
Lan SC, Liu YH (2015) TDDFT study on the excited-state proton transfer of 8-hydroxyquinoline: key role of the excited-state hydrogen-bond strengthening. Spectrochim Acta A 139:49–53
Lee HJ et al (2016) A two-photon fluorescent probe for lysosomal zinc ions. Chem Commun 52:124–127
Li D et al (2015) Genetic targeting of a small fluorescent zinc indicator to cell surface for monitoring zinc secretion. ACS Chem Biol 10:1054–1063
Lim NC, Bruckner C (2004) DPA-substituted coumarins as chemosensors for zinc(II): modulation of the chemosensory characteristics by variation of the position of the chelate on the coumarin. Chem Commun:1094–1095
Lim NC et al (2004) Illuminating zinc in biological systems. Chem Eur J 11:38–49
Liu ZP et al (2014) In vivo ratiometric Zn^{2+} imaging in zebrafish larvae using a new visible light excitable fluorescent sensor. Chem Commun 50:1253–1255
Liu HW et al (2017) Molecular engineering of two-photon fluorescent probes for bioimaging applications. Meth Appl Fluoresc 5:012003
Los GV et al (2008) HaloTag: a novel protein labeling technology for cell imaging and protein analysis. ACS Chem Biol 3:373–382
Loudet A, Burgess K (2007) BODIPY dyes and their derivatives: syntheses and spectroscopic properties. Chem Rev 107:4891–4932

Lu Q et al (2016) Intracellular zinc distribution in mitochondria, ER and the Golgi apparatus. Int J Physiol Pathophysiol Pharmacol 8:35–43
Mahadevan IB et al (1996) The synthesis of Zinquin ester and Zinquin acid, zinc(II)-specific fluorescing agents for use in the study of biological zinc(II). Aust J Chem 49:561–568
Makarov NS et al (2008) Two-photon absorption standards in the 550-1600 nm excitation wavelength range. Opt Express 16:4029–4047
Maret W (2015) Analyzing free zinc(II) ion concentrations in cell biology with fluorescent chelating molecules. Metallomics 7:202–211
Marszalek I et al (2016) Revised stability constant, spectroscopic properties and binding mode of Zn(II) to FluoZin-3, the most common zinc probe in life sciences. J Inorg Biochem 161:107–114
Marszalek I et al (2018) Ternary Zn(II) complexes of FluoZin-3 and the low molecular weight component of the exchangeable cellular zinc pool. Inorg Chem 57:9826–9838
Maruyama S et al (2002) A novel, cell-permeable, fluorescent probe for ratiometric imaging of zinc ion. J Am Chem Soc 124:10650–10651
Masanta G et al (2011) A mitochondrial-targeted two-photon probe for zinc ion. J Am Chem Soc 133:5698–5700
McRae R et al (2009) In situ imaging of metals in cells and tissues. Chem Rev 109:4780–4827
Meeusen J et al (2012) Reaction of metal-binding ligands with the zinc proteome: zinc sensors and *N,N,N′,N′*-tetrakis(2-pyridylmethyl)ethylenediamine. Inorg Chem 51:3625–3632
Morgan MT et al (2019) Ratiometric two-photon microscopy reveals attomolar copper buffering in normal and Menkes mutant cells. Proc Natl Acad Sci U S A 116:12167–12172
Mujumdar RB et al (1993) Cyanine dye labeling reagents: Sulfoindocyanine succinimidyl esters. Bioconjug Chem 4:105–111
Mütze J et al (2012) Excitation spectra and brightness optimization of two-photon excited probes. Biophys J 102:934–944
Nasir M et al (1999) The chemical cell biology of zinc: structure and intracellular fluorescence of a zinc-quinolinesulfonamide complex. J Biol Inorg Chem 4:775–783
Nolan EM, Lippard SJ (2009) Small-molecule fluorescent sensors for investigating zinc metalloneurochemistry. Acc Chem Res 42:193–203
Nowakowski A et al (2015) Chemical-biological properties of zinc sensors TSQ and Zinquin: formation of sensor-Zn-protein adducts versus $Zn(sensor)_2$ complexes. Inorg Chem 54:11637–11647
Pearce DA et al (2001) Derivatives of 8-hydroxy-2-methylquinoline are powerful prototypes for zinc sensors in biological systems. J Am Chem Soc 123:5160–5161
Qin Y et al (2011) Measuring steady-state and dynamic endoplasmic reticulum and Golgi Zn^{2+} with genetically encoded sensors. Proc Natl Acad Sci U S A 108:7351–7356
Qin Y et al (2013) Direct comparison of a genetically encoded sensor and small molecule indicator: implications for quantification of cytosolic Zn^{2+}. ACS Chem Biol 8:2366–2371
Radford RJ et al (2013) Peptide-based targeting of fluorescent zinc sensors to the plasma membrane of live cells (vol 4, pg 3080, 2013). Chem Sci 4:4532–4532
Ricard C et al (2018) Two-photon probes for in vivo multicolor microscopy of the structure and signals of brain cells. Brain Struct Funct 223:3011–3043
Royzen M et al (2005) Ratiometric displacement approach to Cu(II) sensing by fluorescence. J Am Chem Soc 127:1612–1613
Rumi M et al (2000) Structure-property relationships for two-photon absorbing chromophores: Bis-donor diphenylpolyene and bis (styryl) benzene derivatives. J Am Chem Soc 122:9500–9510
Sarkar AR et al (2014) Two-photon fluorescent probes for metal ions in live tissues. Inorg Chem 53:1794–1803
Schießl IM, Castrop H (2016) Deep insights: intravital imaging with two-photon microscopy. Pflugers Arch 468:1505–1516
Sensi SL et al (2003) A new mitochondrial fluorescent zinc sensor. Cell Calcium 34:281–284
Sfrazzetto GT et al (2016) Synthetic fluorescent probes to map metallostasis and intracellular fate of zinc and copper. Coord Chem Rev 311:125–167

Singh H et al (2015) A Golgi-localized two-photon probe for imaging zinc ions. Chem Commun 51:12099–12102
Sjoback R et al (1995) Absorption and fluorescence properties of fluorescein. Spectrochim Acta A 51:L7–L21
Staszewska A et al (2013) Ternary complex formation and competition quench fluorescence of ZnAF family zinc sensors. Metallomics 5:1483–1490
Sumalekshmy S, Fahrni CJ (2011) Metal-ion-responsive fluorescent probes for two-photon excitation microscopy. Chem Mater:823–830
Sumalekshmy S et al (2007) Design of emission ratiometric metal-ion sensors with enhanced two-photon cross section and brightness. J Am Chem Soc 129:11888–11889
Taki M et al (2004) Emission ratiometric imaging of intracellular zinc: design of a benzoxazole fluorescent sensor and its application in two-photon microscopy. J Am Chem Soc 126:712–713
Thompson K et al (2012) Predicting and avoiding subcellular compartmentalization artifacts arising from acetoxymethyl ester calcium imaging probes. The case of fluo-3 AM and a general account of the phenomenon including a problem avoidance chart. Biotech Histochem 87:468–483
Tomat E et al (2008) Organelle-specific zinc detection using zinpyr-labeled fusion proteins in live cells. J Am Chem Soc 130:15776–15777
Tragardh J et al (2015) Exploration of the two-photon excitation spectrum of fluorescent dyes at wavelengths below the range of the Ti:Sapphire laser. J Microsc 259:210–218
Tsien RY (1981) A non-disruptive technique for loading calcium buffers and indicators into cells. Nature 290:527–528
Tsien RY (1988) Fluorescence measurement and photochemical manipulation of cytosolic free calcium. Trends Neurosci 11:419–424
Tsien R, Pozzan T (1989) Measurement of cytosolic free Ca^{2+} with Quin2. Methods Enzymol 172:230–262
Umezawa K et al (2009) Bright, color-tunable fluorescent dyes in the Vis/NIR region: establishment of new "tailor-made" multicolor fluorophores based on borondipyrromethene. Chemistry 15:1096–1106
Vergnano AM et al (2014) Zinc dynamics and action at excitatory synapses. Neuron 82:1101–1114
Vinkenborg JL et al (2009) Genetically encoded FRET sensors to monitor intracellular Zn^{2+} homeostasis. Nat Meth 6:737–740
Walkup GK et al (2000) A new cell-permeable fluorescent probe for Zn^{2+}. J Am Chem Soc 122:5644–5645
Wang C, Yeh AT (2012) Two-photon excited fluorescence enhancement with broadband versus tunable femtosecond laser pulse excitation. J Biomed Opt 17: 025003
Wokosin DL et al (2004) Characterization of a range of fura dyes with two-photon excitation. Biophys J 86:1726–1738
Wommer S et al (2002) Substrate-activated zinc binding of metallo-beta -lactamases: physiological importance of mononuclear enzymes. J Biol Chem 277:24142–24147
Xu C, Webb WW (1996) Measurement of two-photon excitation cross sections of molecular fluorophores with data from 690 to 1050 nm. J Opt Soc Am B 13:481–491
Xu ZC et al (2010) Fluorescent chemosensors for Zn^{2+}. Chem Soc Rev 39:1996–2006
Yao S, Belfield KD (2012) Two-photon fluorescent probes for bioimaging. Eur J Org Chem:3199–3217
Zalewski PD et al (1993) Correlation of apoptosis with change in intracellular labile Zn(II) using zinquin [(2-methyl-8-*p*-toluenesulphonamido-6-quinolyloxy)acetic acid], a new specific fluorescent-probe for Zn(II). Biochem J 296:403–408

Chapter 19
Zinc Signals in Biology

Hak Chung and Amanda J. Bird

Abstract Zinc is an essential nutrient that is required for the structure and catalytic activity of many different proteins. In addition to its well-established role as a cofactor, a growing amount of evidence suggests that zinc ions also function as signaling molecules. In some organisms, zinc ions bind to extracellular receptors which in turn triggers an intracellular response, whereas in other organisms, zinc ions are rapidly released from intracellular compartments into the cytosol, where they inhibit or stimulate the activity of downstream targets. As previous chapters have focused on zinc signaling in mammals, here we summarize our current knowledge of intracellular and extracellular zinc signals in bacteria, fungi, nematodes, and fish. As the mechanisms that maintain cellular zinc homeostasis are critical for zinc to function as a signaling molecule, these mechanisms as well as those that facilitate zinc signals are discussed.

Keywords Two-component signaling systems · Zinc homeostasis · Zinc signaling · Microorganisms · Metallo-regulatory

19.1 Introduction

Zinc is required for the function of a wide range of proteins in all living creatures (Andreini et al. 2006). As a consequence, all organisms rely on a variety of mechanisms to tightly control the levels of zinc inside of cells to maintain sufficient levels of zinc for the function of these zinc-binding proteins. In addition to being a structural

H. Chung
OSU Interdisciplinary Ph.D. Program in Nutrition, The Ohio State University, Columbus, OH, USA
e-mail: chung.745@osu.edu

A. J. Bird (✉)
Departments of Human Sciences and Molecular Genetics, The Ohio State University, Columbus, OH, USA
e-mail: bird.96@osu.edu

T. Fukada, T. Kambe (eds.), *Zinc Signaling*,
https://doi.org/10.1007/978-981-15-0557-7_19

and catalytic cofactor, a growing amount of evidence suggests that zinc ions also function as extracellular and intracellular signaling molecules in many different organisms. Studies in bacteria have revealed two-component signaling systems that sense and respond to subtoxic levels of zinc in the periplasm, whereas studies in nematodes, fish, and potentially yeast suggest that zinc ions can be rapidly released from zinc intracellular stores in response to environmental or developmental cues. In this chapter, we review what is known about the underlying mechanisms that control zinc homeostasis in bacteria, fungi, nematodes, and fish and also highlight new roles of zinc as a signaling molecule.

19.2 Zinc Homeostasis and Signaling in Bacteria

Bacteria are a diverse group of unicellular organisms that are able to occupy and thrive in a broad range of habitats, including those which have exceptionally high levels of zinc and those where zinc ions in the environment are scarce (Fones et al. 2010; Cox and Saito 2013). Many bacteria also live in habitats where the levels of zinc constantly change. For example, in humans the host immune response to microbial pathogens includes nutritional immunity, a condition where the host restricts access to essential metals such as zinc to intentionally limit growth of invading microbes (Corbin et al. 2008; Kehl-Fie and Skaar 2010; Capdevila et al. 2016). Invading bacterial pathogens also run the risk of being engulfed by macrophages and exposed to lethal cocktails of high levels of metals, including zinc (Wagner et al. 2005; Botella et al. 2011; Ong et al. 2014). To survive in this ever-changing battleground, bacterial pathogens require mechanisms to rapidly adapt and survive in both low- and high-zinc conditions. As we will outline below, bacteria primarily maintain zinc homeostasis at the transcriptional level using zinc-responsive metallo-regulatory proteins and also rely on extra help from two-component signal transduction systems to survive conditions of extreme zinc toxicity (Fig. 19.1).

19.3 Zinc Homeostasis in Bacteria

Zinc is critical for the growth and survival of bacteria as it is required for the function of a large number of proteins, most of which are enzymes (Andreini et al. 2006). Bacteria also have to tightly monitor zinc availability as excessive concentrations of zinc can compete for other types of metal-binding sites leading to mismetalation (Foster et al. 2014). To maintain a level of zinc that is both sufficient and not inhibitory for growth, bacteria tightly control zinc influx and efflux, and how they allocate zinc inside of cells. In bacteria, these processes are controlled primarily at the transcriptional level with changes in intracellular zinc levels being rapidly sensed by zinc-responsive metallo-regulatory proteins.

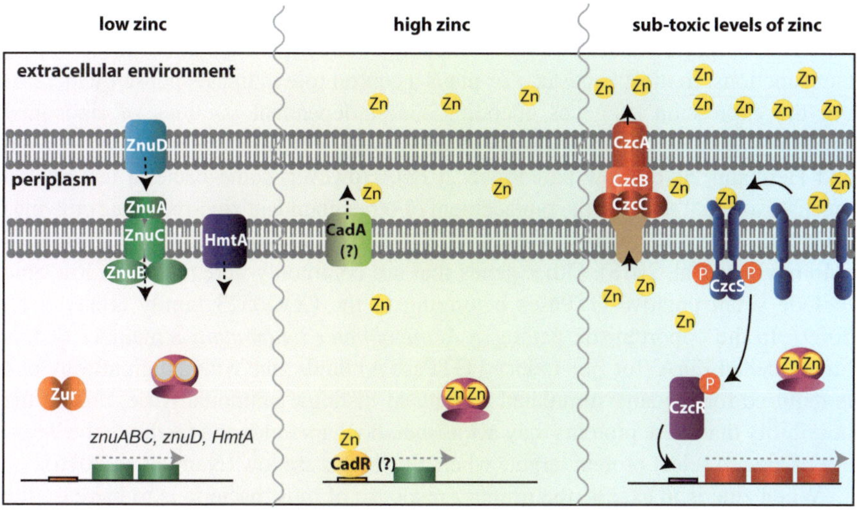

Fig. 19.1 Zinc homeostasis and signaling in *P. aeruginosa*. In low zinc, the Zur repressor is inactive leading to increased expression of gene required for zinc acquisition and zinc sparing. *P. aeruginosa* also expresses CadR and CadA which may protect cells from high-zinc conditions. In the presence of subtoxic levels of zinc, zinc binds to CzcS triggering dimerization and autophosphorylation. This phosphate is then transferred to CzcR, which then increases the expression of genes encoding the CzcCBA efflux pump

In most bacteria, pairs of zinc-responsive metallo-regulatory proteins work together with one triggering changes in gene expression that protect against zinc deficiency and the other triggering changes that protect against zinc toxicity (Capdevila et al. 2016; Choi and Bird 2014; Mikhaylina et al. 2018). For example, in the well-characterized gram-negative bacterium *Escherichia coli*, the metallo-regulatory proteins Zur and ZntR control the expression of genes required for zinc import and efflux, respectively (Outten and O'Halloran 2001). Whereas, in the gram-positive bacterium *Streptococcus pneumoniae* the factors AdcR and SczA control these processes (Kloosterman et al. 2007; Panina et al. 2003). While the precise mechanisms by which these factors activate or repress transcription vary, bacterial zinc-responsive metallo-regulatory proteins share two common features: (1) they bind zinc ions in a manner that is dependent upon intracellular zinc ion availability, and (2) zinc binding leads to a conformational change that allosterically enhances or inhibits their affinity for a specific DNA regulatory element (Reyes-Caballero et al. 2011). By directly sensing intracellular zinc levels, these sensors ensure that the expression of genes that protect against zinc deficiency is increased when zinc is limiting, whereas the expression of genes that protect against zinc toxicity is increased when zinc levels are high (Guerra and Giedroc 2012).

In addition to genes required for zinc import and efflux, bacteria tightly control the expression of other genes that play important roles in zinc homeostasis. One of the most widely used strategies to adapt to conditions of zinc deficiency is called the zinc sparing response. The basic principle behind this mechanism is that abundant zinc-binding proteins are replaced with functionally equivalent proteins that do not bind zinc during conditions of zinc deficiency – thus sparing zinc for more important functions. In most bacteria, Zur plays a central role in this response by increasing the expression of genes encoding zinc-independent paralogs of ribosomal subunits that replace abundant zinc-requiring paralogs (Mikhaylina et al. 2018; Shin and Helmann 2016; Dow and Prisic 2018). However, some bacteria use related mechanisms that result in the replacement of other abundant zinc-requiring enzymes such as alcohol dehydrogenases with alternative non-zinc-requiring substitutes (Mikhaylina et al. 2018). Other genes that are commonly upregulated in low zinc include small putative GTPases belonging to the COG0523 family (Haas et al. 2009). In the opportunistic pathogen *Acinetobacter baumannii*, a member of this family called ZigA (for Zur-Induced GTPase A) binds zinc with a high affinity and is required for the zinc-stimulated activity of histidine ammonia lyase, raising the possibility that these proteins may act as metallochaperones and facilitate the delivery of zinc to select protein targets when zinc levels are low (Nairn et al. 2016).

When zinc is in excess, the primary response of most bacteria is to increase the expression of genes required for zinc efflux. However, some bacteria and cyanobacteria rely on increased expression of small cysteine-rich proteins called metallothioneins to bind and buffer excess zinc in the cytoplasm (Blindauer et al. 2001, 2002). In the absence of zinc efflux systems, a variety of small metabolites are important for zinc tolerance including glutathione in *E. coli* and bacillithiol in *Bacillus subtilis* (Helbig et al. 2008; Ma et al. 2014). As these molecules are abundantly found within the cytoplasm and can form complexes with zinc, it is thought that these low molecular weight thiols, along with other small molecules such as histidine, play an important role in buffering labile zinc ions within the cytosol (Capdevila et al. 2016).

19.4 Zinc-Responsive Two-Component Signal Transduction Systems in Bacteria

The tight control of zinc homeostasis genes by zinc-responsive transcription factors enables bacteria to survive in low- and high-zinc conditions. Although these responses are critical for cell survival over a broad range of environmental zinc stress, a growing amount of evidence suggests that two-component signaling systems are important for survival upon exposure to subtoxic levels of zinc.

Two-component signaling systems are widely used by prokaryotes where they play an important role in sensing and responding to changes in environmental conditions (Stock et al. 2000). Bacterial two-component signaling systems consist of a membrane-bound histidine protein kinase (HK) containing a conserved intracellular kinase core and an intracellular response regulator protein (RR). In response to environmental stimuli, the HK undergoes autophosphorylation at a histidine residue. This phosphate is then transferred to the cognate RR triggering a specific cellular output response, which in most cases is a change in gene expression (Stock et al. 2000).

Two-component signaling systems have been identified that sense and respond to many essential metals including: Fe, Cu, Mn, Ni, Zn, and Co, and toxic metals such as Pb, Ag, and Cd. Here we specifically review what is known about two-component signaling systems that are triggered by excess zinc, which include the CzcRS, ColRS, BaeSR, and ZraSR systems. More information about the regulation of two-component systems by other metals can be found elsewhere (Singh et al. 2014).

One of the first zinc-responsive two-component signaling systems to be characterized was the CzcRS system in *Cupriavidus metallidurans* CH34 (previously *Alcaligenes eutrophus* CH34) (Grosse et al. 1999; van der Lelie et al. 1997). This gram-negative bacterium has been widely used as a model organism to study metal toxicity as it harbors multiple plasmids which enable it to survive under conditions of severe heavy metal stress (von Rozycki and Nies 2009). One of these plasmids, pMOL30, contains the cobalt-zinc-cadmium (*czc*) resistance determinant, a region containing at least nine genes including the *czcCBA* Co, Zn, and Cd efflux system, and the *czcRS* two-component signaling system (Grosse et al. 1999, 2004). While the *czcRS* genes within this resistance determinant are not critical for zinc tolerance or the expression of *czcCBA* in response to zinc excess, deletion of *czcR* or *czcS* leads to a lag in the zinc-dependent induction of these genes (Grosse et al. 1999). As CzcR binds to the 5′ region of one of the genes (*czcN*) in the determinant, which then results in the expression of a longer *czcNICBA* transcript, it is thought that the primary function of this two-component signaling system is to provide an extra level of control by facilitating the rapid transcription of genes upon exposure to zinc (Grosse et al. 1999, 2004).

The role of the CzcRS two-component regulatory system has also been studied in the opportunistic pathogen *Pseudomonas aeruginosa*. In strains of *P. aeruginosa* isolated from metal-polluted rivers, plasmids expressing the *czcCBA* and *czcRS* genes are important for tolerance to excess zinc and cadmium (Hassan et al., 1999). However, most studies in this organism have focused on the function of the chromosomally located *czcR* and *czcS* genes. In the chromosome, the *czcRS* operon is divergently transcribed from the *czcCBA* operon. As treatment with sublethal zinc concentrations results in increased expression of *czcRS* and *czcCBA*, it is thought that the CzcRS regulatory system directly controls the transcription of these genes in response to zinc (Perron et al. 2004). Consistent with this hypothesis, a V194 L substitution in the CzcS sensor results in the constitutive expression of the *czcCBA* genes and resistance to normally lethal zinc concentrations (Perron et al. 2004).

Recent studies characterizing the structure of the periplasmic sensory domain from CzcS have provided additional evidence that CzcS directly senses and responds to zinc. In other HKs, metal binding to the sensory domain triggers homodimerization which in turn leads to increased activity of the cytosolic kinase domain (Affandi et al. 2016; Affandi and McEvoy 2019). In the presence of micromolar levels of zinc, the sensory domain of CzcS forms a homodimer, with a single zinc ion being coordinated to His55 and Asp60 from each monomer at the homodimerization interface (Wang et al. 2017). As genetic mutations that substitute His55 and Asp60 with

Ala prevent CzcS from sensing and responding to zinc in vivo, it is thought that CzcS is regulated by a related mechanism where zinc binding induces CzcS homodimerization triggering autophosphorylation and the subsequent activation of CzcR (Wang et al. 2017).

The structure/function analysis of the sensory domain from CzcS has also revealed important differences between changes in gene expression that are mediated by two-component signaling systems vs. zinc-responsive metallo-regulatory factors, which in *Pseudomonas* include Zur and potentially CadR (Ellison et al. 2013; Pederick et al. 2015; Lee et al. 2001). These differences include (1) that the sensory domain of CzcS is located within the periplasm, whereas Zur and potentially CadR sense changes in cytoplasmic zinc availability, and (2) that CzcS binds zinc with a micromolar affinity, whereas Zur from *P. aeruginosa* binds zinc with a nanomolar affinity (Wang et al. 2017; Pederick et al. 2015). Together these different properties result in the set point for "labile zinc" in the cytosol being low and two-component regulatory systems only operating when cells are exposed to subtoxic levels of zinc (Fig. 19.1).

Studies of the ColRS two-component signal transduction system from *Pseudomonas* have provided additional evidence suggesting that zinc-induced dimerization is critical for zinc sensing and signal transduction in zinc-regulated HKs. The ColRS system senses and confers tolerance to multiple metals including Zn, Fe, Mn, and Cd (Ainsaar et al. 2014). The periplasmic sensory domain from ColRS contains a conserved ExxE motif that is critical for the ColS HK to sense zinc and iron (Ainsaar et al. 2014). As zinc ions are typically coordinated with a tetrahedral geometry, zinc could drive dimerization and activation of ColS with a single zinc ion being coordinated to the ExxE motif from each monomer. In the presence of sublethal levels of zinc, this two-component signaling system plays an important role in protecting cells from membrane stress by upregulating the expression of genes that alter the lipopolysaccharide composition of the outer membrane and polysaccharide content of the cell wall (Mumm et al. 2016; Nowicki et al. 2015).

The BaeSR two-component regulatory system has been extensively studied in *E. coli* where it senses and responds to a broad range of environmental stimuli ranging from exposure to indole, to excessive levels of cations (Leblanc et al. 2011; Raffa and Raivio 2002). Genes that are regulated by this system include the *mdtABC*, *acrD*, and *mdtD* multidrug efflux pumps, the *spy* periplasmic chaperone, as well as *baeRS* (Leblanc et al. 2011; Nishino et al. 2005). In *Salmonella* and *E. coli*, *baeS* and *baeR* are required for tolerance to zinc excess, and exposure to sublethal zinc concentrations leads to increased expression of *baeRS* and other genes of the BaeR regulon (Lee et al. 2005; Nishino et al. 2007). In addition to protecting cells from zinc toxicity, studies in *E. coli* have revealed that genes of the BaeR regulon are also important for survival during zinc shock.

Zinc shock refers to a growth condition where zinc-starved cells are suddenly exposed to high concentrations of zinc in the extracellular environment (MacDiarmid et al. 2003). This transition is potentially hazardous for bacteria because in low zinc they transcriptionally increase the expression of genes required for zinc uptake and at the same time decrease the expression of genes required for zinc export (Outten and O'Halloran 2001). As a consequence, if cells are suddenly exposed to high levels of zinc in the environment, zinc ions will rapidly enter the cell via the uptake systems resulting in the cytosol being flooded with excessive zinc ions. By using a genetically encoded zinc sensor to monitor changes in the labile pools of zinc inside of the cytosol, Fierke and colleagues found that deletion of *mdtA*, *mdtC*, *mdtD*, *acrD*, or *spy* led to the rapid, transient saturation of the sensor during zinc shock (Wang and Fierke 2013). They also found that deletion of *spy* or *mdtC* led to higher levels of total cellular zinc, suggesting that these genes protect zinc-limited cells from environmental zinc shock by facilitating the removal of labile zinc from the cytosol. Although the BaeSR two-component system is not required for the initial response to a zinc shock, it presumably establishes a positive feedback loop to increase the expression of these genes upon prolonged exposure to zinc (Wang and Fierke 2013).

E. coli and *Salmonella* also express a second two-component signaling system, ZraRS (hydH/G), which senses and responds to Zn and Pb (Leonhartsberger et al. 2001; Appia-Ayme et al. 2012). In both bacteria the ZraRS system regulates the expression of a divergently transcribed gene *zraP*, which encodes a periplasmic protein that can bind zinc (Appia-Ayme et al. 2012). In *E. coli*, ~70% of the cellular zinc is bound to ZraP under conditions of zinc stress, suggesting that the primary function of ZraP is to trap or store zinc in the periplasm and protect the cytoplasm from zinc toxicity (Sevcenco et al. 2011). Consistent with this hypothesis, deletion of *zraP* leads to increased expression of *zraP* and *zraS*, suggesting that ZraP is part of a negative feedback loop to tightly regulate the abundance of ZraP in a manner that is dependent upon periplasmic zinc concentrations (Appia-Ayme et al. 2012). While other two-component systems are controlled by related negative feedback mechanisms (Raivio et al. 1999), the precise function of ZraP is still debated because ZraP is not required for zinc tolerance (van der Weel et al. 2019). While it remains possible that other protective mechanisms compensate in the absence of ZraP, other studies have found that ZraP has chaperone activities, and that it binds copper with higher affinity than zinc (Appia-Ayme et al. 2012; van der Weel et al. 2019; Petit-Hartlein et al. 2015). Recent studies have also revealed that ZraR regulates the expression of 25 additional genes including many stress response genes (Rome et al. 2018). Further analyses of these other targets may therefore help to shed light on the precise functions of ZraSR and ZraP during zinc stress.

While many lines of evidence indicate that zinc functions as an extracellular signaling molecule in bacteria, its role as an intracellular signaling molecule is less clear. In mammals, zinc ions can function as an intracellular signaling molecule in part because zinc ions are stored in intracellular compartments such as the endoplasmic reticulum and then can be rapidly released with the appropriate stimuli (Maret 2017). As bacteria lack organelles, it is not possible for them to rapidly release zinc ions from an intracellular store to trigger a response. Although the absence of intracellular stores suggests that zinc is unable to function as an intracellular signaling molecule in bacteria, an interesting situation arises when zinc-limited bacteria are exposed to high levels of zinc. As outlined earlier, this condition leads to a rapid influx of zinc into the cytoplasm, resembling that of the rapid release of zinc seen in mammalian cells during a signaling event. While it is unclear if zinc could have additional regulatory roles under these conditions, there are precedents for the activity of some bacterial enzymes being dependent on intracellular zinc concentration. As one example, zinc allosterically inhibits the activity of the *E. coli* DgcZ diguanylate cyclase, which leads to reduced levels of the signaling molecule cyclic di-GMP and decreased biofilm production (Zahringer et al. 2013; Yeo et al. 2018). The ability of zinc to modulate the activity of specific enzymes in vivo suggests that rapid influxes of zinc during zinc shock, or extreme changes of the concentration of zinc ions in the environment, could function as the trigger to alter the activity of zinc-binding proteins to elicit a specific cellular response.

19.5 Zinc Homeostasis and Signaling in Fungi

The fungal kingdom contains a diverse range of organisms including yeasts, molds, mildews, and mushrooms. However, much of what is known about zinc homeostasis in fungi is based on studies with Ascomycota, or sac fungi, which include the yeasts *Saccharomyces cerevisiae*, *Schizosaccharomyces pombe*, and *Candida albicans*, and the filamentous fungi *Aspergillus fumigatus* (Wilson 2015; Wilson and Bird 2016; Walencik et al. 2016). In contrast to bacteria that tightly regulate zinc efflux, when zinc is excess, sac fungi typically store excess zinc in intracellular compartments. Studies in the budding yeast *S. cerevisiae* have revealed how the processes of zinc acquisition and compartmentalization can be tightly regulated by transcriptional and posttranslational mechanisms in response to cellular zinc status (Fig. 19.2).

In *S. cerevisiae* proteins belonging to the Zrt1- and Irt1-like Proteins (ZIP) family and cation diffusion facilitator (CDF) family play important roles in zinc transport (Fig. 19.2). In this yeast, ZIP family members facilitate zinc uptake into the cytoplasm (Zrt1 and Zrt2), and the release of zinc from the vacuolar stores (Zrt3) (Zhao and Eide 1996a, b; MacDiarmid et al. 2000). A fourth ZIP family member, Yke4, also facilitates the bi-directional transport of zinc to and from the endoplasmic reticulum (Kumanovics et al. 2006). In addition to these proteins, an unrelated transporter called Fet4 plays an important role in zinc acquisition, particularly under anaerobic growth conditions (Waters and Eide 2002). Transport of zinc out of the cytosol is primarily mediated by four CDF family members including Msc2 and Zrg17, which form a complex that transports zinc into the endoplasmic reticulum, and Zrc1 and Cot1, which both transport zinc ions into the vacuolar stores (MacDiarmid et al. 2000; Ellis et al. 2005; Miyabe et al. 2001). Consistent with these proteins having a primary role in controlling where labile zinc ions are sent to inside of cells, the expression or activities of nearly all of these zinc transporters are dependent on cellular zinc status.

At the transcriptional level, a zinc-responsive transcription factor called Zap1 plays a primary role in zinc homeostasis by activating gene expression when zinc levels are low (Wilson and Bird 2016). Genes that are activated by Zap1 under this

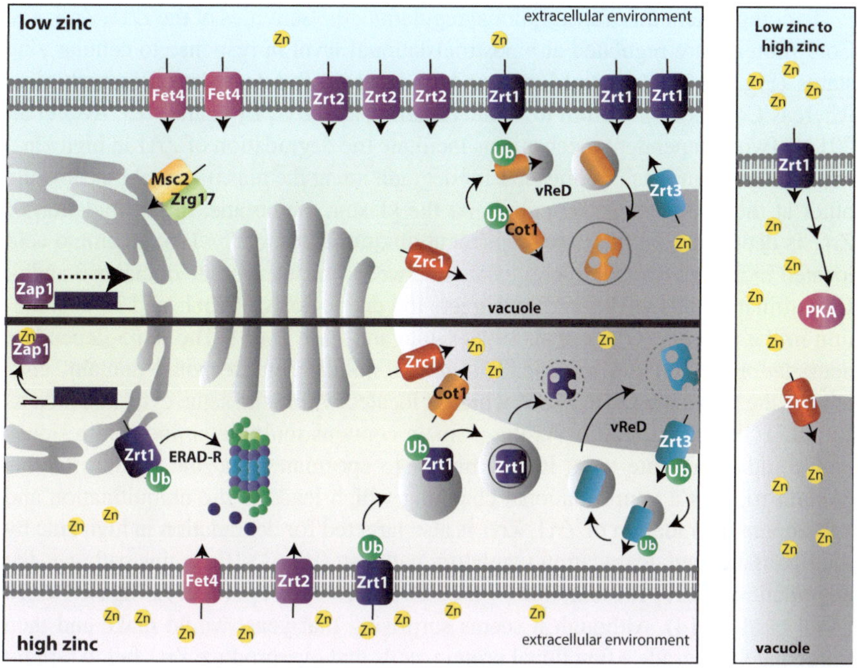

Fig. 19.2 Zinc homeostasis and signaling in *S. cerevisiae*. Left upper panel: When zinc is limiting, the transcriptional activator Zap1 increases the expression of genes encoding protein that facilitate zinc uptake (Zrt1, Zrt2, and Fet4) and the release of zinc from the vacuolar stores (Zrt3). Zap1 target genes also include *ZRG17*, *ZRC1*, and *ZAP*1. As a posttranslational level, Cot1 is targeted for degradation by the vReD pathway. Lower left panel: When zinc is in excess, Zap1 is inactive leading to reduced expression of the above target genes. Newly synthesized Zrt1 is targeted for degradation in the endoplasmic reticulum by the ERAD-R pathway, whereas Zrt1 proteins at the plasma membrane undergo endocytosis and degradation in the vacuole. Zrt3 is also targeted for degradation by the vReD pathway. Right panel: When zinc-limited cells are exposed to zinc, Zrc1 plays a central role in removing excess zinc from the cytoplasm. Under these conditions, Zrt1 may function as a transceptor for the activation of the PKA pathway

condition include the zinc uptake genes *ZRT1*, *ZRT2*, and *FET4* (Waters and Eide 2002; Zhao and Eide 1997). Zap1 also controls the expression of *ZRT3*, *ZRG17*, and *ZRC1* (MacDiarmid et al. 2000, 2003; Wu et al. 2011). While at first it seems counterintuitive that Zap1 would regulate the expression of *ZRG17* and *ZRC1*, the Zap1-dependent regulation of *ZRG17* protects against ER stress in low-zinc conditions, whereas increased expression of *ZRC1* is a proactive mechanism to prevent toxic levels of zinc accumulating in the cytosol during a zinc shock (MacDiarmid et al. 2003; Wu et al. 2011). Recent studies have also found that *MSC2* mRNA levels are higher in low zinc, suggesting that it is also a zinc-regulated gene (Wu et al. 2016). In addition to controlling the expression of zinc transport genes, Zap1 also regulates the expression of a large number of other genes, many of which enable yeast to adapt and survive during longer conditions of zinc deficiency (Wu et al. 2008).

To complement this transcriptional regulation, the activities of the Zrt1, Zrt3, and Cot1 proteins are regulated at a posttranslational level in response to cellular zinc status (Fig. 19.2). Zrt1 and Zrt3 are both targeted for degradation in high zinc, whereas Cot1 is degraded in low zinc (Gitan et al. 1998; Li et al. 2015; Avci et al. 2014). Two independent mechanisms facilitate the degradation of Zrt1 in high zinc, with one mechanism targeting Zrt1 for degradation at the plasma membrane and the other at the endoplasmic reticulum. At the plasma membrane, the degradation of Zrt1 is dependent on the Rsp5-dependent ubiquitination of Lys195, an amino acid located in a large cytosolic loop located between transmembrane domains 3 and 4. Ubiquitination of Lys195 in turn triggers the endocytosis of Zrt1, and its degradation in the vacuole (Gitan et al. 1998; Gitan and Eide 2000). The Rsp5-dependent degradation of Zrt1 is also dependent upon a small "metal response domain" containing the sequence DATSMDV, which is located within the same cytosolic loop as Lys195 (Gitan et al. 2003). As this domain contains multiple amino acid residues that could coordinate zinc, it is tempting to speculate that zinc binding to this domain triggers a conformational changing which leads to the ubiquitination and subsequent degradation of Zrt1. Zrt1 is also targeted for degradation in high zinc by the ER-associated degradation regulatory pathway (ERAD-R). In this pathway, the degradation of Zrt1 is facilitated by the ERAD protease Ypf1 and E3 ligase Doa10 (Avci et al. 2014). Although it seems surprising that yeast would make and then immediately degrade a functional protein, cells that overproduce Zrt1 fail to rapidly sense decreases in cytosolic zinc concentration (Levy et al. 2011). As the ERAD-R-dependent turnover of Zrt1 ensures the absence of Zrt1 on the plasma membrane when there are adequate levels of zinc for growth, this mechanism enables the early sensing of zinc deficiency when zinc levels start to decline (Avci et al. 2014).

The degradation of Zrt3 and Cot1 is dependent upon the vacuole membrane recycling and degradation (vReD) pathway (Li et al. 2015). While the precise signal that triggers the ubiquitination and degradation of Zrt3 in high zinc, and Cot1 in low zinc, is unclear, the ubiquitination of both proteins is dependent upon the RING domain-containing E3 ligase Tul1 and the Defective for SREBP cleavage (Dsc) complex. A second E3 ligase Rsp5 also facilitates the ubiquitination of Zrt3 (Li et al. 2015). Following ubiquitination, Zrt3 and Cot1 are both sorted into the vacuolar lumen for degradation with the aid of the ESCRT machinery (Zhu et al. 2017).

In most other fungi, orthologs of Zap1 tightly control the expression of genes required for zinc uptake, highlighting the importance of zinc-dependent transcriptional control in the fungal kingdom (Wilson 2015). In addition to controlling the expression of zinc uptake genes, the Zap1 orthologs in other fungi often regulate genes which are important for survival in their own environmental habitat. As an example, in pathogenic fungus *C. albicans*, Csr1 (the Zap1 ortholog) induces the expression of the pH-regulated antigen 1 gene *PRA1*, which encodes a secreted zincophore (Citiulo et al. 2012). *PRA1* gene expression is also increased at neutral to alkaline pHs (Sentandreu et al. 1998). In the absence of Pra1, the ability of invasive hyphae to sequester zinc from the host during endothelial invasion is reduced (Citiulo et al. 2012), suggesting that this system helps *C. albicans* to scavenge zinc from the host in neutral to alkaline environments – growth conditions where the solubility of zinc is reduced. While Zap1-like proteins are widely used in the fungal kingdom, there are exceptions. For example, the fission yeast *S. pombe* uses a different zinc-responsive transcription factor called Loz1 to control gene expression in response to zinc (Corkins et al. 2013; Ehrensberger et al. 2014). Fission yeast also produce zinc metallothioneins in high zinc to sequester excess zinc in the cytosol (Borrelly et al. 2002). These differences highlight that studies with other fungi will further our knowledge of zinc homeostasis by revealing novel mechanisms and alternative strategies to survive zinc limitation and zinc excess.

While most studies in fungi have focused on zinc homeostasis, a few recent studies have explored whether zinc can function as a signaling molecule. In yeast, the cAMP-Protein Kinase A (PKA) pathway controls the expression and activity of a wide range of proteins in response to nutrient status; typically acting positively on processes associated with rapid growth and negatively on processes that drive slower growth (Conrad et al. 2014). A recent study in *S. cerevisiae* has shown the addition of zinc to severely zinc-limited cells (i.e., zinc shock) leads to a rapid surge (within 2–3 min) in trehalase activity, which is a widely used reporter to indicate activation of the PKA pathway (Schothorst et al. 2017). While further experiments are required to demonstrate that this increase is a result of a signaling event, as this increase is not observed in a *ZRT1* deletion strain, it has been proposed that Zrt1 functions as a transceptor for activation of the PKA pathway. A link between the PKA pathway and zinc has also been observed in *C. albicans*. In this yeast, the intracellular zinc reporter Zinbo-5 localizes to the endoplasmic reticulum, with glucose stimulation triggering changes in Zinbo-5 fluorescence in a manner that is dependent upon the PKA pathway (Kjellerup et al. 2018). It remains to be tested whether these changes in fluorescence are a direct result of the rapid flux of zinc out of the endoplasmic reticulum into the cytosol.

19.6 Zinc Signals and Homeostasis in *Caenorhabditis elegans*

Caenorhabditis elegans is a member of the phylum Nematoda and has been extensively used as a model system to study animal development and behavior (Corsi et al. 2015). As it is relatively simple to control the levels of essential metals and metallo-cofactors in the diet of *C. elegans*, and to perform genetic screens to identify mutants that have increased or decreased tolerance to low or high levels of these elements, a growing number of researchers are also using this roundworm to study mechanisms of metal homeostasis (Anderson and Leibold 2014; Sinclair and Hamza 2015). So far, studies with *C. elegans* have shown that the expression of many zinc homeostasis genes is regulated at a transcriptional level in response to zinc (Fig. 19.3) and that zinc ions function as an important signaling molecule to activate sperm.

Most studies examining zinc homeostasis in *C. elegans* have focused on the role of the nuclear receptor protein HIZR-1 (high-zinc-activated nuclear receptor 1). HIZR-1 plays a central role in the response to zinc excess by regulating the

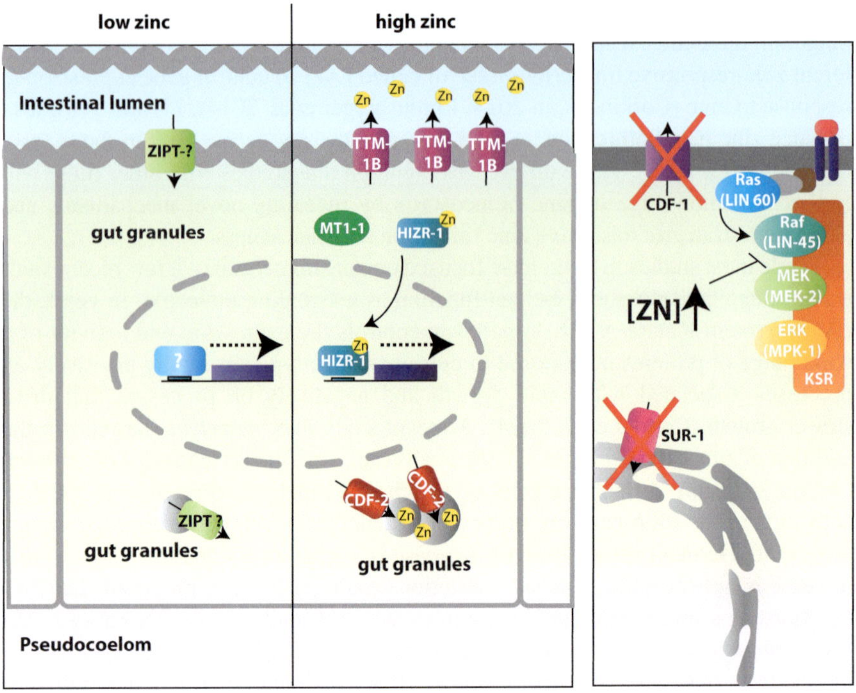

Fig. 19.3 Zinc homeostasis and signaling in *C. elegans*. Left Panel: When zinc is limiting the expression of genes encoding multiple ZIP family members is increased, which may facilitate zinc uptake or release of zinc from intracellular stores. Excess zinc triggers the nuclear localization of HIZR-1 and activation of genes required for zinc export, storage, and compartmentalization. Right panel: Loss-of-function mutations in SUR-1 and CDF-1 both increase intracellular zinc levels and inhibit Ras-mediated signaling

expression of genes required for zinc storage and zinc efflux (Warnhoff et al. 2017). HIZR-1 facilitates these changes by binding to HZA elements that are located in the promoters of its target genes, with binding leading to increased activation of gene expression in high zinc (Warnhoff et al. 2017; Roh et al. 2015). Characterized HIZR-1 target genes include *mtl–1*, which encodes a metal-binding metallothionein, and the CDF family members, *ttm-1b* and *cdf-2*. In *C. elegans* TTM-1b facilitates the excretion of excess zinc ions by transporting them back into the intestinal lumen, whereas CDF-2 transports zinc into gut granules, a lysosome-related organelle where zinc ions can be stored and released when needed (Warnhoff et al. 2017; Roh et al. 2013; Davis et al. 2009). HIZR-1 also autoregulates the expression of its own gene, which established a positive feedback loop to maximize the transcriptional response to high zinc (Warnhoff et al. 2017).

Recent studies in *C. elegans* have also shown that the expression of three genes belonging to the ZIP family is induced at a transcriptional level in response to zinc deficiency (Dietrich et al. 2017). While the precise identity of the regulatory factor that controls these changes is unknown, the zinc-dependent regulation of all of these genes is dependent upon the presence of a conserved low zinc-activation element (LZA). As related DNA elements are found in the promoters of the *ZIP2* and *ZIP13* zinc transporter genes in humans, and the expression of both of these genes is enhanced under conditions of zinc deficiency, the regulatory factor that is required for the low-zinc response in worms may also be conserved in vertebrates (Dietrich et al. 2017).

The above mechanisms play a critical role in cellular zinc homeostasis by adjusting zinc uptake, storage, and efflux according to cellular zinc status. Studies in *C. elegans* have also found that during sperm activation, zinc ions are rapidly released into the cytosol, suggesting that zinc also functions as a signaling molecule in worms.

C. elegans is a useful model organism to study sperm activation as it has two sexes, males and hermaphrodites, both of which make sperm (L'Hernault, 1997). In both sexes, spermatocytes undergo multiple meiotic divisions to generate spermatids, which are round, immotile, transcriptionally silent haploid cells. In a process called sperm activation, these cells then undergo a series of rearrangements which leads to the formation of the pseudopod which enables them to crawl to and fertilize oocytes (L'Hernault 1997). In hermaphrodites, sperm activation is dependent upon the SPE-8 signaling pathway and zinc (Liu et al. 2013; Muhlrad et al. 2014).

The first hint that zinc was required for sperm activation in *C. elegans* was obtained in studies identifying compounds that led to the activation of sperm in vitro. These analyses revealed that extracellular zinc could trigger the sperm activation in a manner that was dependent upon the SPE-8 signaling pathway (Liu et al. 2013). They also demonstrated that the addition of extracellular zinc, or other types of activating signals, promoted the rapid intracellular redistribution of labile zinc ions, consistent with zinc ions functioning as secondary messengers during sperm activation. More recent studies have shown that the *zipt-7.1* zinc transporter gene is expressed in developing spermatocytes, and that deletion of the *zipt-7.1* gene leads to sterility and lower levels of zinc accumulating inside of spermatids (Zhao et al. 2018). As the ZIPT-7.1 transporter is localized to internal membranous organelles, it has been proposed that ZIPT-7.1 facilitates the rapid release of zinc from intracellular stores, which in turn may alter the activity of proteins required for sperm activation (Zhao et al. 2018).

While the above hypothesis has yet to be tested, it is noteworthy that loss-of-function alleles of *cdf-1* and *sur-7* were both discovered in genetic screens to identify factors that affect Ras-mediated signaling (Bruinsma et al. 2002; Yoder et al. 2004). In *C. elegans*, the Ras-mediated signaling pathway controls multiple developmental processes including the development of the vulva, which is a hermaphrodite specific organ that connects the external environment with the internal reproductive system (Sternberg 2005). As outlined in more detail in the following review (Dietrich et al. 2016), loss-of-function alleles of *cdf-1* and *sur-7* were both isolated in genetic screens searching for suppressors of the multi-vulva phenotype of hyperactive alleles of the *let-60 ras* gene (Bruinsma et al. 2002; Yoder et al. 2004). As *cdf-1* and *sur-7* both encode CDF family members that are predicted to transport zinc out of the cytosol, these results are consistent with higher concentrations of zinc in the cytosol inhibiting Ras-mediated signaling (Fig. 19.3). While it is currently unclear how zinc ions inhibit this pathway, potential mechanisms that have been proposed include that CDF proteins directly interact with and control the activity of the kinase Raf1 and that zinc ions modulate the activity of the KSR scaffold protein (Yoder et al. 2004; Jirakulaporn and Muslin 2004).

19.7 Zinc Homeostasis and Signaling in Fish

Danio rerio (zebrafish) is by far the most extensively studied fish model system, particularly for the research of developmental biology (Lele and Krone 1996; Dahm and Geisler 2006). As the nutrient content of the surrounding water can be easily manipulated, a growing number of studies are also using zebrafish as a model system to study nutrient homeostasis and metabolism (Zhao et al. 2014). Studies so far have revealed multiple similarities in the mechanisms that maintain zinc homeostasis in fish to those in mammals.

Phylogenic studies of zinc transporters belonging to the CDF (called ZnTs in fish) and ZIP families have shown that of the ten ZnT and fourteen ZIP human genes, all but ZnT10, ZnT3, ZIP2, and ZIP5 have orthologs in the zebrafish (Feeney et al. 2005). Global genome profiling has also revealed that the expression of many of the genes encoding zinc transporters is subject to transcriptional regulation in the gill, the intestine, and the kidney, in response to altered aquatic zinc levels (Feeney et al. 2005; Zheng et al. 2010a, b). While the mechanisms by which zinc levels affect mRNA level in response to zinc deficiency are largely unclear, the ability of zebrafish to sense zinc in excess is dependent upon the metal-regulatory transcription factor 1 (MTF1) (Dalton et al. 2000). As in mammals, zebrafish MTF1 is activated by increases in intracellular zinc concentration. It also facilitates changes in gene expression by binding to metal-response elements (MREs) located within the promoters of its target genes (Hogstrand et al. 2008). Genes that are upregulated by this MTF1/MRE transcriptional control include *MT2* and *ZNT1*, which encode a metal-binding metallothionein and a zinc export protein, respectively (Hogstrand et al. 2008; Zheng et al. 2008). MTF1 also regulates the expression of the *ZIP10* gene. Due to differences in the transcriptional start site and MRE promoter position, MTF1 functions as an activator of *ZIP10* expression in the gills, while a repressor of *ZIP10* expression in the kidney (Zheng et al. 2008).

Multiple studies in zebrafish have also revealed an important role for zinc during embryogenesis. Riggio et al. (2003) reported a dramatic increase of zinc content after fertilization, followed by a slow reduction before reaching a plateau. Whereas, Ho and colleagues (Ho et al. 2012) noted that each zinc transporter mRNA manifests a distinct expression pattern over the course of embryonic development. A more direct role for zinc was observed in studies investigating the role of Zip7 during vertebrate development. By using morpholino antisense oligonucleotides to downregulate *zip7* expression, Yan et al. (2012) found that this decrease had no effect on the levels of zinc in the whole embryo, but did lead to morphological defects and reduced concentrations of zinc in the brain, eyes, and skeletal structure, which could be compensated for by the supplementation of exogenous zinc. Together these results suggest that Zip7 plays a vital role in the distribution of zinc ions in the developing embryo.

Multiple studies have obtained evidence which suggests that zinc ions may also act as an important signaling molecule in fish. Yamashita et al. (2004) identified that LIV1, a human ZIP6 homolog, is the downstream target of STAT3, a transcription factor that plays an important role in the epithelial-mesenchymal transition (EMT) during gastrulation. They also found that LIV1 is crucial for the nuclear localization of the transcriptional repressor Snail, the master regulator of the EMT. Together these results suggest a pathway where the STAT3-dependent induction of LIV1 results in increased flux of zinc into the cytosol which in turn triggers the nuclear localization of Snail. While additional studies are needed to test this exciting hypothesis, pyrithione zinc was identified in a high-content screening assay as a factor that increased expression of fibroblast growth factor (FGF) reporters (Saydmohammed et al. 2011). Together these studies suggest that the changes in intracellular zinc may affect multiple signaling pathways that are important for normal embryonic development.

The Atlantic croaker (*Micropogonias undulatus*) is another teleost model organism useful for the study of endocrine control of reproduction (Thomas 1990). Despite the limited amount of literature compared to the zebrafish model system, a series of recent studies in Atlantic croaker has revealed another signaling pathway in which zinc is likely involved as a secondary messenger. While ZIP9 belongs to the ZIP family of zinc transporters, Berg et al. (2014) found that in the Atlantic croaker it also functions as an androgen receptor that is activated specifically by testosterone. They also found that the ZIP9-dependent signaling pathway was required for the testosterone-induced apoptosis of ovarian follicle cells, and that testosterone treatment caused an acute increase in the intracellular free zinc concentration. A follow-up paper by the same group further proved the importance of zinc by showing that the absence of zinc in the culture media blocked the testosterone-induced apoptosis (Converse et al. 2017). These results suggest that ZIP9 functions as both an androgen receptor and a zinc transporter, and that increases in the cytosolic zinc concentration may mediate the cell death signal. More recent studies have shown that zinc is also an important signaling molecule in zebrafish oocyte maturation (Li et al. 2019). Although it is unclear which signaling cascade is triggered by increased cytosolic zinc, the protein tyrosine phosphatase b (PTPb)-insulin-like growth factor receptor (Igf1r) pathway has been suggested.

19.8 Conclusions

A growing amount of evidence suggests that zinc ions function as extracellular and intracellular signaling molecules in a wide range of organisms. In this chapter, we highlighted the current knowledge of zinc signaling mechanisms in bacteria, yeast, nematodes, and fish. However, it should be appreciated that examples of zinc signals are found throughout nature, suggesting that zinc ions may function as a signaling molecule in many living creatures (Maret 2017; Rousset et al. 2002; Chang et al. 2005). Despite a wealth of new studies in this emerging field, many questions remain unanswered. For instance, relatively little is known about the factors that trigger the rapid release of zinc from intracellular stores, and the extent to which zinc may function as a signaling molecule. It is also largely unclear what proteins are activated by zinc signals and how these proteins are inactivated. As zinc signals potentially control cellular processes throughout biology, further studies of zinc signals in a wide range of organisms will likely provide answers to some of these questions.

References

Affandi T, McEvoy MM (2019) Mechanism of metal ion-induced activation of a two-component sensor kinase. Biochem J 476:115–135

Affandi T, Issaian AV, McEvoy MM (2016) The structure of the periplasmic sensor domain of the histidine kinase CusS shows unusual metal ion coordination at the dimeric interface. Biochemistry 55:5296–5306

Ainsaar K, Mumm K, Ilves H, Horak R (2014) The ColRS signal transduction system responds to the excess of external zinc, iron, manganese, and cadmium. BMC Microbiol 14:162

Anderson CP, Leibold EA (2014) Mechanisms of iron metabolism in *Caenorhabditis elegans*. Front Pharmacol 5:113

Andreini C, Banci L, Bertini I, Rosato A (2006) Zinc through the three domains of life. J Proteome Res 5:3173–3178

Appia-Ayme C, Hall A, Patrick E, Rajadurai S, Clarke TA, Rowley G (2012) ZraP is a periplasmic molecular chaperone and a repressor of the zinc-responsive two-component regulator ZraSR. Biochem J 442:85–93

Avci D, Fuchs S, Schrul B, Fukumori A, Breker M, Frumkin I, Chen CY, Biniossek ML, Kremmer E, Schilling O, Steiner H, Schuldiner M, Lemberg MK (2014) The yeast ER-intramembrane protease Ypf1 refines nutrient sensing by regulating transporter abundance. Mol Cell 56:630–640

Berg AH, Rice CD, Rahman MS, Dong J, Thomas P (2014) Identification and characterization of membrane androgen receptors in the ZIP9 zinc transporter subfamily: I. discovery in female Atlantic croaker and evidence ZIP9 mediates testosterone-induced apoptosis of ovarian follicle cells. Endocrinology 155:4237–4249

Blindauer CA, Harrison MD, Parkinson JA, Robinson AK, Cavet JS, Robinson NJ, Sadler PJ (2001) A metallothionein containing a zinc finger within a four-metal cluster protects a bacterium from zinc toxicity. Proc Natl Acad Sci U S A 98:9593–9598

Blindauer CA, Harrison MD, Robinson AK, Parkinson JA, Bowness PW, Sadler PJ, Robinson NJ (2002) Multiple bacteria encode metallothioneins and SmtA-like zinc fingers. Mol Microbiol 45:1421–1432

Borrelly GP, Harrison MD, Robinson AK, Cox SG, Robinson NJ, Whitehall SK (2002) Surplus zinc is handled by Zym1 metallothionein and Zhf endoplasmic reticulum transporter in *Schizosaccharomyces pombe*. J Biol Chem 277:30394–30400

Botella H, Peyron P, Levillain F, Poincloux R, Poquet Y, Brandli I, Wang C, Tailleux L, Tilleul S, Charriere GM, Waddell SJ, Foti M, Lugo-Villarino G, Gao Q, Maridonneau-Parini I, Butcher PD, Castagnoli PR, Gicquel B, de Chastellier C, Neyrolles O (2011) Mycobacterial p(1)-type

ATPases mediate resistance to zinc poisoning in human macrophages. Cell Host Microbe 10:248–259
Bruinsma JJ, Jirakulaporn T, Muslin AJ, Kornfeld K (2002) Zinc ions and cation diffusion facilitator proteins regulate Ras-mediated signaling. Dev Cell 2:567–578
Capdevila DA, Wang J, Giedroc DP (2016) Bacterial strategies to maintain zinc Metallostasis at the host-pathogen interface. J Biol Chem 291:20858–20868
Chang HB, Lin CW, Huang HJ (2005) Zinc-induced cell death in rice (*Oryza sativa L.*) roots. Plant Growth Regul 46:261–266
Choi S, Bird AJ (2014) Zinc'ing sensibly: controlling zinc homeostasis at the transcriptional level. Metallomics 6:1198–1215
Citiulo F, Jacobsen ID, Miramon P, Schild L, Brunke S, Zipfel P, Brock M, Hube B, Wilson D (2012) *Candida albicans* scavenges host zinc via Pra1 during endothelial invasion. PLoS Pathog 8:e1002777
Conrad M, Schothorst J, Kankipati HN, Van Zeebroeck G, Rubio-Texeira M, Thevelein JM (2014) Nutrient sensing and signaling in the yeast *Saccharomyces cerevisiae*. FEMS Microbiol Rev 38:254–299
Converse A, Zhang C, Thomas P (2017) Membrane androgen receptor ZIP9 induces croaker ovarian cell apoptosis via stimulatory G protein alpha subunit and MAP kinase signaling. Endocrinology 158:3015–3029
Corbin BD, Seeley EH, Raab A, Feldmann J, Miller MR, Torres VJ, Anderson KL, Dattilo BM, Dunman PM, Gerads R, Caprioli RM, Nacken W, Chazin WJ, Skaar EP (2008) Metal chelation and inhibition of bacterial growth in tissue abscesses. Science 319:962–965
Corkins ME, May M, Ehrensberger KM, Hu YM, Liu YH, Bloor SD, Jenkins B, Runge KW, Bird AJ (2013) Zinc finger protein Loz1 is required for zinc-responsive regulation of gene expression in fission yeast. Proc Natl Acad Sci U S A 110:15371–15376
Corsi AK, Wightman B, Chalfie M (2015) A transparent window into biology: a primer on *Caenorhabditis elegans*. Genetics 200:387–407
Cox AD, Saito MA (2013) Proteomic responses of oceanic *Synechococcus* WH8102 to phosphate and zinc scarcity and cadmium additions. Front Microbiol 4:387
Dahm R, Geisler R (2006) Learning from small fry: the zebrafish as a genetic model organism for aquaculture fish species. Mar Biotechnol (NY) 8:329–345
Dalton TP, Solis WA, Nebert DW, Carvan MJ 3rd. (2000) Characterization of the MTF-1 transcription factor from zebrafish and trout cells. Comp Biochem Physiol B Biochem Mol Biol 126:325–335
Davis DE, Roh HC, Deshmukh K, Bruinsma JJ, Schneider DL, Guthrie J, Robertson JD, Kornfeld K (2009) The cation diffusion facilitator gene *cdf-2* mediates zinc metabolism in *Caenorhabditis elegans*. Genetics 182:1015–1033
Dietrich N, Tan CH, Cubillas C, Earley BJ, Kornfeld K (2016) Insights into zinc and cadmium biology in the nematode *Caenorhabditis elegans*. Arch Biochem Biophys 611:120–133
Dietrich N, Schneider DL, Kornfeld K (2017) A pathway for low zinc homeostasis that is conserved in animals and acts in parallel to the pathway for high zinc homeostasis. Nucleic Acids Res 45:11658–11672
Dow A, Prisic S (2018) Alternative ribosomal proteins are required for growth and morphogenesis of *Mycobacterium smegmatis* under zinc-limiting conditions. PLoS One 13:e0196300
Ehrensberger KM, Corkins ME, Choi S, Bird AJ (2014) The double zinc finger domain and adjacent accessory domain from the transcription factor loss of zinc sensing 1 (loz1) are necessary for DNA binding and zinc sensing. J Biol Chem 289:18087–18096
Ellis CD, Macdiarmid CW, Eide DJ (2005) Heteromeric protein complexes mediate zinc transport into the secretory pathway of eukaryotic cells. J Biol Chem 280:28811–28818
Ellison ML, Farrow JM 3rd, Parrish W, Danell AS, Pesci EC (2013) The transcriptional regulator Np20 is the zinc uptake regulator in *Pseudomonas aeruginosa*. PLoS One 8:e75389

Feeney GP, Zheng D, Kille P, Hogstrand C (2005) The phylogeny of teleost ZIP and ZnT zinc transporters and their tissue specific expression and response to zinc in zebrafish. Biochim Biophys Acta 1732:88–95

Fones H, Davis CA, Rico A, Fang F, Smith JA, Preston GM (2010) Metal hyperaccumulation armors plants against disease. PLoS Pathog 6:e1001093

Foster AW, Osman D, Robinson NJ (2014) Metal preferences and metallation. J Biol Chem 289:28095–28103

Gitan RS, Eide DJ (2000) Zinc-regulated ubiquitin conjugation signals endocytosis of the yeast *ZRT1* zinc transporter. Biochem J 346(Pt 2):329–336

Gitan RS, Luo H, Rodgers J, Broderius M, Eide D (1998) Zinc-induced inactivation of the yeast *ZRT1* zinc transporter occurs through endocytosis and vacuolar degradation. J Biol Chem 273:28617–28624

Gitan RS, Shababi M, Kramer M, Eide DJ (2003) A cytosolic domain of the yeast Zrt1 zinc transporter is required for its post-translational inactivation in response to zinc and cadmium. J Biol Chem 278:39558–39564

Grosse C, Grass G, Anton A, Franke S, Santos AN, Lawley B, Brown NL, Nies DH (1999) Transcriptional organization of the *czc* heavy-metal homeostasis determinant from *Alcaligenes eutrophus*. J Bacteriol 181:2385–2393

Grosse C, Anton A, Hoffmann T, Franke S, Schleuder G, Nies DH (2004) Identification of a regulatory pathway that controls the heavy-metal resistance system Czc via promoter czcNp in *Ralstonia metallidurans*. Arch Microbiol 182:109–118

Guerra AJ, Giedroc DP (2012) Metal site occupancy and allosteric switching in bacterial metal sensor proteins. Arch Biochem Biophys 519:210–222

Haas CE, Rodionov DA, Kropat J, Malasarn D, Merchant SS, de Crecy-Lagard V (2009) A subset of the diverse COG0523 family of putative metal chaperones is linked to zinc homeostasis in all kingdoms of life. BMC Genomics 10:470

Hassan MT, van der Lelie D, Springael D, Romling U, Ahmed N, Mergeay M (1999) Identification of a gene cluster, *czr*, involved in cadmium and zinc resistance in *Pseudomonas aeruginosa*. Gene 238:417–425

Helbig K, Bleuel C, Krauss GJ, Nies DH (2008) Glutathione and transition-metal homeostasis in *Escherichia coli*. J Bacteriol 190:5431–5438

Ho E, Dukovcic S, Hobson B, Wong CP, Miller G, Hardin K, Traber MG, Tanguay RL (2012) Zinc transporter expression in zebrafish (*Danio rerio*) during development. Comp Biochem Physiol C Toxicol Pharmacol 155:26–32

Hogstrand C, Zheng D, Feeney G, Cunningham P, Kille P (2008) Zinc-controlled gene expression by metal-regulatory transcription factor 1 (MTF1) in a model vertebrate, the zebrafish. Biochem Soc Trans 36:1252–1257

Jirakulaporn T, Muslin AJ (2004) Cation diffusion facilitator proteins modulate Raf-1 activity. J Biol Chem 279:27807–27815

Kehl-Fie TE, Skaar EP (2010) Nutritional immunity beyond iron: a role for manganese and zinc. Curr Opin Chem Biol 14:218–224

Kjellerup L, Winther AL, Wilson D, Fuglsang AT (2018) Cyclic AMP pathway activation and extracellular zinc induce rapid intracellular zinc mobilization in *Candida albicans*. Front Microbiol 9:502

Kloosterman TG, van der Kooi-Pol MM, Bijlsma JJ, Kuipers OP (2007) The novel transcriptional regulator SczA mediates protection against Zn^{2+} stress by activation of the Zn^{2+}-resistance gene *czcD* in *Streptococcus pneumoniae*. Mol Microbiol 65:1049–1063

Kumanovics A, Poruk KE, Osborn KA, Ward DM, Kaplan J (2006) *YKE4* (*YIL023C*) encodes a bidirectional zinc transporter in the endoplasmic reticulum of *Saccharomyces cerevisiae*. J Biol Chem 281:22566–22574

L'Hernault SW (1997) Spermatogenesis. In: Riddle DL, Blumenthal T, Meyer BJ, Priess JR (eds) *C. elegans* II, 2nd edn. Cold Spring Harbor, New York

Leblanc SK, Oates CW, Raivio TL (2011) Characterization of the induction and cellular role of the BaeSR two-component envelope stress response of *Escherichia coli*. J Bacteriol 193:3367–3375
Lee SW, Glickmann E, Cooksey DA (2001) Chromosomal locus for cadmium resistance in *Pseudomonas putida* consisting of a cadmium-transporting ATPase and a MerR family response regulator. Appl Environ Microbiol 67:1437–1444
Lee LJ, Barrett JA, Poole RK (2005) Genome-wide transcriptional response of chemostat-cultured *Escherichia coli* to zinc. J Bacteriol 187:1124–1134
Lele Z, Krone PH (1996) The zebrafish as a model system in developmental, toxicological and transgenic research. Biotechnol Adv 14:57–72
van der Lelie D, Schwuchow T, Schwidetzky U, Wuertz S, Baeyens W, Mergeay M, Nies DH (1997) Two-component regulatory system involved in transcriptional control of heavy-metal homoeostasis in *Alcaligenes eutrophus*. Mol Microbiol 23:493–503
Leonhartsberger S, Huber A, Lottspeich F, Bock A (2001) The *hydH/G* genes from *Escherichia coli* code for a zinc and lead responsive two-component regulatory system. J Mol Biol 307:93–105
Levy S, Kafri M, Carmi M, Barkai N (2011) The competitive advantage of a dual-transporter system. Science 334:1408–1412
Li M, Koshi T, Emr SD (2015) Membrane-anchored ubiquitin ligase complex is required for the turnover of lysosomal membrane proteins. J Cell Biol 211:639–652
Li J, Huang D, Sun X, Li X, Cheng CHK (2019) Zinc mediates the action of androgen in acting as a downstream effector of luteinizing hormone on oocyte maturation in zebrafishdagger. Biol Reprod 100:468–478
Liu Z, Chen L, Shang Y, Huang P, Miao L (2013) The micronutrient element zinc modulates sperm activation through the SPE-8 pathway in *Caenorhabditis elegans*. Development 140:2103–2107
Ma Z, Chandrangsu P, Helmann TC, Romsang A, Gaballa A, Helmann JD (2014) Bacillithiol is a major buffer of the labile zinc pool in *Bacillus subtilis*. Mol Microbiol 94:756–770
MacDiarmid CW, Gaither LA, Eide D (2000) Zinc transporters that regulate vacuolar zinc storage in *Saccharomyces cerevisiae*. EMBO J 19:2845–2855
MacDiarmid CW, Milanick MA, Eide DJ (2003) Induction of the *ZRC1* metal tolerance gene in zinc-limited yeast confers resistance to zinc shock. J Biol Chem 278:15065–15072
Maret W. (2017) Zinc in cellular regulation: the nature and significance of "zinc signals". Int J Mol Sci 18
Mikhaylina A, Ksibe AZ, Scanlan DJ, Blindauer CA (2018) Bacterial zinc uptake regulator proteins and their regulons. Biochem Soc Trans 46:983–1001
Miyabe S, Izawa S, Inoue Y (2001) The Zrc1 is involved in zinc transport system between vacuole and cytosol in *Saccharomyces cerevisiae*. Biochem Biophys Res Commun 282:79–83
Muhlrad PJ, Clark JN, Nasri U, Sullivan NG, LaMunyon CW (2014) SPE-8, a protein-tyrosine kinase, localizes to the spermatid cell membrane through interaction with other members of the SPE-8 group spermatid activation signaling pathway in *C. elegans*. BMC Genet 15:83
Mumm K, Ainsaar K, Kasvandik S, Tenson T, Horak R (2016) Responses of *Pseudomonas putida* to zinc excess determined at the proteome level: pathways dependent and independent of ColRS. J Proteome Res 15:4349–4368
Nairn BL, Lonergan ZR, Wang J, Braymer JJ, Zhang Y, Calcutt MW, Lisher JP, Gilston BA, Chazin WJ, de Crecy-Lagard V, Giedroc DP, Skaar EP (2016) The response of *Acinetobacter baumannii* to zinc starvation. Cell Host Microbe 19:826–836
Nishino K, Honda T, Yamaguchi A (2005) Genome-wide analyses of *Escherichia coli* gene expression responsive to the BaeSR two-component regulatory system. J Bacteriol 187:1763–1772
Nishino K, Nikaido E, Yamaguchi A (2007) Regulation of multidrug efflux systems involved in multidrug and metal resistance of *Salmonella enterica serovar Typhimurium*. J Bacteriol 189:9066–9075
Nowicki EM, O'Brien JP, Brodbelt JS, Trent MS (2015) Extracellular zinc induces phosphoethanolamine addition to *Pseudomonas aeruginosa* lipid A via the ColRS two-component system. Mol Microbiol 97:166–178

Ong CL, Gillen CM, Barnett TC, Walker MJ, McEwan AG (2014) An antimicrobial role for zinc in innate immune defense against group A *streptococcus*. J Infect Dis 209:1500–1508
Outten CE, O'Halloran TV (2001) Femtomolar sensitivity of metalloregulatory proteins controlling zinc homeostasis. Science 292:2488–2492
Panina EM, Mironov AA, Gelfand MS (2003) Comparative genomics of bacterial zinc regulons: enhanced ion transport, pathogenesis, and rearrangement of ribosomal proteins. Proc Natl Acad Sci U S A 100:9912–9917
Pederick VG, Eijkelkamp BA, Begg SL, Ween MP, McAllister LJ, Paton JC, McDevitt CA (2015) ZnuA and zinc homeostasis in *Pseudomonas aeruginosa*. Sci Rep 5:13139
Perron K, Caille O, Rossier C, Van Delden C, Dumas JL, Kohler T (2004) CzcR-CzcS, a two-component system involved in heavy metal and carbapenem resistance in *Pseudomonas aeruginosa*. J Biol Chem 279:8761–8768
Petit-Hartlein I, Rome K, de Rosny E, Molton F, Duboc C, Gueguen E, Rodrigue A, Coves J (2015) Biophysical and physiological characterization of ZraP from *Escherichia coli*, the periplasmic accessory protein of the atypical ZraSR two-component system. Biochem J 472:205–216
Raffa RG, Raivio TL (2002) A third envelope stress signal transduction pathway in *Escherichia coli*. Mol Microbiol 45:1599–1611
Raivio TL, Popkin DL, Silhavy TJ (1999) The Cpx envelope stress response is controlled by amplification and feedback inhibition. J Bacteriol 181:5263–5272
Reyes-Caballero H, Campanello GC, Giedroc DP (2011) Metalloregulatory proteins: metal selectivity and allosteric switching. Biophys Chem 156:103–114
Riggio M, Filosa S, Parisi E, Scudiero R (2003) Changes in zinc, copper and metallothionein contents during oocyte growth and early development of the teleost *Danio rerio* (zebrafish). Comp Biochem Physiol C Toxicol Pharmacol 135:191–196
Roh HC, Collier S, Deshmukh K, Guthrie J, Robertson JD, Kornfeld K (2013) *ttm-1* encodes CDF transporters that excrete zinc from intestinal cells of *C. elegans* and act in a parallel negative feedback circuit that promotes homeostasis. PLoS Genet 9:e1003522
Roh HC, Dimitrov I, Deshmukh K, Zhao G, Warnhoff K, Cabrera D, Tsai W, Kornfeld K (2015) A modular system of DNA enhancer elements mediates tissue-specific activation of transcription by high dietary zinc in *C. elegans*. Nucleic Acids Res 43:803–816
Rome K, Borde C, Taher R, Cayron J, Lesterlin C, Gueguen E, De Rosny E, Rodrigue A (2018) The two-component system ZraPSR is a novel ESR that contributes to intrinsic antibiotic tolerance in *Escherichia coli*. J Mol Biol 430:4971–4985
Rousset R, Wharton KA Jr, Zimmermann G, Scott MP (2002) Zinc-dependent interaction between dishevelled and the Drosophila Wnt antagonist naked cuticle. J Biol Chem 277:49019–49026
von Rozycki T, Nies DH (2009) *Cupriavidus metallidurans*: evolution of a metal-resistant bacterium. Antonie Van Leeuwenhoek 96:115–139
Saydmohammed M, Vollmer LL, Onuoha EO, Vogt A, Tsang M (2011) A high-content screening assay in transgenic zebrafish identifies two novel activators of fgf signaling. Birth Defects Res C Embryo Today 93:281–287
Schothorst J, Zeebroeck GV, Thevelein JM (2017) Identification of Ftr1 and Zrt1 as iron and zinc micronutrient transceptors for activation of the PKA pathway in *Saccharomyces cerevisiae*. Microb Cell 4:74–89
Sentandreu M, Elorza MV, Sentandreu R, Fonzi WA (1998) Cloning and characterization of *PRA1*, a gene encoding a novel pH-regulated antigen of *Candida albicans*. J Bacteriol 180:282–289
Sevcenco AM, Pinkse MW, Wolterbeek HT, Verhaert PD, Hagen WR, Hagedoorn PL (2011) Exploring the microbial metalloproteome using MIRAGE. Metallomics 3:1324–1330
Shin JH, Helmann JD (2016) Molecular logic of the Zur-regulated zinc deprivation response in *Bacillus subtilis*. Nat Commun 7:12612
Sinclair J, Hamza I (2015) Lessons from bloodless worms: heme homeostasis in *C. elegans*. Biometals 28:481–489
Singh K, Senadheera DB, Cvitkovitch DG (2014) An intimate link: two-component signal transduction systems and metal transport systems in bacteria. Future Microbiol 9:1283–1293

Sternberg PW (2005) Vulval development. WormBook:1–28
Stock AM, Robinson VL, Goudreau PN (2000) Two-component signal transduction. Annu Rev Biochem 69:183–215
Thomas P (1990) Teleost model for studying the effects of chemicals on female reproductive endocrine function. J Exp Zool Suppl 4:126–128
Wagner D, Maser J, Lai B, Cai Z, Barry CE 3rd, Honer Zu Bentrup K, Russell DG, Bermudez LE (2005) Elemental analysis of Mycobacterium avium-, Mycobacterium tuberculosis-, and Mycobacterium smegmatis-containing phagosomes indicates pathogen-induced microenvironments within the host cell's endosomal system. J Immunol 174:1491–1500
Walencik PK, Watly J, Rowinska-Zyrek M (2016) Fungal zinc homeostasis – a tug of war between the pathogen and host. Curr Med Chem 23:3717–3729
Wang D, Fierke CA (2013) The BaeSR regulon is involved in defense against zinc toxicity in *E. coli*. Metallomics 5:372–383
Wang D, Chen W, Huang S, He Y, Liu X, Hu Q, Wei T, Sang H, Gan J, Chen H (2017) Structural basis of Zn(II) induced metal detoxification and antibiotic resistance by histidine kinase CzcS in *Pseudomonas aeruginosa*. PLoS Pathog 13:e1006533
Warnhoff K, Roh HC, Kocsisova Z, Tan CH, Morrison A, Croswell D, Schneider DL, Kornfeld K (2017) The nuclear receptor HIZR-1 uses zinc as a ligand to mediate homeostasis in response to high zinc. PLoS Biol 15:e2000094
Waters BM, Eide DJ (2002) Combinatorial control of yeast *FET4* gene expression by iron, zinc, and oxygen. J Biol Chem 277:33749–33757
van der Weel L, As KS, Dekker WJC, van den Eijnden L, van Helmond W, Schiphorst C, Hagen WR, Hagedoorn PL (2019) ZraP, the most prominent zinc protein under zinc stress conditions has no direct role in in-vivo zinc tolerance in *Escherichia coli*. J Inorg Biochem 192:98–106
Wilson D (2015) An evolutionary perspective on zinc uptake by human fungal pathogens. Metallomics 7:979–985
Wilson S, Bird AJ (2016) Zinc sensing and regulation in yeast model systems. Arch Biochem Biophys 611:30–36
Wu CY, Bird AJ, Chung LM, Newton MA, Winge DR, Eide DJ (2008) Differential control of Zap1-regulated genes in response to zinc deficiency in *Saccharomyces cerevisiae*. BMC Genomics 9:370
Wu YH, Frey AG, Eide DJ (2011) Transcriptional regulation of the Zrg17 zinc transporter of the yeast secretory pathway. Biochem J 435:259–266
Wu YH, Taggart J, Song PX, MacDiarmid C, Eide DJ (2016) An *MSC2* promoter-*lacZ* fusion gene reveals zinc-responsive changes in sites of transcription initiation that occur across the yeast genome. PLoS One 11:e0163256
Yamashita S, Miyagi C, Fukada T, Kagara N, Che YS, Hirano T (2004) Zinc transporter LIVI controls epithelial-mesenchymal transition in zebrafish gastrula organizer. Nature 429:298–302
Yan G, Zhang Y, Yu J, Yu Y, Zhang F, Zhang Z, Wu A, Yan X, Zhou Y, Wang F (2012) Slc39a7/zip7 plays a critical role in development and zinc homeostasis in zebrafish. PLoS One 7:e42939
Yeo J, Dippel AB, Wang XC, Hammond MC (2018) In vivo biochemistry: single-cell dynamics of cyclic Di-GMP in *Escherichia coli* in response to zinc overload. Biochemistry 57:108–116
Yoder JH, Chong H, Guan KL, Han M (2004) Modulation of KSR activity in *Caenorhabditis elegans* by Zn ions, PAR-1 kinase and PP2A phosphatase. EMBO J 23:111–119
Zahringer F, Lacanna E, Jenal U, Schirmer T, Boehm A (2013) Structure and signaling mechanism of a zinc-sensory diguanylate cyclase. Structure 21:1149–1157
Zhao H, Eide D (1996a) The yeast *ZRT1* gene encodes the zinc transporter protein of a high-affinity uptake system induced by zinc limitation. Proc Natl Acad Sci U S A 93:2454–2458
Zhao H, Eide D (1996b) The *ZRT2* gene encodes the low affinity zinc transporter in Saccharomyces cerevisiae. J Biol Chem 271:23203–23210
Zhao H, Eide DJ (1997) Zap1p, a metalloregulatory protein involved in zinc-responsive transcriptional regulation in *Saccharomyces cerevisiae*. Mol Cell Biol 17:5044–5052

Zhao L, Xia Z, Wang F (2014) Zebrafish in the sea of mineral (iron, zinc, and copper) metabolism. Front Pharmacol 5:33

Zhao Y, Tan CH, Krauchunas A, Scharf A, Dietrich N, Warnhoff K, Yuan Z, Druzhinina M, Gu SG, Miao L, Singson A, Ellis RE, Kornfeld K (2018) The zinc transporter ZIPT-7.1 regulates sperm activation in nematodes. PLoS Biol 16:e2005069

Zheng D, Feeney GP, Kille P, Hogstrand C (2008) Regulation of ZIP and ZnT zinc transporters in zebrafish gill: zinc repression of *ZIP10* transcription by an intronic MRE cluster. Physiol Genomics 34:205–214

Zheng D, Kille P, Feeney GP, Cunningham P, Handy RD, Hogstrand C (2010a) Dynamic transcriptomic profiles of zebrafish gills in response to zinc supplementation. BMC Genomics 11:553

Zheng D, Kille P, Feeney GP, Cunningham P, Handy RD, Hogstrand C (2010b) Dynamic transcriptomic profiles of zebrafish gills in response to zinc depletion. BMC Genomics 11:548

Zhu L, Jorgensen JR, Li M, Chuang YS, Emr SD (2017) ESCRTs function directly on the lysosome membrane to downregulate ubiquitinated lysosomal membrane proteins. elife 6

Index

A
Adaptive and innate immunity, 211, 246, 249, 283, 287, 289, 291, 293, 296, 297, 320
Ageing, 210, 215–217
Aggregation, 193, 203, 205–208, 212, 213, 310, 311
Aging, 133, 140, 141, 143–145, 147, 148, 153–155, 188, 190, 203, 238, 289, 314
Aging and cachexia, 128
Alzheimer's disease (AD), 3, 148, 188, 192, 202, 205, 206, 212, 289, 297
Amyotrophic lateral sclerosis (ALS), 188, 208, 214
Autophagy, 129, 191, 193, 194, 196, 203–205

B
B cells, 3, 58, 64, 80, 82, 85, 86, 92, 211, 248, 257, 263–266, 285, 294, 296, 311, 323
Barrier function, 351, 352, 356, 357
Basophils, 60, 61, 64, 79–93
Beige adipocytes, 237
Browning, 237–239, 313

C
Calcium channels, 15, 83–85, 92, 125, 126, 169, 210, 324, 353
Cancer, 2, 25, 86, 132, 152, 203, 259, 283, 307, 320
Cation diffusion facilitator (CDF), 12, 25, 36, 254, 396, 400–402
CDF, *see* Cation diffusion facilitator (CDF)
Cellular zinc homeostasis, 5–18, 205, 216, 245, 298, 384, 401
Cytokines, 62–66, 85, 87–89, 91, 128, 211, 212, 234, 248, 251, 252, 254, 280, 282, 283, 288–294, 297, 314, 350, 352, 354

D
Dermis, 15, 306, 308, 310, 311, 313
Diabetes, 2, 3, 38, 128, 132, 142, 152, 203, 215, 229–239, 259, 285, 289, 292
Diseases, 2, 44, 58, 88, 102, 129, 140, 141, 176, 188, 202, 238, 244, 285, 307, 319, 348

E
Emission-ratiometric probes, 381
Epidermis, 306–309, 311, 312
Exchangeable zinc, 366, 373, 375, 384
Excitation-contraction coupling, 124–126, 143, 147
Excitotoxicity, 209, 210, 214, 216, 297

F
Fertilization, 3, 100–105, 107, 108, 402
Fluorescence imaging, 366, 372, 377
Fluorogenic probes, 366, 368–371, 374–376, 380, 381, 384
Free zinc ion, 6–8, 10, 12, 14–17, 90, 295

H
Heart, 3, 140–155, 244, 266
Host defense, 69, 80, 280, 292, 296

T. Fukada, T. Kambe (eds.), *Zinc Signaling*,
https://doi.org/10.1007/978-981-15-0557-7

I

Infections, 3, 58–62, 64–67, 111, 152, 176, 203, 211, 212, 215, 247, 248, 251, 252, 255, 258, 265, 266, 279, 286, 287, 289, 293, 298
Inflammation, 3, 61, 64, 65, 67, 90, 152, 153, 203, 211, 212, 234, 247, 279–297, 348, 350, 351, 354
Innate and adaptive immunity, 211, 246, 249, 283, 287, 289, 291, 293, 296, 297, 320
Innate immunity, 3, 60, 65, 92, 211, 246, 249, 251–257, 289, 291, 296, 297
Insulin, 12, 38, 127, 140, 229, 230, 233, 236, 252, 353
Intestine, 3, 38, 154, 244, 308, 309, 348, 349, 352–355, 357, 402
Irt-like protein (zip), 2, 13, 24, 26, 81, 232, 246, 255, 348

L

L-type calcium channels (LTCCs), 83, 85, 92, 125, 126
Lysosomes, 34, 38, 44, 145, 188–196, 203, 204, 250, 260, 400

M

Macrophages, 59, 62, 63, 65, 66, 68, 211, 250–252, 254, 255, 264, 281, 283, 284, 288, 291, 292, 390
Mast cells, 3, 38, 61, 82–86, 88–92, 126, 141, 249–251, 256, 257, 281, 288, 294, 320
Membrane transporters, 10, 23, 350
Metallo-regulatory, 8, 13, 176, 390, 391, 394, 402
Metallothioneins (MTs), 2, 8, 9, 37, 58, 80, 81, 171, 175–177, 192, 204, 207, 208, 210, 215, 245, 280, 309, 321, 365, 392, 400, 402
MG53, 124, 129, 130, 132, 133
Microorganisms, 58
Mitochondria, 10, 14, 36, 102, 140, 175, 203, 286
MT3, 59, 66, 67, 171, 188–196, 204
MTs, *see* Metallothioneins (MTs)
Muscle RING Fingers (MuRF-1), 124, 129

N

Neurotransmission, 12, 17, 166, 168–170, 209, 280

O

Obesity, 3, 132, 133, 229–239
Oocytes, 100–108, 401, 403
Oxidative stress, 13, 61, 67, 129, 144, 146–149, 151, 153, 154, 176, 188, 203, 204, 207, 210, 216, 232, 254, 284, 285, 287, 289, 297, 350, 351

P

Parkinson's disease (PD), 3, 188, 190, 191, 202, 205, 209, 213, 214, 289, 297
Phosphorylation, 3, 14, 15, 63, 84–87, 127, 129, 131, 132, 144, 174–177, 204, 206, 213, 232, 250, 252, 254, 256, 259–265, 281–283, 285, 295, 296, 310, 320, 321, 323, 327, 328, 331–336, 351, 353, 355
Physiology, 2, 5, 39, 44, 132, 177, 322, 350
Potassium channels, 175, 209, 210
Proteolytic cleavage, 230, 266, 324, 332, 333

R

Reproduction, 3, 100–114, 403
Ryanodine receptor (RyR), 125, 126

S

Signal transduction, 6, 9, 15, 85, 86, 132, 144, 245, 246, 251, 252, 259, 260, 263, 265, 267, 280–282, 291, 296, 310, 390, 392–396
Skeletal muscle repair, 124, 128, 130
Skin, 3, 15, 247, 293, 305–314
SLC, *see* Solute carrier family (SLC)
Slc30a, 24, 232, 306
Slc30a8, 38, 231, 234, 235, 237
Slc39a, 24, 169, 306, 307, 321, 348
Slc39a6, 82, 329
Slc39a7, 82, 131
Slc39a10, 82
Solute carrier family (SLC), 12, 13, 24, 246

T

T cells, 3, 60, 63, 64, 79–93, 247, 256, 259, 260, 281, 306
Toll-like receptor (TLR-4), 82, 91, 249, 252–255, 263, 283, 291
Transporters, 2, 6, 10, 24, 68, 80, 106, 124, 151, 167, 195, 205, 232, 233, 245, 280, 282, 305–307, 321, 348, 365, 396

Traumatic brain injury (TBI), 3, 202, 209, 216, 217, 290
Two-component signalling systems, 15, 390, 392–395
Two-photon excitation microscopy (TPEM), 368, 377, 378, 380–383

Z

Zinc
  absorption, 27, 61, 112, 113, 127, 244, 248, 297, 309, 348, 351, 353, 354, 357, 368, 376, 380
  deficiency, 18, 26, 58, 80, 110, 127, 142, 204, 215, 232, 244, 260, 279, 285, 305, 320, 348, 391
  homeostasis, 3, 7, 46, 58, 80, 124, 154, 202, 232, 244, 280, 282, 308, 320, 348, 384, 390, 397, 400
  signalling, 1–4, 6–18, 42, 46, 62, 79–92, 100–114, 123–133, 140–155, 166–178, 190, 210, 233, 239, 244–267, 279–298, 305–314, 319–336, 347–357, 366, 389–404
  transporters, 2, 6, 24, 68, 80, 106, 124, 141, 167, 195, 205, 232, 233, 245, 280, 282, 321, 348, 365, 396
  wave, 80, 82–86, 91, 92, 126, 142, 249, 250, 257, 281, 294, 320
Zinc-finger proteins (ZNFs), 79, 128, 129, 245, 280
Zip, 10, 24, 81, 106, 124, 148, 169, 216, 232, 246, 280, 306, 321, 348, 396, 400
ZIP channels, 321–336
Zip13, 18, 27, 42, 43, 45, 152, 233, 237–239, 313, 322, 323, 401
Zip14, 27, 42–45, 114, 128, 133, 151–154, 233, 284, 292, 295, 314, 322, 353
ZNFs, *see* Zinc-finger proteins (ZNFs)
Znr/gpr39, 3, 209, 284, 351, 353, 355–357
Znt8, 12, 27, 34, 36, 37, 130, 151, 152, 231, 233–235, 237, 238, 333
Zn transporters (Znts), 2, 10, 24, 81, 106, 124, 148, 196, 216, 232, 246, 280, 296, 306, 321, 348, 402
Znts, *see* Zn transporters (Znts)
Zrt, 2, 13, 24, 26, 81, 86, 232, 246, 280, 348